ESSENTIALS OF
Molecular Biology

The "Universal" Genetic Code

First position (5' end)	Second Position				Third position (3' end)
	U	C	A	G	
U	Phe Phe Leu Leu	Ser Ser Ser Ser	Tyr Tyr Stop Stop	Cys Cys Stop Trp	U C A G
C	Leu Leu Leu Leu	Pro Pro Pro Pro	His His Gin Gin	Arg Arg Arg Arg	U C A G
A	Ile Ile Ile Met	Thr Thr Thr Thr	Asn Asn Lys Lys	Ser Ser Arg Arg	U C A G
G	Val Val Val Val	Ala Ala Ala Ala	Asp Asp Glu Glu	Gly Gly Gly Gly	U C A G

NOTE: The boxed codons are used for initiation.

Basic Structure of an α-Amino Acid

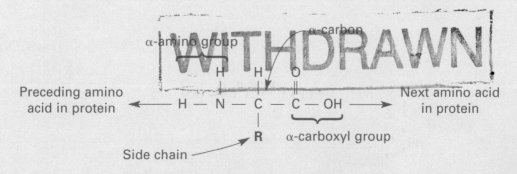

ESSENTIALS OF
Molecular Biology

FOURTH EDITION

George M. Malacinski
Indiana University

JONES AND BARTLETT PUBLISHERS

Sudbury, Massachusetts

BOSTON   TORONTO   LONDON   SINGAPORE

World Headquarters

Jones and Bartlett Publishers
40 Tall Pine Drive
Sudbury, MA 01776
978-443-5000
info@jbpub.com
www.jbpub.com

Jones and Bartlett Publishers Canada
2406 Nikanna Road
Mississauga, ON L5C 2W6
CANADA

Jones and Bartlett Publishers International
Barb House, Barb Mews
London W6 7PA
UK

PRODUCTION CREDITS
Chief Executive Officer: Clayton Jones
Chief Operating Officer: Don W. Jones, Jr.
Executive V.P. & Publisher: Robert W. Holland, Jr.
V.P., Design and Production: Anne Spencer
Director of Sales and Marketing: William J. Kane
V.P., Manufacturing and Inventory Control: Therese Bräuer
Editor in Chief–College: J. Michael Stranz
Executive Editor: Stephen L. Weaver
Associate Managing Editor, College Editorial: Dean W. DeChambeau
Senior Production Editor: Louis C. Bruno, Jr.
Senior Marketing Manager: Nathan J. Schultz
Text Design: Anne Spencer
Cover Design: Kristin E. Ohlin
Copy Editor: Ellice Gerber
Proofreader: Jan Cocker
Illustrator: Elizabeth Morales
Printing and Binding: Courier Westford
Cover Printing: John Pow Company

Library of Congress Cataloging-in-Publication Data
Malacinski, George M.
    Essentials of molecular biology / George M. Malacinski.—4th ed.
        p. cm.
    Includes bibliographical references and index.
ISBN 0-7637-2133-6 (alk. paper)
1. Molecular biology. I. Title.

QH506 .M368 2002
572.8—dc21   2002072756

Printed in the United States of America
06 05 04 03 02   10 9 8 7 6 5 4 3 2 1

TRIBUTE TO DAVID FREIFELDER

David Freifelder taught biochemistry and molecular biology first at Brandeis University and later at the University of California, San Diego. His research interests and expertise were in a broad range of subjects; therefore, he was qualified to write both general and specialized textbooks. He was a gifted writer, and he devoted extensive time and energy to preparing a collection of textbooks and monographs. From his teaching and writing experiences, he developed a remarkable understanding of the *ways in which students learn* and organized his textbooks, including the first edition of *Essentials of Molecular Biology,* using a layering approach for coping with biological complexity. That is, David communicated insight into complex phenomena to his students by helping them build, one piece at a time, a personal *knowledge house,* which made up their understanding of molecular biology.

Succeeding editions, including this fourth edition of *Essentials of Molecular Biology,* have followed that strategy. Students appreciate that approach. This acknowledgment is, therefore, offered as a tribute to David's memory—a symbol of gratitude on behalf of the thousands of students who have been introduced to the field of molecular biology by reading one of the editions of *Essentials of Molecular Biology.*

B R I E F CONTENTS

CHAPTER 1   **Welcome to Molecular Biology!** 1

Part I   **The Structure of Proteins, Nucleic Acids, and Macromolecular Complexes**   17

CHAPTER 2   **Macromolecules** 18
CHAPTER 3   **Nucleic Acids** 32
CHAPTER 4   **The Physical Structure of Protein Molecules** 56
CHAPTER 5   **Macromolecular Interactions and the Structure of Complex Aggregates** 76

Part II   **Function of Macromolecules**   97

CHAPTER 6   **The Genetic Material** 98
CHAPTER 7   **DNA Replication** 118
CHAPTER 8   **Transcription** 146
CHAPTER 9   **Translation** 168
CHAPTER 10   **Mutations, Mutagenesis, and DNA Repair** 192

Part III   **Coordination of Macromolecular Function in Cells**   221

CHAPTER 11   **Regulation of Gene Activity in Prokaryotes** 222
CHAPTER 12   **Regulation of Gene Activity in Eukaryotes** 248
CHAPTER 13   **Genomics and Proteomics Drive Information-Age Biology** 284

Part IV   **Experimental Manipulation of Macromolecules**   303

CHAPTER 14   **Transposons, Plasmids, and Bacteriophage** 304
CHAPTER 15   **Recombinant DNA and Genetic Engineering: Molecular Tailoring of Genes** 346
CHAPTER 16   **Molecular Biology Is Expanding Its Reach** 370
POSTSCRIPT   **Postscript to Your Review of Molecular Biology** 403
APPENDIX   **Chemical Principles Important for Understanding Molecular Biology** 411
                List of Essential Concepts of Molecular Biology 435
                Glossary 439
                Answers to Questions and Problems 455
                Index 479

CONTENTS

About the Author xiii

Preface xiv

Prologue xviii

CHAPTER 1 **Welcome to Molecular Biology!** 1
Goals of Molecular Biology   1
The Early Years   1
Model Biological Systems   2
Methodology of Molecular Biology   7
Rapid Progress in Molecular Biology   9
Putting the Details of Molecular Biology in Perspective   11
Concepts of Molecular Biology   12
Progression Diagram   13
Rewards from Studying Molecular Biology   16

Part I **The Structure of Proteins, Nucleic Acids, and Macromolecular Complexes** **17**

CHAPTER 2 **Macromolecules** **18**
Chemical Structures of the Major Classes
    of Macromolecules   19
Noncovalent Interactions That Determine the
    Three-Dimensional Structures of Proteins and Nucleic
    Acids   23
Macromolecule Isolation and Characterization   26

CHAPTER 3 **Nucleic Acids** **32**
Physical and Chemical Structure of DNA   33
Alternate DNA Structures   36
Circular and Superhelical DNA   38
Renaturation   42
Hybridization   45
The Structure of RNA   46
Hydrolysis of Nucleic Acids   47
Sequencing Nucleic Acids   48
Synthesis of DNA   50
A Future Practical Application?   53

CHAPTER 4 **The Physical Structure of Protein Molecules** **56**
Basic Features of Protein Molecules   57
The Folding of a Polypeptide Chain   58
The α Helix and β Secondary Structures   60
Protein Structure   62
Proteins with Subunits   64
Enzymes   68
A Future Application   73

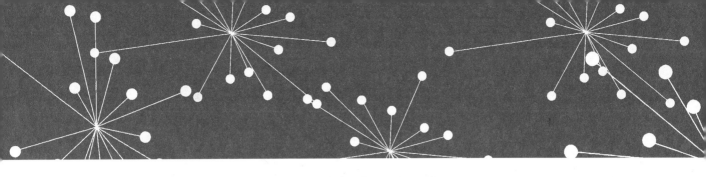

CHAPTER 5   **Macromolecular Interactions and the Structure of Complex Aggregates**. **76**

A Complex DNA Structure: The *E. coli* Chromosome   77

Chromosomes and Chromatin   80

Interaction of DNA and Proteins That Recognize a Specific Base Sequence   85

Biological Membranes   87

Cytoskeletal Elements   92

A Future Practical Application?   93

Part II   **Function of Macromolecules**   **97**

CHAPTER 6   **The Genetic Material**. **98**

Early Observations on the Mechanism of Heredity   99

Identification of DNA As the Genetic Material   101

Identification of RNA As the Genetic Material of Certain Viruses   109

Properties of the Genetic Material   110

RNA As the Genetic Material   115

CHAPTER 7   **DNA Replication** **118**

Semiconservative Replication of Double-Stranded DNA   120

Untwisting of Highly Coiled DNA Is Required for DNA Replication   121

Initiation of DNA Replication   123

Unwinding of DNA for Replication   126

Elongation of Newly Synthesized Strands   126

DNA Polymerase III Consists of Multiple Subunits   130

Antiparallel DNA Strands and Discontinuous Replication   132

The Complete DNA Replication System   137

Replication of Eukaryotic Chromosomes   141

A Future Practical Application?   143

CHAPTER 8   **Transcription** **146**

Enzymatic Synthesis of RNA   147

Transcription Signals   149

Classes of RNA Molecules   154

Transcription in Eukaryotes   156

Means of Studying Intracellular RNA   161

A Future Practical Application?   163

CHAPTER 9   **Translation**. **168**

Outline of Translation   169

The Genetic Code   170

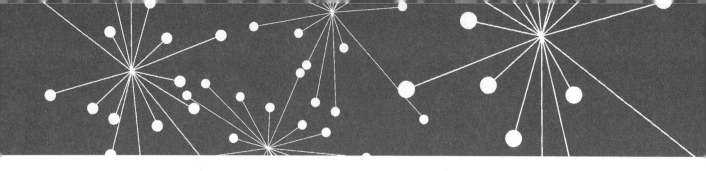

The Wobble Phenomenon 176
Polycistronic mRNA 177
Overlapping Genes 177
Polypeptide Synthesis 179
Stages of Polypeptide Synthesis in Prokaryotes 181
Complex Translation Units 185
Antibiotics 188
A Future Practical Application? 188

CHAPTER 10 **Mutations, Mutagenesis, and DNA Repair** **192**
Types of Mutations 193
Biochemical Basis of Mutants 195
Mutagenesis 196
Induced Mutations 198
Reversion 201
Reversion As a Means of Detecting Mutagens
    and Carcinogens 205
DNA Repair Mechanisms 207
Spontaneous Mutations and Their Repair 207
Repair by Direct Reversal 210
Excision Repair 211
Recombinational Repair 212
The SOS Response 214
Future Practical Applications? 216

Part III **Coordination of Macromolecular Function in Cells** **221**

CHAPTER 11 **Regulation of Gene Activity in Prokaryotes** **222**
Principles of Regulation 224
Transcriptional Regulation 224
Posttranscriptional Control 241
Feedback Inhibition and Allosteric Control 243
A Future Practical Application? 245

CHAPTER 12 **Regulation of Gene Activity in Eukaryotes** **248**
Important Differences in the Genetic Organization of
    Prokaryotes and Eukaryotes 249
The Regulation of Transcription Initiation 251
The Regulation of RNA Processing 260
Regulation of Nucleocytoplasmic mRNA Transport 267
Regulation of mRNA Stability 268
Regulation of Translation 270
Regulation of Protein Activity 272

Gene Rearrangement: Joining of Coding Sequences in the
    Immune System 275
A Future Practical Application? 280

CHAPTER 13   **Genomics and Proteomics Drive Information-
Age Biology** **284**
Genomics—The Use of DNA As a Starting
    Point for Discovery 286
Bioinformatics—Using DNA Sequence
    Information to Build Knowledge 289
Proteomics Focuses on the Totality of Proteins: Their Numbers,
    Structures, Interactions, Locations, and Functions 294
The Emergence of a New "Logic of Molecular Biology"? 298
A Future Practical Application? 299

Part IV   **Experimental Manipulation of Macromolecules**   **303**

CHAPTER 14   **Transposons, Plasmids, and Bacteriophage** **304**
Transposable Elements—Their Discovery Surprised Molecular
    Biologists 306
Transposable Elements in Eukaryotes 312
Plasmids 314
Plasmid-Borne Genes 316
Plasmid Transfer 317
Plasmid DNA Replication 322
Bacteriophage 324
Stages in the Lytic Life Cycle of a Typical Phage 326
Specific Phage 328
Transducing Phage 339
Transposons, Plasmids, and Bacteriophage in Genetic
    Engineering 341
A Future Practical Application? 341

CHAPTER 15   **Recombinant DNA and Genetic Engineering:
Molecular Tailoring of Genes** **346**
Plasmids Act As Nature's Interlopers 347
Restriction Enzymes Function As Nature's Pinking Shears 348
Genetic Interlopers: Vectors Function As Vehicles for
    Transferring Genes 357
Detection of Recombinant DNA Molecules 362
Site-Specific Mutagenesis Using
    Bacteriophage M-13 Vector 363
Applications of Genetic Engineering 366

CHAPTER 16   **Molecular Biology Is Expanding Its Reach** **370**
Uses of Recombinant DNA Technology in Research 371

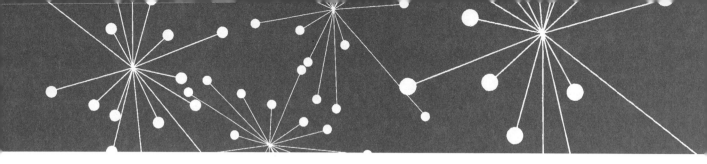

Uses of Recombinant DNA Technology in Medicine   377
Uses of Recombinant DNA Technology in Agriculture   387
Other Commercial and Industrial Applications   390
Molecular Biology: On the Front Line
  of the Battle Against AIDS   391
Social and Ethical Issues   393

POSTSCRIPT   **Postscript to Your Review of Molecular Biology 403**
Molecular Biology Is Enjoying a Golden Era!   403
Speculation—Let's Anticipate a Few
  Discoveries in Molecular Biology!   405
Enhance Your Ability to Learn Molecular Biology!   407
Consider Becoming a Molecular Biologist!   409

APPENDIX   **Chemical Principles Important for Understanding
Molecular Biology 411**
Structure of the Atom   411
Chemical Bonds   412
The Ionization of Water—The pH Scale   417
Organic Chemistry   420
Concluding Note   434

List of Essential Concepts 435

Glossary 439

Answers to Questions and Problems 455

Index 479

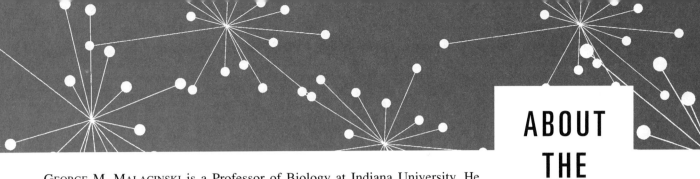

ABOUT THE AUTHOR

GEORGE M. MALACINSKI is a Professor of Biology at Indiana University. He earned his undergraduate degree at Boston University and his graduate degrees at Indiana University. After completing his postdoctoral training at the University of Washington (Seattle), he returned to Indiana University and established a research program in amphibian developmental genetics. His early years were devoted to the analysis of mechanisms that control patterning in early embryogenesis. More recently, he has been researching heart and skeletal muscle development to understand the regulatory mechanisms that establish the differentiation patterns in those muscle types.

For several years Dr. Malacinski has been teaching molecular biology to undergraduates. His course is organized in a collaborative learning format. Students attend traditional lectures three times per week and meet in small groups (under the supervision of a teaching assistant) once per week to work on problem sheets. During lectures he encourages students to enhance their ability for critical (analytical) thinking by engaging in problem solving exercises, becoming proficient at analyzing data (e.g., graphs and charts,), and learning how to design experiments. He uses *Essentials of Molecular Biology* as a template for organizing his course—students are expected to read the entire book! Often, however, he digresses in lectures to model thought processes or explain the logic/rationale that underlies an experimental design. For example, the logic described in Chapters 1 and 13 is reviewed from time to time in the context of a specific data set/concept/illustration during the course of the semester.

Students have responded to Dr. Malacinski's teaching methods by nominating him for various teaching awards.

PREFACE

THE REMARKABLE GROWTH experienced by the discipline of molecular biology is unprecedented in any area of science. Indeed, several commentators have stated that *molecular biology is perhaps the fastest progressing of all of our intellectual endeavors!* Initially, most efforts in the early days of this discipline were devoted to characterizing the physical and chemical properties of biological macromolecules such as DNA, RNA, and protein. Once those parameters were generally understood, attention was given to elucidating the manner in which those macromolecules function in the living cell. Recently, the pyramiding of that knowledge has led to the development of a robust biotechnology industry, one that regularly heralds advances in molecular medicine.

As successes accumulated, molecular biologists developed high levels of confidence that problems that only a decade or so ago were considered "formidable" could now be solved. Their efforts have contributed to an ever increasing understanding of the origin of the human species; to the creation of new, genetically engineered drugs; and to the sequencing of the entire genome of a variety of species, including disease-causing microorganisms, agriculturally important plants, and, of course, the human organism itself.

The scope of molecular biology is now so broad as to have an impact on virtually all areas of biology, from the more traditional disciplines of taxonomy and systematics, to the more contemporary courses of study such as gene function. To ensure that *Essentials of Molecular Biology* captures the distinctive features of this field, a distinguished group of contributors was enlisted to edit the second and third editions of this book. They brought to this textbook project high levels of research, achievement, and perspectives that validate the winnowing processes necessary to derive essential features, key concepts, and general principles. For this fourth edition, reviewers were engaged to ensure that the text is accurate and up-to-date. I gratefully acknowledge their efforts—a complete listing of contributors and reviewers to the recent editions can be found in the Acknowledgments section.

Because of the broad extent to which molecular concepts have been integrated into most biology disciplines (including cell biology, developmental biology, genetics, and even evolutionary biology), the faculty at Indiana University, where I teach, revised its basic curriculum to position Molecular Biology as the first course beyond the freshman level. To provide an effective learning experience for entry-level biology majors, an appropriate textbook was required—one that would not intimidate students by its depth, or overpower them with its quantity of detail. *Essentials of Molecular Biology* was selected because it emphasizes the fundamental features of various aspects of DNA, RNA, and protein structure, function, and expression without delving into an inordinate amount of detail on the related molecular processes.

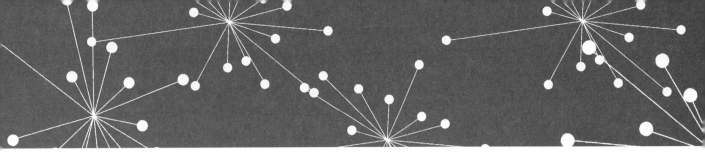

The first and second editions of *Essentials of Molecular Biology* were hailed by students at Indiana University and at other institutions at which it was used for their success in making molecular biology understandable. One of the key features of these editions was their "layering" approach to knowledge construction. Complexity was developed chapter by chapter, rather than having the information presented all at once as a series of detail-laden descriptions. Students appreciated this approach, which was retained in the third and, now, the fourth edition. Recently, the increased emphasis on a "critical thinking" approach to the undergraduate biology learning experience at Indiana University encouraged revisions such as the explicit discussion of the logic and rationale that facilitate progress in molecular biology found in Chapters 1 and 13 of this edition.

The Preface explains how this edition differs from the previous editions and, I hope, how it improves on them. Each new feature was designed with the help of undergraduate students to ensure that this textbook continued to be "user friendly."

Note to Instructors

Instructors who use this textbook prefer *brevity of presentation* and emphasis on *fundamental features* of macromolecular structure and function. My own experience reveals that undergraduate biology students who are beginning to learn molecular biology prefer a concise text, rather than an encyclopedic one, which they often find intimidating. An earlier edition of *Essentials of Molecular Biology* was recognized by the National Health Museum (NHM) *as one of the most readable texts in the field.* The NHM differentiated it from one of the more comprehensive textbooks, which it indexed as a *veritable encyclopedia of molecular biology and cell biology.*

This new edition continues to embrace the "anti-encyclopedic" theme. Nevertheless, despite suggestions from some reviewers that I suppress enlargement or avoid expansion, this edition is slightly enlarged, for it contains two new features: A chapter on genomics, bioinformatics, and proteomics (Chapter 13), which explains the growing importance of the study of nucleotide sequence information for understanding cell function, and a Postscript, which offers beginning students a glimpse into the research realm of molecular biology. Otherwise, the succinctness of previous editions has been maintained.

Essentials of Molecular Biology, therefore, differs from the "mega-books" in purpose, in scope, and in depth. Because the big books contain more background information and substantially more detail (often in the form of color-coded illustrations), they are—by nature—not well-suited for beginning undergraduate courses, which emphasize intellectual development (the main goal of an undergraduate education). Those mega-books may give students the erroneous impression that *science is about facts,* when, indeed, practitioners gener-

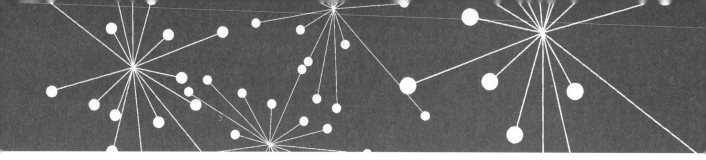

ally agree that science is really about *using experimentation to collect the evidence that establishes truths.*

To enhance the user-friendly features of earlier editions, the present edition was extensively edited by undergraduate students. Frequently, sentences were shortened, and the tendency to cram too much information into a phrase or paragraph was subdued. The use of "stick diagrams" as illustrations, in the style that an instructor would draw on the blackboard or on an overhead transparency, was maintained in accordance with student preferences. Because the discipline of molecular biology is currently in a data-collection mode, I anticipate that the "key concepts" that punctuate the text will provide meaning to the collection of details that define a reductionist discipline such as molecular biology. These key concepts should help alleviate the tendency for students to become frustrated with what they often refer to as "information overload."

I have found that many undergraduates are aware of the distinction between what the discipline of molecular biology actually represents—an intellectually sophisticated experimental strategy (reductionism) for learning about cell behavior—and what it is often portrayed as in encyclopedic textbooks—merely another body of knowledge (collection of detailed facts).

I encourage instructors to contact me for access to the supplementary material that I use in my L211 Molecular Biology course at Indiana University. Old examinations, weekly worksheets used in our collaborative learning discussion groups, and access to a course Internet site are available on request.

Acknowledgments

I gratefully acknowledge my L211 Molecular Biology undergraduate students who helped improve the this text's readability. Susan Duhon assisted with the editing and prepared the index and is hereby thanked. Graduate teaching assistants, who edited and improved the questions and problems at the ends of chapters, are also acknowledged.

For each revision expert reviewers have been consulted. Their impact has been substantial, and cumulative. In many instances their contributions have defined the scope and content of various chapters. Those specialists include Sankar L. Adhya, Developmental Genetics Division, National Cancer Institute; Gerald Becker, Lilly Laboratories, Indianapolis, IN; Robert Cedergren, Department of Biochemistry, University of Montreal; Larry Gold, SomaLogic, Inc., Boulder, CO; Philip C. Hanawalt, Department of Biological Sciences, Stanford University; Pamela L. Hanratty, Department of Biology, Indiana University; Christie A. Holland, Children's Medical Center, Washington, DC; Leonard Holmes, Department of Chemistry and Physics, University of North Carolina, Pembroke; Masao Kawakita, Department of Applied Chemistry, Kogakuin University, Tokyo, Japan; Kenneth Marians, Department of Molecular Biology, Memorial Sloan-Kettering

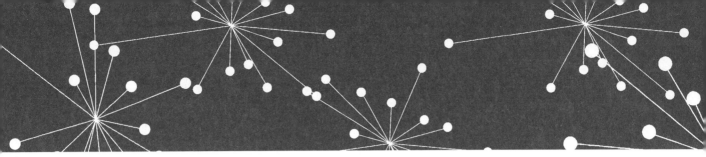

Cancer Institute; Hans-Peter Müller, Institute of Molecular Biology, University of Zurich; Alexandra Cynthia Newton, Department of Pharmacology, University of California, San Diego; David Parma, SomaLogic, Inc., Boulder, CO; Barry Polisky, ThermoBioStar, Inc., Boulder, CO; John Richardson, Department of Chemistry, Indiana University; Walter Schaffner, Institute of Molecular Biology, University of Zurich; Robert L. Sinsheimer, Department of Biological Sciences, University of California, Santa Barbara; and Dorothy M. Skinner, Biology Division, Oak Ridge National Laboratory.

GEORGE M. MALACINSKI
Bloomington, Indiana

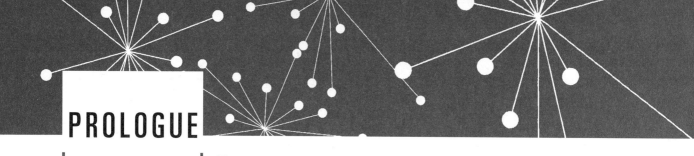

PROLOGUE

THIS FOURTH EDITION OF ESSENTIALS OF MOLECULAR BIOLOGY is the third revision of David Freifelder's highly successful original edition. It is intended to serve as the textbook for a traditional one semester first course in molecular biology. Given the broad extent to which the concepts of molecular biology have been integrated into most of the other disciplines of biology (including cell biology, developmental biology, genetics, and evolutionary biology), the faculty at Indiana University revised its basic curriculum to position molecular biology as the first course beyond the freshman level. As a result, this textbook is used by several hundred students at Indiana University each year.

For students beginning their study of molecular biology, these are exciting times. No discipline in biology has ever experienced the explosion in growth and popularity that molecular biology is enjoying. Most scholars acknowledge that the discipline began its rapid growth in the early 1950s with the simple yet elegant descriptions of the structure of the DNA double helix. Intense intellectual curiosity provided the initial driving force for progress. Then, a series of technological advances propelled the discipline to the point where its horizons now appear to be virtually unlimited.

During the early years, the focus was on the purification and characterization of the key macromolecules in DNA, RNA, and protein synthesis. From information on the biochemical and physical properties of the key macromolecules, functional roles were deduced. Aided by discoveries in microbial genetics, the general blueprint for gene structure and expression was elucidated in a remarkably short time. More recently, success with gene isolation (recombinant DNA technology, or genetic engineering) has provided all the confidence molecular biologists needed to believe that a complete molecular understanding of virtually any biological phenomenon is within reach. That attitude, right or wrong, has in turn spawned an unprecedented interest in the commercial applications of biology. Molecular biology has led to the establishment of novel ties between universities and industries and has powered the development of new disciplines such as molecular medicine.

Molecular biology is, by its very nature, *reductionist;* that is, it deals with ever smaller pieces (i.e., molecules) of the whole (i.e., cell). To provide an effective learning experience for a biology student such as yourself, an experience that does not intimidate you with depth or overpower you with detail is desirable. *Essentials of Molecular Biology* strives to enhance your learning experience by providing you with a reasonable amount of detail, as well as an overview of the importance of various macromolecules to the functioning of living cells. It is designed to guide you through a rewarding learning experience.

The content and format of this book has been field tested extensively by the hundreds of undergraduates who use it each year at Indiana University. This fourth edition has features that should help you avoid getting lost in a sea of details:

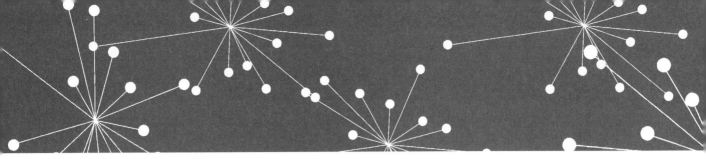

The text is divided into sections (Parts I–IV), which should help you focus your learning strategies and efficiently gain an understanding of the subject matter.

The relationship of the parts to the whole should easily be grasped with the progression diagram and the list of specific aims that begin each chapter.

Key concepts are highlighted in each chapter, so the generalization or principle being developed is stated in "nutshell" form.

Genomics, bioinformatics, and proteomics—the newest frontiers in molecular biology are described in a new chapter (13). The logic and rationales that drive those "information-age" endeavors are reviewed therein.

The Postscript guides you through the research realm of contemporary molecular biology and offers a list of suggestions for enhancing your ability to learn molecular biology.

One of the key features of this text is the "layering" approach to knowledge construction. Complexity is developed chapter by chapter, rather than presented all at once as a series of detail-laden descriptions. Students at Indiana University appreciate this approach! Indeed, several hundred of our biology undergraduates have been surveyed to ascertain how they think and, consequently, how they learn. Results have clearly demonstrated that our biology students are linear thinkers; therefore, the layering approach is much appreciated.

To improve your critical (i.e., analytical) thinking skills, graphs and tables of data are interspersed throughout the text. Together with the questions and problems at the end of each chapter, suitable opportunities are thereby readily available to accelerate your intellectual development.

Please welcome yourself to molecular biology by reading Chapter 1. Enjoy your learning experience! Molecular biology offers a myriad of opportunities for your personal gain—intellectual development, new knowledge, and a stepping stone for further study in other areas of biology.

GMM

Welcome to Molecular Biology!

Goals of Molecular Biology

The ultimate goal of molecular biology is ambitious: to understand the five basic cell behavior patterns (growth, division, specialization, movement, and interaction) in terms of the various molecules that are responsible for them. That is, molecular biology wants to generate a complete description of the structure, function, and interrelationships of the cell's macromolecules, and thereby to understand why living cells behave the way they do.

This goal might appear overly zealous. Yet the rate at which progress is being made often astonishes even the most optimistic scientists. One might, in fact, consider these years to represent a golden era for biology, with the field of molecular biology providing the main driving force.

Significant discoveries are emerging from research laboratories nearly every day and the front pages of national newspapers frequently herald exciting announcements of the identification of disease-causing genes, or promising biotechnology products, or new agricultural processes.

A few decades ago the most important discoveries in molecular biology were made using the simplest organisms (e.g., viruses and bacteria). Now-adays, however, equally important findings are regularly reported for both plants and mammals. A few key discoveries and the efforts of a small group of pioneering scientists have set the stage for the present era.

The Early Years

The term molecular biology was first used in 1938 by Warren Weaver. As Director of the Natural Sciences Section of the Rockefeller Foundation, he advocated that financial support be given to this "new branch of science—a new biology—**Molecular Biology.**" By that time, biochemists began to discover many fundamental intracellular chemical reactions, and to appreciate the importance of specific reactions and of protein structure in defining the numerous properties of cells. However, the development of molecular biology itself could not begin until this realization was reached: The most productive advances would be made by

Molecular Biology

studying "simple" systems such as bacteria and bacteriophages (bacterial viruses). Although bacteria and bacteriophages are still quite complicated, they are far simpler than animal cells. In fact, they enabled scientists to identify DNA as the molecule that contains most, if not all, of the genetic information of a cell.

Although DNA was first described in 1869 by F. Miescher, its significance to cell function and the definitive proof that it is responsible for inherited traits did not come until almost a century later. The experimental evidence that led to the assignment of genes to DNA depended heavily on the use of bacteria and their viruses (described in Chapter 6).

Once it became clear that DNA contained the chemical basis of heredity, it was not long before J. D. Watson and F. H. C. Crick offered a model for the physical structure of DNA. That model (described in Chapter 3) also proposed a mechanism for DNA replication and the spontaneous origin of mutations. Shortly thereafter, RNA was revealed to be an intermediate for the synthesis of enzymes and other proteins.

Following these discoveries, the new field of molecular genetics progressed rapidly in the late 1950s and early 1960s. It provided new concepts at a rate matched only by the development of quantum mechanics in the 1920s. The initial success and the accumulation of an enormous body of information enabled researchers to apply the techniques and powerful logical methods of molecular genetics to a variety of subjects: muscle and nerve function, membrane structure, the mode of action of antibiotics, cellular differentiation and development, immunology, and others. Faith in the basic uniformity of life processes was an important factor in this rapid growth. That is, it was believed that the fundamental biological principles that govern the activity of simple organisms, such as bacteria and viruses (organisms that lack an organized nucleus), must apply to more complex cells; only the details should vary. This faith has been amply justified by experimental results.

In this book, prokaryotes and eukaryotes will often be discussed separately and compared and contrasted. Usually prokaryotes will be discussed first, because they are simpler. In keeping with this practice, we begin by briefly reviewing the properties of several model living systems, including bacteria and their viruses.

Model Biological Systems

virus

The simplest living organism is, of course, the **virus.** As a minimalist life form, it consists of a DNA (or, in some cases, an RNA) inner core surrounded by a protein coat. The key to the virus's simplicity is its parasitic nature. It borrows functions from its host cell. The host is, for some kinds of viruses, a bacterial cell, while for others it is a plant cell, and for yet others it is an animal cell. Of these hosts, the bacterial cell is the simplest. We will review its general features here briefly, then return to viruses.

bacteria

Bacteria are free-living unicellular organisms. They have a single chromosome, which is not enclosed in a nucleus (they are prokaryotes), and, compared to eukaryotes, they are simple in their physical organization. For all practical purposes, a bacterium can be thought of as consisting of several thousand chemicals and a few organized particles, all in liquid solution, enclosed in a rigid cell wall.

Bacteria have many features that make them suitable objects for the study of fundamental biological processes. For example, they can be grown easily and rapidly and, compared to cells in multicellular organisms, they are relatively simple in their needs. The bacterium that has served the field of molecular biology best is *Escherichia coli* (usually referred to as *E. coli*), which divides every 20 minutes at 37°C under optimal conditions. Thus, a single cell becomes 10^9 bacteria in about 20 hours!

Bacteria can be grown in a **liquid growth medium** or on a solid surface. A population growing in a liquid medium is called a bacterial **culture.** If the liquid is a complex extract of biological material, it is called a **broth.** If the growth medium is a simple mixture containing no organic compounds other than a carbon source, such as a sugar, it is called a **minimal medium.** A typical minimal medium contains each of the ions Na^+, K^+, Mg^{2+}, Ca^{2+}, NH_4^+, Cl^-, HPO_4^{2-}, SO_4^{2-} and a source of carbon (such as glucose, glycerol, or lactate). If a bacterium can grow in a minimal medium—that is, if it can synthesize *all* necessary organic substances, such as amino acids, vitamins, and lipids—the bacterium is said to be a **prototroph.** If any organic substances other than a carbon source must be added for growth to occur, the bacterium is termed an **auxotroph.**

Bacteria are commonly grown on solid surfaces. The earliest surface used for growing bacteria was a slice of raw potato. This was eventually replaced by **agar,** a jelling agent obtained from seaweed. Agar is resistant to the action of bacterial enzymes and, hence, is considered biologically inert. **Figure 1-1** illustrates growth of bacteria on an agar surface.

Metabolism in bacteria is precisely regulated. Thus, bacteria represent the most efficient free-living organisms yet discovered. They rarely synthesize substances that are not needed. For example, the amino acid tryptophan is not formed if tryptophan is present in the growth medium, but when the tryptophan in the medium is used up, the tryptophan-synthesizing enzymatic system will be quickly activated. The systems responsible for the utilization of various energy sources are also efficiently regulated. A well-studied example is the metabolism of the sugar lactose as an alternate carbon source to glucose. Control of both tryptophan

minimal medium

auxotroph

agar

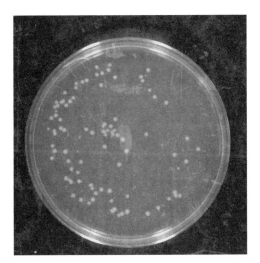

Figure 1-1 A glass dish displaying colonies of *E. coli* grown on the surface of an agar-containing growth medium. Each colony contains a cluster, or clone, of cells that represents the progeny of a single cell that divided many times. For instance, if 100 cells were spread on the agar surface, 100 colonies would appear the next day.

synthesis and lactose degradation are two examples of **metabolic regulation.** This very general phenomenon will be explored extensively throughout the book, especially in Part III. Both simple and complex regulatory systems will be described, all of which act to determine how much of a particular compound is utilized and how much of each intracellular compound is synthesized at different times and in different circumstances. We will learn about the lengths to which the so-called simple cells go to utilize limited resources efficiently and to optimize their metabolic pathways for efficient growth.

The ease with which bacteria can be grown in the laboratory on agar surfaces or even in liquid growth medium in large, industrial-scale tanks has facilitated rapid progress in the physical and chemical characterization of macromolecules. For example, the original descriptions of nucleic acids (Chapter 3) and proteins (Chapter 4) were largely made with components fractionated and isolated from mass quantities of bacterial cells. In addition, knowledge of metabolic regulation (Part III) can be traced back to early studies of bacteria, and more recently to plant and animal cells, grown under conditions similar to those illustrated in Figure 1-1.

Bacteriophage

bacteriophage
phage

Once the culture conditions for growing bacteria were established and many of the metabolic processes of normal bacterial life were known, bacterial viruses (**bacteriophage,** or the shortened form **phage**) were studied in earnest. Being much simpler than bacteria (unlike bacteria, they are not "free-living"), they could be studied in a relatively straightforward fashion. Since they indeed represent the simplest form of life, as their biochemical features became known, several physicists began to study them in the hopes of discovering new laws or first principles of physics!

Figure 1-2 illustrates a bacteriophage that is relatively complex. It contains a protein coat, or phage head, to which is attached a tail. Some phages are simpler yet, and lack well-defined tail structures.

Molecular biologists have utilized viruses as simple model systems for many kinds of studies. One of the most significant studies used the minimalist protein

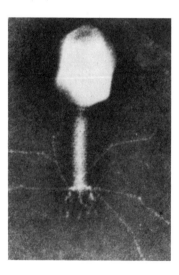

Figure 1-2 An *E. coli* T4 phage. The DNA is contained in the head. Tail fibers come from the pronged plate at the tip of the tail and serve to attach the virus to the host bacterium's surface.

coat/DNA core composition of phage to establish whether protein or DNA carries heredity information. Chapter 6 will explain the life cycle of a typical bacteriophage and review how phage, like the one illustrated in Figure 1-2, were used in pioneering experiments that helped prove that DNA rather than protein contains genetic information.

Archaebacteria

In addition to those two broad categories of organisms, another group of prokaryotes has recently been designated—the **Archaebacteria.** They are probably modern descendants of a very ancient type of prokaryote. Features that distinguish them include gene expression and cell division mechanisms, which are intermediate between typical prokaryotes and eukaryotes. Microbiology texts should be consulted to gain an understanding of the ways in which they differ from both prokaryotes and eukaryotes, such as energy metabolism, environmental tolerance, and ribosomal RNA sequence.

archaebacteria

Yeasts

Another favorable model system that shares many of the advantages of bacteria is yeast [**Figure 1-3**(a, b)]. They are, however, eukaryotes. Having a true nuclear membrane surrounding their chromosomes, they represent a higher level of organization than bacteria. That level approaches the complexity of animal (e.g., human) cells. Yet being a microorganism, yeast can be grown and manipulated much like *E. coli.*

Yeasts have been used for millennia for producing wine and beer. A great deal of early biochemical research was carried out with yeasts rather than bacteria, work stimulated mainly by interest in understanding and improving beer.

In contemporary molecular biology, mutant strains of yeast are often employed to discover genes that control growth, division, and cell behavior patterns. In addition, yeasts are presently employed as tools for producing large numbers of copies of human chromosome fragments. So-called "yeast artificial chromosomes" represent miniature chromosomes that contain foreign (e.g., human) DNA. They are propagated within yeast cells, providing the molecular biologist with opportunities for genetic engineering.

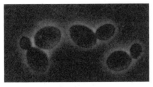

(a)

(b)

Figure 1-3 (a) A light micrograph of the yeast *Saccharomyces cerevisiae.* Many cells are budding by outgrowth from the cell wall of the mother. (Courtesy of Breck Byers.) (b) A fluorescence micrograph of a single cell stained with a fluorescent dye that binds to nucleic acids. The bright spot is the nucleus. The dark region is the vacuole, a liquid-filled sac that is free of nucleic acids.

Animal Cells (and Embryos)

Many types of animal cells, including several kinds of human cells, can be cultured using methods based on bacteria and yeast culture techniques. **Primary cell cultures** represent normal animal tissue that is usually derived from either skin cells (**Figure 1-4**) or embryos. They grow well initially, but eventually die off. Tumor cells, however, grow indefinitely, and are often easier to propagate in culture. Such tumor (or "transformed") cells are therefore frequently used for routine laboratory experimentation.

Cells from early-stage mammalian (e.g., mouse) embryos provide an especially versatile model system. When cultured in an appropriate medium, they specialize in various ways that can be conveniently studied by the molecular biologist. These so-called **embryonal stem cells** can, in some instances, be grown in the presence of various culture medium supplements and thereby specialize into one or another cell type (e.g., muscle, nerve, liver, etc.). As it holds great promise as a starting point for the production of tissues and organs as human replacement parts, embryonal stem cell research is attracting great interest among molecular biologists. Molecular biologists also can introduce foreign genes into isolated embryonal stem cells and then reintroduce them into a growing (mouse) embryo. Often, the foreign genes will be expressed as the embryo develops, providing novel insights into the ways gene expression is regulated.

In addition, molecular biologists have succeeded in injecting foreign genes directly into animal eggs and thereby generating **transgenic animals.** This experimental procedure will be described in detail in Chapter 16. In the case of some animals, improved agricultural productivity, such as enhanced milk production from dairy cows, has been achieved. Some scientists speculate that similar procedures will one day be applicable to humans. Attempts will be made to inject foreign genes into human cells in order to correct genetic defects. Such **gene therapy** strategies are also briefly discussed in Chapter 16.

Plant Cells

The discovery that the cells of many plants can be nurtured and cultivated into whole, complete plants expanded the horizons of molecular biologists. Studies on

embryonal stem cells

gene therapy

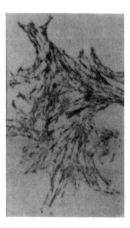

Figure 1-4   A microcolony of Chinese hamster fibroblasts that have been growing for a few days on a glass surface. (Courtesy of Theodore Puck.)

the mechanism of action of plant hormones such as auxin were given a tremendous boost when cultivation methods for plant cells were perfected.

More recently, a small simple plant, the common wall cress (*Arabidopsis*), which can be grown in a test tube, is being established as a tool for molecular genetics. It has a small amount of nuclear DNA, and exhibits several advantageous features (such as the ease with which genetic variants can be produced) that will be exploited by molecular biologists to learn about gene function in higher plants.

Methodology of Molecular Biology

For certain, the choice of appropriate research organisms was important for advancing the knowledge base of molecular biology. Especially advantageous was the fact that, despite the vast array of bacteria, viruses, and animal cells that were potentially available for laboratory experimentation, a consensus among researchers quickly emerged. A limited selection, which included *E. coli,* its bacteriophages, and only a small number of animal cell lines, were chosen as focal points. Thus, during the early years, researchers were able to conveniently exchange information and collaborate on achieving common goals, such as preparing genetic maps and understanding metabolic regulation.

Physical and Chemical Characterization Methods

The development of a vast array of laboratory methods was, however, equally important. In several cases, the application of those methods provided great leaps in the research scientist's ability to discover new features and functions of macromolecules. The originators of several of those breakthrough methods were awarded Nobel Prizes. Some of those methods will be described in this textbook as we encounter the phenomena to which they have been applied. For now, a summary of many of them is included in **Table 1-1**.

In the review of those methods, it should quickly become apparent that they provide ways to learn mainly about physical and chemical features of macromolecules. Most of them represent procedures for measuring or visualizing eversmaller quantities or tinier features of molecules, or parts of macromolecules.

Especially when applied in combination, the application of these methods has led to enormous advances in understanding the structural features of proteins and nucleic acids. For example, electrophoresis of nucleic acids and hybridization have been coupled (e.g., Southern blotting described in Figure 8-13) to learn about gene expression patterns in cells that are undergoing specialization. The polymerase chain reaction and DNA sequencing are often coupled to gain knowledge about the structure of a particular gene. From this knowledge, researchers can investigate the way cells regulate genes during growth.

Genetic Methods for Molecular Biology

Many of the most significant advances in molecular biology have come about through the application of genetic analysis to molecular phenomena. Initially, genetics provided an abstract, logical system for deducing relationships between specific genes and the metabolic processes for which they are responsible. Mutant

TABLE 1-1	HISTORICAL REVIEW OF SEVERAL IMPORTANT LABORATORY METHODS FOR MOLECULAR BIOLOGY		
Method	**Approximate Introduction**	**Brief Description**	**Sample Textbook Reference**
Ultracentrifuge*	1920s	Estimation of size and shape of molecule in centrifugal field	
Electrophoresis*	1930s	Separation of proteins or nucleic acids based on size and/or charge in electric field	Figure 2-11
Electron microscope	1930s	Direct visualization of cell ultra-structure, including nucleic acids	Figure 1-2
Radioisotope tracers*	1940s	Tracking flow of molecule through a metabolic pathway	Figure 7-17
Column chromatography	1940s	Separation of macro-molecules through a resin column	
X-ray diffraction*	1950s	Provides precise measurement of three-dimensional structure of protein or nucleic acid	
Amino acid analysis* (and sequencing)	1950s	Determination of order of amino acids in proteins	
Nucleic acid hybridization	1960s	Quantitative assessment of similarity of RNAs or DNAs	Figure 3-10
Nucleotide sequencing*	1970s	Determination of sequence of bases in DNAs	Figure 3-15
Recombinant DNA technology*	1970s	Genetic engineering of novel genes	Figure 15-9
DNA synthesis	1980s	Custom synthesis of desired nucleotide sequences	Figure 3-18
"Monoclonal" antibody histo-chemistry*	1980s	Highly specific reagents for detecting proteins	
Polymerase chain reaction*	1980s	Production of large numbers of copies of small amount of DNA starting material	Figure 15-5
Automated DNA sequencing	1990s	Coupled processing and reading of dideoxynucleo-tide fragments	
Gene knockouts*	1990	Specific genes (in mouse) are inactivated	Figure 16-1

*Denotes associated Nobel Prize

organisms have traditionally provided the tools that allowed geneticists to develop models or schemes to explain how a metabolic process works, or how a structural component contributes to cell function.

In some cases, mutants have allowed the molecular biologist to recognize processes previously not known to exist (e.g., steps in DNA replication that are reviewed in Chapter 7). In other instances, mutants have helped scientists unravel a complex metabolic pathway (e.g., carbohydrate metabolism, reviewed in Chapter 11). Perhaps most importantly, mutants often allow scientists to establish a direct correlation between an identified gene and its previously unknown function (e.g., termination step in protein synthesis, described in Chapter 9).

More recently, with the availability of pure genes prepared using methods described in Table 1-1, nucleotide sequences can be altered. Those alterations can be directed to specific nucleotides, either through simple pinpoint substitutions or through deletion. In these ways, the role particular nucleotide sequences play in determining protein structure and/or function can be directly tested. No longer is it necessary to rely exclusively on naturally occurring or laboratory-induced mutations in whole organisms. Beginning with a pure gene, a researcher can work back to a function in the intact organism ("reverse genetics").

Rapid Progress in Molecular Biology

Direct experimentation using the model systems and laboratory methods just described has proven to be a powerful driving force in molecular biology. A variety of thought processes has emerged as the discipline has matured, and they guide today's molecular biologist. Although certainly not exclusive to molecular biology, these thought processes are employed by most molecular biologists on an almost daily basis as data are collected, interpretations are formulated, conclusions are drawn, and additional experiments are designed. The following are examples of what constitutes the **logic of molecular biology.**

The Efficiency Argument

During the hundreds of millions of years that living organisms have been evolving, the rigors of competition for survival in an environment where resources are limited has led to a natural selection for efficiency. Accordingly, when attempting to understand how a mechanism or process works, molecular biologists intuitively favor the simplest, most efficient scheme over more complex, contrived, or cumbersome schemes. This approach has been especially useful for understanding the molecular mechanisms involved in regulating bacterial metabolism, RNA, DNA, and protein synthesis, as well as phage life cycles. However, analogous processes in the cells of higher plants and animals lack this intense competitive pressure between individual cells for survival. Simple, efficient mechanisms, therefore, do not always exist in the cells of higher organisms.

Phylogenetic History

When attempting to explain why a particular mechanism works the way it does, researchers often make comparisons with similar mechanisms in organisms that are either more or less advanced on the evolutionary scale. Species divergence is often due to changes in molecular processes. By comparing these processes among evolutionarily diverse species, biologists frequently gain a deeper understanding of molecular mechanisms.

Quantitative Assessment

The molecular biologist is constantly searching for quantitative relationships. In the process of unraveling a molecular mechanism, an assessment of the timing, chemical balance, and flow of components in space is routinely made in order to establish whether a proposed mechanism or hypothesis is feasible. Often, proposed pathways or schemes can be either approved or disregarded based on quantitative assessments.

Conceptual Model Building

Models attempt to explain complex hypotheses by casting them in a readily understandable form (often in the form of a diagram or cartoon). In this context, they often represent conceptual entities rather than the three-dimensional hardware associated with molecular model kits for constructing larger-than-life representations of the structure of one or another macromolecule (e.g., a protein). For example, models have been developed to explain how an antigen stimulates the immune system to produce antibodies that react specifically with the challenge antigen.

Conceptual models are generally regarded as being good only if they can be subjected to a direct experimental test. That situation generally applies to the usefulness of almost any hypothesis. Molecular biologists are, therefore, constantly revising and refining their models as the interpretations of new experimental results call for modification.

Parallelism

Molecular biologists have great faith in the universality of basic biological processes. As a starting point in an attempt to understand a molecular mechanism or process, explanations that are satisfactory for simple systems (e.g., bacteriophages) are often scaled up to fit similar phenomena in more complex systems (e.g., mammalian cells). This approach does not always work, as has been demonstrated for messenger RNA processing, which is inherently different in eukaryotic cells. Nevertheless, newly discovered phenomena in molecular biology are often initially explained in terms of similar processes that had been analyzed earlier.

Strong Inference

Like all scientific enterprises, molecular biology is a human endeavor. Accordingly, intuition, which we might define here as a sort of comprehension based more on experience and common sense than on a conscious and elaborate line of logic argument, is often employed to develop a set of explanations or possible models for a particular molecular event. Strong inference is based on the exclusion of all but one final alternative explanation for a phenomenon. Initially, one states all the reasonable explanations for a particular phenomenon. Then, by direct experimentation, various possible explanations are ruled out one by one. The final alternative, which cannot be ruled out, and which satisfies our common sense, is considered most likely to be correct. This strong inference approach accounts for the relatively common use of the phrase "it is likely that" when experimental data are being interpreted.

Despite the importance of intuition in "strong inference," do not forget the critical role played by experimental evidence!

Optimism

Molecular biology thought processes enthusiastically endorse "optimism" as a key feature. That the use of the reductionist approach, which breaks a process down to its simplest components (e.g., molecules), will continue to be productive is a basic tenet of the discipline. Indeed, many molecular biologists believe that not only cellular functions but also increasingly higher-order biological phenomena (including perhaps even human behavior) will be comprehended in terms of molecular processes.

Rapid progress in molecular biology has generated literally thousands of volumes of detailed information about the structure and function of cellular components. Organizing those details in a comprehensive way represents a challenge for both the beginning student and the accomplished professional scientist. In order to develop a working knowledge of the discipline, details about molecular structure and function require attention, and generalizations need to be made.

Putting the Details of Molecular Biology in Perspective

Molecular biology has profited greatly from the use of the so-called "reductionist" approach to the study of biological phenomena. That is, many molecular researchers believe the whole (organism) can only be fully understood when it is dissected and analyzed. Eventually, the lowest common denominators, such as the nature of the bonding between DNA and proteins, are revealed and general theories are constructed. Often, the analysis of one or another aspect of a single molecule (e.g., the enzyme lysozyme) has provided the focus of a large number of research laboratories, each following a similar reductionist theme.

That approach represents a sharp contrast to the "holistic" approach, which gives special emphasis to interactions between components. The old adage—"the whole is greater than the sum of its parts"—applies to the holistic approach. This approach emphasizes the integration of various biological phenomena into higher-order systems. In some cases, those systems are represented by individual types of organisms (e.g., the physiology of amphibia; the human embryo), while in other cases several organisms are integrated into even higher orders of organization, such as the disciplines of animal or plant ecology.

Because of the spectacular success of the reductionist approach in molecular biology, modern researchers instinctively favor reductionist strategies. However, beginning students need to appreciate the fact that individual molecules do not function in a vacuum. They function rather as components of complex gene expression mechanisms, in metabolic pathways, or as structural elements, always in concert with other molecules.

Those various mechanisms, pathways, and structural elements themselves also operate in the context of other systems, which are themselves usually organized into cells. In eukaryotic organisms, those cells cluster to form tissues or organs that in turn comprise a single whole organism. Finally, whole organisms interact with one another and their environment to form an ecosystem.

Individual molecules are, of course, the basis of all living activities. Nevertheless, in order to fully comprehend the function and significance of any single molecule, an excursion beyond the test tube and perhaps even out of the laboratory and on to biological field stations will eventually be necessary.

In several instances, such a "vertical" integration of knowledge has generated spectacular results. One classic example is the protein hormone prolactin found in virtually all vertebrates. It was first discovered to stimulate the growth of the cells that line the crop in the pigeon. Later it was found to stimulate milk secretion in mammals. Then it was discovered to promote tail growth in aquatic amphibia. More recently, it was discovered to regulate salt balance in fish.

The boundaries of the holistic approach are even being extended. For example, data generated in the molecular biology laboratory from isolation and characterization procedures have been integrated into the research activities of other disciplines. Mitochondrial DNA research provides an interesting case—it has been discovered that mitochondrial DNA is inherited separately from chromosomal (nuclear) DNA. It also evolves quickly, since mitochondria lack DNA repair enzymes, and mutations, therefore, arise relatively often. The nucleotide sequence of mitochondrial DNA thus changes rapidly. Molecular biologists and anthropologists have joined forces to use mitochondrial DNA sequence variation to trace lineages within human populations. Their studies have pinpointed Africa as the location where the human species originated. Since mitochondrial DNA is inherited only through the egg (sperm contribute no mitochondria to the embryo), recent research has revealed that in all likelihood each member of the human race (including you and I!) can be traced back to a single ancestral mother.

In order to keep the rapidly increasing volume of molecular biology details in proper perspective, *Essentials of Molecular Biology* will endeavor to highlight the various concepts that unify the many subdisciplines of the field.

Concepts of Molecular Biology

As the discipline of molecular biology has grown and matured, numerous notions, ideas, and general understandings have emerged. These represent the so-called "concepts of molecular biology." These concepts are valuable to beginning students, for they serve to unify the numerous details that often overwhelm a novice. These concepts also provide the beginner with touchstones for connecting or interrelating the disparate aspects of the structure and function of large numbers of seemingly different molecules.

For the accomplished, professional molecular biology research scientist, these same concepts are still useful, providing a starting point for developing higher levels of understanding. As scientists attempt to formulate unifying theories to account for such phenomena as rates of genetic change within large populations, or as they endeavor to devise rules or principles to understand issues such as the origin of life, scientists constantly evaluate the conceptual underpinnings of the discipline of molecular biology.

Summarized below are a few examples of concepts that will emerge from a review of the information included in the chapters comprising Part I of *Essentials of Molecular Biology.*

The "Biopolymer" Concept: Polymers of small molecules, such as amino acids (proteins) or nucleotides (nucleic acids), are extraordinarily versatile, and hence have evolved in complexity to generate the diverse functional repertoires of living cells (p. 23).

The "Shape/Size" Concept: The specific function a macromolecule is best suited to perform is largely dictated by its size and shape. Size and shape set distinct limits on the number and types of interactions and relationships in which macromolecules can participate (p. 64).

In Part II other concepts will emerge. Two examples are:

The "Gene" Concept: Genes contain all the information for the synthesis and functioning of cellular components (p. 115).

The "RNA Processing" Concept: Eukaryotic mRNAs are extensively processed ("cut and shut") prior to functioning as templates for protein synthesis (p. 161).

In Parts III and IV, more concepts will be presented. Five examples follow:

The "Regulation" Concept: A multitude of regulatory mechanisms function in cells to insure that the supply of various macromolecules matches up with the cell's ever-changing needs (p. 224).

The "Instability → Flexibility" Concept: The inherent instability of many macromolecules (e.g., some proteins and mRNAs) in the living cell facilitates adapting to changing metabolic conditions (p. 274).

The "Changing Genome" Concept: The sequence or arrangement of nucleotides in genes is constantly evolving, either by random mutation, by occasional recombination events, or from the action of transposable elements (p. 313).

The "Simple Genome" Concept: Viral genomes are extraordinarily simple, in part because viruses borrow functions from their host cells (p. 328).

The "Levels of Sophistication" Concept: Prokaryotes, such as viruses and bacteria, under heavy natural selection pressure, have evolved more highly streamlined mechanisms for controlling gene expression than higher organisms (e.g., humans) (p. 341).

In the Appendix, additional concepts are presented, including:

The "Surface Interaction" Concept: Interactions between surfaces, facilitated by a variety of complementary weak forces, permit macromolecules to serve as templates for ordering and/or synthesizing other cellular components (p. 434).

In several chapters, additional concepts will be highlighted to help students construct a "big picture" understanding of the meanings and relationships of the details, methods, and thought processes dominating the field of modern molecular biology.

Progression Diagram

In addition to presenting key concepts throughout the text, a "progression diagram," illustrated in **Figure 1-5**, will be referred to from time to time to help

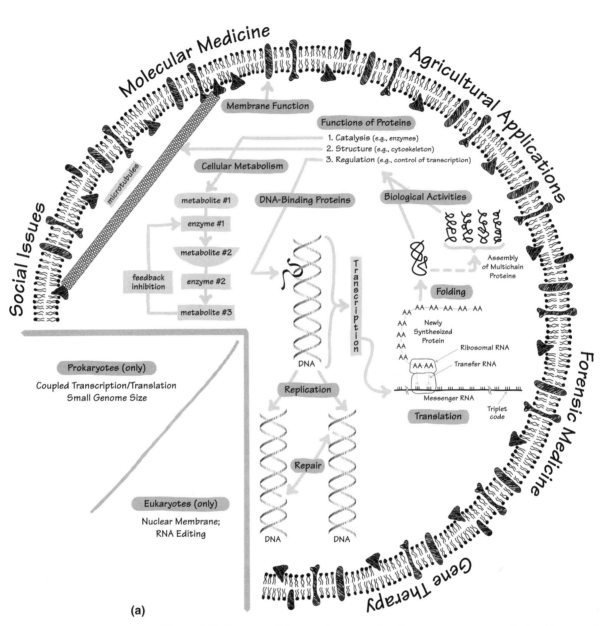

Figure 1-5 Summary of the use of a progression diagram to construct a unified and integrated understanding of the molecular components of a living cell: (a) represents the basic diagram;

(continued)

students construct a knowledge framework that incorporates these various concepts and their underlying details into an integrated whole. Using this diagram as a guide, students should be able to develop a broad understanding of the discipline, and be more prepared to cope with the increasing complexity of the later chapters.

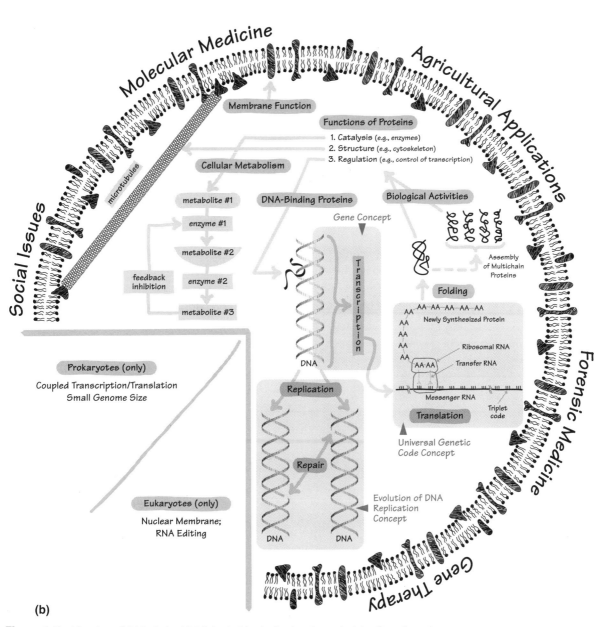

(b)

Figure 1-5 (*Continued*) (b) includes highlights in blue indicating the topical details reviewed in one or another part of the textbook. In this case, Part II (Macromolecular Function) details are highlighted. Concepts presented in Part II are indicated by the shaded patches. At the beginning of each chapter, the progression diagram will be presented. The processes reviewed and concepts discussed in that chapter will be highlighted.

Rewards from Studying Molecular Biology

Serious and diligent attention to learning molecular biology from this textbook offers several rewards to the beginning student. First, the use of the "layering approach" to constructing a knowledge base, which is used in this textbook (see the Preface), provides a learning tool that, once mastered, can be applied to other disciplines, ranging from poetry to physics. Second, studying molecular biology provides an excellent opportunity for enhancing one's analytical thought processes (so-called "critical thinking skills"). From time to time, experimental data are encountered. Learning how to interpret it usually involves analyzing entries in tables or graphs, and thereby improves problem-solving skills. Third, by learning the concepts and details of molecular biology, the beginning biology student is provided a gateway to virtually all other disciplines in biology, ranging from cell biology to genetics to population biology.

Finally, should you, the beginning student, develop an interest in molecular biology, you might consider choosing a career in a field related to this discipline. The Postscript section at the end of this textbook and the accompanying Student Manual can be consulted for advice on preparing for lifelong work in this field.

The Structure of Proteins, Nucleic Acids, and Macromolecular Complexes

in this chapter you will learn

1. The nature of the components of nucleic acids and proteins, and the chemical linkages by which they are joined to form larger molecules.

2. The role that nonco-valent interactions play in determining the three-dimensional structure of macro-molecules.

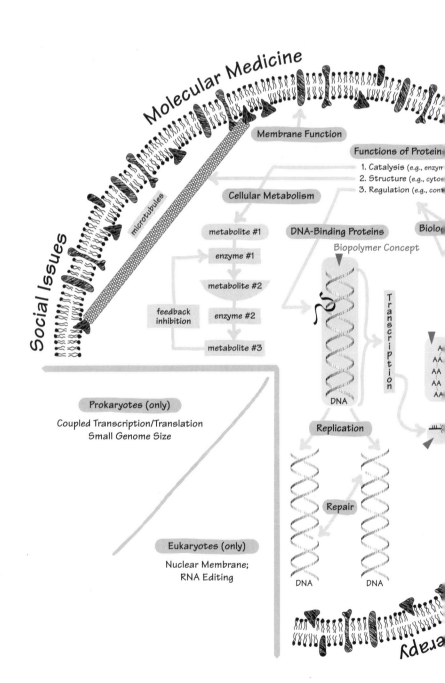

Molecular Medicine

Social Issues

Membrane Function

microtubules

Cellular Metabolism

Functions of Protein
1. Catalysis (e.g., enzym
2. Structure (e.g., cytos
3. Regulation (e.g., cont

metabolite #1

enzyme #1

metabolite #2

enzyme #2

metabolite #3

feedback inhibition

DNA-Binding Proteins

Biopolymer Concept

Biolo

Transcription

A
AA
AA
AA
AA

DNA

Replication

Repair

DNA   DNA

Prokaryotes (only)

Coupled Transcription/Translation
Small Genome Size

Eukaryotes (only)

Nuclear Membrane;
RNA Editing

erapy

Macromolecules

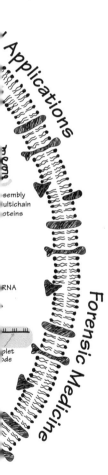

A typical cell contains 10^4–10^5 different kinds of molecules. Roughly half of these are small molecules—namely, inorganic ions and organic compounds whose molecular weights usually do not exceed several hundred. The others are polymers so massive (molecular weights from 10^4 to 10^{12}) that they are called **macromolecules.** These molecules are of three major classes: proteins, nucleic acids, and polysaccharides. They are polymers of amino acids, and sugars, respectively.

Our approach in these introductory chapters is to review the general properties of macromolecules. Macromolecules represent good examples of the axiom "The whole is greater than the sum of its parts." The properties of macromolecules, we will soon realize, are very different from the properties of their component amino acids, nucleotides, and sugars. Whereas these individual components alone contribute little to the functioning of the living cell, their polymers have extraordinary functions. Some of them (e.g., proteins) catalyze chemical reactions, while others (e.g., nucleic acids) store information and transfer it from one generation to the next. Indeed, a living cell's inventory of macromolecules gives it its identity. Cells differ from one another in size, shape, and function. Those features are largely generated by macromolecules.

Knowledge of the properties of macromolecules is essential for understanding living processes. This chapter and the next three chapters will emphasize nucleic acids and proteins. Information about other types of macromolecules will be given as needed in other chapters. The Appendix includes additional information about basic chemistry, which serves as a background for Chapters 3 and 4.

Chemical Structures of the Major Classes of Macromolecules

In this section we examine the chemical structure of proteins, nucleic acids, and polysaccharides—in particular, the monomers and the chemical linkages by which they are joined. Physical properties of these macromolecules and the interactions that exist between various parts of a single macromolecule are described in later sections.

Proteins

A protein is a polymer consisting of several amino acids (a **polypeptide**). Each amino acid can be thought of as a single carbon atom (the α carbon) to which there is attached one carboxyl group, one amino group, and a side chain denoted R (**Figure 2-1**). The side chains are generally carbon chains or rings to which var-

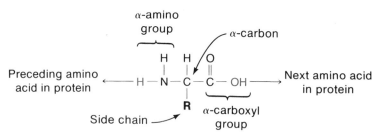

Figure 2-1   Basic structure of an α-amino acid. The NH$_2$ and COOH groups are used to connect amino acids to one another. The blue OH of one amino acid and the blue H of the next amino acid are removed when two amino acids are linked together (see Figure 2-3).

ious functional groups are attached. The simplest side chains are those of glycine (a hydrogen atom) and alanine (a methyl group). The complete chemical structures of each amino acid are given in **Figure 2-2**. The ionization state each would have within a living cell is shown.

To form a protein the amino group of one amino acid reacts with the carboxyl group of another; the resulting chemical bond is called a **peptide bond** (**Figure 2-3**). Amino acids are joined together in succession to form a linear **polypeptide chain** (**Figure 2-4**). When the number of peptide bonds exceeds about 15 (the number is arbitrary), the polypeptide is called a **protein.**

The two ends of every protein molecule are distinct. One end has a free —NH$_2$ group and is called the **amino terminus;** the other end has a free —COOH group and is the **carboxyl terminus.** The ends are also called the N (or NH$_2$) and C termini, respectively.

The amino-acid side chains usually do not engage in covalent bond formation; an exception is the —SH group of cysteine, which often forms a disulfide (S—S) bond by reaction with a second cysteine in the vicinity (within either the same or different polypeptide chain).

amino terminus

carboxyl terminus

Nucleic Acids

A nucleic acid is a polynucleotide—that is, a polymer consisting of nucleotides. Each nucleotide has the three following components (**Figure 2-5**):

1. A cyclic five-carbon sugar. This is ribose, in the case of ribonucleic acid (RNA), and deoxyribose, in deoxyribonucleic acid (DNA). The difference in the structure of ribose and 2′-deoxyribose is shown in Figure 2-5.
2. A purine or pyrimidine base attached to the 1′-carbon atom of the sugar by an *N*-glycosylic bond. The bases, which are shown in **Figure 2-6**, are the purines, adenine (A) and guanine (G), and the pyrimidines, cytosine (C), thymine (T), and uracil (U). DNA and RNA contain A, G, and C; however, T is found only in DNA and U is found only in RNA.
3. A phosphate attached to the 5′ carbon of the sugar by a phosphoester linkage. This phosphate is responsible for the strong negative charge of both nucleotides and nucleic acids.

nucleoside

A base linked to a sugar is called a **nucleoside;** thus *a nucleotide is a nucleoside phosphate.*

Hydrophobic (neutral) R Groups

Glycine   Alanine   Valine   Leucine   Isoleucine

Phenylalanine   Proline   Methionine   Tryptophan

Hydrophilic (neutral) R Groups

Serine   Threonine   Asparagine   Glutamine   Tyrosine   Cysteine

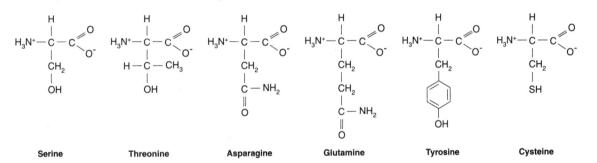

Acidic R Groups

Aspartic acid   Glutamic acid

Basic R Groups

Lysine   Arginine   Histidine

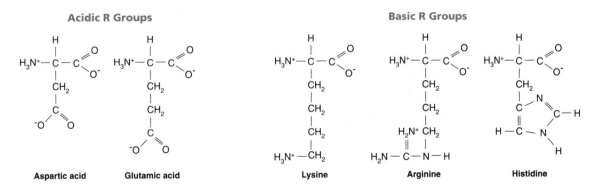

Figure 2-2 Chemical structures of the amino acids. In this illustration (and also in Figure 2-4) they are shown at physiological pH (approx. 7.2–7.6). Thus, the α-amino and α-carboxyl groups are charged.

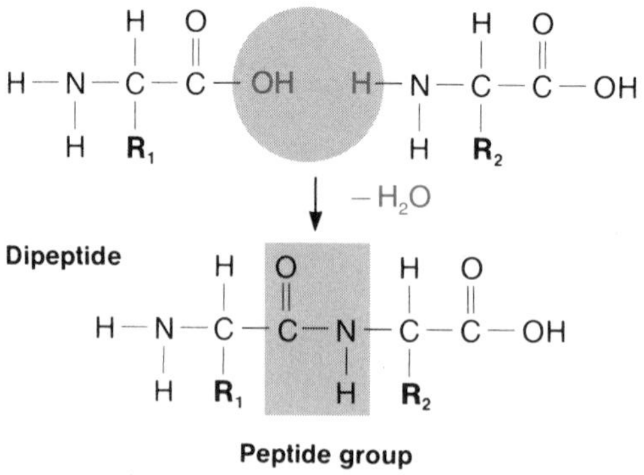

Figure 2-3 Formation of a dipeptide from two amino acids by elimination of water (shaded circle) to form a peptide group (shaded rectangle).

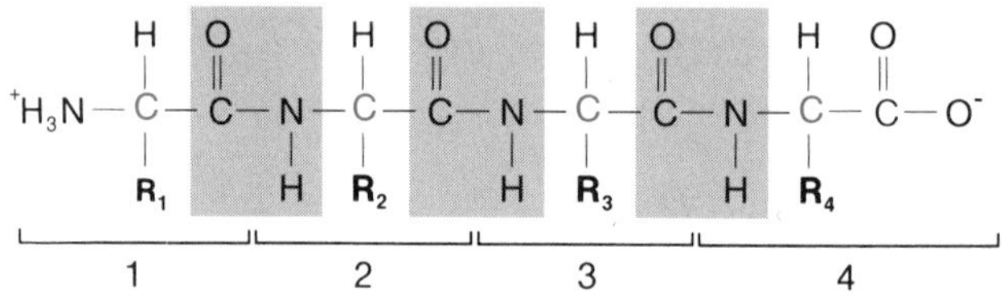

Figure 2-4 A tetrapeptide showing the alternation of α-carbon atoms (unshaded) and peptide groups (shaded). The four amino acids are numbered below.

The nucleotides in nucleic acids are covalently linked by a second phosphoester bond that joins the 5′-phosphate of one nucleotide and the 3′-OH group of the adjacent nucleotide (**Figure 2-7**). Thus, the phosphate is esterified to both the 3′- and 5′-carbon atoms; this unit is often called a **phosphodiester group** (see Appendix).

Polysaccharides

Polysaccharides are polymers of sugars (most often glucose) or sugar derivatives. These are very complex molecules because sometimes covalent bonds occur between many pairs of carbon atoms. This has the effect of allowing one sugar unit to be joined to more than two other sugars, which results in the formation of highly branched macromolecules. These branched structures are sometimes so enormous that they are almost macroscopic. For example, the cell walls of many bacteria and plant cells are single gigantic polysaccharide molecules.

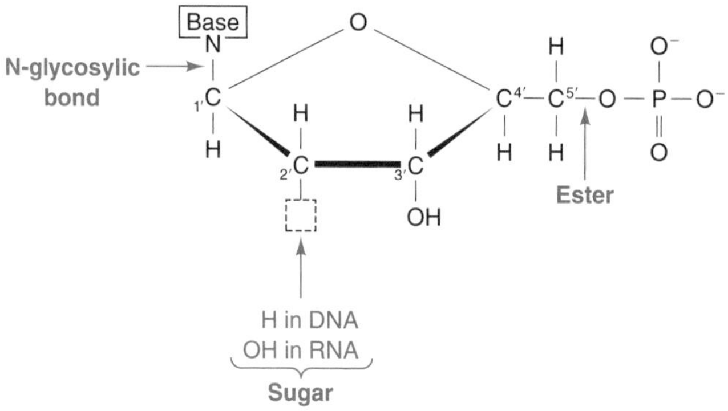

Figure 2-5 Structure of a mononucleotide. (The carbon atoms in the sugar ring are numbered.) The OH in RNA at the 2′C position facilitates chemical breakdown of RNA. Thus, DNA is more stable than RNA.

Adenine

Guanine

Cytosine

Thymine

Uracil

Figure 2-6 The bases found in nucleic acids. The weakly charged groups are shown in blue.

The brief review included in the preceding section, which emphasized the basic chemical features of macromolecules, provides an opportunity to present the first of many concepts that will be featured through the text.

Noncovalent Interactions That Determine the Three-Dimensional Structures of Proteins and Nucleic Acids

The biological properties of macromolecules are mainly determined by weak interactions that result in each molecule acquiring a unique three-dimensional structure. These noncovalent interactions are much weaker than typical covalent bonds, and, for any given type of macromolecule, several of these weaker inter-

KEY CONCEPT

The "Biopolymer" Concept

Polymers of small molecules, such as amino acids (proteins) or nucleotides (nucleic acids), are extraordinarily versatile, and hence have evolved in complexity to generate the diverse functional repertoires of living cells.

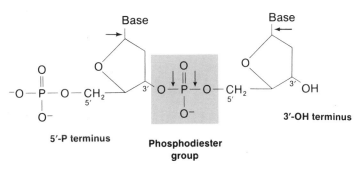

Figure 2-7 The structure of a dinucleotide. The vertical arrows show the bonds in the phosphodiester group about which there is free rotation. The horizontal arrows indicate the N-glycosylic bond about which the base can freely rotate. A polynucleotide would consist of many nucleotides linked together by phosphodiester bonds.

actions may play a role in determining three-dimensional structure. In this section, the noncovalent interactions that are important and the chemical groups responsible for them are described. The Appendix should be consulted for further information on noncovalent bonds.

The Random Coil

random coil

Linear polypeptide and polynucleotide chains contain several bonds about which there is free rotation. In the absence of any intrastrand interactions,* each monomer would be free to rotate with respect to its adjacent monomers; this is limited only by the fact that more than one atom cannot occupy the same space. The three-dimensional configuration of such a freely rotating chain is called a **random coil;** it is a somewhat compact and globular structure that *changes shape continually,* owing to constant bombardment by solvent molecules.

Few, if any, nucleic acid or protein molecules exist in nature as random coils because there are many interactions between regions of the chains. These interactions are hydrogen bonds, hydrophobic interactions, ionic bonds, and van der Waals interactions.

Hydrogen-Bonding

The most common hydrogen bonds found in biological systems are shown in **Figure 2-8**. In RNA, int*ra*strand hydrogen bonds cause the polynucleotide strand to be folded. Thus, it is more compact than would be expected for long, single-stranded molecules. In DNA, int*er*strand hydrogen bonds are responsible for the double-stranded helical structure. In proteins, intrastrand hydrogen-bonding occurs between a hydrogen atom on a nitrogen adjacent to one peptide bond and an oxygen atom adjacent to a different peptide bond. This interaction gives rise to several particular polypeptide chain configurations, which will be discussed shortly.

The Hydrophobic Interaction

A hydrophobic interaction is an interaction between two molecules (or portions of molecules) that are somewhat insoluble in water. These two molecules (*which may be different*) are poorly soluble in water, which tends to repulse them. In response to that natural repulsion, they tend to associate.

Many components in nucleic acids and proteins have the hydrophobic property just described. For example, the bases of nucleic acids are planar organic rings carrying localized weak charges (Figure 2-6). The localized charges are sufficient to

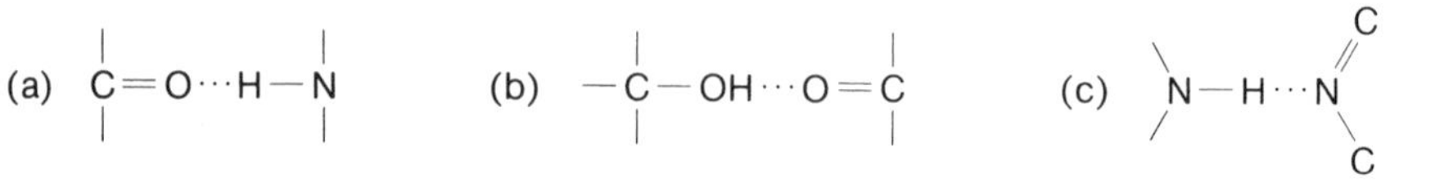

Figure 2-8 Structures of three types of hydrogen bonds (indicated by three dots). (a) A type found in proteins and nucleic acids. (b) A weak bond found in proteins. (c) A type found in DNA and RNA.

---

*An int*ra*strand interaction is an interaction between two regions of the same strand. An int*er*strand interaction is between different strands. It is important to remember the distinction between these similar words.

maintain solubility, but the large, poorly soluble organic ring portions of the bases tend to cluster, minimizing the contact with water molecules. The most efficient kind of clustering is one in which the faces of the rings are in contact, an array known as **base-stacking** (**Figure 2-9**). Note that since the bases are adjacent in the chain, stacking gives some rigidity to single polynucleotide strands. Therefore, instead of a random coil forming, the strands are extended. In the following chapter we will see how stacking is of major importance in determining nucleic acid structure. Many amino-acid side chains are very poorly soluble, and this causes (1) the benzene ring of phenylalanine to stack and (2) the hydrocarbon chains of alanine, leucine, isoleucine, and valine to form tight unstacked clusters. Since these amino acids are not necessarily adjacent, *hydrophobic interactions tend to bring distant hydrophobic parts of a polypeptide chain together*. Figure 2-2 indicates the hydrophobic amino acids.

Ionic Bonds

An ionic bond results from the attraction between unlike charges. At physiological pH several amino-acid chains are ionized: negatively charged carboxyl groups (aspartic and glutamic acids) and positively charged amino groups (lysine, histidine, and arginine). These five amino acids thus can form ionic bonds.

Ionic bonds tend to bring together distant parts of the chain. Ionic interactions can also be repulsive—namely, between two like charges; it would therefore be unlikely for a polypeptide chain to fold in such a way that two aspartic acids are very near each other. Ionic bonds are the strongest of the noncovalent interactions. However, they are destroyed by extremes of pH, which can change the charge of the groups, and by high concentrations of salt, the ions of which shield the charged groups from one another.

Van der Waals Attraction

Van der Waals forces exist among all molecules and are a result of both permanent dipoles and the circulation of electrons. The dependence of the attractive force between two atoms is proportional to $1/r^6$, in which r is the distance between their nuclei. The attraction is a very weak force and is significant only if two atoms are very near one another (1–2 Å apart).

Since van der Waals forces are very weak, they are easily overcome by thermal motion; in general, the force between two atoms will not keep them in proximity. However, if interactions of *several* pairs of atoms are combined, the cumulative

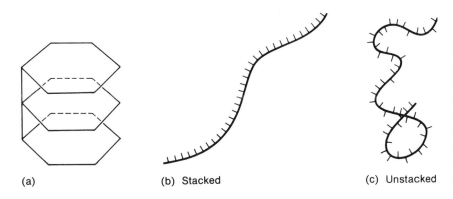

(a)   (b) Stacked   (c) Unstacked

Figure 2-9 The meaning of stacking. (a) Three stacked bases. (b) and (c) Structures of stacked and unstacked polynucleotides. The stacked polynucleotide is more extended because the stacking tends to decrease the flexibility of the molecule.

attractive force can be great enough to withstand being disrupted by thermal motion. Two molecules thus can attract one another if several of their component atoms can mutually interact. However, because of the $1/r^6$-dependence, the intermolecular fit must be nearly perfect. This means that *two molecules can bind to one another by van der Waal forces if their shapes are complementary.* This is also true of two separate regions of a polymer—that is, the regions can hold together if their shapes match. Sometimes the van der Waals attraction between two regions is not large enough to cause this; however, it can significantly strengthen other weak interactions (such as the hydrophobic interaction) if the fit is good.

Summary of the Effects of Noncovalent Interactions

The effect of noncovalent interactions is to constrain a linear chain to fold in such a way that different regions of the chain, which may be quite distant from each other in a chain (if the chain were linear), are brought together. An example of a molecule showing the four kinds of noncovalent folding that have been described is shown in **Figure 2-10**.

In Chapters 3, 4, and 6 we will discuss the specific structures brought about by these interactions and the physical properties of nucleic acids and proteins.

Macromolecule Isolation and Characterization

Several methods for studying macromolecules are practically standard in molecular biology laboratories. Proteins are usually isolated from extracts of living cells with molecular sieves and chromatography on supports that are either positively or negatively charged. The former method takes advantage of the heterogeneity in molecular weight displayed by different proteins. The latter exploits the fact that most proteins differ from one another in the charge they carry, which is due, of course, to the unique amino-acid composition of individual proteins.

electrophoresis

The most convenient and widely employed method for monitoring a protein purification scheme, and also for initial characterization studies, is **electrophoresis.** This method consists of setting the protein sample on a solid support (usually a gel) and establishing an electric field. Depending on the net charge of specific proteins,

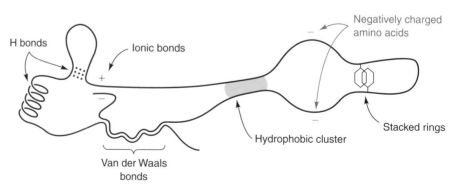

Figure 2-10 A hypothetical polypeptide chain showing attractive (black) and repulsive (blue) interactions.

positively charged proteins will migrate toward the cathode (negative pole), and negatively charged proteins toward the anode (positive pole). The gel consists of a semisolid sheet with pores through which the migrating proteins pass on their way toward either pole. Acrylamide, an inert synthetic polymer, is most commonly employed as the gel. By formulating the gel so as to have relatively small pores, smaller proteins will migrate much faster than larger proteins. In fact, the most sophisticated gel electrophoresis methods for proteins employ a sequence of charge and size separation methods, as shown in **Figure 2-11**.

Other protein isolation and characterization methods include ultracentrifugation, amino acid sequencing, x-ray crystallography, and occasionally, for larger proteins, electron microscopy (see Table 1-1 in Chapter 1).

For nucleic acid isolation prior to characterization, **gene cloning** methods are routinely employed. The simplest protocol is illustrated in **Figure 2-12**. Pieces of DNA are inserted into a plasmid that is then replicated inside a bacterial host. The duplicated plasmid is then harvested, and the replicated foreign DNA is isolated from the plasmid.

Due to the unusually large size of most nucleic acids, a molecular sieve with larger pores than normally used for electrophoresis of proteins is employed (**Figure 2-13**). Agarose, a plant polysaccharide with various desirable qualities, is commonly employed to make the semi-solid gel support. After electrophoresis,

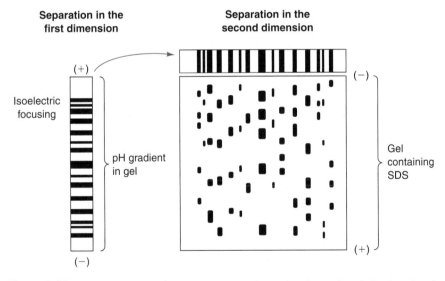

Figure 2-11 Two-dimensional gel electrophoresis of proteins. A protein solution (e.g., crude extract of cells) is first subjected to electrophoresis in a tube with a semisolid gel inside it. The gel contains a pH gradient. This first step separates proteins according to their net charges (isoelectric focusing). Next, that gel is set on top of a rectangular slab containing an acrylamide gel that includes the anionic detergent sodium dodecyl sulfate (SDS). SDS molecules interact with amino acids (one SDS molecule/amino acid) to give all proteins a similar charge-to-mass ratio. In this second dimension, once an electric field is applied proteins are separated during electrophoresis solely according to their molecular size. Using this method, several hundred of the proteins in a crude extract of cells can be identified as individual spots on the rectangular second dimension gel.

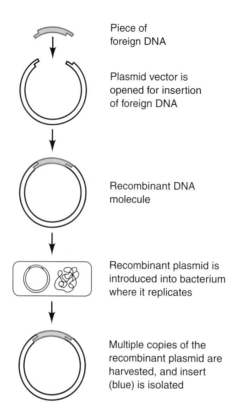

Figure 2-12 Procedure for cloning foreign DNA in a plasmid vector.

the location of the nucleic acids is easily visualized by soaking the gel in an appropriate dye solution (**Figure 2-14**).

Additional nucleic acid isolation and characterization methods include nucleotide sequencing, x-ray crystallography, and electron microscopy. Technological developments associated with these isolation and characterization methods have

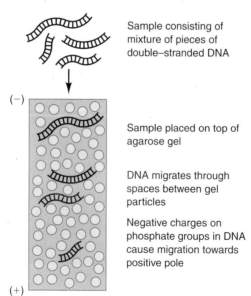

Figure 2-13 Gel electrophoresis of nucleic acids. Double-stranded DNA is shown. RNA (normally single stranded) can also be characterized on agarose gels.

(−)

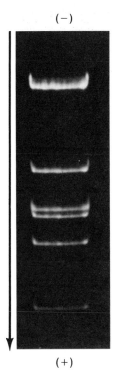

(+)

Figure 2-14   A gel electrophoregram of six fragments of *E. coli:* phage DNA. The direction of movement is from top to bottom. The DNA is made visible by the fluorescence of bound ethidium bromide. (Courtesy of Arthur Landy and Wilma Ross.)

been the driving force behind the recent rapid accumulation of information on macromolecule structure. From this information on structure, ideas about cellular function of specific macromolecules can often be developed.

SUMMARY

About half of all molecules found in living cells can be termed macromolecules because of their massive molecular weights. The linear structure of the three major classes of macromolecules—proteins, nucleic acids, and polysaccharides—is determined by polymerization of specific monomers. Proteins are built from amino acids joined by peptide bonds that result from the reaction of the carboxyl group of one amino acid with the amino group of another amino acid. Nucleic acids are formed by phosphoester bonds linking the 5′-phosphate of one nucleotide to the 3′-OH group of another. Sugars and their derivatives are the building blocks, forming highly balanced polysaccharides. The three-dimensional molecular properties of these macromolecules, however, are determined by noncovalent interactions, including hydrogen bonds, hydrophobic interaction, ionic bonds, and van der Waals attractions. The joining of

cysteine residues through the formation of disulfide bonds also affects molecular properties of proteins. All of these intrastrand interactions act to constrain a chain so that regions that are often distant from one another in the linear chain can be brought together.

Electrophoresis, a standard method used for studying macromolecules in molecular biology laboratories, allows charged macromolecules to be separated in an electric field while traveling on a solid support, such as an acrylamide or agarose gel. In the case of protein separation, sequential electrophoresis can be performed. The first electrophoretic run separates proteins on the basis of their differing net charges, while the second run causes a separation by size. Nucleic acids are also routinely separated using gel electrophoresis, followed by visualization using appropriate dyes.

DRILL QUESTIONS

1. What are the monomers of which proteins, nucleic acids, and polysaccharides are composed?
2. What groups in an amino acid become linked in forming a polypeptide chain? What is the linkage called?
3. What chemical groups are at the ends of a protein molecule or a polypeptide chain? Do polypeptide chains or proteins contain these groups at positions other than their termini?
4. Which nucleic acid base is unique to DNA? to RNA?
5. What chemical groups on nucleotides become covalently linked to form a nucleic acid? What is the linkage created called?
6. What is the difference between a nucleoside and a nucleotide?
7. In simple terms, what is the nature of a hydrophobic interaction?
8. What is the strongest and the weakest noncovalent force involved in stabilizing macromolecules?
9. All DNA molecules move in the same direction when electrophoresed. Why? Is this also generally true of proteins? Explain.
10. In SDS-electrophoresis of proteins, what features of the technique cause protein molecules to move in a single direction?

PROBLEMS

1. A mixture of different proteins is subjected to electrophoresis in three polyacrylamide gels, each having a different pH value. In each gel, five bands are seen. (a) Can one reasonably conclude that there are only five proteins in the mixture? Explain. (b) Would the conclusion be different if a mixture of linear DNA fragments were being sized?
2. Name the kinds of bonds in proteins in which each of the following amino acids might participate: cysteine, lysine, isoleucine, glutamic acid?
3. (a) If a particular macromolecule that is very compact in 0.01 M NaCl expands considerably in 0.5 M NaCl, what forces are probably important in determining the overall size and shape of the molecule? (b) If a macromolecule that is a near-random coil in 0.2 M NaCl becomes very extended and rigid in 0.01 M NaCl, what forces are probably involved in the extension?
4. A protein consisting of one polypeptide chain contains four cysteines. If all were engaged in disulfide bonds and if all possible pairs of cysteines could be joined, how many different theoretical structures would be possible? What would determine which was the naturally occurring structure?
5. Amino acids can be separated by electrophoresis on paper. Suppose that a pH is chosen to make some amino acids positively charged and others negatively charged. Thus, some amino acids will move toward the anode and others toward the cathode, their charge principally determining the direction and rate of movement. Often amino acids having the same charge (such as alanine and valine) also separate; in other words, they move at different rates, though in the same direction. What determines the difference in rate, and would alanine or valine move faster?

CONCEPTUAL QUESTIONS

1. With an understanding of the various types of noncovalent interactions that determine protein three-dimensional structure, predict which of these interactions would be most prevalent in proteins located in different body tissues. For example, what noncovalent interactions would most likely be found in proteins spanning lipid cell membranes? In proteins that transport iron throughout the body?

2. When studying a protein, one can gain information on (a) the protein's three-dimensional structure and the interactions that determine it, as well as (b) the order of amino acids comprising the protein. How could each set of information be used to determine an unknown protein's function?

3. Upon isolating a new microorganism, what information might you gain from studying its macromolecules that would help you compare it with previously known microorganisms? How could you discover this information?

CHAPTER

3

in this chapter you will learn

1. About the physical and chemical nature of DNA and RNA molecules, the differences between them, and the conformations in which they may exist.

2. Denaturation, hybridization, and sequencing of nucleic acids.

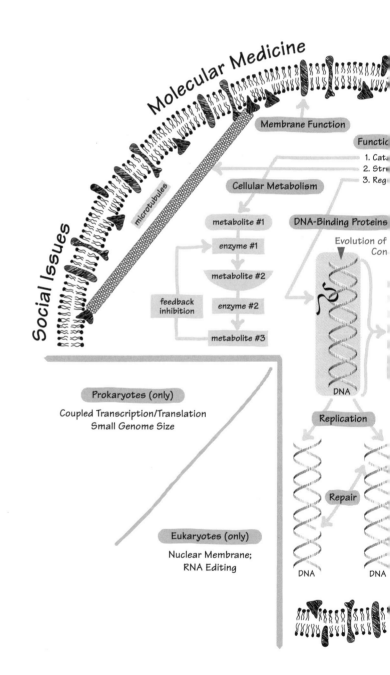

Molecular Medicine

Social Issues

Membrane Function

microtubules

Cellular Metabolism

metabolite #1

enzyme #1

metabolite #2

feedback inhibition

enzyme #2

metabolite #3

Functic
1. Cata
2. Stru
3. Reg

DNA-Binding Proteins

Evolution of
Con

DNA

Replication

Repair

Prokaryotes (only)

Coupled Transcription/Translation
Small Genome Size

Eukaryotes (only)

Nuclear Membrane;
RNA Editing

DNA   DNA

Nucleic Acids

Deoxyribonucleic acid (DNA) is the single most important molecule in living cells and contains all of the information that a cell needs to live and to propagate itself. In this chapter, its remarkable structure and many of its physical and chemical properties are described, together with those of ribonucleic acid (RNA), a class of molecules intimately linked to the use or expression of genetic information.

We will focus in this chapter on the structural properties of nucleic acids. Realizing that genes consist of simple polymers of nucleotide sequences, we will first learn how the individual atoms in DNA are covalently linked to form the double helix. Our study of the structural properties of DNA in this chapter will then reveal how nucleic acids store information. Later (Chapters 9 and 10) we will learn how that information is converted into the functional units (i.e., proteins) that give various cells their individual characteristics.

Along the way, we will learn several modern techniques that are routinely used in molecular biology laboratories to study nucleic acids. Soon we will appreciate why such spectacular advances have been made in understanding gene structure and function. Not only is it possible to establish the nucleotide sequence of virtually any gene, it is possible to biochemically synthesize genes using special chemical methods.

The development of those methods is in large part responsible for such sensational breakthroughs as the gene cloning and transgenic animals described later in this textbook (e.g., Chapters 15 and 16).

Physical and Chemical Structure of DNA

In the previous chapter, the structure of nucleotides and how they are joined to form a polynucleotide, or a nucleic acid, were described. In this section, we show how nucleotides in a polynucleotide chain interact with one another to produce complex three-dimensional structures.

In early physical studies of DNA, many experiments indicated that the molecule is an extended chain having a highly ordered structure. The most important technique was x-ray diffraction analysis, by which information was obtained about the arrangement and dimensions of various parts of the molecule. The most significant observations were that the molecule is helical and that the bases of the nucleotides are stacked, with their planes separated by a spacing of 0.34 nanometers (nm).

Chemical analysis of the molar content of the bases (generally called the **base composition**) adenine, thymine, guanine, and cytosine in DNA molecules isolated from many organisms provided the important fact that [A] = [T] and [G] = [C], in which [ ] denotes molar concentration, from which followed the corollary [A + G] = [T + C] or [purines] = [pyrimidines].

James Watson and Francis Crick combined chemical and physical data for DNA with a feature of the x-ray diffraction diagram that suggested that two helical strands are present in DNA and showed that the two strands are coiled about one another to form a **double-stranded helix** (**Figure 3-1**). In this model, the sugar-phosphate backbones follow a helical path at the outer edge of the molecule and the bases are in a helical array in the central core. The bases of one strand are hydrogen-bonded to those of the other strand to form the purine-to-pyrimidine base pairs A · T and G · C (**Figure 3-2**).

The two bases in each base pair lie in the same plane, and the plane of each pair is perpendicular to the helix axis. The base pairs are rotated 36° with respect to each adjacent pair, so that there are ten pairs per helical turn. The diameter of the double helix is 2.0 nm.

The helix creates two external helical grooves, a deep wide one (the **major groove**) and a shallow narrow one (the **minor groove**); both of these grooves are large enough to allow protein molecules to come in contact with the bases, as will be discussed in Chapter 5.

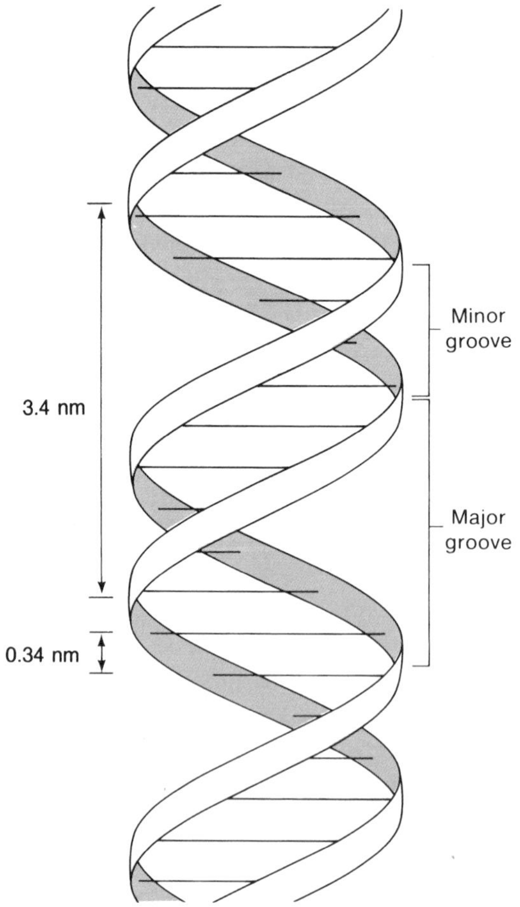

double-stranded helix

major groove
minor groove

Minor groove

Major groove

3.4 nm

0.34 nm

Figure 3-1 Diagrammatic model of the DNA double helix in the common B form. Parallel lines represent the base pairs, of which there are 10 per complete turn. The base pairs are perpendicular to the 10 long axis of the helix.

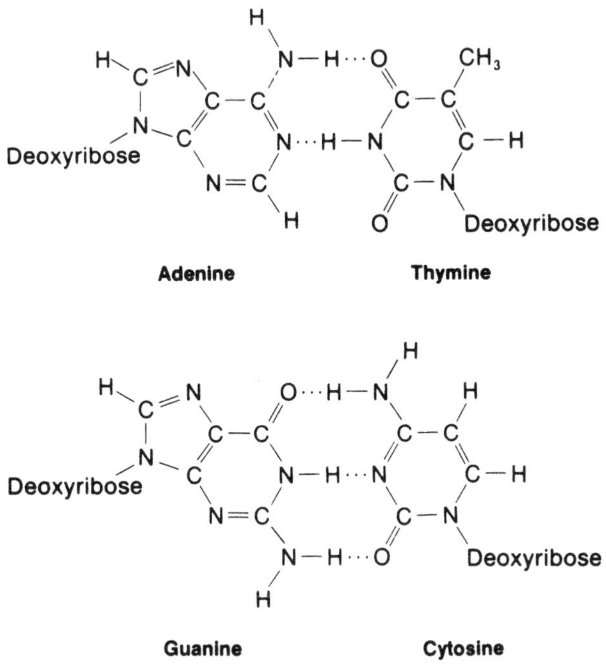

Adenine   **Thymine**

Guanine   **Cytosine**

Figure 3-2 The two common base pairs of DNA. Note that hydrogen bonds (dotted lines) are between the weakly charged polar groups noted in Figure 2-6.

The double helix shown in Figure 3-1 is a right-handed helix. A right-handed helix is the arc described by your right hand when you touch your nose with your thumb; a left-handed helix is that described by your left hand. Naturally occurring DNA molecules are generally right-handed helices.

Base-pairing is one of the most important features of the DNA structure because it means that the base sequences of the two strands are complementary (**Figure 3-3**); thus, an A in one strand is paired with a T in the other. Likewise, a C in one strand is paired with a G in the other. This has deep implications for the mechanism of DNA replication because, in this way, the replica of each strand has the base sequence of its complementary strand, and from one strand, the other can be made.

The two polynucleotide strands of the DNA double helix are antiparallel—that is, the 3′-OH terminus of one strand is adjacent to the 5′-P (5′-phosphate) terminus of the other (**Figure 3-4**). The significance of this is twofold. First, in a linear double helix there is one 3′-OH and one 5′-P terminus at each end of the helix. The second and generally more significant point is that the orientations of the two strands are different. That is, if two nucleotides are paired, the sugar portion of one nucleotide lies upward along the chain, whereas the sugar of the other nucleotide lies downward. We will see in Chapter 7 that this structural feature poses an interesting constraint on the mechanism of DNA replication.

Because the strands are antiparallel, a convention is needed for stating the sequence of bases of a single chain. The convention is to write a sequence with the 5′-P terminus at the left; for example, ATC denotes the trinucleotide 5′-P—ATC—3′-OH.

It is always true that [A] = [T] and [G] = [C], but there is no rule governing the relationship between the concentrations of guanine and cytosine and those of

base-pairing

Figure 3-3 A diagram of a DNA molecule showing complementary base-pairing of the individual strands. The light lines represent hydrogen bonds. The numbers orient the strands. 5′ and 3′ ends reveal the antiparallel nature of the strands.

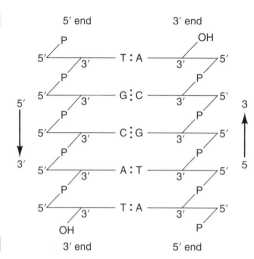

Figure 3-4   A stylized drawing of a
segment of a DNA duplex showing the
antiparallel orientation of the complementary chains. The arrows indicate the
5′ to 3′ direction of each strand. The
phosphates (P) join the 3′ carbon of one
deoxyribose (horizontal line) to the
5′ carbon on the adjacent deoxyribose.
The dots between the pairs indicate the
number of hydrogen bonds between
the bases.

adenine and thymine in a DNA molecule; in fact, there is enormous variation in
this ratio. The usual way that the base composition of the DNA molecule from a
particular organism is expressed is not by a ratio, but by the percent of all bases
that are G · C pairs; that is ([G] + [C]) × 100/[all bases]. This is termed the G + C
content, or percent G + C. The base composition of hundreds of organisms has
been determined. Generally speaking, the percent of G + C content is near 50%
for most complex organisms and has a very small range from one species to the
next. For instance, for most plants and animals, the extremes are 48% and 52%,
with 49%–51% for primates. The percent G + C varies widely, however, from one
genus to another in single-cell organisms. For example, in bacteria the extremes
are 27% for the genus *Clostridium* and 76% for the genus *Sarcina; E coli* is
around 50% G + C.

Alternate DNA Structures

DNA structure is very rich in variety. Although the double-stranded B helix is the
predominant form, other helical and some completely unexpected forms can be
found. Among these structural variants is the **A helix** (the predominant form of
double-stranded RNA), which has 11 bases per turn rather than 10, and has the
plane of the bases tilted 30° with respect to the helical axis. This helix, favored
under conditions of dehydration, is wider and shorter than the B helix, and the distinction between the major and minor grooves is reduced.

    The **Z helix** (so-named because the backbone has a zigzag shape) is a more
radical departure from the B helix theme. Although double stranded, the Z helix is
left-handed (**Figure 3-5**) and has 12 base pairs per turn. Since the length of a turn
is 4.5 nm rather than 3.4 nm (B helix), Z DNA appears longer and slimmer-looking
than the B helix. In the very long DNA molecules found in cells, different conformations can coexist in the same molecule. Regions of the Z conformation can be
found interspersed among regions of essentially B conformation; this fact raises
the question of what determines the conformation of DNA. In the case of Z DNA,

A helix

Z helix

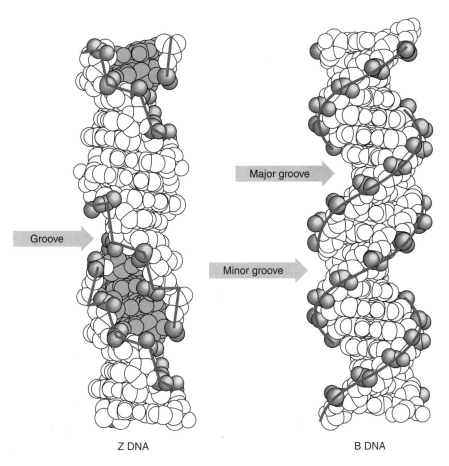

Groove

Major groove

Minor groove

Z DNA

B DNA

Figure 3-5 Space filling models of Z DNA and B DNA. The dark spheres are the phosphate groups, and the heavy lines link adjacent phosphates. The zigzag shape of the Z DNA backbone is clearly visible. To verify if the helices are right- or left-handed, rotate the book 180° and look at the helices. (From Rich et al. *Am. Rev. Biochem.* 53:795. Copyright 1984: Annual Reviews Inc. [Palo Alto])

a primary structure or sequence of G alternating with C favors the Z DNA conformation. The presence of regions of Z DNA near genes on the same molecule can influence their expression.

Further variations of DNA structure involve the "strandedness" of the DNA. For example, single-stranded DNA occurs in some viruses. In these DNA molecules, the rule of double-stranded complementarity [G] + [A] = [C] + [T] does *not* hold. The structure of this DNA is much more irregular—the single strand folds back on itself. Short, double-stranded helical regions are formed between complementary regions of the molecule. These duplexes are separated by loop structures, and residual single-stranded regions. Single-stranded DNA is more dense than double-stranded DNA as a result of its more contorted and compacted structure. Alternating helix-loop structures are also a feature of single-stranded RNA molecules.

Triple- and even four-stranded DNA helices have also been observed. Triple-stranded helices are found in regions of DNA that have repeating stretches of purines alternating with stretches of pyrimidines complementary to the purine stretches. Under these circumstances, one strand of a repeating unit folds back into the major groove of the preceding repeating unit. This conformation leaves one strand of the repeating unit unpaired. The four-stranded helix is characterized by G-rich regions of DNA. This structure is often found at the end of chromosomes in the telomeres, and may be important in the process of meiosis.

DNA structure can thus be best described as a number of variations on the B helix theme. Quite clearly the structure adopted by a particular region of a DNA molecule depends on the base sequence in that region. This is a simple example of how the actual sequence of its monomers (in this case, nucleotides) influences the three-dimensional structure of a biological polymer. In the next chapter on proteins, similar features of three-dimensional structure will be traced to the actual sequence of the amino acid monomers.

Circular and Superhelical DNA

The rather crude techniques originally used to isolate and study DNA suggested that DNA had the form of a linear double helix. The overall length of DNA makes it extremely sensitive to shear forces during preparation. Typical DNA preparations are thus often fragmented (both strands broken) or nicked (single strand broken). However, as techniques improved, it became clear that many DNAs, especially prokaryotic chromosomes and mammalian mitochondrial DNA, are actually circular, with the ends of the helix covalently joined. This observation raised novel topological questions concerning the effect of the circular state on the helical structure.

Because of the polarity of the strands, the 5′ terminus of one strand can only join its own 3′ end to close the circle. Thus, circular, double-stranded DNA is two circles of single-stranded DNA twisted around each other. You might imagine this as pieces of twine joined to their own ends.

However, consider the result of giving the double-helix exactly one 360° twist before joining the ends. (Keep in mind that the double helix already has a complete turn every 10–11 bases.) If the twist is in the same direction as the helix, it would have the effect of tightening the double helix. Alternatively, a twist in the opposite direction of the helix would loosen the helix. Circular DNA that has incorporated one or more twists to increase the number of times one strand crosses the other is said to be positively **supercoiled** or **superhelical.** A negative superhelix results from decreasing the number of times the strands cross.

Be sure to realize that DNA's "natural" one turn per 10–11 bases is *not* supercoiling. Supercoiling is *additional* twists (whether in the "tightening" or the "loosening" direction). Also, note that "positive" and "negative" supercoiling can refer to different directions of twisting under different DNA conformations. In a right-handed helix (B and A conformation), "positive" supercoiling would add additional right-handed twists. However, in a left-handed helix (Z conformation), "positive" supercoiling would entail left-handed twists. DNA compensates for the higher energy state of supercoiling by modifying the dimensions (e.g., pitch) of the preferred conformations of the double helix.

One major structural effect of positive or negative supercoiling is that the helical axis curves slightly to maintain the 10-base periodicity of the B helix, which would otherwise be modified. The bent helical axis can in fact be observed by electron microscopy (**Figure 3-6**). Other effects are more dramatic: highly neg-

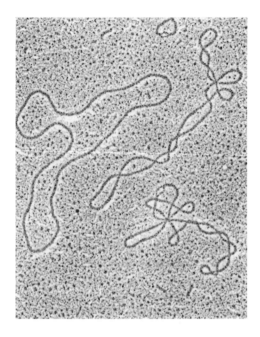

Figure 3-6 Circular and supercoiled DNA of phage PM2 as imaged by an electron microscope. (Courtesy of K. G. Murti.)

atively superhelical DNA can be accommodated by conformational changes in the molecule. The ultimate accommodation is the switch to a Z DNA structure, thus compensating for two negative turns per left-handed turn.

The formation of short bubbles of unpaired, single-stranded DNA from severe negative supercoiling can be further stabilized by **cruciform** structures (**Figure 3-7**). Cruciforms are conformations that can be adopted by highly symmetrical regions of DNA. Because of the 5′ to 3′ polarity of the strands and the comple-

cruciform

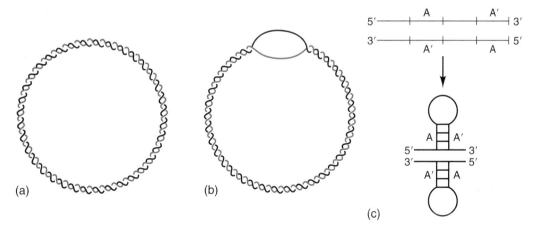

Figure 3-7 (a) A nonsupercoiled covalent circle of DNA; (b) a negatively supercoiled circle of DNA. The single-stranded bubble allows the remainder of the molecule to adopt the standard B helix; (c) a cruciform structure of DNA existing in a palindromic region (an inverted repeat). Region A is complementary to A′, so, due to the symmetry of an inverted repeat, A can base-pair with A′ of the second strand or with the A′ sequence of its own strand.

palindromic

mentarity of bases, cruciforms occur in regions of **palindromic** (inverted repeat) DNA. Such cruciform structures serve as "docking sites" for proteins involved in DNA function. Supercoiling thus favors large, local deviations from the standard conformation. However, since DNA structures are dynamic during the lifetime of a cell, many regions undergo a constant conformational flux between a normal B helix and other forms.

The existence of superhelical DNA suggests that the structural variations previously discussed have some biological function. One of the first surprising research developments was the discovery of enzymes called **topoisomerases** that modify the superhelicity of DNA. Of course, the only way to modify the superhelicity of closed circular DNA is to break one or both strands and twist the ends with respect to each other before linking them together again; topoisomerases are able to accomplish this.

topoisomerases

In addition, considerable information is accumulating that relates the presence of DNA palindromes to protein binding sites. Since most DNA in its natural state is negatively supercoiled, it is likely that some palindromes normally exist in the cruciform conformation, and are therefore accessible to binding by other molecules. Likewise, single-stranded bubbles generated by negative supercoiling can serve as regions for protein binding.

Denaturation of DNA

The helical structure of nucleic acids is determined by thermodynamically favored stacking between adjacent bases in the same strand (Figure 2-9), and the double-stranded structure of the helix is maintained by hydrogen-bonding between the bases of the base pairs (Figure 3-2). This conclusion has come from studies of the thermal stability of DNA molecules.

The free energies of the noncovalent interactions described in Chapter 2 are not much greater than the energy of thermal motion at room temperature. Thus, at elevated temperatures, the three-dimensional structures of both proteins and nucleic acids are disrupted. A macromolecule in a disrupted state, in which the molecules are in a nearly random conformation, is said to be **denatured;** the ordered state, which is presumably that originally present in nature, is called **native.** A transition from the native to the denatured state is called **denaturation.** When double-stranded DNA or native DNA is heated, the bonding forces between and within the strands are disrupted, and the two strands can separate; thus, *denatured* DNA is single stranded.

denaturation

A great deal of information about structure and stabilizing interactions has been obtained by studying nucleic acid denaturation. This is done by measuring some property of the molecule that changes as denaturation proceeds—for example, the absorption of ultraviolet light. For DNA, denaturation can be detected by observing the increase in the ability of a DNA solution to absorb ultraviolet light at a wavelength of 260 nm. A standard measure of such absorption is a logarithmic ratio of transmitted light and incident light called the **absorbance** $A;$ at 260 nm, this is written A_{260}. Nucleic acid bases absorb 260-nm light strongly, and the amount of light absorbed by a given number of bases depends on their proximity to one another. When the bases are highly ordered (very near one another), as in double-stranded DNA, A_{260} is lower than that for the less ordered state in single-stranded DNA. In particular, if a solution of double-stranded DNA has a value of $A_{260} = 1.00$, a solution of single-stranded DNA at the same concentration has a value of $A_{260} = 1.37$. This relation is often described by stating that a solution of double-stranded DNA becomes **hyperchromic** when heated. The corresponding value for free bases is 1.60, so that absorbance can also be used to measure degradation of nucleic acids.

hyperchromic

If a DNA solution is slowly heated and the A_{260} is measured at various temperatures, a melting curve, such as that shown in **Figure 3-8**, is obtained. The following features of this curve should be noted:

1. The A_{260} remains constant up to temperatures well above those encountered by most living cells in nature.
2. The rise in A_{260} occurs over a relatively narrow range of 6°–8°C.
3. The maximum A_{260} is about 37% higher than the starting value.

The state of a DNA molecule in different regions of the melting curve is also illustrated in Figure 3-8. Before the rise begins, the molecule is fully double-stranded. In the rise region, base pairs in various segments of the molecule are broken; the number of broken base pairs increases with temperature. In the initial part of the upper plateau, a few base pairs remain to hold the two strands together until a critical temperature is reached, at which the last base pair breaks and the strands separate completely. Please note that during melting all covalent bonds, including phosphodiester bonds, remain intact. Only hydrogen bonds and stacking interactions are disrupted. A convenient parameter to characterize a melting transition is the temperature at which the rise in A_{260} is half complete. This temperature is called the **melting temperature**; it is designated T_m.

melting temperature
T_m

A great deal of information is obtained by observing how T_m varies with base composition and experimental conditions. For example, it has been observed that T_m increases with percent $G + C$, which has been interpreted in terms of the greater number of hydrogen bonds in a $G \cdot C$ pair (three) than in an $A \cdot T$ pair (two). That is, a higher temperature (more energy) is required to disrupt a $G \cdot C$ pair than an

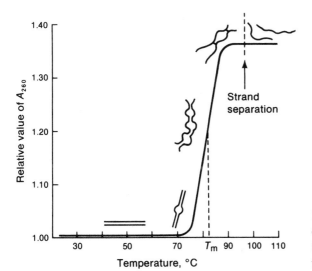

Figure 3-8 A melting curve of DNA showing T_m and possible molecular configurations for various stages of melting.

A · T pair because the double-stranded structure is stabilized, at least in part, by hydrogen bonds. In addition, reagents that increase the solubility of the bases and therefore decrease hydrophobic interactions, such as base stacking, also lower T_m. These results imply that a polynucleotide would tend to have a three-dimensional structure that maximizes the contact of the highly soluble phosphate group with water and minimizes contact between the hydrophobic bases and water. This explains why, in DNA, the sugar-phosphate chain is on the outside in contact with water, and the bases are on the inside in a stacked array. Clearly, stacked bases are more easily hydrogen-bonded, and correspondingly, hydrogen-bonded bases, which are oriented by the bonding, stack more easily. Thus, the two interactions act cooperatively to form a very stable structure. If one of the interactions is eliminated, the other is weakened: This explains why T_m drops so markedly after the addition of a reagent that destroys either type of interaction.

A very effective denaturant is NaOH, for the net charge of several groups engaged in hydrogen-bonding is changed at high pH, and a base bearing such a group loses its ability to form these bonds. At a pH greater than 11.3, all hydrogen bonds are eliminated and DNA is completely denatured. Since the phosphodiester backbone of DNA is quite resistant to alkaline hydrolysis, this procedure is frequently the method of choice for denaturing DNA because high temperatures often promote phosphodiester bond cleavage. Recall that denaturation is designed to break hydrogen bonds, while leaving covalent bonds intact.

A repulsive force is also present in DNA—namely, that between the negatively charged phosphate groups in complementary strands. These charges are neutralized by bound cations such as Na^+ and Mg^{2+}. As a result, the repulsion is so great that when DNA is placed in distilled water, strand separation occurs.

Renaturation

renaturation
reannealing

A solution of denatured DNA can be treated in such a way that native DNA reforms. The process is called **renaturation** or **reannealing** and the re-formed DNA is called **renatured DNA.** Renaturation is a valuable tool in molecular biology since it can be used to demonstrate genetic relatedness between different organisms, to detect particular species of RNA, to determine whether certain sequences occur more than once in the DNA of a particular organism, and to locate specific base sequences in a DNA molecule.

Two requirements must be met for renaturation to occur:

1. The salt concentration must be high enough that electrostatic repulsion between the phosphates in the two strands is eliminated—usually a concentration in the range of 0.15–0.50 M NaCl is used.
2. The temperature must be high enough to disrupt the random intrastrand hydrogen bonds present in single-stranded DNA. However, the temperature cannot be too high, or stable interstrand base-pairing will not occur and be maintained. The optimal temperature for renaturation is 20°–25° below the value of T_m.

Since strands can separate during denaturation, renatured DNA does not contain its own original single strand; renaturation is, thus, a random mixing process.

For example, if DNA labeled with the nonradioactive isotope ^{15}N is mixed with DNA containing the normal isotope ^{14}N and then denatured and renatured, three kinds of double-stranded DNA molecules will result: 25% will have ^{14}N in both strands, 25% will have ^{15}N in both strands, and 50% will have one ^{14}N strand and one ^{15}N strand.

Let us now consider how renaturation actually takes place. After denaturation, the DNA solution is composed of a mixture of essentially single-stranded fragments. In order to renature, a fragment must find its complementary partner in the solution. If the concentrations of the fragment and its partner are low, they will take much more time to find each other than if the two fragments are in high concentration. The process of finding a partner, or renaturation, can be followed by observing the absorption of the solution at 260 nm (A_{260}), since during renaturation the bases order themselves and the A_{260} will decrease with time. The **rate of decrease** is the rate of renaturation, which is related to the concentration of individual fragments in the solution. The value of A_{260} in the solution can be used to estimate the overall concentration of DNA. Renaturation kinetics offers a powerful tool for the analysis of the presence of repeated sequences in a DNA sample.

A hypothetical 100 base pair DNA molecule is shown in **Figure 3-9**. The sequence ABC/A′B′C′ represents 3% of the total number of base pairs in the sample; however, since it is repeated five times this sequence makes up 15% of the total weight of the DNA sample. When the molecule is broken into smaller pieces, there are five times more of the fragments originating from the repetitive sequence than from the remainder of the DNA that is not repetitive. Because of its higher concentration in the solution, the repetitive DNA renatures faster than single copy

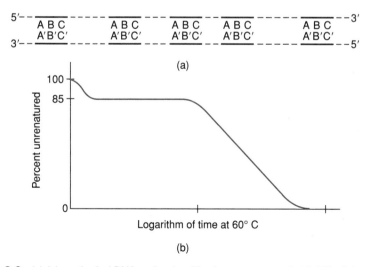

Figure 3-9 (a) A hypothetical DNA molecule with a base sequence that is 3% of the total length of the DNA and is repeated five times. The dashed lines represent the nonrepetitive sequences; they account for 85% of the total length. (b) A renaturation curve for the DNA shown in (a). Time is logarithmic in order to keep the curve on the page. The *y* axis—percent unrenatured DNA—is equivalent to the unit, relative value of A_{260}, used in previous figures with $A_{260} = 1.37$ equal to 100% (percent unrenatured-percent single stranded).

DNA. Such a hypothetical DNA molecule would have a renaturation curve such as that shown in Figure 3-9(b). Thus, this curve, which is obtained with DNA from a single organism, contains two components. The more rapidly renaturing component accounts for 15% of A_{260}, from which one can conclude that 15% of the DNA (by weight) contains a repeating sequence. From the kinetics one can determine the number of bases in the repeating sequence (called its **complexity**), and thus the number of copies of this sequence and the number of bases in a single nonrepeating sequence.

Repeated sequences are found in most eukaryotes, but not generally in unicellular organisms. Most eukaryotes' chromosomes are composed of regions of varying levels of repetitiveness. Distinctions of "repetitiveness" can be made to renaturation analysis. **Figure 3-10** shows a comparison between the renaturation kinetics of a bacterial and mouse DNA. Close examination of these curves from many species shows that there are four classes of sequences: **unique** (single copy), **slightly repetitive** (1–10 copies), **middle repetitive** (10 to several hundred copies), and **highly repetitive** (several hundred to several million copies). The boundaries of these classes are arbitrary, so one must be careful when making generalizations about the classes.

The unique sequences account for most of the genes of the cell, but often for only a small percentage of the DNA. This is because most eukaryote DNA is made up of highly repetitive sequences that do not code for proteins. The slightly repetitive class consists of genes that exist in a few copies; in some cases the sequences are not perfectly repetitive: the gene products may have slight differences in amino acid composition—for example, the various forms of hemoglobin. The middle-repetitive sequences are of two types: clustered and dispersed. The clustered sequences usually represent genes that exist in multiple copies in order to increase the amount or the rate of synthesis of the gene product. Some of the dispersed sequences may have a function in regulating gene expression. However, the major fraction consists of 20–60 copies of sequences, containing 2,000–7,000 nucleotide pairs that are able to move from one location to another in a chromosome and be-

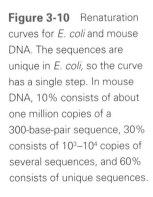

Figure 3-10 Renaturation curves for *E. coli* and mouse DNA. The sequences are unique in *E. coli,* so the curve has a single step. In mouse DNA, 10% consists of about one million copies of a 300-base-pair sequence, 30% consists of 10^3–10^4 copies of several sequences, and 60% consists of unique sequences.

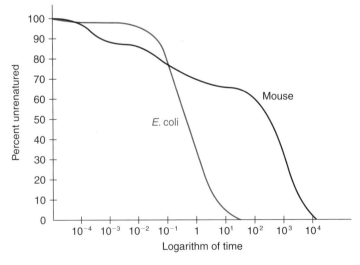

tween chromosomes. These are called transposable elements and are discussed in detail in Chapter 14. The highly repetitive sequences include the short sequences in satellite DNA and sequences of normal length in certain genes that exist in very large numbers, such as those making ribosomal RNA, which is needed for all protein synthesis. There are, in fact, very few different highly repetitive sequences, but the number of copies is so great that these sequences may account for 20% or more of the mass of the DNA. Generally, the highly repetitive sequences are relatively short in length. Chapter 13 will expand on this topic, for nucleotide sequence data is helping to understand the functional significance of various base sequence arrangements.

Hybridization

The technique wherein renatured DNA is formed from separate single-stranded samples is called **hybridization.** This renaturation technique can be done on thin paper filters (**membrane filters**) made of nitrocellulose, which bind single-stranded DNA very tightly but fail to bind either double-stranded DNA or RNA. They provide an important technique for measuring hybridization.

hybridization

Renaturation is carried out on a filter, as shown in **Figure 3-11**. A sample of denatured DNA is first filtered. The single strands bind tightly to the filter along the sugar-phosphate backbone, but the bases remain free. The filter is then placed in a vial with a solution containing a small amount of radioactive denatured DNA and a reagent that prevents additional binding of single-stranded DNA to the filter. After a period of renaturing, the filter is washed. Radioactivity is found on the filter only if renaturation has occurred.

Another important use of filter hybridization is the detection of sequence complementarity between a single strand of DNA and an RNA molecule—this is called either **DNA-RNA hybridization** or Northern hybridization, and it is the method of choice to detect an RNA molecule that has been copied from a particular DNA molecule. In this procedure, a filter to which single strands of DNA have been bound, as mentioned above, is placed in a solution containing radioactive RNA. After renaturation the filter is washed and hybridization is detected by the presence of radioactive RNA on the filter.

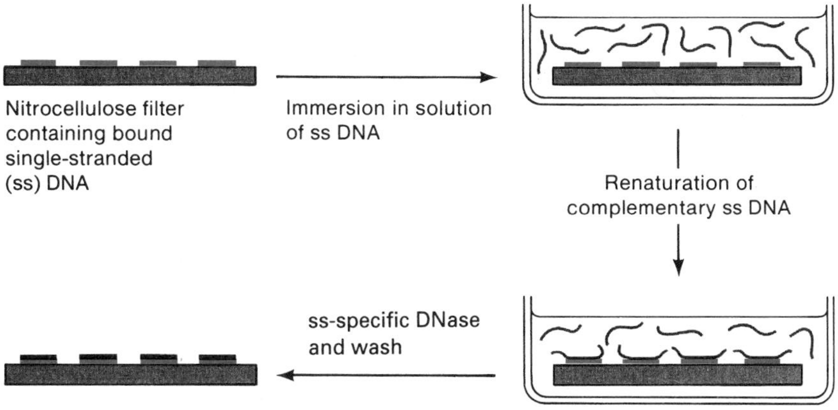

Nitrocellulose filter containing bound single-stranded (ss) DNA

Immersion in solution of ss DNA

Renaturation of complementary ss DNA

ss-specific DNase and wash

Figure 3-11 Method of hybridizing DNA to nitrocellulose filters containing bound, single-stranded DNA. In the final step the filter is treated with DNase to remove unhybridized DNA and washed.

The Structure of RNA

ribosomal RNA (rRNA)
transfer RNA (tRNA)
messenger RNA (mRNA)

A typical cell contains about ten times as much RNA as DNA. The high RNA content is due mainly to the large variety of roles played by RNA in the cell. There are three primary types of RNA—**ribosomal RNA (rRNA),** by far the most abundant, **transfer RNA (tRNA),** and **messenger RNA (mRNA).** Each of these general classes is actually composed of several unique species: 3 or 4 rRNAs, up to 50 tRNAs, and over 1,000 mRNAs. In fact, that abundance of RNA types has been attributed by some researchers to represent a carryover from an RNA "world" which may have preceded the appearance of DNA polymers.

RNA has many of the structural characteristics of DNA. That is, both molecules are linear polynucleotide chains. However, RNA has the following distinctions: ribose replaces deoxyribose, U replaces T, and, with the exception of some viruses, RNA is single-stranded. Most RNAs therefore resemble the conformation of single-stranded DNA, that is, they have stem-loop or hairpin structures because the chains fold back on themselves, creating loops and small base-paired stretches between complementary regions (**Figure 3-12**). RNA is defined in terms of its primary structure (the order of the nucleotides in the chain), its secondary structure (the pattern of base pairing), and its tertiary structure (the conformation of the molecule in three dimensions).

Due to the essential relationship of RNA structure to its functions, considerable research has been devoted to the prediction of the base-pairing pattern. This is a necessary step in determining its three-dimensional form. Since the techniques involved in prediction are empirical, secondary structures, although extremely useful, should only be considered as a model. Two methods of prediction (avail-

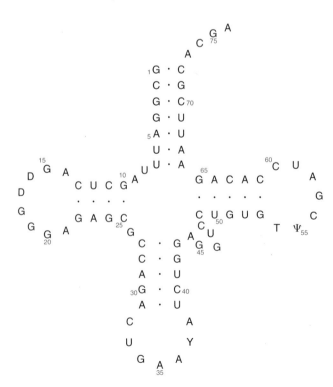

Figure 3-12 The proposed secondary structure of a tRNA. D, T, Y, and ψ are modified nucleotides often found in tRNA.

able as computer programs) maximize either (1) the number of base pairs in the molecule or (2) the overall free energy of the molecule using assigned energy values for different base pairs. The fact that G-C base pairs are held together by three hydrogen bonds favors their presence in the model over the presence of A–U pairs. Another approach is based on the availability of sequences of the same molecule from several organisms. Base pairs are proposed between positions in sequences that have undergone compensatory base changes—potential base-paired positions in one sequence must be base paired in a second, even though the identity of the nucleotides at these positions are different (**Figure 3-13**).

Hydrolysis of Nucleic Acids

With few exceptions (e.g., tRNA), the structure of nucleic acids are of such large size that they are difficult to characterize. The strategy that has emerged to circumvent this problem involves breaking down the polymers into smaller fragments. The phosphodiester bonds of both DNA and RNA can be broken by hydrolysis (addition of a H_2O molecule) either chemically or enzymatically.

At very low pH (1 or less), phosphodiester bond hydrolysis of both DNA and RNA occurs. This is also accompanied by breakage of the *N*-glycosylic bond between the base and the sugar (see Figure 2-5) so that free bases are produced. At high pH the behavior of DNA is strikingly different from that of RNA. DNA remains very resistant to hydrolysis at pH 13 (about 0.2 phosphodiester bonds broken per million bonds per minute at 37°C), whereas the presence of the 2′-OH groups renders RNA very sensitive to alkaline hydrolysis; at 37°C free ribonucleotides are produced within minutes at pH 11.

A variety of enzymes called **nucleases** hydrolyze nucleic acids. They usually show chemical specificity and are classified as either deoxyribonucleases (DNase)

nucleases

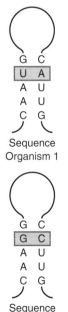

Sequence
Organism 1

Sequence
Organism 2

Figure 3-13 Compensatory base changes that support secondary structure. In sequences from organisms 1 and 2, it was found that potential base pairs U–A and G–C occupy similar positions. The fact that the sequences are different at these positions but the bases are complementary supports the hypothesis that these are indeed paired positions.

exonucleases
endonucleases

or ribonucleases (RNase). Many DNases act only on single-stranded or only on double-stranded DNA, although some act on both kinds. Furthermore, nucleases that act only at the end of a nucleic acid, removing a single nucleotide at a time, are called **exonucleases,** and may also be specific for the 3′ or 5′ end of the strand. Most nucleases act within the strand and are **endonucleases;** some of these are even more specific in that they cleave only between particular bases.

The availability of base-specific endonucleases opened the door to the determination of the primary structure of many RNAs. RNase T1, which cleaves an RNA chain 3′ to a G, and pancreatic RNase, which cleaves 3′ to a U or C, were particularly useful. Since the tertiary structures of many of these ribonucleases are now known, these enzymes also provide valuable insight into how proteins interact with nucleic acids (**Figure 3-14**).

In contrast, work on DNA structures lagged far behind RNA structures because of a lack of base-specific DNases. This situation has dramatically changed, however. As we shall see in the next section, it is now much easier to determine the sequence of DNA than RNA. The breakthrough for DNA occurred when restriction endonucleases were discovered. Restriction nucleases cleave double-stranded DNA at specific sequences that are several nucleotides long. They are also largely responsible for the development of the genetic engineering techniques discussed in Chapter 15.

ribozymes

One of the most remarkable and unexpected discoveries of recent years is that some RNAs have a transphosphorylation, catalytic activity. Previously, the dogma of biochemistry was that only proteins could have the catalytic activity for making or breaking bonds. The RNA enzymes, called **ribozymes,** are able to cleave and form specific phosphodiester bonds in a manner analogous to protein enzymes. An example is provided later (see Chapter 8, Figure 8-9).

Sequencing Nucleic Acids

Throughout this book, nucleotide sequences of DNA molecules will be described as a means of understanding the kinds of physical and chemical signals that are used by cells. This section describes how these sequences are determined.

Sanger procedure

DNA sequence determination is based on the high-resolution separation of polynucleotide fragments on polyacrylamide gels (Chapter 2). In the **Sanger procedure,** which is outlined here, polynucleotide fragments are produced by an ingenious use of the enzyme DNA polymerase. Since the activity of DNA polymerase is detailed in Chapter 7 only a superficial description of its mechanism of action is given here.

DNA polymerase uses triphosphate derivatives of deoxyribonucleotides to construct polynucleotides. This enzyme also requires the presence of a short primer

KEY CONCEPT

The "RNA Enzyme" Concept

Although most enzymes are proteins, some RNAs are catalytically active.

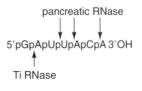

Figure 3-14 Hydrolysis of RNA. The position of digestion by T1 RNase and pancreatic RNase are shown. The sequence includes a "p" for the phosphate between nucleosides to illustrate which bond is cleaved.

polynucleotide and a longer single-stranded template fragment (the DNA to be sequenced) [**Figure 3-15**(a)]. Normally, double-stranded DNA is made continuously in this solution until all the nucleotide triphosphates are consumed. Consider the same reaction where a dideoxyribonucleotide (a nucleotide derivative lacking hydroxyl groups at both the 2′ and 3′ carbons) is added; these nucleotide analogs are incorporated into the growing nucleotide chain, but the resulting chain cannot be extended since there is no 3′ terminal hydroxide upon which the following nucleotide can be attached [Figure 3-15(b)]. When dideoxy G is used at 1/100 of the concentration of G, an average of one analog is randomly incorporated into the newly synthesized strand for every 100 "normal" Gs [Figure 3-15(b)].

If three additional reactions are prepared where each contains one of the remaining dideoxy nucleotides (i.e., A, T, or C), then four reaction mixtures will exist—one for each nucleotide—and each will contain a complex mixture of polynucleotide fragments differing in length, depending upon the point at which a dideoxynucleotide is incorporated. After the reactions are completed, each mixture can be analyzed by gel electrophoresis. The fragments will separate based on their lengths—the shortest migrating the farthest. As can be seen in **Figure 3-16**, each lane has a different pattern that represents the positions of the nucleotide whose dideoxy derivative was used. In this example, the G lane has the shortest fragment, and, therefore, the first position in the newly made sequence is G. The second fastest moving band is in the A track, so A is the second nucleotide. Continuation of this reading allows the definition of the entire sequence of the DNA. A word of caution: The determined sequence is the complement of the template sequence—a G in the template directs the incorporation of a C in the growing chain, and the determined sequence has the opposite 3′-5′ orientation from the template sequence, so that when the gel pattern tells us that there is a G at the 5′ terminus of the newly synthesized strand, we know that there is a C at the 3′ terminus of the DNA strand we are sequencing.

In a typical experiment, many copies of an isolated single strand of DNA are divided into four samples. Primers, the four types of nucleotide triphosphates, and

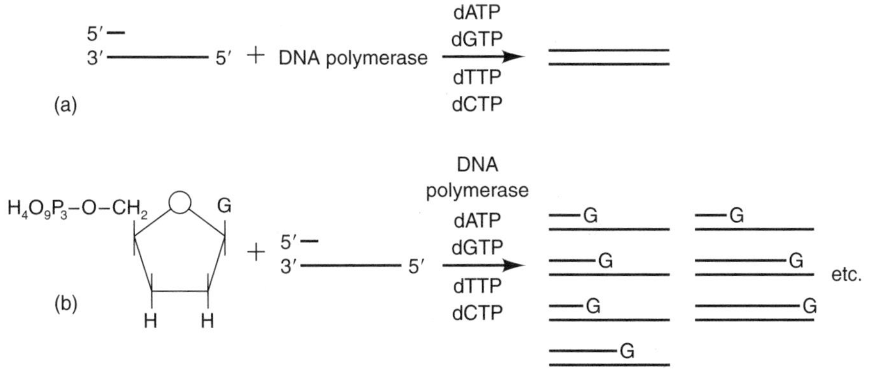

Figure 3-15 (a) DNA polymerase reactions. ATP–adenosine triphosphate, GTP–guanosine triphosphate, TTP–thymidine triphosphate, CTP–cytidine triphosphate. (b) The same reaction with added dideoxy G. Chain termination takes place at each position where dideoxy G is incorporated.

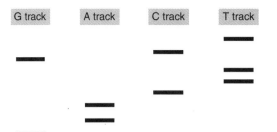

Figure 3-16 The gel pattern of DNA sequencing reactions. The bottom of the gel contains the 5′ end of the synthesized sequence. That is because the shortest fragments are closest to the 5′ end of the sequence. Recall, polymerization always goes 5′→ 3′. The determined sequence is 5′GAACTTGCT3′. This means that the template sequence (which is what we are interested in knowing) is 5′AGCAAGTTC3′.

one type of dideoxy nucleotide are added to each sample. In order to analyze the very small amounts that are generally available, either radioactive nucleotides or primers are used; analysis is then done with x-ray films, an example of which is shown in **Figure 3-17**. This technique is obviously limited by the capacity of a gel to separate fragments differing in length by one nucleotide; however, with care, a sequence of over 600 nucleotides can be determined on one gel.

Happily, automated versions of the Sanger methods have been devised for sequencing long DNAs, including whole genomes. Dideoxy nucleotides tagged with fluorescent dyes are employed. As the collection of dideoxy fragments are separated, lasers scan them, automatically recording both the fragment type (e.g., ddG; ddA, etc.) and length. Computer programs then interpret that information in terms of the linear sequence of the test DNA. In addition, pieces of DNA prepared by endonuclease digestion of whole genomes—once sequenced in this manner—can be aligned with computer programs, should individual pieces contain ends that overlap with other ends of other (contiguous) pieces.

As will be discussed in Chapter 13, the genomes of a variety of organisms have been sequenced using such automated methods. Indeed, the complete nucleotide sequence (3 billion base pairs) of human DNA is now available!

Synthesis of DNA

One of the techniques responsible for the explosive progress made in molecular biology in recent years is the chemical synthesis of DNA. Years of research on the optimal way to synthesize DNA have produced a method now available in most laboratories. The technique is based on two chemical principles: the activation of the phosphate group so that it will combine with another nucleotide at a reasonable speed, and the protection of reactive groups on the nucleotide to prevent unwanted side reactions (**Figure 3-18**). Activation is accomplished using the condensing agent tetrazole on highly reactive nucleoside phosphoramidites.

Protection is the process wherein a chemical group is temporarily added to critical positions of the nucleotide to ensure that during the formation of the internucleotide bond only the correct 3′-5′ phosphodiester bonds are made. Amino groups on the bases, for example, react under conditions of phosphoramidite activation so they must be made chemically inert. In addition, 5′-OH groups must

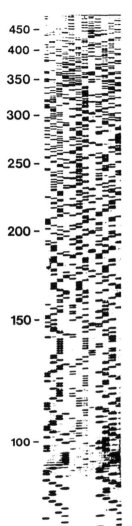

450 -
400 -
350 -
300 -
250 -
200 -
150 -
100 -

Figure 3-17 Sequences of DNA are stored and freely available from DNA Databanks in Europe, the United States, and Japan via international computer networks.

be modified to insure that it is the 3'OH atoms that are used for the phosphodiester bond. Later in the synthesis, protecting groups need to be removed to produce the desired unmodified DNA. In addition, the example in Figure 3-18 makes use of an insoluble support to facilitate purification of intermediates and harvesting of the final product.

The synthesizing strategy starts when the first nucleotide is added to an insoluble support (generally made of porous glass) via its 3'-OH. To ensure that only this OH reacts, the 5'-OH is protected with a trityl group (MMT in Figure 3-18). This group is acid sensitive so that it can be removed easily to make the 5'-OH available for subsequent steps. The resin is a porous glass preparation; the trityl group is generally the mono- or dimethoxy derivative, and reactive versions of all four nucleosides are commercially available.

After removal of the trityl group by an acid wash, the next nucleotide (in the form of its protected phosphoramidite derivative) is added in the presence of tetrazole. The internucleotide bond formation is virtually instantaneous; however, the

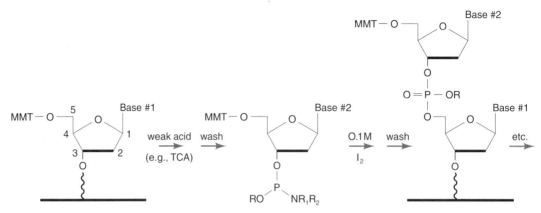

Figure 3-18 Chemical synthesis of DNA. The first protected nucleoside is linked to the in-soluble support via the 3'-OH. Acid hydrolysis of the monoethoxytrityl group (MMT) is done in 5% trichloroacetic acid (TCA). The next nucleotide in the sequence is added as its phospho-ramidite derivative (R = methyl; R_1 and R_2 are isopropyl groups). Iodine is used to oxidize the phosphite to phosphates. The process is repeated with a third nucleoside, and a fourth, and so on to produce a long chain oligonucleotide. Eventually, methyl groups (R) are removed at alka-line pH. Notice that this chemical synthesis is done in the 3'-to-5' direction, which is opposite from the natural (biological) process.

product is a *phosphite* rather than a *phosphate* ester. The phosphite bond is then oxidized to phosphate using iodine, the trityl group is removed, and the product is ready for the next cycle.

Now the advantage of the resin can be exploited. Since the product is insoluble in the reaction medium at all times during the synthesis, purification is achieved by simple filtration of the support. The support can be packed into a vertical column and various reagents can be passed through it, one after the other, creating a continuous process. At the termination of the synthesis reaction, the DNA is cleaved from the support and all the remaining protective groups are removed.

This process has now been completely automated. Reagents are stored in bottles connected to the column with tubing and a series of valves. A computer controls which valves are opened and shut—the chemist simply keys in the desired sequence. The cycle of nucleotide addition has been reduced to around five minutes, and the synthesizing machines (so-called "gene machines") are capable of producing oligonucleotides over 100 units long. By using nucleotide sequence information to synthesize altered versions of specific regions of a gene, it is now possible to modify naturally occurring genes (e.g., human insulin gene) and thereby produce novel versions of particular proteins.

Recently, a scheme for the chemical synthesis of RNA has been implemented. RNA synthesis is complicated by the fact that RNA is much more chemically la-bile than DNA, so the additional 2'-OH must be protected during the synthesis. More recently, phosphoramidites of not only ribo- and deoxyribonucleotides, but also of various modified nucleotides have become available to the researcher. By synthesizing nucleic acids that contain such substituted, modified bases, it is possible to analyze which chemical groups of a nucleotide (such as the amino, 2-hydroxyl, or oxo groups) are involved in a particular biological function.

A Future Practical Application?

Once it became possible to sequence large pieces of DNA, several laboratories joined forces to sequence the entire genomes from plants, bacteria, and humans. Although the human genome consists of approximately three billion base pairs, automation techniques reduce the time it takes to complete this huge project. With the availability of immense data banks of sequence information, a new era of molecular biology and biochemistry is emerging (see Chapter 13). Up until now, a molecule was characterized by its biological activity, from which chemical and structural studies could be analyzed in this context. In the future, this strategy will be turned around. That is, sequence databases will be searched for structural domains or special features from which biological activity will eventually be inferred. Progress in identifying new proteins and establishing relationships between specific genes and disease states will, therefore, be greatly facilitated.

SUMMARY

DNA consists of deoxyribonucleotides linked together by phosphodiester bonds connecting the 5′-carbon of one nucleotide with the 3′-carbon of the next nucleotide. DNA commonly exists as an antiparallel double strand. In double-stranded DNA, every base of one strand is hydrogen bonded to a base of the other strand, and since purines pair only with pyrimidines, the concentration of purines should equal that of pyrimidines—which can be expressed as $[A + G] = [C + T]$. Double-stranded DNA usually assumes the B form of a double helix, with ten base pairs per turn, 0.34 nm between each adjacent base pair, and bases oriented perpendicular to the helical axis. In addition to the B form, DNA also exists in an A form (a right-handed helix with 11 base pairs per turn, and bases tilted 30° with respect to the helical axis), and a Z form (a left-handed helix with 12 base pairs per turn). DNA can also form a double-stranded, covalently closed circle. These circular molecules are often coiled into a superhelix, the formation of which is catalyzed by enzymes called topoisomerases.

The hydrophobic properties and hydrogen-bonding abilities of the bases serve to hold the double-stranded molecule together. However, this double-stranded conformation is sensitive to the temperature and ion concentration of its environment. At a sufficiently high temperature, the hydrogen bonds between bases are broken and the molecule is denatured. DNA made single stranded will spontaneously renature if the temperature is high enough to disrupt any intrastrand base-pairing and if the salt concentration is sufficient to counteract the electrostatic repulsion between the phosphates. Renaturation kinetics can be used to determine the number of bases in a sequence that is repeated within a molecule and the number of copies of that sequence that are present.

Three primary classes of RNA exist: rRNA, mRNA, and tRNA. RNA is usually single stranded, although some molecules—tRNA, for example—have extensive secondary structure. Enzymes called nucleases catalyze the hydrolysis of phosphodiester bonds linking adjacent nucleotides in either RNA or DNA.

DNA molecules can be sequenced using the Sanger method, in which synthesis of a copy of the strand to be sequenced is halted at specific positions by the incorporation of dideoxy derivatives of specific nucleotides. The fragments are separated by gel electrophoresis, and the sequence is determined from the banding pattern of the gel. Also, nucleic acid molecules can be synthesized to include virtually any nucleotide sequence. Such oligonucleotides are routinely employed for Sanger sequencing (as primers) and for genetic engineering (see Chapter 15).

DRILL QUESTIONS

1. What is the difference between the major and minor grooves of double-stranded DNA?
2. A new virus is discovered, the genome of which is found to be 28% A, 23% T, 24% C, and 25% G. What does this tell you about the genome of the virus?
3. In a double-stranded DNA molecule, what is the relationship between [A] + [G] and [C] + [T], and between [A] + [T] and [G] + [C]?
4. In what circumstances does a double-stranded DNA molecule form an A helix? a B helix? a Z helix?
5. Could a supercoiled DNA molecule be formed by:
   (a) joining the ends of a linear DNA molecule and twisting the resulting circle?
   (b) twisting the ends of a linear DNA molecule and then linking the ends together?
   (c) linking together the ends of a linear DNA molecule?
6. What noncovalent interactions are involved in maintaining the double-helical conformation of DNA?
7. Why are salt concentration and temperature important to the renaturation of DNA?
8. What would happen if double-stranded DNA were placed in distilled water?
9. Suppose that a nonsupercoiled DNA molecule is composed of 4,800 base pairs. How many helical turns are present? How long is the linear, double-helical form of this molecule?
10. Could an exonuclease remove nucleotides from a circular DNA molecule?

PROBLEMS

1. The DNA molecules below are denatured and then allowed to reanneal. Which of the two is least likely to re-form the original structure? What would prevent that molecule from doing so?
   (a) ATATGTATATATAGAT
       TATACATATATATCTA
   (b) GCCTATACGTGCACCA
       CGGATATGCACGTGGT
2. Which of the two molecules shown above would have the highest T_m? Why? Which would require a higher temperature for renaturation?
3. The following gel pattern was obtained in an attempt to sequence a DNA molecule using the Sanger method. What is the sequence of the DNA molecule in question?

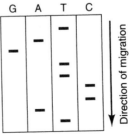

4. Suppose, by mistake, too much dideoxy T was used in the "T reaction mixture" for the gel described in the previous question. How would the bands in the T lane appear?

5. Which of the following types of RNA would likely be more stable (i.e., resistant to nuclease digestion) in the living cell: ribosomal, mRNA, or tRNA?

6. The melting curve below was obtained from a solution of DNA. What does this curve tell you about the composition of the solution?

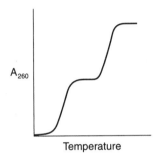

Temperature

CONCEPTUAL QUESTIONS

1. What features of the DNA double helix facilitate its replication? What features might make its replication complicated?

2. In what ways does RNA represent the ideal molecule for converting the information encoded in a nucleotide sequence into the structural and catalytic macromolecules of a cell?

3. With modern DNA sequencing techniques it is now feasible to sequence all the DNA in a human cell. Social, moral, and ethical considerations can, however, be raised. Should such a major sequencing endeavor nevertheless be encouraged?

in this chapter you will learn

1. About polypeptide folding and various configurations proteins adopt.

2. How several poly-peptides associate to form a larger unit.

3. The formation and structure of enzyme-substrate complexes and antibodies, and the models put forth to represent them.

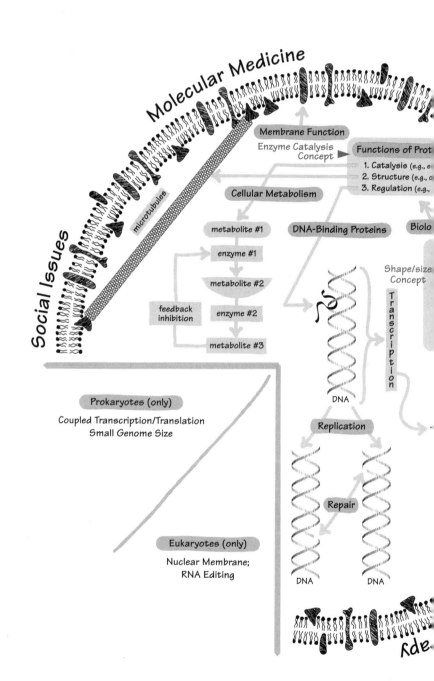

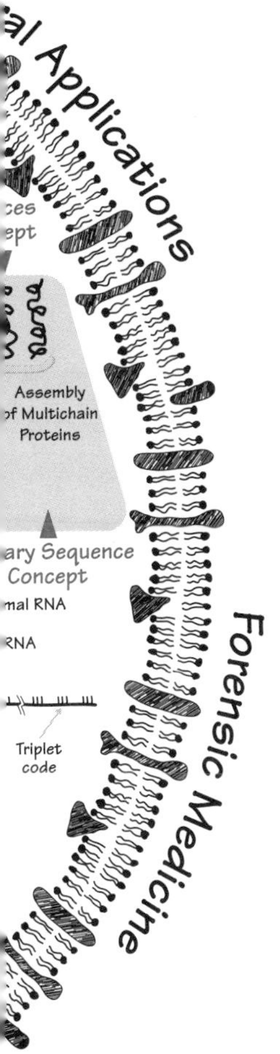

al Applications

ces
ept

Assembly
of Multichain
Proteins

ary Sequence
Concept
mal RNA

RNA

Triplet
code

Forensic Medicine

The Physical Structure of Protein Molecules

Like nucleic acids, all proteins are polymers, but instead of nucleotides, proteins are built from amino acids. Each species of protein molecule has a unique three-dimensional structure determined principally by the amino acid sequence of that protein. This contrasts sharply with DNA, which has a universal structure—the double helix. This makes the study of proteins very complex but, on the other hand, the diversity of protein structures enables these molecules to carry out the thousands of different processes required by a cell and makes proteins fascinating objects to study.

Proteins have been called the "footsoldiers" of the living cell. Their functions are the most diverse of all types of macromolecules. As we shall learn in this chapter, size, shape, and macromolecular composition vary tremendously among proteins. Those diverse features generate a wide range of functional properties. Some proteins carry out structural functions (e.g., stabilize a living cell's shape), while others speed up the rate of chemical reactions (e.g., enzymes).

In order to understand the manner in which proteins manage to carry out such diverse functions, we will learn how individual amino acids contribute a structural feature to a protein. Beginning with the amino acid sequence, we will review twisting, folding, and aggregation, each of which contributes to a protein's function. Then we will examine several examples of proteins that display widely different properties and, hence, carry out diverse functions.

The study of the detailed structure of proteins is beyond the scope of this book, being heavily dependent on a variety of optical techniques, especially the mathematically complex technique of x-ray diffraction. For this reason, the discussion of proteins in this chapter will be a survey.

Basic Features of Protein Molecules

In Chapter 2, the basic chemical features of protein molecules were described. That is, a protein is a polymer of amino acids in which carbon atoms and peptide

primary structure

groups alternate to form a linear polypeptide chain, while specific groups—the amino-acid side chains—project from the α-carbon atom (Figure 2-1). This linear sequence of amino acids is called the **primary structure** of the protein. The term "linear" requires careful consideration, for, as will be seen in the following sections, a polypeptide chain is highly folded and can assume a variety of three-dimensional shapes; each shape in turn consists of several standard elementary three-dimensional configurations and other configurations that may be unique to that molecule.

In Chapter 3, it was often seen that nucleic acid molecules are very large, having molecular weights often as high as 10^{10}. Protein molecules are much smaller; in fact, the molecular weight of a typical protein molecule is comparable to that of the smallest nucleic acid molecules, the transfer RNA molecules. The molecular weights of hundreds of different proteins have been measured. Typical polypeptide chains have molecular weights ranging from 15,000 to 70,000. The average molecular weight of an amino acid is 110, which means that typical polypeptide chains contain some 135–635 amino acids.

The length of a typical polypeptide chain, if it were fully extended, would be 1,000–5,000 Å. A few of the longer fibrous proteins, such as myosin (from muscle) and tropocollagen (from tendon) are in this range, with lengths of 1,600 Å and 2,800 Å, respectively. Most proteins, however, are highly folded and their longest dimension usually ranges from 40 Å to 80 Å. This folding constitutes the **secondary structure** of a protein and is described in the next section.

secondary structure

The Folding of a Polypeptide Chain

A fully extended chain, if it were to exist, would have the configuration shown in **Figure 4-1**. This stretched out form has no biological activity. Rather, protein function arises because polypeptides fold in complex ways to yield precise three-dimensional structures. The folded structure is stabilized by many noncovalent interactions (e.g., H-bonding, ionic interactions, van der Waals attractions, de-

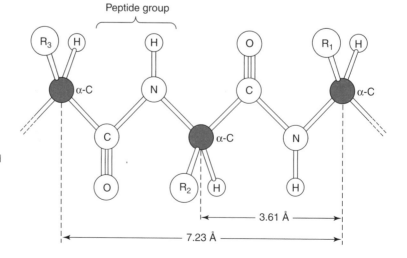

Figure 4-1 The configuration of a hypothetical fully extended polypeptide chain. The length of each amino acid residue is 3.61 Å; the repeat distance is 7.23 Å. The α-carbon atoms are shown in blue. Side chains are denoted by R.

scribed in Chapter 2 and the Appendix). Physical properties of the peptide bond and the amino acids in the polypeptide chain govern the manner of this folding. Three rules based on these physical properties can be described:

1. The peptide bond has a partial double-bond character (**Figure 4-2**) and hence is constrained to be planar and rigid. Free rotation occurs only between the α carbon and the peptide unit. The polypeptide chain is thus flexible but not as flexible as would be the case if there were free rotation about all of the bonds.
2. The side chains of the amino acids cannot overlap, so the folding, therefore, can never be truly random because certain orientations (especially those with bulky side chains) are forbidden.
3. Two charged groups having the same sign will not be very near one another. Thus, like charges tend to cause extension of the chain.

However, in addition to these rules, folding behaves according to the following general tendencies:

1. Amino acids with polar side chains tend to be on the surface of the protein in contact with water.
2. Amino acids with nonpolar side chains tend to be internal. Very hydrophobic side chains tend to cluster. Figure 2-2 in Chapter 2 indexes the nonpolar amino acids.
3. Hydrogen bonds tend to form between the carbonyl oxygen of one peptide bond and the hydrogen attached to a nitrogen atom in another peptide bond.
4. The sulfhydryl group of the amino acid cysteine tends to react with an —SH group of a second cysteine to form a covalent S—S (**disulfide**) **bond.** Such bonds pose powerful constraints on the structure of a protein (**Figure 4-3**).

disulfide bond

These tendencies should not be considered invariable because there are many exceptions. Nevertheless, they indicate that the structure of a protein may be changed markedly by a single amino acid substitution—for example, a polar amino acid for a nonpolar one; similarly, the change in structure might be minimal if one nonpolar amino acid replaces another nonpolar one. This notion will be encountered again in Chapter 10 when mutations are considered.

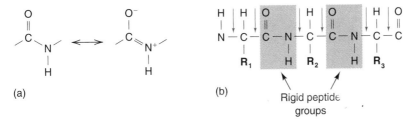

Figure 4-2 The planarity of the peptide group. (a) Two forms of the peptide bond in resonance. Owing to the resonance, the peptide group has a partial double-bond character and is thus rigid. (b) The peptide groups in a protein molecule. They are planar and rigid, but there is freedom of rotation about the bonds (blue arrows) that join peptide groups to α-carbon atoms (blue).

KEY CONCEPT

The "Primary Sequence" Concept

The primary sequence of building blocks (amino acids in proteins and nucleotides in nucleic acids) dictates the folding pattern and final shape, and, therefore, activity of a macromolecule.

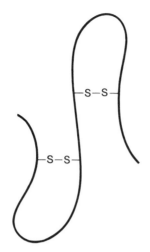

Figure 4-3 A polypeptide chain in which four cysteines are engaged in two disulfide bonds.

The three-dimensional shape of a polypeptide chain is a result of a balance between all of the rules and tendencies just described and can be very complex. However, in examining many polypeptide chains, it has become apparent that certain geometrically regular arrays of the chain are found repeatedly in different polypeptide chains and in different regions of the same chain. These are the arrays resulting from hydrogen-bonding between different peptide groups. These arrays are described in the next section.

The α Helix and β Secondary Structures

In the absence of any interactions between different parts of a polypeptide chain, a random coil would be the expected configuration. Hydrogen bonds, however, easily form between the H of the N—H group and the O of the carbonyl group. This H-bonding between atoms along the peptide backbone results in a variety of ordered structures of the polypeptide chain called secondary structures. The most common of these secondary structures are the α helix and the β structures.

α helix

In an **α helix,** the polypeptide chain follows a helical path that is stabilized by hydrogen-bonding between peptide groups. Each peptide group is hydrogen-bonded to two other peptide groups, one three units ahead and one three units behind in the chain direction (**Figure 4-4**). All α helices have a pitch of 5.4 Å, which is the repeat distance, a diameter of 2.3 Å, and contain 3.6 amino acids per turn. So it is a much tighter helix than the DNA helix. The side chains stick out from the rod-shaped core of the α helix.

In the absence of all interactions other than the hydrogen-bonding just described, the α helix is the preferred form of a polypeptide chain because, in this structure, all monomers are in identical orientation and each forms the same hydrogen bonds as any other monomer. Polyglycine, which lacks side chains and hence cannot participate in any interactions other than those just described, is an α helix.

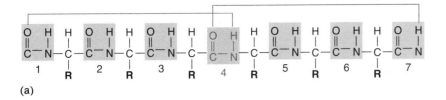

(a)

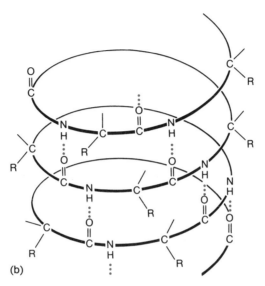

(b)

Figure 4-4 Properties of an α helix. (a) The two hydrogen bonds in which peptide group 4 (blue) is engaged. The peptide groups are numbered below the chain. (b) An α helix drawn in three dimensions, showing how the hydrogen bonds stabilize the structure. The blue dots represent the hydrogen bonds. The hydrogen atoms that are not in hydrogen bonds are omitted for the sake of clarity.

If all monomers are not identical, or if there are secondary interactions that are not equivalent, the α helix is not necessarily the most stable structure. While some amino acid side chains can stabilize the α helix, others are responsible for preventing α-helicity. A striking example of the disruptive effect of a side chain on an α helix is evident with the synthetic polypeptide polyglutamic acid, a polypeptide containing only glutamic acid. At a pH below about 5, the carboxyl group of the side chain is not ionized and the molecule is almost purely an α helix. However, above pH 6, when the side chains are ionized, electrostatic repulsion totally destroys the helical structure. (Please refer to the Appendix [Table A-3] if you would like further discussion of the ionization behavior of amino acids at various pH values.)

If the amino acid composition of a real protein is such that the helical structure is extended a great distance along the polypeptide backbone, the protein will be somewhat rigid and fibrous (although not all rigid fibrous proteins are α helical). This structure is common in many structural proteins, such as the α-keratin in hair.

Another hydrogen-bonded configuration is the **β structure.** In this form, the molecule is almost completely extended (repeat distance = 7 Å) and hydrogen bonds form between peptide groups of polypeptide segments lying adjacent and parallel with one another (**Figure 4-5**(a)). The side chains lie alternately above and below the main chain.

β structure

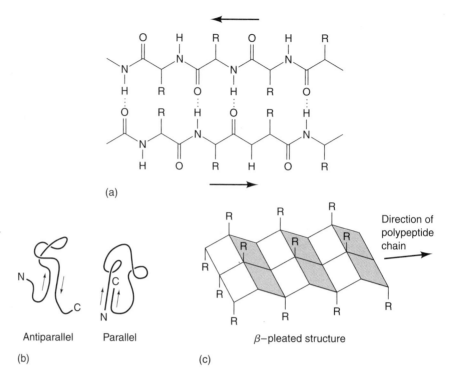

Figure 4-5　β structures. (a) Two regions of nearly extended chains are hydrogen-bonded (blue dots) in an anti-parallel array (arrows). The side chains (R) are alternately up and down. (b) Antiparallel and parallel β structures in a single molecule. (c) A large number of adjacent chains forming a β pleat.

(a)

Antiparallel　　Parallel

β–pleated structure

Direction of polypeptide chain

(b)　　　　　　　　　　　(c)

Two segments of a polypeptide chain (or two chains) can form two types of β structure, which depend on the relative orientations of the segments. If both segments are aligned in the N-terminal-to-C-terminal direction or in the C-terminal-to-N-terminal direction, the β structure is **parallel.** If one segment is N-terminal-to-C-terminal and the other is C-terminal-to-N-terminal, the β structure is **antiparallel.** Figure 4-5(b) shows how both the parallel and antiparallel β structures can occur within a single polypeptide chain.

When many polypeptides interact in the way just described, a pleated structure results called the **β-pleated sheet** (Figure 4-5(c)). These sheets can be stacked and held together in rather large arrays by van der Waals attractions and are often found in fibrous structures such as silk.

Protein Structure

Few proteins are pure α helix or β structure; usually regions having each structure are found within a protein. Since these configurations are rigid, a protein in which most of the chain has one of these forms is usually long and thin and is called a **fibrous protein.** In contrast are the quasispherical proteins called the **globular proteins,** in which α helices and β structures are short and interspersed with randomly coiled regions and compact structures.

The fibrous proteins are typically responsible for the structure of cells, tissues, and organisms. Some examples of structural proteins are collagen (the protein of tendon, cartilage, and bone), elastin (a skin protein), and spectrin (a protein that contributes to cell shape). Some of the fibrous proteins are not soluble in water—examples are the proteins of hair and silk.

The catalytic and regulatory functions of cells are performed by proteins that have a well-defined but deformable structure. These are the globular proteins, of which the catalytic proteins, or enzymes, are the most widely studied. Globular proteins are compact molecules having a generally spherical or ellipsoidal shape. Large segments of the polypeptide backbone of a typical globular protein are α helical. However, the molecule is extensively bent and folded. Usually, the stiffer α-helical segments alternate with very flexible randomly coiled regions, which permit bending of the chain without excessive mechanical strain. Numerous segments of the chain, which might be quite distant along the backbone, form short parallel and antiparallel β structures (Figure 4-5(b)); these are also responsible in part for the folding of the backbone (Figure 4-5(c)). The extensive folding of the backbone is usually called the **tertiary structure** or **tertiary folding.**

tertiary structure

A very important distinction can be made between secondary and tertiary structure, namely,

> Secondary structure results from hydrogen-bonding between peptide groups that are close together in the sequence of the polypeptide chain. Tertiary structures are formed from the folding of the secondary structures (α helices, β sheets) to yield three-dimensional structures.

The most prevalent interactions responsible for tertiary structure are the following:

1. Ionic bonds between oppositely charged groups in acidic and basic amino acids, for example, glutamic acid forming an ionic interaction with lysine.
2. Hydrogen bonds between H-bond donors and acceptors in amino acid side chains. For example, the hydroxyl group in tyrosine and a carboxyl group of aspartic or glutamic acids.
3. Hydrophobic clustering between the hydrocarbon side chains of nonpolar residues such as phenylalanine, leucine, isoleucine, and valine.
4. Metal-ion coordination complexes between amino, hydroxyl, and carboxyl groups, ring nitrogens, and pairs of SH groups.

Hydrophobic clustering (see preceding item 3) is the most important stabilizing feature (please refer to the Appendix).

Figure 4-6 shows a schematic diagram of a hypothetical protein (in two dimensions) in which several of these interactions determine the structure. This figure should be examined carefully, for it indicates the role of different features of a protein molecule in determining the overall conformation of the molecule. One can see the following:

1. Disulfide bonds bring distant amino acids together.
2. Hydrophobic interactions bring distant amino acids together.
3. Hydrogen bonds sometimes bring distant amino acids together, but usually a single hydrogen bond makes a more subtle change in position.
4. Electrostatic interactions bring amino acids together or keep them apart, depending on the signs of the charges.

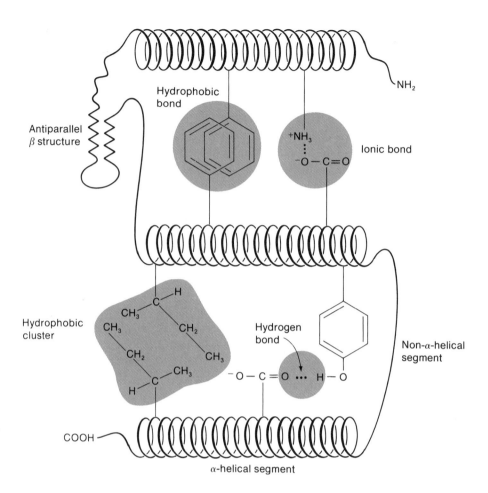

Figure 4-6 A hypothetical globular protein having several types of side-chain interactions.

5. A β structure brings distant segments of the polypeptide backbone together and creates rigidity.

6. An α helix makes adjacent regions of a polypeptide backbone stiff and linear.

7. Van der Waals attractions produce specific interactions between clusters of amino acids that may or may not be nearby in the polypeptide chain.

Figure 4-7 shows the structures of a number of globular proteins; some are predominantly α helical (a), some are predominantly β sheet (b), and some are mixed α helix and β sheet (c).

Proteins with Subunits

A polypeptide chain usually folds so that nonpolar side chains are internal—that is, isolated from water. However, it is rarely possible for a polypeptide chain to fold in such a way that *all* nonpolar groups are internal. It is, thus, often the case that nonpolar amino acids on the surface form clusters in an effort to minimize contact with water. A protein molecule having a large hydrophobic patch can fur-

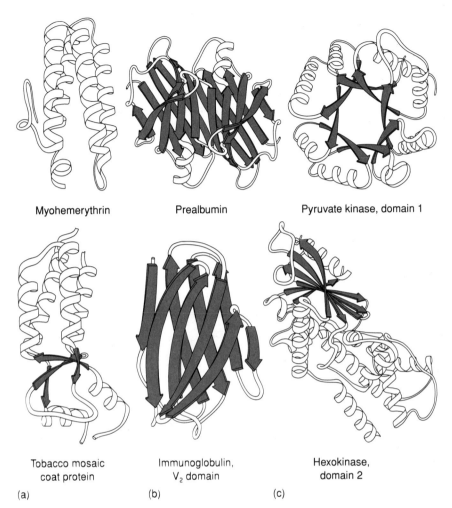

Myohemerythrin   Prealbumin   Pyruvate kinase, domain 1

Tobacco mosaic   Immunoglobulin,   Hexokinase,
coat protein   V₂ domain   domain 2

(a)   (b)   (c)

Figure 4-7 Idealized drawings of the tertiary structures of several globular proteins. Coiled region represents α helix and flat arrows indicate β structure. (a) Myohemerythrin and tobacco mosaic coat protein are mostly α helical. (b) Prealbumin and the immunoglobulin V_2 domain are predominantly β sheet. (c) Pyruvate kinase, domain 1, and hexokinase, domain 2, have α helical and β sheet structures. (Adapted and redrawn from Mathews and van Holde, *Biochemistry.* Menlo Park, CA: Benjamin Cummings, 1990.)

ther reduce contact with water by pairing with a hydrophobic patch on another protein molecule. Similarly, if a molecule has several distantly located hydrophobic patches, a structure consisting of several protein molecules in contact effectively minimizes contact with water. The protein is then said to consist of identical **subunits;** this is, in fact, a very common phenomenon, with two, three, four, and six subunits occurring most frequently. A multisubunit protein may also contain unlike subunits, and this is quite common. For example, hemoglobin, the oxygen carrier of blood, consists of four subunits, two each of two different types; likewise, RNA polymerase, which catalyzes synthesis of RNA, has five subunits of which four are different; and DNA polymerase III, which synthesizes DNA in *E. coli,* consists of ten different subunits.

Multisubunit proteins have certain advantages, particularly with respect to economy of synthesis and regulation of enzymatic activity. An example of a multisubunit protein, whose physical and biological properties are well known, is immunoglobulin G. It will be described in some detail in the following paragraphs; its synthesis is discussed in Chapter 12.

antibodies

antigens

The immunoglobulins (**antibodies**) are the proteins of the immune system. Their function is to interact with specific foreign molecules (**antigens**) and thereby render them inactive. This interaction is called the **antigen–antibody reaction.** The best-understood immunoglobulin is immunoglobulin G (**IgG**); other classes of immunoglobulins are IgA, IgM, IgD, and IgE. In this section, we shall examine only the IgG class, itself comprising several subclasses defined by slight structural differences.

Cleavage of the disulfide bonds of IgG yields two polypeptide chains whose molecular weights are about 25,000 and 50,000. The lighter polypeptide is called an **L chain;** the heavier one is an **H chain.** IgG is a tetramer containing two L chains and two H chains. A schematic structure of IgG is shown in **Figure 4-8.** Experimentally, IgG can be cleaved with the proteolytic enzyme papain, which causes each of the H chains to break, as shown in the figure (arrows), thus producing three separate subunits. The two units that consist of an L chain and a fragment of the H chain equal in mass to the L chain are called F_{ab} fragments (the subscript stands for antigen binding). The third unit, consisting of two equal segments of the H chain, is called the F_c (the subscript stands for "constant" or "common") fragment.

Two sites on an IgG molecule can bind antigen. Each site is at the end of an F_{ab} segment, as shown in **Figure 4-9.** The F_c segment is not involved in antigen–antibody binding, but is used in later processes needed to destroy the antigen.

The amino acid sequences of many subclasses of IgG molecules, each capable of combining with only a single kind of antigen molecule, have been determined.

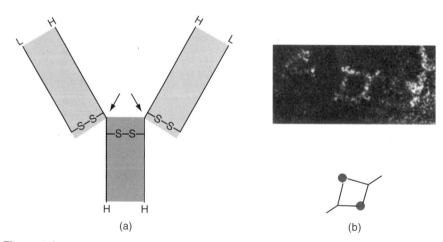

(a)   (b)

Figure 4-8   (a) The subunit structure of immunoglobulin G. There are two L subunits and two H subunits. Disulfide bonds join each L strand to an H strand and join the two H strands to one another. Treatment with papain cleaves the H strands at the arrows, yielding two F_{ab} units (shaded black) and one F_c unit (shaded blue). (b) An electron micrograph showing two immunoglobulin G molecules joined at the antigen binding site. The joining is a result of binding two antigen molecules not visible in the micrograph but shown as red dots in the interpretive drawing. (Courtesy of R. C. Valentine.)

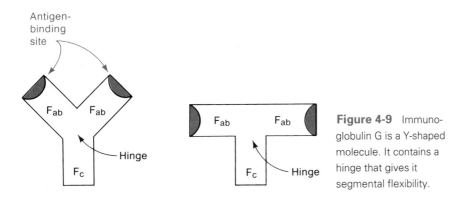

Figure 4-9 Immunoglobulin G is a Y-shaped molecule. It contains a hinge that gives it segmental flexibility.

All immunoglobulins are structurally similar inasmuch as L and H chains both have what is called a **variable (V)** and a **constant (C)** region.

The V and C regions have separate functions in IgG. The V regions confer antigen-binding specificity, whereas the C regions are responsible for the overall structure of the molecule and for its recognition by other components of the immune system.

The C regions of the L chains (C_L) of all types of IgG molecules have an identical amino acid sequence. Likewise, the C regions of all H chains (C_H) are identical for all IgG, although different from the C_L sequences. The V regions differ from one type of IgG to the next, however. The arrangement of the C_L, V_L, C_H, and V_H regions is shown in **Figure 4-10**.

Some of the similarities between amino acid sequences in different parts of an IgG molecule are quite striking. For example, the C region of the H chain can be divided into three segments—C_H1, C_H2, and C_H3—whose amino acid sequences are quite similar, but not identical, to one another. In addition, the sequence of C_L resembles the C_H sequences, although generally they are different. The V_L and V_H sequences for a particular IgG molecule also are nearly

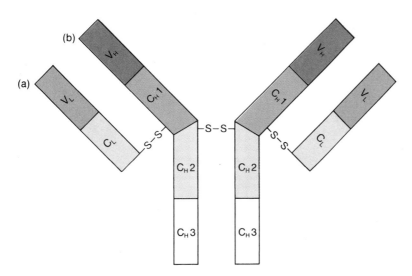

Figure 4-10 Model of an antibody molecule showing arrangement of the variable (V) and constant (C) regions of the IgG (a) L chain and (b) H chain.

KEY CONCEPT

The "Shapes and Surfaces" Concept

Subtle features of the shape and surface characteristics of a macromolecule allow it to interact with complementary surfaces on a neighboring macromolecule.

the same. This means that the IgG molecule, which already has twofold symmetry, consists of four domains, each of which has twofold symmetry, as shown in Figure 4-10, in which the intensity of the shading indicates two homologous regions. The symmetry of the molecules is made especially evident in **Figure 4-11**, which shows the three-dimensional structure of IgG. A feature of IgG, not clearly shown in the figure, is that the antigen-binding site is formed by joining parts of different subunits, namely, the V_H and V_L regions. Formation of a binding site by a subunit interaction is a property of many multisubunit proteins.

The structures of some multisubunit proteins are quite complex. For example, collagen, the protein of tendon, cartilage, bone, and skin, is formed by the mutual binding of three polypeptide chains as a triple helix and the interaction of numerous triple helices to form a tough fiber.

Enzymes

Enzymes are special proteins able to catalyze chemical reactions. Their catalytic power exceeds all manmade catalysts. A typical enzyme accelerates a reaction 10^8- to 10^{10}-fold, although there are enzymes that increase the reaction rate by a factor of 10^{15}. This is accomplished by lowering the energy level required for conversion of the substrate(s) or reactants to product(s). Enzymes are also highly

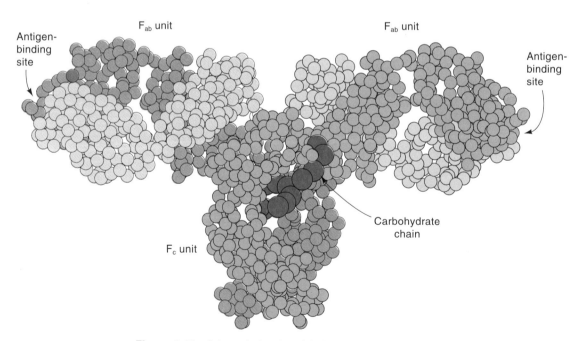

Figure 4-11 Schematic drawing of the three-dimensional structure of an IgG molecule. One of the H chains is shown in dark blue, the other in dark gray. One of the L chains is shown in light blue, the other in light gray. (After E. W. Silverton, M. A. Navia, and D. R. Davies. *Proc. Nat. Acad. Sci.*, 1944; 74: 5142.)

specific in that each catalyzes only a single reaction or set of closely related reactions. Furthermore, only a small number of reactants, often only one, can participate in a single catalyzed reaction. Since nearly every biological reaction is catalyzed by an enzyme, these clearly require a very large number of distinct enzyme molecules.

The detailed mechanism of catalysis by particular enzymes is beyond the scope of this book. All enzymes, however, have certain general features, the knowledge of which is important for understanding molecular biological phenomena. These features are described in this unit.

The Enzyme-Substrate Complex

In any reaction that is enzymatically catalyzed, one reactant always forms a tight complex with the enzyme. This reactant is called the **substrate** of the enzyme; in descriptive formulas, it is denoted S. The complex between the enzyme E and the substrate is called the enzyme-substrate or **ES complex.**

The substrate binds to a region of the enzyme called the **active site,** and it is here that the chemistry occurs. The active site is often a cleft in the enzyme, with some side chains of amino acids contributing to the binding of the substrate and others to the catalysis of the reaction. The extraordinary selectivity in enzyme catalysis is almost entirely a result of specificity of enzyme-substrate binding. After the ES complex forms, the substrate is usually altered in some way that facilitates further reaction. When the substrate is in the altered state, the ES complex is said to be active and is usually denoted **(ES)*.** The **(ES)*** complex then engages in one or a series of transformations, which result in conversion of the substrate to the product and dissociation of the product from the enzyme. The extent to which ES forms is determined by the strength of binding between E and S; this is called the **affinity** of E and S.

Theories of Formation of the Enzyme-Substrate Complex

Catalysis by an enzyme occurs in several steps—binding of the substrate, conversion to the product, and release of the product. The initial step, formation of the ES complex, is, in principle, easy to understand and is often considered in thinking about molecules. The subsequent rearrangements are chemical phenomena and will not be discussed in this book.

There are two major theories of enzyme binding, the **lock-and-key** and **induced-fit** models (**Figure 4-12**). In the lock-and-key model, the shape of the active site of the enzyme is complementary to the shape of the substrate. In the induced-fit model, the enzyme changes shape upon binding the substrate, and the active site has a shape that is complementary to that of the substrate only *after* the substrate is bound. For every enzyme-substrate interaction examined to date, one of these two models applies. Although some enzymes employ the lock-and-key model, most enzymes act through an induced-fit mechanism. It is often the case, though, that the substrate itself undergoes a small change in shape; in fact, the strain to which the substrate is subjected is often the principal

KEY CONCEPT

The "Enzyme Catalysis" Concept

Enzymes lower the activation energies of the chemical groups that participate in a reaction, and thereby speed up the reaction.

substrate

ES complex
active site

(a) Lock-and-key model

(b) Induced-fit model

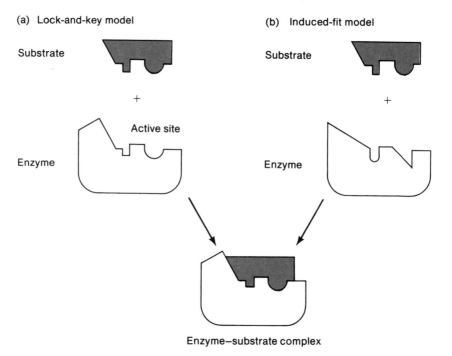

Figure 4-12 Two models for enzyme-substrate binding. (a) The lock-and-key model. The active site of the enzyme by itself is complementary in shape to that of the substrate. (b) The induced-fit model. The enzyme changes shape upon binding a substrate. The active site has a shape complementary to that of the substrate only after the substrate is bound.

mechanism of catalysis—that is, the substrate is held in an enormously reactive configuration.

Molecular Details of an Enzyme-Substrate Complex

The first detailed analysis of enzyme-substrate binding was carried out using hen egg-white lysozyme. This enzyme cleaves certain bonds between sugar residues in some of the polysaccharide components of bacterial cell walls and is responsible for maintaining sterility within eggs. The amino acid sequence of lysozyme is shown in **Figure 4-13**. The 19 amino acids that are part of the active site are printed in red; it should be noticed that they form widely separated clusters along the chain. Only when the chain is folded do they come into proximity and form the active site. The folding of the chain is shown in **Figure 4-14**. The deep cleft indicated by the arrow is the active site. This is seen more clearly in the space-filling

Figure 4-13 Schematic diagram of the amino acid sequence of lysozyme showing that the amino acids (blue) in the active site are separated along the chain. Folding of the chain brings these amino acids together.

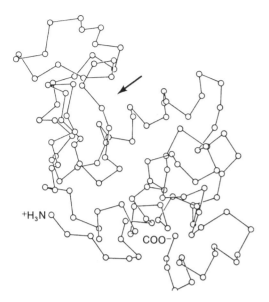

+H₃N
COO⁻

Figure 4-14 Three-dimensional structure of lysozyme. Only the α-carbon atoms are shown. The active site is in the cleft indicated by the arrow. (Courtesy of Dr. David Phillips.)

model shown in **Figure 4-15**. The substrate is a hexasaccharide segment that fits into the cleft and is distorted upon binding. The enzyme itself changes shape when the substrate is bound.

Included in the cleft-like active site of lysozyme are glutamic and aspartic acid side chains. For functioning of this enzyme, and virtually all others, the acid/base balance (pH—see Appendix) needs to be optimal. The so-called pH optimum varies from enzyme to enzyme (**Figure 4-16**), and reflects the nature of the tertiary structure of the enzyme, as well as the specific amino acids included in the active site.

Another enzyme, yeast hexokinase A, has been studied in order to examine what structural changes occur on substrate binding; these changes are now well doc-

Figure 4-15 A space-filling molecular model of the enzyme lysozyme. The arrow points to the cleft that accepts the polysaccharide substrate. (C atoms are black; H, white; N, gray; O, gray with slots.) (Courtesy of John A. Rupley.)

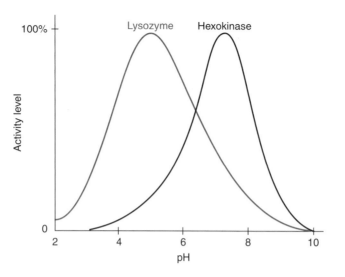

Figure 4-16 Experimental determination of the optimal acid/base balance, as expressed by pH (H ion concentration) for two enzymes.

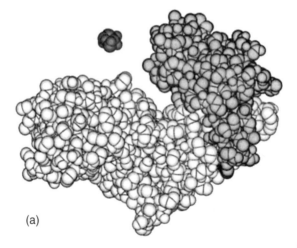

(a)

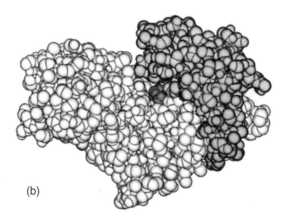

(b)

Figure 4-17 Structure of yeast hexokinase A. All atoms except hydrogen are shown separately. (a) Free hexokinase and its substrate, glucose. (b) Hexokinase complexed with glucose. Note that glucose binding causes two domains to move together and close the binding site cleft. (With permission, from J. Bennett and T. A. Steitz, *J. Mol. Biol.,* 140: 211. Copyright 1980: Academic Press Inc. [London] Ltd.)

umented. This is shown in the pair of space-filling models shown in **Figure 4-17**. It represents another example of "induced fit."

A Future Application

With increasing knowledge of the mechanism of action and substrate specificity of various enzyme proteins, it is now feasible to custom-design inhibitors to shut down the activity of some of those enzymes. For example, researchers are presently attempting to design inhibitors to protein kinases and protein phosphatases. These proteins play key roles in allowing cells to translate information received from the environment. Their "language" involves modifying the function of specific proteins by attaching (kinases) or removing (phosphatases) phosphates. Because uncontrolled phosphorylation or dephosphorylation has been implicated in cancer, many researchers are interested in developing specific inhibitors to kinases and phosphatases. The design of such inhibitors is based on the recently elucidated three-dimensional structures and catalytic mechanisms of these phospho-transfer enzymes.

SUMMARY

The primary structure of a polypeptide is the linear sequence of amino acids connected by peptide bonds. The secondary structure involves H-bonding between nearby peptide bonds. In the α helix, each peptide group H-bonds with groups three ahead and behind itself. Each turn contains 3.6 amino acids. The side chains stick out from the helical core. The β sheet is more extended, and the H-bonding occurs between peptide units in two or more adjacent polypeptide chains that can be parallel or antiparallel to one another. The side chains extend alternately above and below the sheet. Tertiary structure is the overall folding of a polypeptide in which secondary structure domains are positioned via long-range interactions that may be covalent—for example, disulfide bonds, ionic (repulsion or attraction of charged side chains), hydrophobic (clustering of nonpolar side chains), or H-bonds. Fibrous proteins have long, uninterrupted sections of α helix or β sheet; globular proteins are more spherical, with short stretches of secondary structure. Generally, polar side chains are on the surface near water, and nonpolar side chains are internal, away from water. Multisubunit proteins contain several polypeptides—for example, antibodies that contain two heavy and two light chains. The antigen-binding site is formed by the interaction of two chains, a common property for binding and active sites. Enzymes are protein catalysts that speed up the rate of cellular reactions. The active site is where substrate is bound, forming an enzyme-substrate complex where chemical reactions take place. These steps are facilitated by amino acid side chains, often from separated regions of the polypeptide. Substrate binding may occur by a lock-and-key mechanism, in which the active site is of complementary shape to the substrate, or, more often, by an induced-fit mechanism, in which both the active site and substrate change shape slightly, thus facilitating catalysis.

DRILL QUESTIONS

1. (a) Which amino acids are polar and which are nonpolar?
   (b) Is isoleucine more nonpolar than alanine? Why?
2. About which bond in the polypeptide backbone is there no free rotation?
3. Name the kinds of bonds in proteins in which the side chains of each of the following amino acids might participate: cysteine, arginine, valine, aspartic acid.
4. What would you guess to be the environment of a glutamate that is internal? What is the environment of an internal lysine?
5. Which of the following sets of three amino acids are probably clustered within a protein? (a) Asn, Gly, Lys; (b) Met, Asp, His; (c) Phe, Val, Ile; (d) Tyr, Ser, Lys; (e) Ala, Arg, Pro.
6. For a particular enzyme, 15 amino acid residues are known to be at the active site of the molecule. What would you expect their relative positions to be in the primary sequence of the protein?
7. What is meant by the primary, secondary, and tertiary structure of a protein?
8. What would be the most likely general tertiary structure for a protein with very long α helical or β structure domains? With many short regions of α helix and/or β structure?
9. What is meant by induced-fit? How does this facilitate catalysis?

PROBLEMS

1. State whether the following polypeptides would aggregate to form a multisubunit protein consisting of identical subunits and give the reason for your conclusion.
   (a) The folded molecule contains distinct surface regions with complementary shapes, and no polar amino acids are nearby.
   (b) There is a large hydrophobic cluster in a crevice just below the surface.
   (c) The protein has a large hydrophobic patch flanked by two lysines.
   (d) The surface has a region where positively and negatively charged amino acids alternate in a linear array.
2. Is an enzyme likely to have a very rigid configuration? Explain.
3. In general, a diploid cell containing the $^+$ and $^-$ alleles encoding an enzyme E has an E^+ phenotype. Suppose you isolated a particular mutant that yields an E^- phenotype even when the $^+$ allele is present. Suggest a possible explanation for the observed phenomenon.
4. Many proteins consist of several identical subunits. Some of these proteins have a single binding site, whereas others have several identical binding sites. What might you predict about the locations of the binding sites in these two classes of multisubunit proteins?
5. Assuming you had a way to look at the exact structure of the active site of an enzyme or binding protein, both before and after binding, how might you distinguish between the lock-and-key and induced-fit models of substrate binding? Explain the expected results.

CONCEPTUAL QUESTIONS

1. How might the various amino acid side chains (functional groups) that are associated with the active site of an enzyme chemically aid in the binding or catalysis of the substrate?

2. Enzymatic reactions are very fast, accelerated up to 10^{15}-fold. Suggest ways that you might slow this down in the laboratory in order to be able to study the mechanism of a particular enzyme activity.

3. It is generally estimated that in a typical cell 1,000–2,000 different chemical reactions are catalyzed by enzymes. Why do you suppose it is necessary that each of those reactions have its own special enzyme?

CHAPTER

5

in this chapter you will learn

1. About the organization of prokaryotic and eukaryotic genomes.

2. Several characteristics of proteins that interact with DNA.

3. About several aspects of biological membranes.

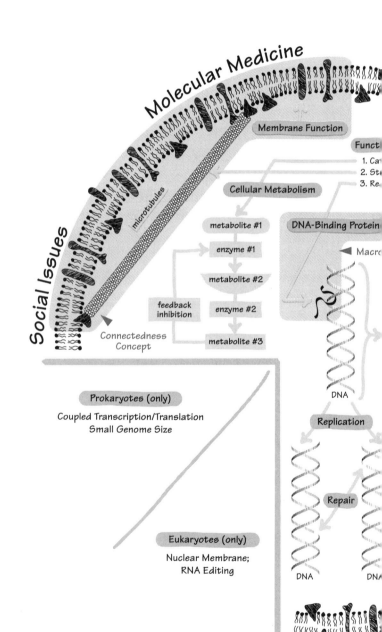

Molecular Medicine

Social Issues

Membrane Function

Funct[
1. Ca
2. St
3. Re

Cellular Metabolism

microtubules

metabolite #1

enzyme #1

metabolite #2

feedback inhibition

enzyme #2

metabolite #3

DNA-Binding Protein

Macr

Connectedness Concept

DNA

Replication

Repair

DNA

Prokaryotes (only)

Coupled Transcription/Translation
Small Genome Size

Eukaryotes (only)

Nuclear Membrane;
RNA Editing

DNA   DN

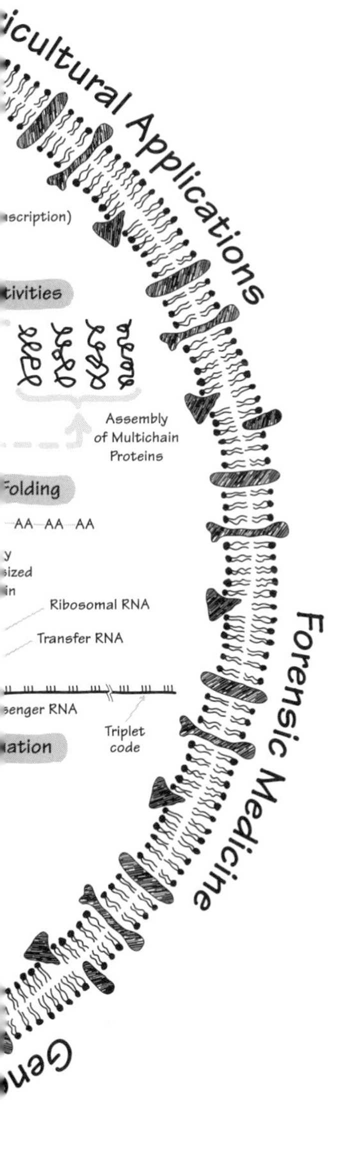

Macromolecular Interactions and the Structure of Complex Aggregates

Interactions between macromolecules underlie most biological phenomena. Structural components both within individual cells and in-between the cells associated with various tissues and organs invariably consist of assemblies of macromolecules. For example, nucleic acids are often associated with proteins, as in chromosomes (which are DNA-protein complexes), viral nucleic acids (which are encased in protein shells), and bone and cartilage (which are complex assemblies of proteins and other macromolecules). Proteins can also interact with lipids to produce membranes such as those that separate the contents of a cell from the environment, and those that separate different intracellular components from one another. Finally, polysaccharides form extraordinarily complex structures, such as the cell walls of bacteria and of plants.

The study of such complex structures and how they are formed is called structural biology. A few structures are almost completely understood and many more are actively being studied. In this chapter, only a few structures will be described. These have been selected by two criteria: They are of general importance, or they illustrate general principles and their structures are reasonably well understood. Examples such as these are typical, for most proteins function in the living cell by interacting with other proteins or other macromolecules, rather than by acting alone. This realization has generated a new research focus—proteomics—which will be reviewed in detail later, in Chapter 13. We will begin by reviewing the structure of one macromolecular complex—the *E. coli* chromosome.

A Complex DNA Structure: The *E. coli* Chromosome

All genes of *E. coli,* and presumably of most bacteria, are contained on a single supercoiled circular DNA molecule. The total length of the circle is about 1,300 μm.

A cylindrical bacterium has a diameter and a length of about 1 and 3 μm, respectively (**Figure 5-1**); clearly the bacterial DNA must be highly folded when it is in a cell.

When *E. coli* DNA is isolated by a technique that avoids both breakage of the molecule and denaturation of proteins, a highly compact structure known as a bacterial chromosome, or **nucleoid,** is found. This structure contains protein and a single supercoiled DNA molecule. Some RNA is also present in the isolated nucleoid, possibly becoming associated with the structure during the isolation procedure, but it is probably not an essential component. An electron micrograph of the *E. coli* chromosome is shown in **Figure 5-2**. The particular feature of the structure to note is that the DNA is in the form of loops that are supercoiled and emerge from a dense protein-containing structure, sometimes called the **scaffold.** The physical dimensions of the isolated chromosome are affected by a variety of factors, and some controversy exists about the state of the nucleoid within the cell. If the nucleoid is treated with an enzyme that degrades proteins or various detergents that break down protein–protein interactions, the chromosome expands markedly, although it remains much more compact than a free DNA molecule. As proteins are removed, the scaffold is disrupted and the chromosome goes through a series of transitions between forms having different degrees of compactness.

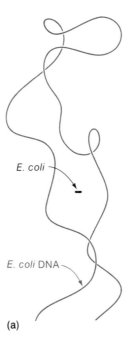

E. coli

E. coli DNA

(a)

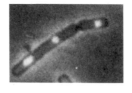

(b)

Figure 5-1   (a) Schematic diagram showing the relative sizes of *E. coli* and its DNA molecule, drawn to the same scale except for the width of the DNA molecule, which is enlarged approximately 10^6 times. (b) An *E. coli* cell whose DNA has bound an added fluorescent dye. (Courtesy of Todd Steck and Karl Drlica.)

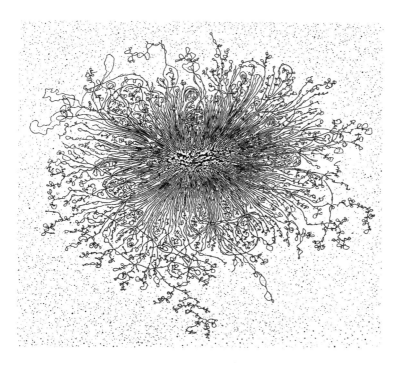

Figure 5-2 An electron micrograph of an *E. coli* chromosome showing the multiple loops emerging from a central region or scaffold. (Bluegenes #1. 1983. All rights reserved by Designer-genes Posters Ltd.)

The conclusion drawn from these and other observations is that the chromosome is held in compact form by the binding of different regions of the DNA molecule to the scaffold. Whether the scaffold has a well-defined organization and is a true "structure" (as opposed to a disorganized aggregate) is not known.

Chromosome Superstructure Is Elucidated by Enzyme Treatment

Treatment of a chromosome with very tiny amounts of a DNase, producing single-strand breaks, followed by sedimentation in a centrifuge of the treated DNA, gives some insight into the physical structure of the chromosome. In Chapter 3, it was explained that if a typical supercoiled naked DNA molecule receives one single-strand break, the strain of underwinding is immediately removed by free rotation about the opposing sugar-phosphate bond, and all supercoiling is lost. Since a nicked circle is much less compact than a supercoil of the same molecular weight, the nicked circle sediments much more slowly. Thus, one single-strand break in a circle causes an abrupt decrease (by about 30%) in the *s* value (a centrifugation measure).

If one single-strand break, however, is introduced by a nuclease into the *E. coli* chromosome, the *s* value decreases by only a few percent. Furthermore, a second break causes a second decrease in the *s* value, and subsequent breaks cause additional stepwise changes. After about 45 breaks, the form of the chromosome remains constant. This clearly indicates that free rotation of the entire DNA molecule does not occur when a single-strand break is introduced. The structure of the *E. coli* chromosome that has been deduced from these data is shown in **Figure 5-3**. The DNA is assumed to be fixed to the scaffold at approximately 100 positions,

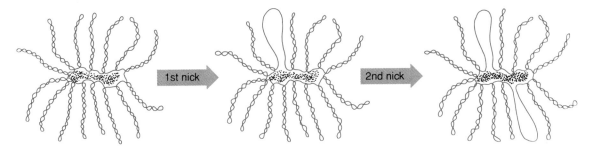

Figure 5-3 A schematic drawing of the highly folded supercoiled *E. coli* chromosome showing only 15 of the approximately 100 loops attached to the scaffold and the opening of a loop by a single-strand break (nick).

each of which prevents free rotation, so there would be about 100 supercoiled loops of DNA. One single-strand break would then cause one supercoiled loop to become open, and each subsequent break would, on average, open another loop. This notion has been confirmed in electron micrographs of chromosomes in which there are one or two nicks; structures such as that in Figure 5-2, but with one or two open loops, are observed.

The enzyme DNA gyrase, which plays an important role in DNA replication (see Chapter 7), is responsible for the supercoiling. If coumermycin, an inhibitor of *E. coli* DNA gyrase, is added to a culture of *E. coli* cells, the chromosome loses its supercoiling in about one generation time. Indirect measurements of the number of binding sites for gyrase on the chromosome indicate that there are roughly 45 sites. The spatial distribution of these sites is not known, but it is tantalizing to think that there may be one binding site in each supercoiled loop.

In the next section, it will be seen that DNA is arranged in a much more complex way in eukaryotic cells.

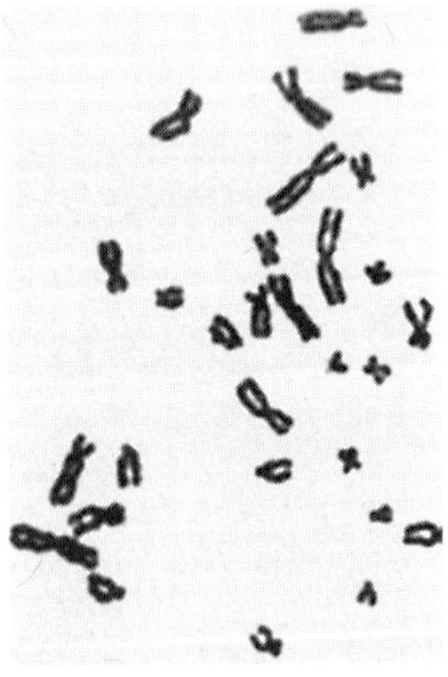

Chromosomes and Chromatin

The DNA of all eukaryotes is organized into morphologically distinct units called **chromosomes** (**Figure 5-4**). Each chromosome contains only a single enormous DNA molecule. For example, the DNA molecule in a single chromosome of the fruit fly *Drosophila* has a molecular weight greater than 10^{10} and a length of 1.2 cm. (Since the width of all DNA molecules is 2.0 nm or 2×10^{-7} cm, the ratio of length to width of *Drosophila* DNA has the extraordinarily high value of 6×10^{6}.) These molecules are much too long to be seen in their entirety by electron microscopy because, at the minimum magnification needed to see a DNA molecule, the field of view is only about 0.01 cm across.

A chromosome is much more compact than a DNA molecule and, in fact, a DNA molecule cannot spontaneously fold to form such a compact structure because the molecule would be strained enormously. Instead, DNA is made compact by a hierarchy of different types of folding, each of which is mediated by one or more protein molecules.

Figure 5-4 Human chromosomes from a cell at metaphase. Each chromosome is partially separated along its long axis prior to separation of the two daughter chromosomes. The constriction is the site of attachment to the mitotic spindle. (Courtesy of Theodore Puck.)

Organization of DNA into Chromosomes Requires Histone Proteins

The DNA molecule in a eukaryotic chromosome is bound to basic proteins called **histones.** The complex comprising DNA and histones is called **chromatin.** There are five major classes of histones—H1, H2A, H2B, H3, and H4—whose properties are listed in **Table 5-1**. Histones have an unusual amino acid composition in that they are extremely rich in the positively charged amino acids lysine and arginine. The lysine-to-arginine ratio differs in each type of histone. The positive charge of the histones is one of the major features of the molecules, enabling them to bind to the negatively charged phosphates of the DNA. This electrostatic attraction is apparently the most important stabilizing force in chromatin since, if chromatin is placed in solutions of high salt concentration (e.g., 0.5 M NaCl), which breaks down electrostatic interactions, chromatin dissociates to yield free histones and free DNA. Chromatin can also be reconstituted by mixing purified histones and DNA in a concentrated salt solution and gradually lowering the salt concentration by dialysis.

Reconstitution experiments have been carried out in which histones from different organisms are mixed. Usually, almost any combination of histones works because, except for H1, the histones from different organisms are very much alike. In fact, the amino acid sequences of both H3 and H4 are nearly identical (sometimes one or two of the amino acids differ) from one organism to the next. Histone H4 from a cow differs by only two amino acids from peas—arginine for lysine and isoleucine for valine—which shows that the structure of histones has not changed in the 10^9 years of evolutionary history since plants and animals diverged. Clearly, histones are very special proteins indeed.

As a cell passes through its growth cycle, chromatin structure changes. In a resting cell the chromatin disperses, but it does not spread throughout the entire nucleus. Rather, the chromatin associated with individual chromosomes appears to remain localized within the nucleus. Later, after DNA replication has occurred, the chromatin condenses about 100-fold and chromosomes form. Chromosomes have been isolated and gradually dissociated, and chromosomes at various degrees of dissociation have been observed by electron microscopy. The most elementary structural unit is a fiber 10 nm wide, which appears like a string of beads (**Figure 5-5**). The beadlike structure, which is chromatin, is also seen when chromatin is isolated from resting cells. The structural hierarchy of a chromosome, which has been deduced from a variety of studies, is shown in **Figure 5-6**. The

histones

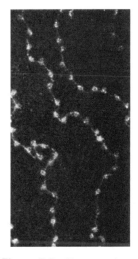

Figure 5-5 Electron micrograph of chromatin. The beadlike particles have diameters of nearly 100 Å. (Courtesy of Ada Olins.)

TABLE 5-1	TYPES OF HISTONES	
Type	**Lys/Arg Ratio**	**Location**
H1	20.0	Linker
H2A	1.25	Core
H2B	2.5	Core
H3	0.72	Core
H4	0.79	Core

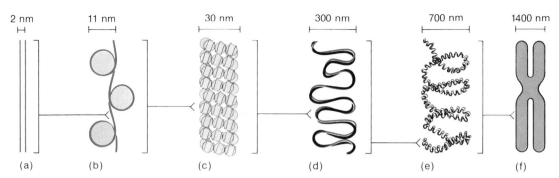

Figure 5-6 (a) As indicated, long linear pieces of DNA (vs circles of many prokaryotic chromosomes) comprise eukaryotic chromosomes. The human genome, for example, contains a total of approximately 2 m DNA distributed among 46 chromosomes. That DNA is coiled into nucleosomes (b), which are condensed (c–e) to form a metaphase chromosome (f). The dimensions indicate known sizes of intermediates, but the detailed structures are hypothetical. Note that each diagram represents a magnified portion of a section of the next diagram in the series.

higher orders of structure are still somewhat speculative, but a great deal of information is known about the beads themselves, each of which is an ordered aggregate of DNA and histones.

Prior to electron microscopic studies, the effect of various DNA endonucleases on chromatin also suggested that chromatin contains repeating units. Treatment of chromatin with pancreatic DNase I (which cannot attack DNA that is in contact with protein) yields a collection of small particles containing DNA and histones (**Figure 5-7**). If, after enzymatic digestion, the histones are removed, DNA fragments having roughly 200 base pairs, or a multiple of 200, are found. After a long period of digestion the multiple-size units are not found, and all of the DNA has the 200-base-pair-unit size. Fragments have also been isolated before removal of the histones and then examined by electron microscopy; it has been found that a fragment containing $200n$ base pairs consists of n connected beads, indicating that there is a fundamental bead unit containing about 200 base pairs.

DNA Is Organized into Nucleosome Core Particles

nucleosomes

The beadlike particles are called **nucleosomes.** Each nucleosome is found to consist of one molecule of histone H1, two molecules each of histones H2A, H2B, H3, and H4, and a DNA fragment. Treatment of the nucleosomes (which have been obtained by digestion of chromatin with pancreatic DNase) with the enzyme micrococcal nuclease gradually removes additional amounts of DNA. All histones remain associated with the DNA until the number of base pairs is less than 160, at which point H1 is lost. More bases can be removed, but the number cannot be reduced to less than 146 base pairs. The structure that remains is called the **core particle.** It contains an octameric protein disc consisting of two copies each of H2A, H2B, H3, and H4, around which the 146-base-pair segment is wrapped like a ribbon (**Figure 5-8**). Thus, a nucleosome consists of a core particle, additional DNA that links adjacent core particles, and one molecule of H1. The H1 binds to the

core particle

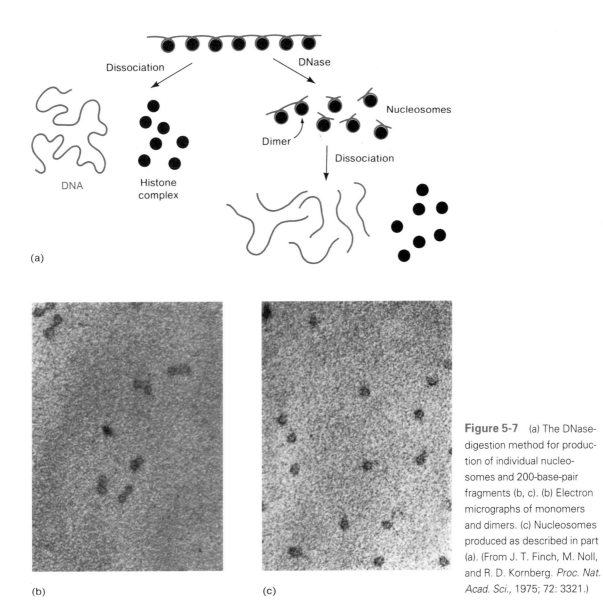

(a)

(b)   (c)

Figure 5-7 (a) The DNase-digestion method for production of individual nucleosomes and 200-base-pair fragments (b, c). (b) Electron micrographs of monomers and dimers. (c) Nucleosomes produced as described in part (a). (From J. T. Finch, M. Noll, and R. D. Kornberg. *Proc. Nat. Acad. Sci.*, 1975; 72: 3321.)

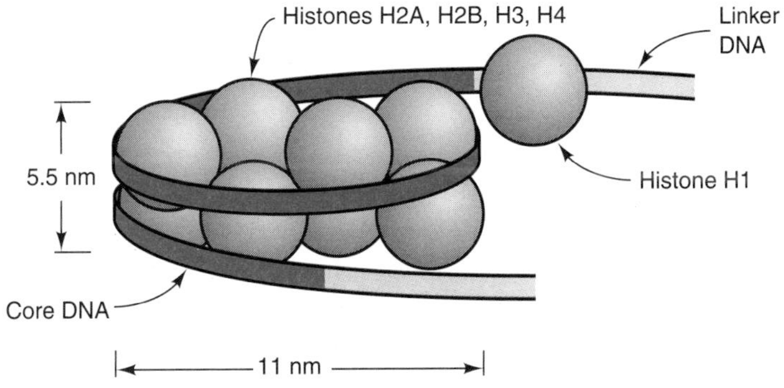

Figure 5-8 Schematic diagram of nucleosome core particle. The DNA molecule is wound 1.75 turns around a histone octamer (two molecules each of histones H2A, H2B, H3, and H4). Histone H1 is bound to the DNA between nucleosomes.

histone octamer and the linker DNA, causing the linkers extending from both sides of the core particle to cross and draw nearer to the octamer, although some of the DNA does not come into contact with any histones. The overall structure of the chromatin fibril is probably a zigzag, as shown in panels (d) and (e) of Figure 5-6. Assembly of DNA and histones is the first stage of shortening of the DNA strand in a chromosome—namely, a sevenfold reduction in length of the DNA and the formation of a beaded flexible fiber 11 nm wide, roughly five times the width of free DNA.

The binding of DNA and histones has been examined and it appears that about 80% of the amino acids in the histones are in α-helical regions. Many of the extended α-helical regions lie in the major groove of the DNA helix, and the complex is stabilized by an electrostatic attraction between the positively charged lysines and arginines of the histones and the negatively charged phosphates of the DNA.

The DNA content of nucleosomes varies from one organism to the next, ranging from 140 to 240 base pairs per unit. The core particles of all organisms have the same DNA content (146 base pairs), so the observed variation results from different sizes of the DNA between the nucleosomes (namely, 10 to 100 nucleotide pairs). Little is known about the structure of DNA, which runs between nucleosomes, or whether it has a special function, and the cause of variation of its length is unknown.

In forming a compact chromosome, the DNA molecule is folded and refolded in several ways, as shown in Figure 5-6. The second level of folding is the shortening of the 11-nm fiber to form a solenoidal supercoil with six nucleosomes per turn, called the **30-nm fiber** (Figure 5-6(c) and **Figure 5-9**); this form has been isolated and well characterized. The remaining levels of organization—folding of the 30-nm fiber and further folding of the folded structure—shown in Figure 5-6(d–f), are less well understood. Electron micrographs of isolated metaphase chromosomes from which histones have been removed indicate that the partially unfolded DNA has the form of an enormous number of loops that seem to extend from a central core or scaffold, composed of nonhistone chromosomal proteins, as was seen for the *E. coli* chromosome (**Figure 5-10**).

It is interesting to compare and contrast prokaryotic and eukaryotic chromosome compaction. Although we have focused on the protein scaffold with prokaryotes and the histones with eukaryotes, the two systems also have similarities.

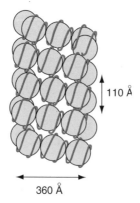

110 Å

360 Å

Figure 5-9 A proposed solenoidal model of chromatin. Six nucleosomes (shaded) form one turn of the helix. The DNA double helix (shown in blue) is wound around each nucleosome. (After J. T. Finch and A. Klug. *Proc. Nat. Acad. Sci.*, 1976; 73: 1900.)

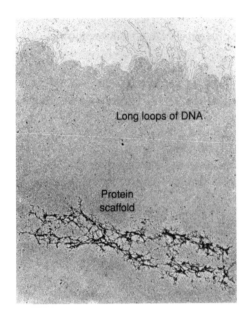

Long loops of DNA

Protein scaffold

Figure 5-10 Electron micrograph of a segment of a human metaphase chromosome from which the histones have been chemically removed. The dense network near the bottom of the figure is the protein scaffold to which the DNA loops are attached and on which the chromosome is assembled. (Courtesy of Ulrich Laemmli.)

Although prokaryotes do not have nucleosomes, when *E. coli* chromosomes are carefully prepared they display areas with a beaded appearance. Thus, prokaryotic DNA is likely complexed, to one extent or another, with a histone-like protein (in addition to the scaffold proteins).

Although histones play a central role in compacting the eukaryotic chromosome, only half of the proteins in the eukaryotic chromosome are true histones. Many of the rest are involved in the formation of the eukaryotic scaffold seen in Figure 5-10.

Interaction of DNA and Proteins That Recognize a Specific Base Sequence

In the formation of chromatin, the histones do not bind to specific base sequences. Rather, they recognize only the general DNA structure. The positive charges of the amino groups on some of the amino acid side chains (e.g., lysine and arginine) in histones electrostatically bond to the negatively charged phosphates on DNA's backbone. There are also other proteins, such as the topoisomerases we encountered in Chapter 3, which likewise bind to the general DNA structure, rather than to specific base sequences.

In addition to those "nonspecific" DNA-binding proteins, there are, however, numerous proteins that bind only to particular sequences of bases. We will encounter many such proteins in later chapters of this book (for example, see Chapters 11 and 12). Now, while on the subject of macromolecular complexes, is an appropriate time to learn about their general features.

These proteins usually form relatively weak hydrogen bonds to certain DNA sequences. They are of three main types: (1) polymerases, which initiate synthesis of DNA and RNA from particular base sequences; (2) regulatory proteins, which turn the activity of particular genes on and off by interacting with specific base sequences; and (3) certain nucleases, which cut phosphodiester bonds between a single pair of adjacent nucleotides contained in unique sequences of four to six nucleotides. A general pattern seems to be emerging from studies of literally dozens of such proteins that applies to sequence-specific binding. This pattern is the following:

1. The base sequence of the DNA frequently has some sort of twofold symmetry, allowing binding of the protein to two sites.
2. The protein–DNA contact region is on only one side of the DNA molecule.
3. The protein usually has one or more α-helical regions and these regions bind to the DNA within two adjacent turns of the major groove of the DNA.
4. The protein is often in the form of a dimer, having twofold symmetry; the monomers are arranged so that the α helices are in a symmetric array that matches the symmetry of the base sequence.
5. Side chains of amino acids of the α-helical regions are in contact with DNA bases in particular positions; the correct bases must be present at those positions. A hydrogen-bonding complementarity between those specific bases and amino acid side chains of the protein provides correct matching.
6. Positively charged amino acids often form ionic bonds with the phosphate groups, which are negatively charged. Although this binding does not confer specificity, it stabilizes the interaction of the sequence-specific protein with DNA.

A detailed analysis of the structure of a DNA-protein complex is beyond the scope of this book. Some of the features just listed, however, will be illustrated by an examination of the complex between the Cro protein of *E. coli* phage λ and the operator sequence in λ DNA. Cro is a small protein consisting of 66 amino acids; it forms a dimer and binds to a DNA sequence containing 17 nucleotide pairs (**Figure 5-11**). Each monomer of the Cro protein contains three α-helical regions, one of which is in direct contact with DNA bases in the Cro-DNA complex. **Figure 5-12** shows the binding region of the DNA molecule, with the bases and phosphate groups that contact the Cro protein shown. These sites have been identified by treating the complex with reagents [DNases, see Figure 5-7(a)] that attack "exposed" DNA but cannot alter a region of the DNA that is in contact with Cro. Analysis of the chemical structure of the DNA after exposure of the complex to these reagents showed that certain bases and phosphates were unaltered and, hence, are in the contact region. **Figure 5-13**(a) shows the Cro-DNA complex. The symmetry of the Cro dimer is apparent; the relevant α helices are in the upper and lower parts of the dimer and can be seen to fit nicely into the major groove of the DNA. This kind of symmetric binding array has been observed for a large number of prokaryotic DNA-binding proteins. Another more complex example is the lambda repressor protein shown in Figure 5-13(b). Common to those proteins is a dimeric subunit structure and a so-called **helix-turn-helix** arrangement of its folds. That folding pattern permits the protein to nest in the major groove and interact with specific bases in the minor groove.

A close look at the complex (for which there is not enough detail in the figures) would show that amino acids in each Cro monomer are situated in a way that they can form hydrogen bonds with the DNA bases, which confers binding specificity. They also interact electrostatically with the phosphate groups, as described

```
TATCACCGCAAGGGATA
ATAGTGGCGTTCCCTAT
```

Figure 5-11 The base sequence of one of the λ Cro protein binding sites. This sequence is called *oR3*. The adenines and guanines shown in Figure 5-12 are printed in blue.

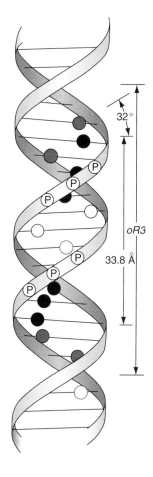

32°

oR3

33.8 Å

Figure 5-12 Points at which the *oR3* region of λ DNA makes contact with Cro. Phosphates are labeled P; guanines and adenines in contact are solid black and blue circles; guanines and adenines not in contact are open black and blue circles, respectively. Note that the bases in contact are only in the wider (major) groove of the DNA. (After W. F. Anderson et al. *Nature,* 1981; 290: 754.)

in point 6 on the previous page. This additional bonding confers further binding strength to the DNA-protein complex.

In eukaryotes, various types of DNA-binding proteins that play similar roles have also been recognized. These include monomer proteins in which intricate folding patterns are stabilized by zinc ions [Figure 5-13(c)]. That family of proteins is referred to as the **zinc finger** DNA-binding proteins. Dimeric versions of DNA-binding proteins also exist in eukaryotes, including—among others—the family of so-called **leucine zipper** proteins [Figure 5-13(d)]. The subunits of these proteins are held together by the interaction of the hydrophobic side chains of the amino acid leucine.

DNA and its associated proteins therefore comprise a true macromolecular complex. In the absence of those proteins, DNA's functional capacity is minimal.

Biological Membranes

Biological membranes are organized assemblies consisting mainly of proteins and lipids. The structures of all biological membranes have many common features, but small differences exist to accommodate the varied functions of different membranes.

KEY CONCEPT

The "Macromolecular Complex" Concept

By acting in concert, the functional capacity and repertoires of proteins and nucleic acids is enhanced.

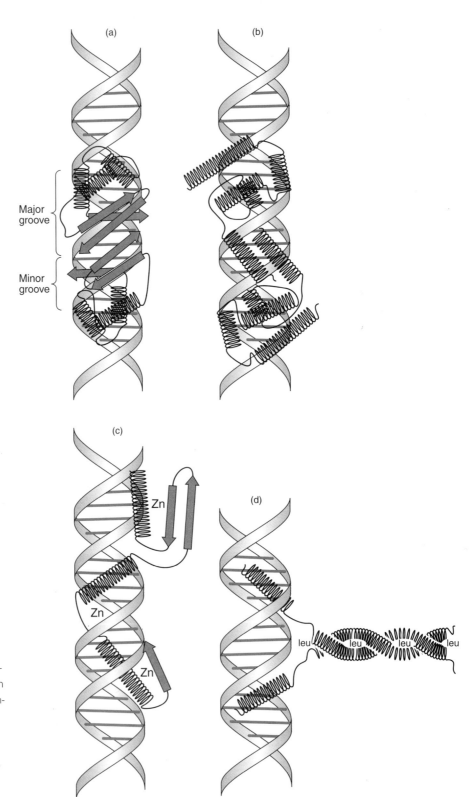

Figure 5-13 Idealized drawings of interaction with DNA of various proteins that play a role in regulating gene expression. All possess α helices that nest in the major grooves of DNA and interact with specific nucleotide bases in DNA (example shown in Figure 5-12). (a) Cro protein dimer. (b) Lambda repressor protein dimer, comprised of two monomers each with five α helices. (c) Zinc finger protein (monomer) from eukaryote. (d) Leucine zipper protein (dimer) from eukaryote. Coils represent α helices and ribbons represent β structures.

A basic function of membranes is to separate the contents of a cell from the environment. However, it is always necessary for cells to take up nutrients from the surroundings; therefore, the enclosing membrane of a cell must be permeable. However, the permeability of the membrane must be selective, so that the concentrations of compounds within a cell can be controlled; otherwise, intracellular concentrations would be inseparable from extracellular concentrations. Selective permeability is attained by restricting free passage of most intra- and extracellular substances and allowing transport of specific substances by molecular pumps, channels, and gates. External membranes also have a signal function. For instance, they often contain receptors for signal molecules such as hormones. The external membrane is also responsible for transmitting electrical impulses, as in nerve cells.

There are also numerous types of intracellular membranes. These membranes serve to compartmentalize various intracellular components, as well as to provide surfaces on which certain molecules can be adsorbed prior to chemical reaction with other adsorbed molecules.

Figure 5-14 shows an electron micrograph in which the membrane enclosing a red blood cell is seen in cross-section. The most notable feature of this membrane is that it consists of two layers—it is called a **membrane bilayer.** This structure is a consequence of the chemical nature of the lipids that form the membrane. There are many membrane lipids, of which the most prevalent are the phospholipids, molecules consisting of a polar phosphate-containing group and two long nonpolar hydrocarbon chains. **Figure 5-15** shows a space-filling model of one phospholipid and a schematic drawing. The essential feature of the molecule is a polar "head" group to which hydrocarbon "tails" are attached. Such a molecule, which has distinct polar and nonpolar segments, is called **amphipathic.** When placed in water, amphipathic molecules tend to spontaneously aggregate. This is because only the polar head is capable of interaction with the polar water molecules. The hydrocarbon tails are brought together by hydrophobic (water-hating) interactions—that is, they cluster because they are unable to interact with water.

Studies of synthetic vesicles formed from a single type of phospholipid and containing a variety of small molecules have shown that the lipid bilayer is impermeable to all ions and highly polar molecules, raising the question of how these molecules get in and out of cells. The answer is that naturally occurring biological membranes contain certain proteins that are responsible for transport of all molecules having polar regions and for most nonpolar molecules as well.

Various Proteins Are Associated with Membranes

There are two general classes of membrane proteins—the **integral membrane proteins,** which are contained wholly or in part within the membrane, and the **peripheral proteins,** which lie on the membrane surface and are bound to the integral proteins. As much as 50% of the mass of a typical cell membrane is comprised of proteins. Each type of cell has some unique proteins associated with its membrane, for the function of a cell is partially defined by its membrane components. Recall, membranes play key roles in such diverse processes as partitioning the cell into compartments, anchoring cytoskeletal elements, screening molecules for passage into and out of the cell–nucleus, connecting cells to each other, and passing

membrane bilayer

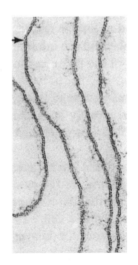

Figure 5-14 Electron micrograph of a preparation of plasma membranes from red blood cells. A single membrane is denoted by an arrow. (Courtesy of Vincent Marchesi.)

(a)

(b)

Polar
(hydrophilic)
head group

Hydrocarbon
(hydrophobic)
tails

Figure 5-15 (a) Space-filling model of phosphatidyl choline, a phospholipid. (b) The essential features of a phospholipid or gycolipid molecule, such as the one illustrated in (a).

signals between cells. Thus, it should be expected that the protein composition of a cell membrane is complex. An electron micrograph of the plasma membrane of a red blood cell in which integral membrane proteins can be seen is shown in **Figure 5-16**. The outer surface of the membrane is smooth, but the inner surface is covered with many globular protein molecules.

Numerous physical measurements have shown that the integral membrane proteins can freely diffuse laterally throughout the bilayer. The degree of motion is determined by the fluidity of the lipid layer, which is in turn a function of the Van der Waals attractions between the hydrocarbon tails of the particular phospholipids in the membrane. The **fluid mosaic model** of membrane structure (**Figure 5-17**) recognizes these features. It was further shown that the proteins do not spontaneously rotate (flip-flop) from one side of the membrane to another. The integral proteins do not spontaneously rotate because they have polar and nonpolar regions, as shown in Figure 5-17. Since the external region is polar and the internal region is nonpolar, rotation would require the passage of the polar region through the nonpolar center of the bilayer.

One subclass of integral membrane proteins, **transmembrane** proteins, is embedded in the lipid bilayer core. Such proteins actually span the membrane. Hence the term "transmembrane." **Figure 5-18** illustrates their orientation within the membrane.

fluid mosaic model

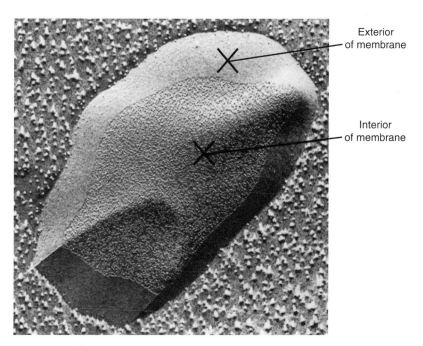

Exterior of membrane

Interior of membrane

Figure 5-16 An electron micrograph of the plasma membrane of red blood cells. The interior of the membrane, which has been exposed by fracture of the membrane, contains numerous globular particles having a diameter of about 7.5 nm. These particles are termed integral membrane proteins. (Courtesy of Vincent Marchesi.)

Since different proteins protrude from the two sides of the membrane, the membrane is said to be an asymmetric structure having (in the case of a cell) both an inside and an outside. It is this asymmetry that determines the direction of movement of molecules entering and leaving a cell.

There are many modes of transport of molecules across the bilayer and only a few will be mentioned. One mode makes use of channels through the membrane. These channels usually are passages through the integral membrane proteins; the passages will have an abundance of polar amino acids if their function

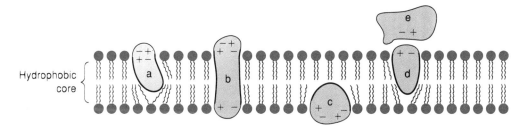

Hydrophobic core

Figure 5-17 The structure of a membrane according to the fluid mosaic model. Four integral proteins (a–d) are embedded to various degrees into a lipid bilayer such that the hydrophobic surface of each protein (heavy lines) is in the membrane and the polar region (indicated by + and −) is external. Peripheral proteins (e.g., (e)) are on the surface and bound to a polar region of an integral protein. Integral proteins can drift laterally but cannot flip-flop.

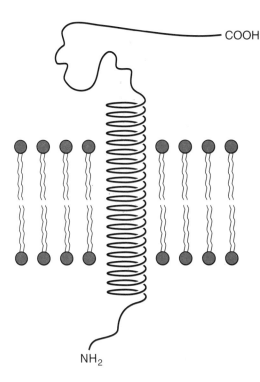

Figure 5-18 Transmembrane proteins such as (b) in Figure 5-17 contain α helices that extend through the lipid bilayer. The α helix contains hydrophobic amino acids that facilitate interactions with the hydrophobic core of the membrane.

is to allow transit of polar substances. Often the channels can be opened and closed by means of conformational changes of the membrane proteins. Such channels are found in nerve cell membranes. When open, they permit sodium ions to enter and thereby change the electrical properties of the cell. Other transport systems utilize chemical reactions that convert the substance to be transported into a molecule that can enter the membrane and then, after transit of the modified molecule, restore the original molecule at the other side of the membrane; these chemical mechanisms are usually very complex and consume a great deal of energy.

Since the types of protein that are included in the lipid bilayer (Figure 5-17) to a large extent define the functional properties of membranes of different kinds of cells, methods for characterizing those proteins have been developed. Typically, detergents are employed to solubilize the membrane and thereby release intramembrane proteins. Those proteins are then initially characterized by using analytical methods such as gel electrophoresis (Chapter 2).

KEY CONCEPT

The "Connectedness" Concept

The cell's need for connectivity and communication to the outside is achieved through the action of various membrane proteins.

Cytoskeletal Elements

Anchored to the interior of the eukaryotic cell's membrane is a complex network of protein filaments. These filaments exist in a dynamic, constantly reorganizing state. They thereby contribute to the cell's shape, its motility, and the movement within the cell of various components (e.g., muscle contractile proteins). Three types of cytoskeletal elements exist in most eukaryotic cells. Each represents a macromolecular complex: **actin filaments** are composed of polymers of actin

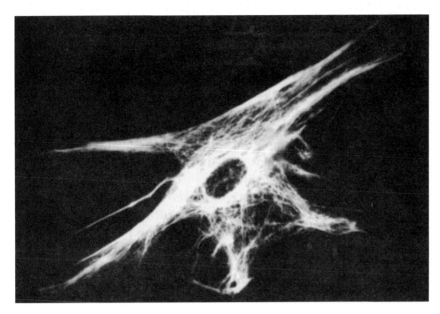

Figure 5-19 Light micrograph of a cell stained by the fluorescent antibody technique showing the distribution of intermediate filaments throughout the cytoplasm. Intermediate filaments are not present in the nucleus. (Courtesy of Michael W. Klymkowsky.)

protein subunits, **microtubules** consist of aggregates of tubulin subunits, and **intermediate filaments** consist of bundles of fibrous proteins.

These cellular components represent the largest protein complexes within a cell. In many instances they can be viewed with various microscopic methods (e.g., **Figure 5-19**).

A Future Practical Application?

The intracellular malaria parasite *Plasmodium* is responsible for the worldwide deaths of approximately two million children each year. This parasite invades erythrocytes (red blood cells: see Figure 5-16) by binding to a specific transmembrane protein (see Figure 5-18) called glycophorin. One component of the parasite's surface, also a transmembrane protein, has recently been identified as being responsible for the attachment of the parasite to glycophorin on the erythrocyte's surface. Thus, a protein–protein interaction mediates the infection process.

Much research is aimed at designing drugs or vaccines that will block that interaction, and thereby interrupt the host–parasite interaction that leads to malaria.

SUMMARY

The *E. coli* chromosome (also called a nucleoid) consists of a single supercoiled, circular DNA molecule that is attached to a dense protein-containing scaffold at various points. Using the enzyme DNase, it has been demonstrated that the DNA is attached to the scaffold at approximately 100 points. The bacterial chromosome also has approximately 100 binding sites for DNA gyrase, an enzyme responsible for supercoiling DNA. The organization of the eukary-

(*continued*)

SUMMARY (*Continued*)

otic genome is quite different, in that more than one chromosome is present in the cell, and the degree of compaction is much higher. The DNA molecule in a eukaryotic chromosome is first complexed with proteins called histones. Histones are rich in lysine and arginine, and bind to DNA through electrostatic interactions with the phosphate groups. The DNA-histone complex (called chromatin) has the appearance of beads on a string. The beads represent the core particle—two molecules each of histones H2A, H2B, H3, and H4—that form an octameric disc wrapped by 146 base pairs of DNA. The DNA connecting the core particles is called linker DNA, and it is to this that histone H1 binds. The chromatin coils to form a solenoid with six nucleosomes per turn. Further folding and re-folding, and complexing with nonhistone chromosomal proteins, results in a highly compact eukaryotic chromosome.

Certain proteins, namely polymerases, some nucleases, and regulatory proteins, bind to DNA molecules at specific base sequences. Although different proteins bind to different sequences, these proteins share several characteristics: the protein and the DNA sequence to which it binds have twofold symmetry, the region of contact is on one side of the DNA molecule, and α helices of the protein lie within the major groove to allow amino acid-base interaction.

Biological membranes are bilayers composed of amphipathic phospholipids. These membranes are impermeable to most molecules in the environment. Associated with the membranes are many protein molecules. Integral membrane proteins are contained, at least in part, within the membrane, while peripheral proteins are found on the surface of the membrane, attached to integral proteins.

DRILL QUESTIONS

1. What are the differences between a nucleosome, a core particle, and an octameric disc? Which directly involves the histone H1?
2. Why is the level of DNA compaction higher in eukaryotes than in prokaryotes?
3. Why is it necessary for DNA to exist at various levels of compaction?
4. At what stage of the eukaryotic cell cycle would you expect DNA to be the most compact? At what stage would you expect it to be the least compact? Why?
5. Referring to the Cro protein, how far apart are the two α helices that contain the amino acids that interact with the bases of the DNA molecule?
6. Where are the polar and the nonpolar regions of a membrane-spanning integral protein?
7. What noncovalent interactions are involved in stabilizing a lipid bilayer?

PROBLEMS

1. If a DNA molecule is 2.4 cm long, how many nucleosomes could be present in its most highly compacted state? (Assume linker DNA has 60 base pairs.)
2. When an isolated DNA-protein complex was treated with 2M NaCl, the protein dissociated from the DNA. Digestion of the complex with a nuclease and a protease yields only free nucleotides and amino acids. What is the main type of interaction involved in binding the protein to the DNA?

3. Another DNA-protein complex is isolated. When treated with 2M NaCl, the complex does not dissociate. The complex was treated with a nuclease and a protease and, in addition to free nucleotides and amino acids, a component was found that was neither a free nucleotide nor amino acid. What is likely to be the identity of this component, and what interaction is responsible for protein-DNA binding?

4. A DNA-binding protein binds tightly to double-stranded DNA, but very poorly to single-stranded DNA. It can bind to all base sequences with equal affinity. Binding is poor in 1M NaCl. What part of the DNA molecule is likely to be the binding site?

5. Acridine molecules bind to DNA by intercalating (inserting) between the base pairs. If the Cro protein and DNA are mixed, the Cro-DNA complex forms. What would happen if acridine were added to the DNA before the Cro protein?

CONCEPTUAL QUESTIONS

1. Although prokaryotes do not possess histones, Archaebacteria do. What are the implications of that fact for understanding the evolutionary history of eukaryotes?

2. What reasons might explain why it is beneficial to a cell that flip-flopping of membrane proteins occurs infrequently?

3. Speculate on why most functions carried out by a cell rely on macromolecular interactions and/or the use of complex aggregates of macromolecules instead of smaller, simpler molecules.

Function of
Macromolecules

in this chapter you will learn

1. Background information on the mechanisms of heredity.

2. Experimental evidence that identified DNA as the genetic material.

3. Properties of DNA as the genetic material, including storage, transmission, stability, and mutation.

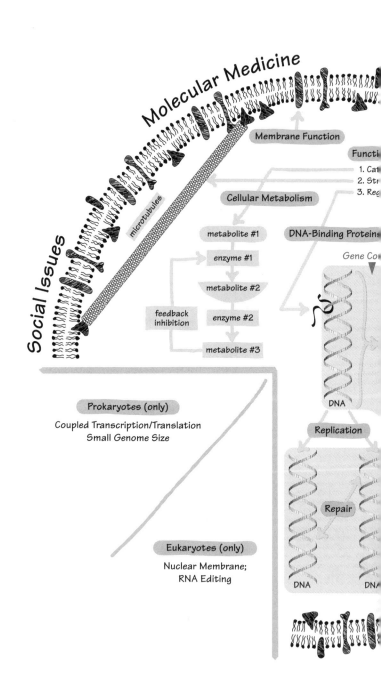

Molecular Medicine

Social Issues

microtubules

Membrane Function

Functi...

1. Cat
2. Str
3. Reg

Cellular Metabolism

metabolite #1

enzyme #1

metabolite #2

enzyme #2

metabolite #3

feedback
inhibition

DNA-Binding Protein...

Gene Co

DNA

Prokaryotes (only)

Coupled Transcription/Translation
Small Genome Size

Replication

Repair

DNA   DNA

Eukaryotes (only)

Nuclear Membrane;
RNA Editing

The Genetic Material

In the first five chapters, we have examined the physical and chemical properties of DNA, RNA, and protein. While this information is critical to our understanding of molecular biology, it alone doesn't supply the missing link between the purely physical properties of macromolecules and their biological functions in living organisms. In this chapter, we will begin to consider the biological functions of macromolecules, starting with the role of nucleic acids as the genetic material of living organisms.

The genetic material of an organism is the substance that contains the information specifying the inherited characteristics of that organism. As we will see shortly, this information must be stable, be capable of being expressed in the cell as needed, undergo accurate replication, and be transmitted, in a largely unchanged form, from parent to progeny. Most organisms utilize DNA as their genetic material. Exceptions occur in the cases of certain bacteriophages, plant and animal viruses, in which the genetic material is RNA. Two important questions arise: (1) How do we know that DNA is the genetic material?; and (2) What properties of DNA make it so ideally suited to function in this capacity? In the following sections, we will discuss several classic experiments that led to the identification of DNA as the genetic material, and, in doing so, laid the foundation for molecular genetics. We will also examine the properties essential in genetic material, and see why DNA is so well suited to this role.

Early Observations on the Mechanism of Heredity

Long before the formal beginnings of molecular biology, scientists were attempting to learn more about the properties of living cells that set them apart from inanimate objects. By the mid-to-late 1800s, many studies began to suggest that the cell nucleus possessed unique properties, and might play a key role in heredity. It is important to remember that these early observations involved eukaryotic cells; their larger size made them more feasible to study under the microscope. A few of these important findings are summarized below:

1. *The nucleus has a unique chemical composition.* As early as 1869, Friedrich Miescher had reported that the nuclei of human white blood cells contain a unique chemical component, *nuclein,* which is rich in phosphorus, but,

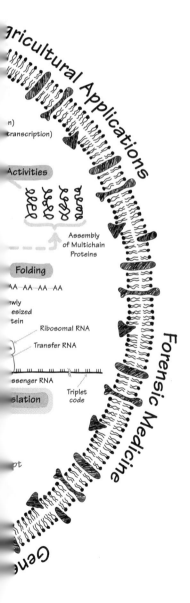

unlike protein, contains no sulfur. This substance would later be identified as DNA.

2. *Fusion of sperm and egg nuclei at the time of fertilization produces a zygote that contains all of the hereditary information necessary to develop into a mature adult organism.*

3. *Cell nuclei contain chromosomes. The precise numbers and types of chromosomes may vary from one species to another, but normal members of a particular species will have a characteristic set of chromosomes.*

4. *Chromosomes are duplicated prior to cell division and one copy of each is precisely apportioned to the daughter cells (via mitosis).* Thus, each daughter cell will possess the same numbers and types of chromosomes as the parent cell that gave rise to it.

5. *To maintain a constant chromosome number in sexually reproducing organisms, chromosome number is reduced by one-half (via meiosis) during gamete formation.* When the egg and sperm fuse at fertilization, normal chromosome number is restored, and each offspring consequently receives half of its chromosomes from each parent.

6. *The patterns of inheritance of chromosomes mirrors the patterns of inheritance of genetic traits described by Gregor Mendel (1865).*

7. *In certain instances, it was possible to correlate the inheritance of specific physical traits, such as gender or eye color (in fruit flies), with the inheritance of specific recognizable chromosomes.*

Thus, scientists knew that the nucleus plays a critical role in heredity, that the nucleus contains chromosomes, and that the inheritance of chromosomes parallels the inheritance of certain traits. But what *was* this mysterious component of the chromosomes that might carry the genetic information? Chemical analyses of nuclei and chromosomes demonstrated that nuclein was associated with proteins, forming complexes called *nucleoproteins*. As improved analytical techniques became available, the nuclein itself was chemically analyzed, and found to be composed of four different nucleotides.

Given a choice between these two candidates for the role of genetic material, namely nucleic acids and proteins, the "common wisdom" of that era held that the genetic material was more likely to consist of proteins. After all, inherited traits exhibit tremendous diversity, and it was assumed that genes would likewise be highly diverse in composition and structure. As you learned in Chapter 4, proteins vary considerably in amino acid composition, size, and shape. On the other hand, nucleic acids, merely consisting of four types of nucleotides, were thought to be lacking in complexity. In fact, by 1910, Phoebus Levene had proposed a widely accepted *tetranucleotide model* of structure of the DNA molecule. According to this model, DNA was depicted as a small, relatively simple molecule, a tetranucleotide composed of one molecule each of the four different nucleotides (**Figure 6-1**). Consequently, many scientists believed that proteins were the logical candidates for storing and transferring genetic information. The DNA was thought to function in more of a structural capacity, holding the "informational proteins" together in the chromosomes. Now, in the modern era of biology, it is, of course, taken for granted that DNA is the genetic material.

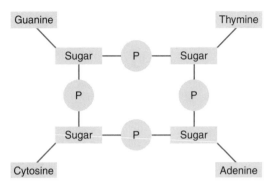

Figure 6-1 Proposed tetra-nucleotide structure of DNA.

A Historical Review Offers an Exercise in Analytical Thinking

In the following sections, you will learn about the landmark experiments, performed, in some cases, over 65 years ago, that paved the way for accepting the fact that genes consist of a "simple" macromolecule such as DNA. Aside from their historical importance, these experiments can help you, the student, gain some insight into the applications of the scientific method and the analysis of experimental data, while helping you develop critical thinking skills.

Identification of DNA As the Genetic Material

The Transformation Experiments

The development of our understanding that DNA is the genetic material began with an observation in 1928 by Fred Griffith, who was studying the pathogenic properties of the bacterium responsible for causing human pneumonia—that is, *Streptococcus pneumoniae,* or *Pneumococcus.* The virulence (disease-causing ability) of this bacterium was known to be dependent upon a polysaccharide capsule that surrounds the bacterial cell and protects it from the defense systems of the body. Strains of *S. pneumoniae* possessing this capsule produce **smooth-edged (S)** colonies when grown on solid medium. Mice injected with these virulent S strains of bacteria normally die, and live S bacteria can be recovered from their blood (**Figure 6-2**(a)). It is important to note that different S strains (called **I-S, II-S, III-S,** and so on) produce capsules that differ in their chemical composition; the ability of the bacterium to produce a capsule at all, as well as the type of capsule produced, are *genetic* (*i.e., inherited*) *traits.* Griffith also isolated mutant strains of *Pneumococcus,* which produced **rough (R)** colonies on solid media. These mutants proved to be both nonencapsulated (lacking a capsule) and nonvirulent when injected into mice (Figure 6-2(b)). Smooth bacteria could undergo mutation to produce R mutants, and vice versa, but the mutation always involved the loss or gain of the ability to make a *particular* type of capsule (i.e., **II-S ↔ II-R; not III-S ↔ II-R**). Griffith next took smooth bacteria and heated them *before* injecting them into mice. As might be expected, the

(a) S strain kills mouse

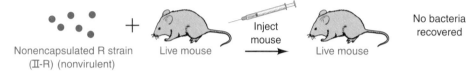

Encapsulated S strain (Ⅱ-S) (virulent) + Live mouse → Inject mouse → Dead mouse → Encapsulated S strain (Ⅱ-S) (virulent recovered)

(b) R strain does not kill mouse

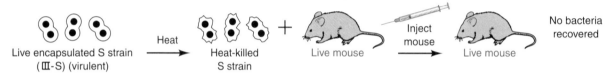

Nonencapsulated R strain (Ⅱ-R) (nonvirulent) + Live mouse → Inject mouse → Live mouse → No bacteria recovered

(c) Heat-killed S strain does not kill mouse

Live encapsulated S strain (Ⅱ-S) (virulent) → Heat → Heat-killed S strain + Live mouse → Inject mouse → Live mouse → No bacteria recovered

(d) R strain plus heat-killed S strain, both of which are separately nonlethal, kill mouse

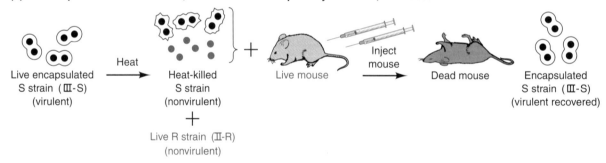

Live encapsulated S strain (Ⅱ-S) (virulent) → Heat → Heat-killed S strain (nonvirulent) + Live R strain (Ⅱ-R) (nonvirulent) + Live mouse → Inject mouse → Dead mouse → Encapsulated S strain (Ⅱ-S) (virulent recovered)

Figure 6-2 The Griffith experiment showing conversion of a nonlethal bacterial strain to a lethal form by a cell extract.

bacteria were killed by the heating process and the mice survived (Figure 6-2(c)). But Griffith then made a very significant observation: When he mixed heat-killed S cells (type **III-S**) with live R cells (type **II-R**) and injected them into mice, the mice died, even though neither batch of cells was lethal by itself (Figure 6-2(d))!

Of even greater significance was the fact that live bacteria isolated from a mouse that had died from such a mixed infection were only of the **S type,** in particular the *same* S type (**III-S**) as the heat-killed S cells, and thus could *not* have arisen via mutation from these particular R cells. The live R bacteria had somehow been replaced by or **transformed** into S bacteria, which then reproduced as S bacteria. Griffith referred to the unknown substance that brought about this transformation as the **transforming principle.** He had no idea of its nature, but mistakenly suspected that it might either be a protein involved in capsule synthesis, or else some substance that acted as a precursor of the bacterial capsule.

transforming principle

Several years later, it was shown that a mouse was not needed to mediate this transformation. When a mixture of live R cells and heat-killed S cells was cultured in nutritive medium, some live S cells were produced.

A possible explanation for this surprising result was that the R cells had somehow revived the heat-killed S cells. This possibility was eliminated when it was observed that live S cells could be obtained when the live R cells were mixed, not with intact heat-killed S cells, but with an extract containing only the molecular contents of these heat-killed S cells. All intact cells, cell debris, and capsular material had been removed from this extract by centrifugation and filtration (**Figure 6-3**). It was concluded that the S cell extract contained a transforming principle, the nature of which was unknown.

DNA Is Identified As the Transforming Principle

The next development occurred some 15 years later when Oswald Avery, Colin MacLeod, and Maclyn McCarty partially purified the transforming principle from S cell extract and demonstrated that the fraction with transforming ability contained DNA. They sequentially purified the cell-free extract, and discovered that removal of proteins, polysaccharides, lipids, or RNA from the extracts *did not* significantly diminish their ability to bring about transformation. By the end of the purification procedure, they had obtained a relatively pure, viscous material that had the physical and chemical properties of DNA. This "purified DNA" was then added to a live culture of R cells. When the cells were plated on agar and allowed to form

Preparation of transforming principle from S strain

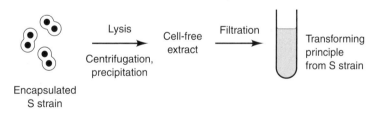

Addition of transforming principle to R strain

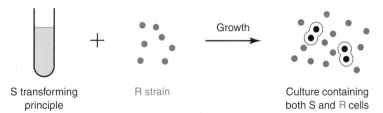

Figure 6-3 The transformation experiment.

colonies, some of the colonies that grew (about 1 in 10^4) were S type (**Figure 6-4**)! To show that this was a permanent genetic change, they dispersed many of the newly formed S colonies and placed them on a second agar surface. The resulting colonies were again S type. On the other hand, when an R type colony arising from the original mixture was dispersed, only R bacteria grew in subsequent generations. Hence, the R colonies retained the R character, whereas the S colonies bred true as S.

At this point, Avery and his coworkers were far from being able to state with absolute certainty that the transforming principle was, in fact, DNA. Although the purified fraction clearly contained DNA, it was also contaminated with small amounts of RNA. The possibility existed that minute amounts of protein might also remain in the extract, and that it might be these proteins, and not the DNA itself, that was responsible for the transforming activity. To address these criticisms, they undertook additional experiments to determine whether the transformation was actually caused by the DNA alone.

This evidence was provided by five sets of data:

1. Chemical analysis disclosed that the major component of the purified transforming principle was a deoxyribose-containing nucleic acid.
2. Physical evidence showed that the sample contained a highly viscous substance having the properties of DNA.
3. Experiments demonstrated that the transforming activity was not lost when the purified transforming principle was treated with either (a) purified proteolytic (protein hydrolyzing) enzymes—trypsin, chymotrypsin, or a mixture of both—or (b) ribonuclease (an enzyme that hydrolyses RNA).
4. Treatment with materials known to contain enzymes that hydrolyze DNA (DNase) *did* inactivate the transforming principle and also destroyed its characteristic physical properties.
5. The DNases used to treat the purified transforming principle were themselves first inactivated by heating or by chemical treatment. These

Figure 6-4 Avery, MacLeod, and McCarty's experiments demonstrated that only DNA-containing fraction of the purified transforming principle was capable of causing the transformation of rough (R) cells into smooth (S) cells. In 1994, a special "public health award" was given by the Lasker Foundation to Maclyn McCarty, professor emeritus at Rockefeller University, to honor his earlier discovery that genes are comprised of DNA!

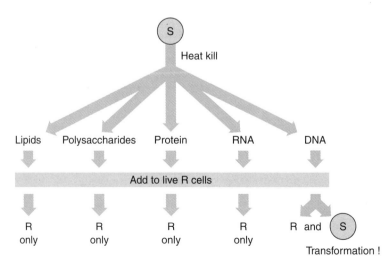

inactivated DNases simultaneously *lost* their ability to degrade DNA and their ability to inactivate the transforming principle.

A Conservative Conclusion Was Drawn

The results of these experiments provided strong evidence that the transforming principles were, in fact, DNA. Nevertheless, Avery, MacLeod, and McCarty were very cautious as they stated in their conclusions: *"The evidence presented supports the belief that a nucleic acid of the desoxyribose type is the fundamental unit of the transforming principle of Pneumococcus Type III."* Notice that they *didn't* conclude that DNA is the genetic material of all living organisms. They had examined only one specific type of organism (*S. pneumoniae*) and, in fact, had limited their transformation studies to a single genetic trait: the ability of the bacterium to produce a polysaccharide capsule. The possibility, therefore, still existed that they might be examining a phenomenon that was peculiar to bacteria or to the production of a specific type of genetic change. They also expressed concern that the biological activity of the purified transforming principle might be due to the presence of minute amounts of some contaminant in their DNA preparation. Thus, their experimental data provided strong but not conclusive evidence that DNA is the universal genetic material.

The transformation experiments also failed to convince many in the scientific community that DNA is the genetic material, largely because the tetranucleotide model of DNA structure was so widely accepted. The tetranucleotide hypothesis was based on chemical analyses that indicated that DNA consisted of equimolar amounts of the four bases. This conclusion can be attributed to (1) the use of inadequate chemical procedures for analyzing base composition, and (2) the use of higher organisms as sources of DNA. (In these organisms the base composition is, in fact, not far from being equimolar.) As techniques improved and a greater range of organisms was examined, it was found that base composition of DNA varies widely from one species to another, and that DNA is neither a simple, small molecule nor a simple repeating structure.

In 1952, Erwin Chargaff published a key paper that would provide crucial evidence that DNA possessed sufficient structural complexity and diversity to allow it to function as the genetic material. Chargaff used the technique of chromatography to analyze the base composition of DNA isolated from a variety of different organisms, and discovered that the DNA from all organisms examined shared certain common properties. (This set of properties would later come to be known as "Chargaff's rule," and would provide important clues concerning base pairing to Watson and Crick as they developed their model of the DNA molecule.) Chargaff's findings are summarized below:

1. In the DNA of all species examined, the amount of A is equal to the amount of T, and the amount of G is equal to the amount of C.
2. The total amount of purines (A + G) present in DNA is equal to the total amount of pyrimidines (T + C).
3. *However,* the ratio of (A + T)/(G + C) *does **not** necessarily equal one, and, in fact, this ratio varies considerably from species to species.* DNA base composition *does* indeed vary from one species to another!

With these results, the idea that DNA is the genetic material *became* acceptable. The following experiment provided additional proof.

The Blender Experiment Provided Further Evidence for DNA

An elegant confirmation of the genetic nature of DNA came from an experiment involving the *E. coli* phage T2. This experiment, known as the **Blender Experiment** because a kitchen blender was used as a major piece of apparatus, was performed by Alfred Hershey and Martha Chase. These scientists exploited the simple features of the bacteriophage system to differentiate between DNA and protein as the genetic material of one particular bacteriophage, **T2.** They demonstrated that the DNA, injected by a phage particle into a bacterium, contained all of the information required to achieve the production of progeny phage particles.

Figure 6-5 shows a diagram of a typical bacteriophage. Recall, as discussed in Chapter 1, that the coat or *capsid* of the bacteriophage (along with other external structures such as its tail and tail fibers, and the like) consists entirely of protein, while the core, which is packaged inside of the protein capsid, contains the DNA (at least in those viruses that use DNA as their genetic material).

Phages are intracellular parasites and cannot multiply, except in a host bacterium. Thus, a phage must be able to enter a bacterium, multiply, and then escape. There are many ways in which this can be accomplished. However, the outline of a basic life cycle follows and is depicted in **Figure 6-6**.

We now know that the life cycle of a phage begins when a phage particle adsorbs (attaches) to the surface of a susceptible bacterium. The phage nucleic acid then leaves the phage particle through the phage tail (if the phage has a tail) and enters the bacterium through the bacterial cell wall. In a complicated but understandable way, the phage essentially converts the bacterium to a phage-synthesizing factory. Within approximately an hour, the infected bacterium bursts or

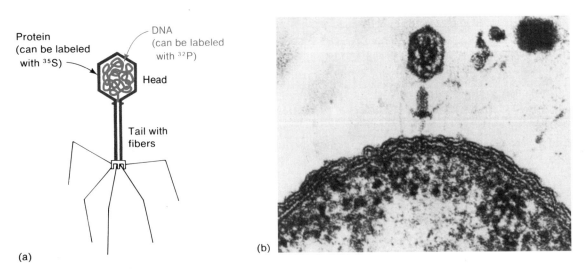

Figure 6-5 (a) A diagram of T2 phage. (b) Phage T2 adsorbed to the surface of *E. coli*. (Courtesy of Lee Simon and Thomas Anderson.)

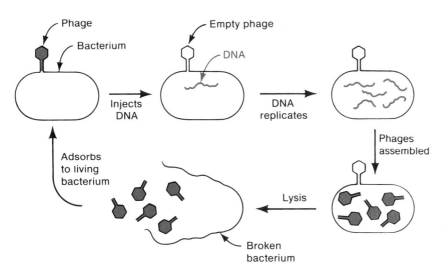

Figure 6-6 A schematic diagram of the life cycle of a typical bacteriophage. The number of phage released usually ranges from 20 to 500.

lyses and several hundred phage progeny are released. The suspension of newly synthesized phage is called a **phage lysate.** (For a more detailed look at phage replication, consult Chapter 14.)

At the time at which Hershey and Chase performed their experiments, it was known that phage infection began with the attachment of the phage to the bacterium, and ended with the lysis of the infected bacterium and the release of the progeny phase. The events occurring in the interim had not yet been defined. Clearly, the genetic material of the phage, whatever it might be, had to enter the bacterium and direct the production of new phage progeny. The simplicity of the T2 phage structure suggested an elegant approach to the question of the identity of the phage genetic material. A single particle of T2 consists of DNA (now known to be one double-stranded molecule) encased in a protein shell (Figure 6-5(a)). The DNA is the only phosphorus-containing substance in the phage particle; the proteins of the shell, which contain, among others, the amino acids methionine and cysteine, have the only sulfur atoms. Thus, by growing phage in a bacterial culture in which the nutrient medium contains radioactive phosphate ($^{32}PO_4^{3-}$) as the sole source of phosphorus, phage containing radioactive DNA can be prepared. If, instead, the growth medium contains radioactive sulfur (as $^{35}SO_4^{2-}$), phage containing radioactive proteins are obtained. If these two kinds of labeled phage are used, in separate experiments, to infect bacterial hosts, the phage DNA and protein molecules can always be located by their specific radioactivity. Hershey and Chase used these radioactive phage to show that ^{32}P, but not ^{35}S, is injected into the bacterium.

Each phage T2 has a long tail by which it attaches to sensitive bacteria (Figure 6-5(b)). Hershey and Chase showed that an attached phage can be torn from the bacterial cell wall (without harming the bacteria) by violent agitation of the infected cells in the kitchen blender. Thus, it was possible to separate the adsorbed phage from a bacterium and determine the component(s) of the phage that could not be shaken free by agitation—presumably those components that had been injected into the bacterium.

In the first experiment (**Figure 6-7**(a)), ^{35}S-labeled phage particles were allowed to adsorb to bacteria for a few minutes. The bacteria (with their attached phages) were separated from unadsorbed phage and phage fragments by centrifuging the mixture and collecting the sediment (the pellet), which consisted of the phage-bacterium complexes. The supernatant, which contained the lighter, unadsorbed phage particles and phage fragments, was poured off and saved. (This was done to eliminate any phage-associated radioactivity that had not become associated with the bacteria.) The pellet (containing the phage-bacterium complexes) was resuspended in liquid and blended to remove the attached phage particles. (By this time, the phage had already injected their genetic material into the bacterial host.) The suspension was again centrifuged, and the pellet (which now consisted almost entirely of bacteria alone) and the supernatant (which contained the detached phage) were collected separately. It was found that 80% of the ^{35}S label was in the supernatant, and 20% was in the pellet. The 20% of the ^{35}S that remained associated with the bacterium was shown some years later to consist mostly of phage-associated tail fragments that adhered too tightly to the bacterial surface to be removed by blending.

A very different result was observed when the phage population was labeled with ^{32}P instead of ^{35}S (Figure 6-7(b)). In this case, 70% of the ^{32}P remained associated with the bacteria in the pellet after blending, and only 30% in the super-

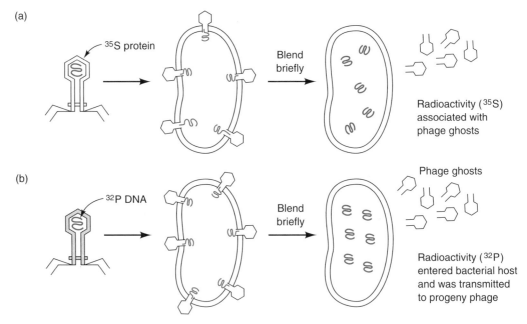

(a)

^{35}S protein

Blend briefly

Radioactivity (^{35}S) associated with phage ghosts

(b)

^{32}P DNA

Blend briefly

Phage ghosts

Radioactivity (^{32}P) entered bacterial host and was transmitted to progeny phage

Figure 6-7 The Hershey-Chase blender experiment demonstrated that DNA, and not protein, is the genetic material of the bacteriophage T$_2$. (a) ^{35}S-labeled T$_2$ phages were used to infect *E. coli*. Most of the radioactivity was removed from the bacterial cells by blending, and appeared in the phage ghosts, indicating that the ^{35}S protein had not entered the bacterial host. (b) ^{32}P-labeled T$_2$ phages were used to infect *E. coli*. After a brief period of adsorption, the cells were agitated in a kitchen blender to remove phage particles. Most of the radioactivity (^{32}P) had entered the bacterial cells and ultimately appeared in the phage progeny.

natant. Of the radioactivity in the supernatant, roughly one-third could be accounted for by breakage of bacteria during the blending; the remainder was shown some years later to be a result of defective phage particles that could attach to the bacterium, but could not inject their DNA. Most importantly, when the pellet material (containing the infected bacteria) was resuspended in growth medium and reincubated, it was found to be capable of phage production, and the progeny phage produced in this instance contained ^{32}P! Thus, the ability of a bacterium to synthesize progeny phage was associated with the transfer of ^{32}P, and, hence, of DNA, from parental phage to the bacterium. At least some of the radioactivity found in the parental DNA was subsequently transmitted to the progeny phage, as would be expected if the *genetic information* was transmitted from parent to progeny! It is equally important to note that ^{35}S, which did not enter the bacterium, *was not transmitted to the phage progeny*—further evidence that DNA, and not protein, is the genetic material.

At Last, DNA Is Proved to Be the Genetic Material

Further proof for DNA as the genetic material came a few years later. It was demonstrated that the naked, purified DNA of the small bacteriophage φX-174 could successfully infect bacterial protoplasts (cells lacking a portion of their cell wall) in the absence of any viral protein, and produce normal, infectious phage progeny.

Identification of RNA As the Genetic Material of Certain Viruses

As we have already anticipated at the beginning of this chapter, most organisms utilize DNA as their genetic material. However, there are exceptions to this rule: Certain bacteriophages, as well as a number of plant and animal viruses, use RNA as their genetic material. The classic experiments described below demonstrated that RNA is the genetic material of one particular plant virus, *tobacco mosaic virus* (TMV).

TMV is a relatively simple virus consisting of two chemical components, RNA and protein. Structurally, the helical RNA core of the TMV virus is surrounded and protected by protein subunits arranged in a helical configuration (**Figure 6-8**). Plants infected by TMV develop characteristic mosaic-like lesions on their leaves. Different strains of TMV exist that vary with respect to the composition of their protein capsid and the ability to infect certain plants. Scientists discovered that it was possible to separate the protein components of the virus from the RNA, thus enabling them to test the ability of each of these components to cause viral infection individually. It was found that inoculating the tobacco plants with purified TMV protein did not result in the production of TMV lesions. Inoculation with TMV RNA *alone*, however, resulted in viral infection. This evidence that RNA was functioning in the capacity of genetic material in the TMV virus was further supported by the observation that treatment of the TMV RNA with ribonuclease prior to inoculation of the tobacco plants destroyed its ability to cause infection.

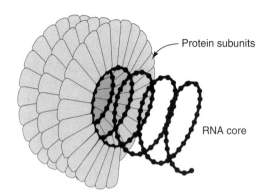

Figure 6-8 Tobacco mosaic virus particle consists of a helical RNA core, surrounded by protein subunits in a helical arrangement.

The most elegant experiment supporting the genetic role of TMV RNA was performed by H. Fraenkel-Conrat and B. Singer in 1957. These scientists isolated the RNA and protein components from two distinct strains of TMV, then made reconstituted virus particles consisting of the RNA of one strain combined with the protein of the other strain. These two reconstituted ("hybrid") viruses were then used to infect tobacco plants. The following question was posed: Which of the two viral components, the RNA or the protein, would determine the properties of the viral progeny produced during infection caused by these hybrid viruses? When progeny viruses were isolated from tobacco leaves infected by the hybrid virus, the progeny always had the same properties as the virus whose **RNA** was present in the hybrid. These results demonstrated conclusively that RNA is, in fact, the genetic material of TMV.

Properties of the Genetic Material

It is often said that "hindsight is 20/20." With the definitive results of the transformation experiments, the Hershey-Chase experiments, and others that followed, it was clearly established that DNA is the genetic material of living organisms. As scientists reviewed the results of Chargaff's studies of base composition, and began to reexamine the physical and chemical properties of DNA, everything began to make sense! It soon became obvious that DNA possessed a number of properties that made it particularly well suited for its role as the genetic material.

1. DNA has the ability to store genetic information, which can be expressed in the cell as needed.
2. This information can be transmitted to daughter cells with minimal error. (This process requires complex enzymes and repair mechanisms.)
3. DNA possesses both physical and chemical stability so information (present in duplicate) is not lost over long periods of time (years).
4. DNA has the potential for heritable change without major loss of parental information.

Let's examine each of these properties of DNA and see how each contributes to its ability to function as the genetic material.

Storage and Transmission of Genetic Information by DNA

DNA possesses and conveys several types of information:

1. The sequence of all RNA molecules synthesized by the cell.
2. The sequence of all amino acids in every protein synthesized by the cell.
3. Start and stop signals for the synthesis of each RNA, the processing of each RNA, and the synthesis of each protein.
4. A set of signals that interacts with the cellular components and determines whether and when a particular RNA or protein is to be made, and how many molecules are to be made per unit time.
5. Signals that serve as origins and terminators for the replication of the DNA.
6. Signals that provide essential features of chromosome structure (centromeres and telomeres).

This information is contained in the sequence of the DNA bases. DNA serves principally as the template for the synthesis of specific RNA molecules, called RNA transcripts. **Transcription,** the process of synthesizing RNA molecules from a DNA template, is discussed in Chapter 8. The base sequence of each RNA is complementary to a specific region on one of the DNA strands. The complementary RNA is made by means of an enzymatic system that adds a ribonucleotide to the growing end of the RNA molecule only if that base can hydrogen-bond with the DNA base being copied. This restriction ensures a copy error rate of less than 1 in 10^4.

After synthesis, the strand of RNA may be "processed"—it may be trimmed or modified at the ends, cut into defined fragments, internal segments may be excised, or individual components may be modified by components of the cellular machinery. Processing is much more extensive in eukaryotic cells than in prokaryotic cells (see Chapters 8 and 12).

The processed RNA is used for varied cellular functions. Much of the processed RNA—the **messenger RNA**—is used to specify the amino acid sequences of cellular proteins. For this purpose, the RNA nucleotide sequence is read (on ribosomes) in sequential groups of three bases called **triplets** or **codons.** Each codon corresponds to a particular amino acid, or to a start or stop signal. (The process of protein synthesis will be discussed in detail in Chapter 9.)

This two-stage process has an advantage—the genetic DNA molecule is sheltered and does not have to be used very often or enter the protein synthesis machinery. In eukaryotes, it is possible for the DNA to remain within the relatively protected environment of the nucleus, where DNA replication and transcription occur, while the RNA molecules exit the nucleus to participate in translation. In addition, since it is possible to use *one* DNA molecule to make *many* copies of an RNA molecule, protein synthesis can occur much more rapidly and efficiently.

Specific regions of the DNA serve, by interaction with cellular components, to regulate the transcription of adjacent messages. Other regions direct cellular components to initiate and terminate DNA replication.

The use of specific patterns of hydrogen-bonding enables the cell to use less genetic material to hold information than if only Van der Waals attractions had been selected in the course of evolution; this is because Van der Waals interactions

transcription

messenger RNA

triplets
codons

are so weak that the error frequency in transmitting information would be unacceptable unless more bases (or different molecules) were used.

Transmission of Information from Parent to Progeny by DNA

When a cell divides, each daughter cell must receive identical genetic information. That is, each DNA molecule must undergo duplication to become two identical molecules, each carrying the information that was contained in the parent molecule. This duplication process is called **replication.** Once again, the importance of the specific hydrogen-bonding capacity of nucleotide bases in DNA is apparent. As with transcription, the DNA replication system only requires that the base being added to the growing end of a new DNA chain is capable of hydrogen bonding to the particular base being copied.

However, the fidelity (accuracy) required for DNA replication is much higher than for transcription. Occasional synthesis of an aberrant protein can be tolerated. After all, additional, nondefective copies of the same protein molecule may well exist within the cell, and furthermore, since most messenger RNA is relatively short-lived, it is likely that "defective" mRNA molecules will eventually be replaced by newly synthesized, "normal" mRNAs. Occasional error in the replication of a gene, however, could be lethal, since the daughter cells receiving this defective gene might lack the ability to carry out some vital function encoded in the gene.

Consequently, cells have developed additional mechanisms to ensure the fidelity of gene duplication. The enzymes involved in DNA replication possess an editing (proofreading) function which re-checks the correctness of the nucleotides that have just been inserted into the growing chain, and excises them if an error has been made. This procedure removes 99.9% of the few errors made in the initial insertions.

In addition, the cells have a means to differentiate, for a period of time, the newly synthesized DNA strand from the older parental strand. Special enzymes monitor the DNA at all times, looking for mismatched base pairs. If such a mismatch is found on a recently duplicated DNA strand, it is corrected using the older DNA strand as a template. This combination of error-correcting mechanisms reduces the copy-error rate in DNA duplication in cells to between 1 in 10^9 and 1 in 10^{10} base pair.

The Physical and Chemical Stability of DNA

In long-lived organisms, a single molecule of the genetic material may have to last 100 years or more. Furthermore, the information contained in the molecule is passed on to succeeding generations over millions of years with only small changes. Thus, DNA molecules must have great stability.

The sugar-phosphate backbone of DNA is extremely stable. The C—C bonds in the sugar are resistant to chemical attack under all conditions except strong acids at very high temperatures. The phosphodiester bond is a little less stable; it can be hydrolyzed at room temperature at pH 2, but this is not a normal physiological condition. In considering the stability of the phosphodiester bond, one can see immediately why 2′-deoxyribose rather than ribose is the constituent sugar in DNA. The phosphodiester bond in RNA is rapidly broken in alkali. The chemical mechanism that breaks the bond requires the presence of an OH group on the

KEY CONCEPT

The "DNA-Genetic Material" Concept

Double-stranded DNA has evolved as the genetic material because it is especially well suited for replication, repair, occasional change, and long-term stability.

$2'$-carbon of the sugar. In DNA, which contains deoxyribose, there is no $2'$-OH group present, so the molecule is extremely resistant to alkaline hydrolysis. (You may wish to refer to Figure 2-5 in Chapter 2.) The N-glycosylic bond that links the base to the 5-carbon sugar (Figure 2-5) is also very stable, although it would not be so if the bases were not hydrophobic rings.

An alteration in the chemical structure of a base would mean a loss of genetic information. In a cell there are certainly a large number of chemical compounds that can attack a free base. In considering this problem, one can immediately see the value of the double helix. The molecule is redundant in the sense that identical information is contained in both strands, that is, the base sequence of one strand is complementary to the other strand and can be derived from it. In fact, there exist in the cell elegant repair systems that can remove an altered base, read the sequence on the complementary strand, and then reinstate the correct base. (DNA repair will be discussed in Chapter 10.)

Another merit of the double-helical structure is that it provides the bases of the DNA with great protection against chemical attack. The bases are hydrophobic rings with charged groups that contain the genetic information. It is these charged groups that specifically need protection. The hydrophobic structure of the bases causes them to stack so tightly that water is almost completely excluded from the stacked array. As a result, water-soluble compounds are often unable to come into close contact with the "dry" stack of bases, and the likelihood of their reaching the hydrogen-bonded charged groups is small.

The bases themselves, with the exception of cytosine, are very stable. At a very low rate, however, cytosine is deaminated to form uracil:

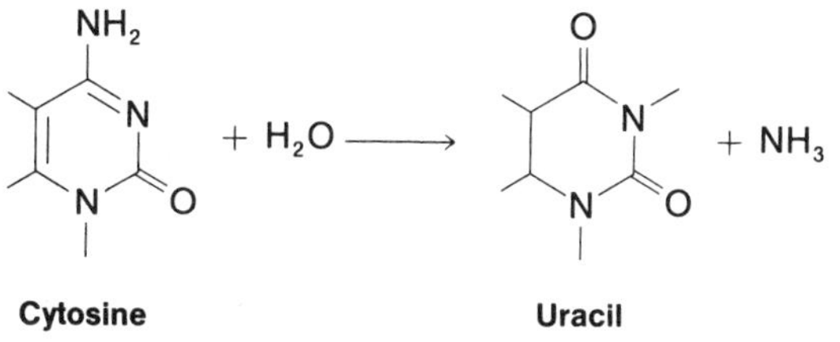

Cytosine   **Uracil**

Deamination is a potentially disastrous change because the deamination product, uracil, pairs with adenine rather than with guanine. This could have two effects: (1) When the DNA containing the deaminated cytosine (uracil) was transcribed, an incorrect base could appear in the RNA, since the erroneous uracil would direct the incorporation of an adenine into the RNA molecule; (2) an adenine, rather than a guanine, would similarly appear in newly replicated DNA strands synthesized using the deaminated DNA strand as a template (**Figure 6-9**). To prevent these effects, there exists an intracellular DNA repair system that removes uracil from DNA and replaces it with cytosine. It will be described more fully in Chapter 10.

The necessity of eliminating uracil formed by deamination of cytosine explains why DNA uses thymine and not uracil as the complementary base for adenine. If uracil were a DNA base, there would be no way to distinguish between a "correct" uracil and a uracil produced by the deamination of cytosine. Because DNA uses thymine, the cell can follow a simple rule:

■ **Always remove uracil from DNA because it is unwanted.**

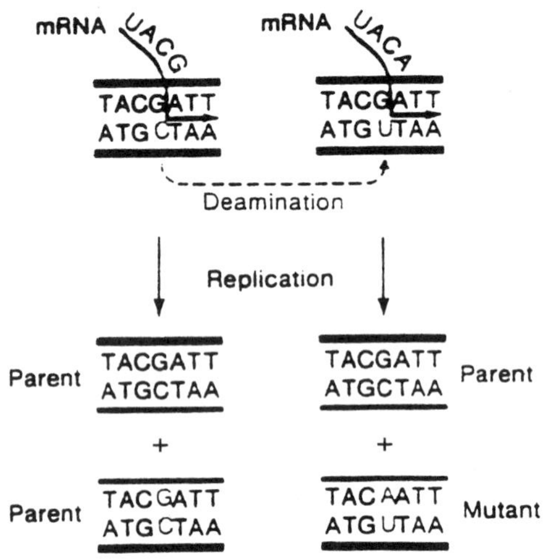

Figure 6-9 The effect of deamination of cytosine to form uracil in the base sequences of mRNA and of two daughter DNA molecules. The C → U transition is shown in blue in the uppermost panel. "Parent" and "mutant" refer to base sequences before and after deamination. Newly replicated DNA is indicated by a thin line.

It is not known why RNA uses uracil (and not thymine) or why DNA evolved with cytosine, rather than with a base that would not deaminate. This may have been an evolutionary accident. It is possible that the original RNA (and DNA?) molecules contained cytosine and uracil because these were the only pyrimidines available in the primordial sea. Cells may have later gained the ability to methylate uracil to form thymine because this would provide a cell with a criterion for eliminating the result of a cytosine–uracil conversion. Perhaps it is not merely a coincidence that the final step in the synthesis of thymidylic acid is the methylation of a deoxyuridylic acid.

The Ability of DNA to Change: Mutation

All of the genetic information contained in a cell resides in its DNA. Thus, if a cell or an organism is to be able to evolve through time, the base sequence of its DNA must be capable of change. Furthermore, the new sequence must persist so that progeny cells or organisms will inherit the new property. The process by which a base sequence changes is called **mutation.** At the nucleotide level, there are two principal mechanisms of mutation:

1. A **chemical alteration** of the base that gives it new hydrogen-bonding properties and, thus, causes a different base to be incorporated into the daughter DNA strand during subsequent replication.
2. A **replication error** by which an incorrect base is erroneously incorporated, or an extra base is accidentally inserted or deleted in the daughter molecule.

On the average, mutational changes are deleterious and may lead to cell death or impaired function. It is important, therefore, that not too many mutations occur in a single DNA molecule at one time because, otherwise, the rare, advantageous alteration would always occur in a cell or virus destined to die by virtue of an accumulation of lethal mutations. Keeping the mutation rate low is accomplished in two ways: (1) The hydrophobic, water-free core of the DNA molecule reduces its accessibility to attacking molecules; and (2) the cell has evolved a number of repair mechanisms for correcting alterations or replication errors.

These repair systems are not completely efficient, so mutations occur at a rate that is very low, but useful in an evolutionary sense.

Mutations are, as we have said, frequently deleterious, so it is important that when such evolutionary experiments take place, the original parental information not be discarded. Such a loss is prevented in two ways: (1) The other members of the species retain the parental base sequence; and (2) the double-stranded DNA molecule is redundant. Normally, only one strand of the DNA molecule is altered. When the DNA replicates, it does so in such a way that, after cell division, the DNA molecule in each daughter cell contains only *one* of the parental strands. Thus, it is possible for one of the daughter DNA molecules to be a copy of the parent, and the other to be a mutant. If the mutant is better equipped to survive and multiply than the parent organisms or any other members of the species, then, after a great many generations, Darwin's principle of survival of the fittest will lead to the ultimate replacement of the parental genotype by the mutant.

Although most replication errors are corrected by the cellular systems described in Chapters 7 and 10, a small fraction of such copy errors do persist, and these give rise to mutations.

In addition to mutations involving minor alterations of nucleotide sequence, mutations may involve larger chromosomal modifications—duplications, deletions, inversions, or translocations—all with varied consequences. Any of these mutations may affect protein or RNA sequences or the control of the expression of a gene or gene cluster.

RNA As the Genetic Material

In the preceding sections, the properties of DNA that particularly suit it to be the genetic material have been described. However, it seems likely that RNA might have been the primordial genetic material. As a single-stranded molecule, its hydrogen-bonding groups are available to perform catalytic functions and perform a catalytic role. DNA presumably evolved as a more stable genetic material. RNA, nonetheless, has survived as the genetic material in some viruses, even though it lacks the beneficial features of base protection and redundancy associated with a double helix (although it should be noted that some viruses have double-stranded RNA).

In viruses, the RNA is protected from the environment by a protein coat. Viral RNA molecules spend most of their time as inert particles, replicating infrequently in the cells of their host organism. When they do replicate, they do so rapidly and a large number of progeny particles are produced in a short period of time. The sheer numbers compensate for the lesser chemical stability of RNA. Additionally, although the lower degree of stability may result in a higher mutation rate among RNA viruses, this may be of some advantage to viruses that must change and evolve relatively rapidly in order to evade the defense mechanisms of their host organism. So it appears that the RNA phages and viruses have evolved special compensatory features that enable them to survive, despite their lesser stability relative to DNA-containing viruses.

KEY CONCEPT

The "Gene" Concept

Genes contain all the information for the synthesis and functioning of cellular components.

SUMMARY

Several classic experiments led to the identification of DNA as the genetic material. The transformation experiments of Griffith and of Avery, MacLeod, and McCarty, using the bacterium *Streptococcus pneumoniae,* led to the discovery that a substance, present in heat-killed bacteria having a Smooth (S) phenotype, could bring about a stable genetic transformation of bacteria having a variant Rough (R) phenotype. This substance, the **transforming principle,** was identified as having the physical and chemical properties of DNA. Hershey and Chase showed that when T2 phage adsorbed to *E. coli,* only DNA was injected into the bacterium, and this DNA provided all of the genetic information necessary for development of progeny phage. Along similar lines, other scientists demonstrated that it is the RNA and not the protein of the RNA-containing virus, *tobacco mosaic virus,* that directs the production of viral progeny.

The central dogma of modern biology is: *DNA makes RNA, makes protein.* DNA has many features that makes it useful as a genetic material: (1) It can store genetic information in its base sequence; (2) it can transfer this information to the protein-synthesizing apparatus of the cell as needed by synthesizing an RNA molecule whose sequence is complementary to the DNA; (3) it has great physical and chemical stability; (4) DNA can transfer genetic information to progeny cells by replication, with each strand serving as a template for synthesis of a complementary strand; and (5) genetic change can occur without a loss of parental information via mutational alterations that result in the production of one daughter molecule identical in base sequence to the parent molecule, and a second daughter molecule with a slightly different base sequence.

DRILL QUESTIONS

1. What is meant by bacterial transformation?
2. In which organisms is RNA the sole genetic material?
3. What term describes the base sequence that corresponds to a single amino acid?
4. In the Blender Experiment, what radioactive atoms were used to label DNA and protein?
5. What is the process whereby DNA is changed?
6. What is the product of deamination of cytosine?
7. How are uracil and thymine chemically related?
8. What are two main mechanisms of mutation?

PROBLEMS

1. Why are viruses able to use RNA as a genetic material?
2. What experimental result showed that transformation consists of a permanent genetic change?
3. Why is RNA (but not DNA) hydrolyzed at alkaline pH?
4. Why was the transformation experiment not accepted by the scientific community as proof that DNA is the genetic material?
5. How was the transforming principle in Griffith's transformation experiment later shown to be DNA?
6. How does a replication error cause mutations?

CONCEPTUAL QUESTIONS

1. What other experiments could be performed to show that DNA is the genetic material?
2. Is it beneficial to a species that most mutational changes are deleterious?
3. Are mutational changes necessary for the various species of living organisms to prosper?

CHAPTER 7

in this chapter you will learn

1. About the individual events that occur at a replication fork, and how these individual events together accomplish the task of DNA replication.

2. The functions of enzymes and other proteins present at the replication fork.

3. The different types of DNA replication used in eukaryotes, prokaryotes, and viruses.

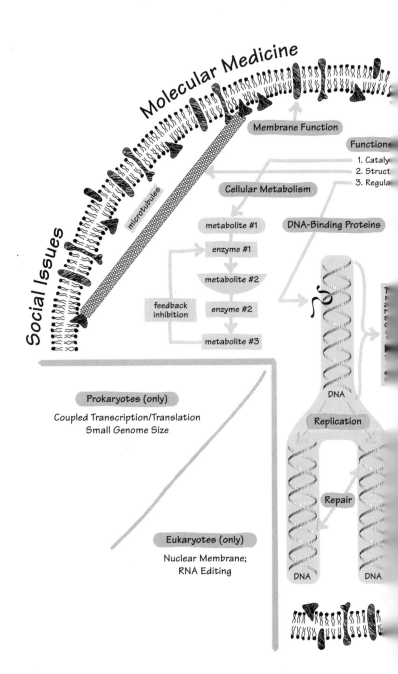

Molecular Medicine

Social Issues

Membrane Function

Functions
1. Cataly
2. Struct
3. Regula

Cellular Metabolism

microtubules

metabolite #1

enzyme #1

metabolite #2

feedback inhibition

enzyme #2

metabolite #3

DNA-Binding Proteins

DNA

Replication

Repair

DNA   DNA

Prokaryotes (only)
Coupled Transcription/Translation
Small Genome Size

Eukaryotes (only)
Nuclear Membrane;
RNA Editing

DNA Replication

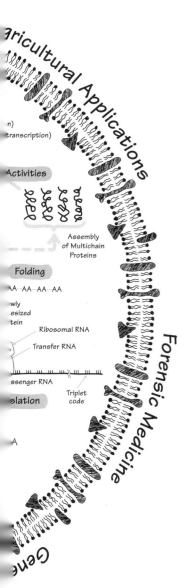

Genetic information is transferred from parent to progeny organisms by a faithful replication of the parental DNA molecules. In most organisms the information resides in one or more double-stranded DNA molecules. Exceptions do, however, exist. Some bacteriophage species contain single-stranded instead of double-stranded DNA. In these systems replication consists of several stages in which single-stranded DNA is first converted to a double-stranded molecule, which then serves as a template for synthesis of single strands identical to the parent molecule.

From a purely abstract perspective, DNA replication might be thought to be relatively uncomplicated. After all, as proposed in **Figure 7-1**, each of the 4 deoxynucleotide bases need only be arranged so that they pair with an appropriate "partner" base (adenine with thymine, cytosine with guanine). Thus, one might expect that a gradual "unzipping" of DNA would allow nucleotide bases to "fit in" the other side, and thereby replicate DNA quickly and efficiently.

Replication of double-stranded DNA is, however, a complicated process that is not yet completely understood. This complexity is, at least in part, due to the following features of the process: (1) The two strands of a double-stranded DNA molecule run in opposite (anti-parallel) directions. New nucleotides need to be added in such a way as to maintain those reverse polarities; (2) Replication requires a supply of energy to unwind the helix; (3) Single-stranded DNA tends to form inter- and intrastrand base pairs when it finds a complementary sequence, so that once unwound, separated strands of double-stranded DNA are likely to quickly re-anneal; (4) A single enzyme can catalyze only a limited number of physical and chemical reactions. Since several reactions are required for DNA replication, a collection of enzymes is involved; (5) Replication errors occasionally occur and mechanisms need to exist to correct them; and (6) Both circularity and the enormous lengths of some DNA molecules impose physical constraints on the replicative system. If those complications aren't enough to baffle the beginning student of molecular biology, one more can be mentioned: throughout the biological kingdom some variations in key steps in DNA replication have been recognized!

Our strategy for learning about DNA replication will be to proceed, stepwise, with a review of the important features of each individual step. First, we will review how twisted, and sometimes circular, DNA is unraveled prior to replication. Then we will learn that DNA replication usually begins at special sites on DNA. Next, we will study how the double helix is unwound to generate a single-stranded template. Later we will discuss the enzymology of the replication process, which is complicated, as mentioned in the introductory remarks of this chapter, by the fact that the two DNA strands run in opposite directions. Finally, we will learn

Figure 7-1 Schematic diagram of DNA replication as conceived by scientists before information on the complications associated with individual physical and chemical reactions was collected.

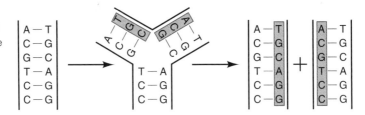

how all those systems and components are integrated into the so-called "replication fork," the site on unwinding DNA where most of the action occurs.

Semiconservative Replication of Double-Stranded DNA

semiconservative replication

DNA polymerase

DNA replication is accomplished by **semiconservative replication.** In this replication mode each parental single strand is a template for the synthesis of one new daughter strand. The enzyme **DNA polymerase** adds nucleotide monomers one by one to the growing strands. As each new strand is formed it is hydrogen-bonded to its parental template (**Figure 7-2**). Thus, as replication proceeds, the parental double helix unwinds and then rewinds again into two new double helices, each of which contains one original parent strand and one newly formed daughter strand.

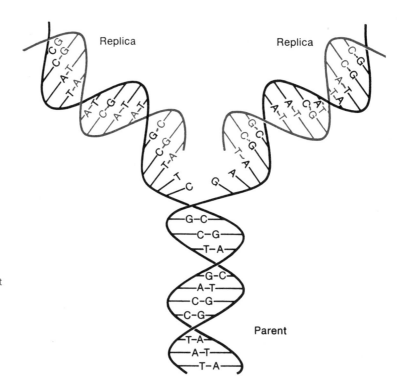

Figure 7-2 The replication of DNA according to the mechanism proposed by Watson and Crick. The two replicas consist of one parent strand (black) plus one daughter strand (blue). Each base in a daughter strand is selected by the requirement that it form a base pair with the parent base.

Untwisting of Highly Coiled DNA Is Required for DNA Replication

For the various replication enzymes to add nucleotide bases to the growing strands, it is necessary that the replicating portion of a DNA molecule exist—momentarily—in the ladder configuration diagrammed in Figure 7-1. As you learned in Chapter 3, double-stranded DNA is twisted, or coiled, 360 degrees for every 10 bases (e.g., Figure 3-1). This twisting must first be uncoiled so that the hydrogen bonds holding the two strands together are accessible to a special enzyme ("helicase"—you will encounter it later in this chapter) that unwinds the two DNA strands from each other. Untwisting of prokaryotic DNAs is complex because they are circular and, therefore, have no free ends that might facilitate easy uncoiling. Also, DNA is usually highly folded, as was described in Chapter 5 (e.g., Figure 5-3), into a compact structure so as to fit conveniently into a cell. Eukaryotic DNA, although long and linear, rather than circular, is similarly twisted and tangled when it is organized into individual chromosomes.

Thus, the twisting and tangling of DNA, overlooked in the simplistic diagram in Figure 7-1, presents a formidable task for cells preparing to replicate DNA.

Replication of Circular DNA Introduces Twists

The circular structures of replicating DNA resemble the Greek letter θ (theta) (**Figures 7-3** and **7-4**). The term **θ replication** is, therefore, used to describe them. Replication of such circles involves a topological problem—a problem common not only to prokaryotic DNA, which is usually circular, but also to eukaryotic DNA. Although not circular, eukaryotic DNA is replicated simultaneously in many local regions resembling long chains of small circles (please look ahead to Figure 7-25). Therefore, it is usually discussed—in conceptual terms—as if it replicates as a series of connected circles.

As replication of the two daughter strands proceeds along the double-stranded DNA helix, as shown in Figure 7-4, in the absence of some kind of swiveling, the nongrowing ends of the daughter strands would cause the entire unreplicated portion of the molecule to become overwound, or "tight." This, in turn, would

θ replication

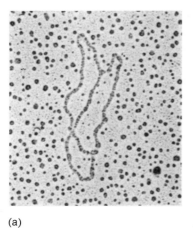

(a)

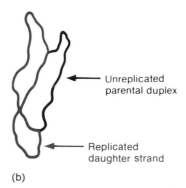

Unreplicated
parental duplex

Replicated
daughter strand

(b)

Figure 7-3 (a) Electron micrograph of a ColE1 DNA molecule (molecular weight = 4.2×10^6) replicating by the θ mode. (b) The parent and daughter segments are shown in the drawing. (Courtesy of Donald Helinski.)

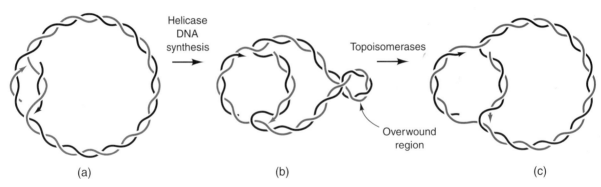

Figure 7-4 Relaxation of DNA overwinding by topoisomerases. DNA replication is shown after it has begun on a circular DNA molecule. In this example a replication fork is working at both sides of the growing "bubble" (a). As the replication bubble grows (b), an overwound region appears ahead of the replication bubble. Through the action of topoisomerases the overwound region is relaxed (c). Illustration provided by C. Ullsperger, A. Vologodskii, and N. Cozzarelli.

cause *positive* supercoiling (see Chapter 3) of the unreplicated portion. This supercoiling obviously cannot increase indefinitely because, if it were to do so, the unreplicated portion would become coiled so tightly that no further advance of the replication fork would be possible. As discussed in Chapter 3, most naturally occurring circular DNA molecules are *negatively* supercoiled. Thus, the overwinding motion initially is no problem because it can be taken up by the underwinding already present in the negative supercoil. However, after about 5% of the circle is replicated, the negative superhelicity is used up, and the topological problem arises.

Special Untwisting/Uncoiling Enzymes (Topoisomerases) Exist

topoisomerase

Separating the tightly twisted DNA requires that the individual strands be rotated about one another. A pair of **topoisomerase** enzymes accomplish this. A type-I topoisomerase working just ahead of the replicating DNA makes a cut in *one* strand of the double-stranded DNA. With the loose DNA ends attached to the enzyme, one of the single strands is passed through the gap generated by the cut. Then the ends are reattached to one another. As a result, a circular double-stranded DNA molecule is uncoiled with each topoisomerase action one full 360-degree turn. As a result of this action, the DNA is "relaxed" somewhat. Later, the helicase enzyme mentioned previously will come into play and undo the hydrogen bonds that hold the 10 base pairs present in each full 360-degree turn. That is, topoisomerase works ahead of the helicase, relieving the overwinding mentioned in Figure 7-4.

Two Topoisomerases Function in Concert

gyrase

As DNA replication continues around a circle, positive supercoiling of the entire circle occurs. A second enzyme, which is also a topoisomerase, acts to introduce negative supercoils into the replicating circle. It is called **gyrase.** This enzyme acts by cutting *both* strands (in contrast to the type-I enzyme, which cuts only one strand). Therefore, it is also sometimes referred to as topoisomerase II. A segment of the cut DNA, which is located away from the actual cut site, is passed through the gap and then the cut is resealed. In this way the replicating circular DNA is maintained in its normal state of negative supercoiling. **Figure 7-5** below illustrates the manner in which one of these enzymes (topoisomerase I) untwists DNA.

Initiation of DNA Replication

Now that we have considered the problems associated with untwisting DNA in order to relieve unwanted supercoiling, let us examine the process of initiation. Initiation occurs at specific base sequences, which comprise the so-called **"origin of replication" (ori).**

"origin of replication" (ori)

Two general types of initiation processes, both acting at the ori, have been discovered. One is referred to as *de novo* "from the new," in which synthesis begins from RNA primers. It is common to most organisms. The other is called the **"covalent extension mechanism."** It involves the synthesis of a new strand as an extension of a preexisting (old) strand. Its use is mostly limited to certain viruses. Those two general mechanisms for initiation are illustrated in **Figure 7-6**. Since those mechanisms are indeed so different, they will be discussed separately.

covalent extension mechanism

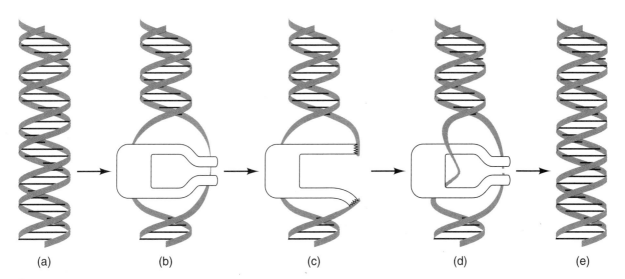

(a) (b) (c) (d) (e)

Figure 7-5 Topoisomerase I binds to the double helix. It unwinds a local region (b) and introduces a nick (single-strand cut) (c). While holding onto the cut ends, the enzyme then passes the other, intact strand through the gap in the cut strand (d). The gap is sealed, and the enzyme is released from the DNA, leaving a double helix (e) which, compared to the starting DNA (a), is identical, except that the extent of negative supercoiling of the whole, circular molecule is *reduced* by one complete turn. Gyrase (topoisomerase II) works in the opposite way—it *introduces* negative supercoiling to compensate for the positive supercoiling in circular DNA.

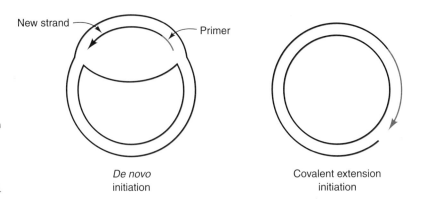

Figure 7-6 Two different mechanisms for the initiation of DNA synthesis have been discovered. *De novo* is most common (primer shown in blue), while covalent extension (new strand shown in blue) is limited to some viruses and bacterial F (fertility) factors (see Figure 14-8 in Chapter 14).

De Novo Initiation

The most common mechanism of initiation (*de novo*) is shared by DNA molecules as diverse as the *E. coli* chromosome, phage λ DNA, and the DNA of the eukaryotic simian virus 40 (SV40). During the initiation of replication of the *E. coli* chromosome (shown diagrammatically in **Figure 7-7**), the DnaA protein binds to a 9 nucleotide sequence that is repeated 4 times within a 250 nucleotide sequence. That region is defined as the *ori*C ("C" for chromosomal). In the presence of ATP, the bound 50-kDa DnaA protein monomers interact to form a complex protein–DNA structure consisting of 15–20 DnaA monomers and about 150 base pairs of DNA. Formation of this structure acts to locally denature an A + T-rich region directly adjacent to it. Then, because of specific interactions between the DnaA protein and the other replication proteins (like DnaB), the replication fork

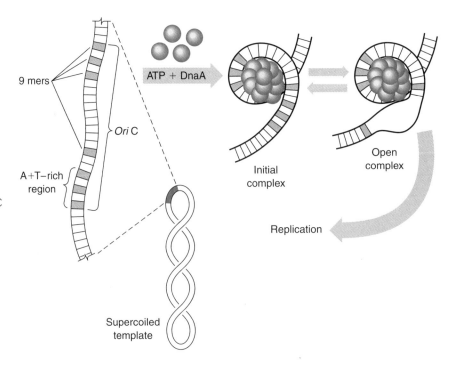

Figure 7-7 Initiation at *ori*C occurs after DnaA protein binds the four 9 mers. The A + T-rich region is then denatured, and this open complex serves as a replication site. (Adapted and redrawn from D. Bramhill and A. Kornberg, *Cell,* 1988; 54: 915–918.)

machinery is recruited to the unwound site on the DNA. Once all the requisite enzymes are present and primers have been synthesized, DNA pol III can begin to synthesize DNA. In the case of *ori*C, two replication forks then proceed in opposite directions away from the origin (see later section on Bidirectional Replication, page 140). The exact mechanism for initiation of the leading and lagging strands is somewhat unclear and currently under investigation.

During phage λ replication in *E. coli,* as well as the replication of the animal virus SV40, a very similar mechanism is used, and a virally encoded protein binds repeated sequences at the *ori* to form the initial complicated nucleoprotein structure. Thus, this mechanism seems to be general.

A different type of *de novo* initiation is employed by other organisms. In this mechanism, an RNA primer is synthesized on one DNA strand through the direct action of an RNA polymerase enzyme.* The eventual extension of that RNA primer by DNA polymerase, during the synthesis of a daughter strand, generates a replication bubble that consists of one double-stranded and one single-stranded branch.

Since the growing daughter strand displaces the unreplicated parental strand, the bubble is called a displacement loop, or, more commonly, a **D loop** (**Figure 7-8**).

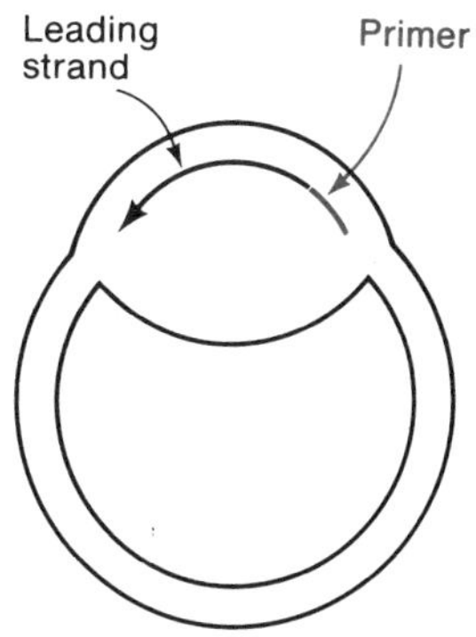

Figure 7-8 A circular DNA molecule with a D loop. The "upper" strand is double because the primer is bound to it and a daughter strand is being synthesized as an extension of the primer.

D loop
concatamer

Covalent Extension Initiation

In some of the very simplest biological systems (e.g., plasmids and viruses), natural selection favors rapid production of large numbers of copies of DNA. Using the "covalent extension" mode of replication (see Figure 7-6) illustrated again in **Figure 7-9**, circular DNA molecules produce a **concatamer,** which is a DNA base sequence tandemly repeated many times. In this mode of initiation, a nick is made in one of the DNA strands, thereby creating a free 3′-OH end (along with a free 5′-P end). The free 3′-OH provides DNA polymerase with a starting point. The synthesis of a daughter strand begins at this free end of the

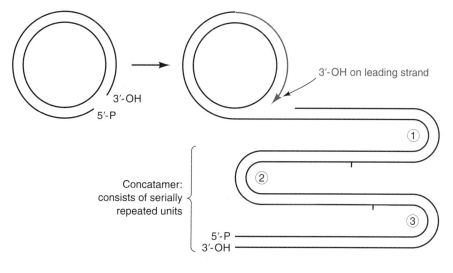

Figure 7-9 Covalent extension initiation of DNA replication begins at the free 3-OH generated when the double-stranded DNA is nicked. Newly synthesized DNA is shown in blue. Three copies of the circular DNA are shown in the concatamer.

*This enzyme polymerizes ribonucleotides, and is discussed in detail in Chapter 8.

nicked DNA strand. That is, the free 3-OH serves the same role as did the RNA primer illustrated for *de novo* initiation in Figure 7-8.

A similar covalent extension process for the initiation of DNA synthesis is employed during the bacterial mating process. As will be described in Chapter 14 (e.g., Figure 14-8), a linear DNA synthesized as shown in Figure 7-9 is transferred from a donor cell to a recipient cell. Replication of virus DNA also employs covalent extension, and is described in Chapter 14 (e.g., see Figure 14-24 in Chapter 14).

Unwinding of DNA for Replication

In order for the enzymes that synthesize new (i.e., daughter) strands of DNA to begin their polymerization functions, the DNA template needs to be readied. This process of template preparation involves at least three steps: unwinding of the double helix (recall, from Figure 3-1 that the double helix is wound 360 degrees every 10 base pairs); breakage of the H-bonds holding the two DNA strands together (see Figure 3-2) to expose the single-stranded templates for new synthesis; and separation of the single-stranded DNA templates so that they will not re-anneal before DNA polymerase has had an opportunity to act.

Although there is one enzyme, DNA polymerase I (pol I), that can accomplish all of those tasks, its role in DNA replication is mostly limited to repair processes. Another enzyme, the DNA polymerase III mentioned earlier, is the main enzyme for polymerizing nucleotides in most cells. It is, however, highly specialized for polymerization and therefore requires the assistance of other proteins to carry out the three steps mentioned previously.

Helicase Enzymes Unwind DNA

helicase

The unwinding of double-stranded DNA is accomplished by the **helicase** enzymes. When helicases encounter double-stranded DNA, they continue to move forward by unwinding and prying open the base pairs. The "unzipping" action of helicase involves breaking hydrogen bonds and eliminating hydrophobic interactions. Energy is required for this process. It is supplied by ATP, which is hydrolyzed in the process, and the energy released is used by the helicase enzyme.

single-strand DNA binding proteins (ssb)

As mentioned previously, the single-stranded DNA exposed by the action of a helicase enzyme could easily re-form hydrogen bonds with other bases on the opposite strand. This is prevented by the action of **single-strand DNA binding proteins (ssb),** which bind to the exposed strands, coating them in their single-stranded form and thereby blocking the re-annealing process. Once the strand has reacted with ssbs, polymerization of nucleotides proceeds. The unwinding action of helicase and the binding of ssbs is diagrammed in **Figure 7-10.**

Elongation of Newly Synthesized Strands

The enzymatic synthesis of new daughter DNA strands is complicated by three additional factors:

1. The need for high fidelity in copying the base sequence.
2. The fact that the two parental strands are antiparallel (one runs $5'$-P $\rightarrow 3'$-OH, and the other $3'$-OH $\rightarrow 5'$-P), and DNA polymerase can only function in one direction.

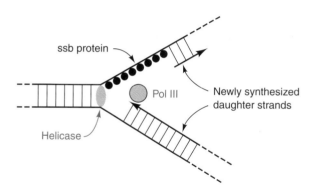

ssb protein

Pol III

Newly synthesized
daughter strands

Helicase

Figure 7-10 The unwinding events in a replication
fork.

3. The need for speed (the replication fork moves at 1,000 nucleotides per second in *E. coli*).

Considering those complications, it is easy to understand that the number of steps that must be completed is far too great to be accomplished by a single enzyme. In fact, twenty or more proteins are known to be necessary. In an effort to provide some understanding of the process with a minimum of confusion, each step will be treated separately. We will consider the basic chemistry of polymerization and the enzymes involved, the problems raised by the chemistry of polymerization, the mechanisms for eliminating replication errors, the synthesis of RNA primers, and, finally, the topology of a complete replication apparatus.

The Polymerization Reaction and the Polymerases

In 1957, Arthur Kornberg showed that in extracts of *E. coli* there exists a DNA polymerase (now called **polymerase I** or **pol I**). This enzyme is able to synthesize DNA from four precursor molecules—namely, the four deoxynucleoside 5′-triphosphates (dNTP), dATP, dGTP, dCTP, and dTTP—as long as a DNA molecule to be copied (a **template** DNA) is provided. Neither 5′-monophosphates nor 5′-diphosphates, nor 3′-(mono-, di-, or tri) phosphates can be polymerized— only the 5′-triphosphates can serve as substrates for the polymerization reaction; soon we will see why this is the case. Some years later, it was found that pol I, although playing an essential role in the replication process, is not the major polymerase in *E. coli;* instead, the enzyme responsible for advance of the replication fork is polymerase III or pol III.* Pol III also exclusively uses 5′-triphosphates as precursors and requires a DNA template before polymerization can occur. Pol** I and pol III have some features in common. The overall chemical reaction catalyzed by both DNA polymerases is:

$$\text{Poly(nucleotide)}_n\text{-3}'\text{-OH} + \text{dNTP} \rightarrow \text{Poly(nucleotide)}_{n+1}\text{-3}'\text{-OH} + \text{PP}$$

in which PP represents pyrophosphate cleaved from the dNTP.

polymerase I
pol I

template

---

*The numbers I and III refer only to the order in which the enzymes were first isolated, not to their relative importance. Another polymerase, pol II, has also been isolated from *E. coli*. It plays no other role in DNA replication. Although mutant cells lacking pol II grow normally, recent studies indicate that the enzyme may be involved in certain DNA damage-induced DNA repair mechanisms. Additional DNA polymerases—also implicated in DNA repair mechanisms—have also recently been discovered.
**For convenience, the abbreviation "pol" will be used in this chapter to refer to specifically named polymerases.

primer

Both pol I and pol III only polymerize deoxynucleoside 5′-triphosphates and can do so only while copying a template DNA. Furthermore, polymerization can only occur by addition to a **primer**—an oligonucleotide hydrogen-bonded to the template strand and whose terminal 3′-OH group is available for reaction (that is, a "free" 3′-OH group). The meaning of a primer is made clear in **Figure 7-11**, which depicts six potential template molecules; of these, only three can be said to be active—(c), (e), and (f)—each of which has a free 3′-OH group. The lack of activity with (d) and the direction of synthesis with (e) and (f) indicate that nucleotides do not add to a free 5′-P group. The lack of any synthesis with (a) or (b) indicates that addition to a 3′-OH group cannot occur if there is nothing to copy. Thus, we draw two conclusions:

1. Both a primer with a free 3′-OH group and a template are needed.
2. Polymerization consists of a reaction between a 3′-OH group at the end of the growing strand and an incoming nucleoside 5′-triphosphate. When the nucleotide is added, it supplies another free 3′-OH group. Since each DNA strand has a 5′-P terminus and a 3′-OH terminus, strand growth is said to proceed in the 5′ → 3′ (5′-to-3′) direction.

These two points are summarized in **Figure 7-12**.

Occasionally polymerases add a nucleotide that cannot hydrogen-bond to the corresponding base in the template strand. This may be purely a mistake, or may result from the tautomerization of adenine and thymine discussed in Chapter 10 (see Table 10-1). In any case, it is important that the unpaired base be removable while its incorrectness is recognizable—namely, as an unpaired base at the 3′-OH terminus of a growing strand.

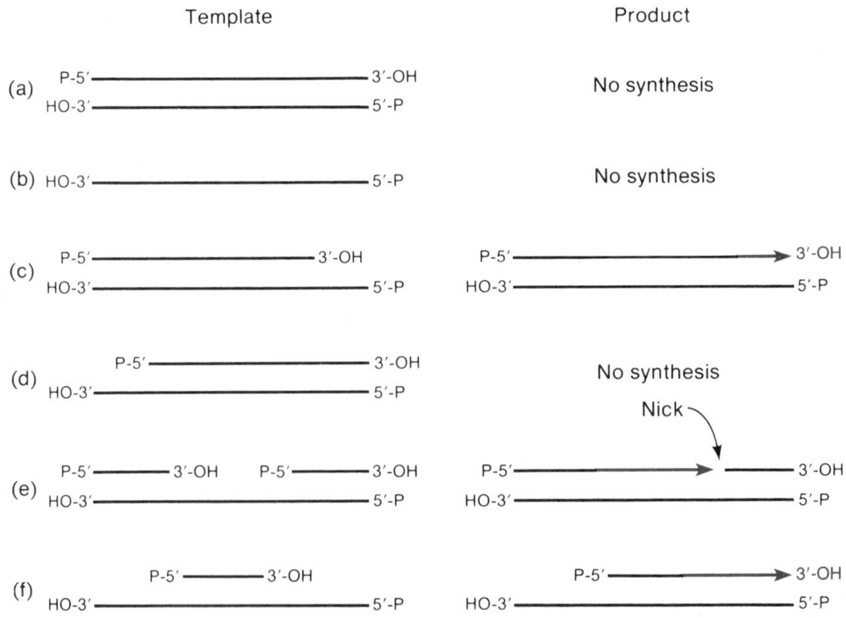

primer

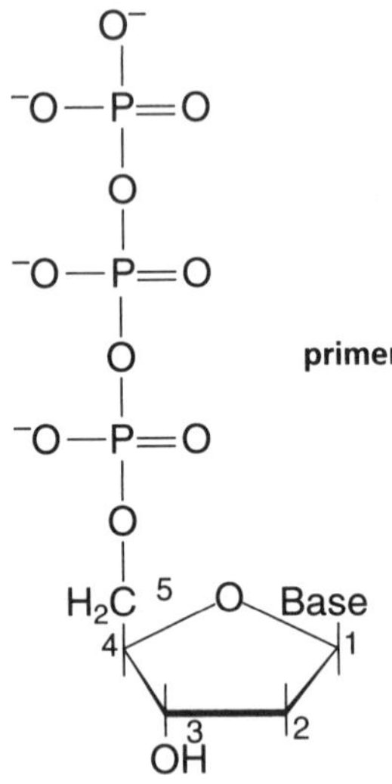

Deoxynucleoside triphosphate

Figure 7-11 The effect of various templates used in DNA polymerization reactions. A free 3′-OH on a hydrogen-bonded nucleotide at the strand terminus and a non-hydrogen-bonded nucleotide at the adjacent position on the template strand are needed for strand growth. Newly synthesized DNA is blue.

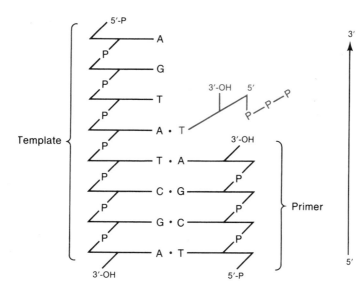

Figure 7-12 Schematic diagram of a replicating DNA molecule showing the distinction between template and primer and the meaning of $5' \to 3'$ synthesis.

Correcting Mismatched Bases

Both pol I and pol III respond to an unpaired terminal base by terminating polymerizing activity. That is because those enzymes require a primer that is correctly hydrogen-bonded. When such an impasse is encountered, a $3' \to 5'$ exonuclease function (which is a separate catalytic activity located at a site distant from the active site for nucleotide polymerization), is stimulated and the unpaired base is removed (**Figure 7-13**). After removal of this base, polymerizing activity resumes and chain growth, therefore, continues. This enzyme activity, because it removes a base from the end of a nucleic acid, is referred to as an *exo*nuclease activity (see Chapter 3). Thus, pol I and pol III are said to possess a **proofreading** or **editing function.**

An additional function of DNA polymerase I is that of an opposite direction, $5' \to 3'$ exonuclease. This activity has the following features:

1. Nucleotides are removed one by one from the 5'-P terminus only. That is, of course, the opposite of the activity illustrated in Figure 7-13.
2. More than one nucleotide can be removed by successive cutting.
3. The nucleotide removed must have been base paired.
4. The nucleotide removed can be either of the deoxy- or the ribo-type.
5. Activity can occur at a nick as long as there is a 5'-P group.

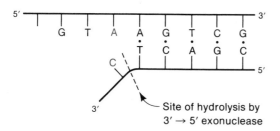

Figure 7-13 The $3' \to 5'$ exonuclease activity of DNA polymerase I showing the site of hydrolysis. That C is removed (blue) because it does not base pair with the A being copied (blue).

We will see shortly that the main function of the 5′ → 3′ exonuclease activity is to remove ribonucleotide primers.

The 5′ → 3′ exonuclease activity at a single-strand break (nick) can occur simultaneously with polymerization. That is, as a 5′-P nucleotide is removed, a replacement can be made by the enzyme's polymerizing activity (**Figure 7-14**). Since pol I cannot form a bond between a 3′-OH group and a 5′-monophosphate, the nick moves along the DNA molecule in the direction of synthesis leaving a gap. This movement is called **nick translation.**

nick translation

DNA Polymerase III Consists of Multiple Subunits

Pol III is a very complex enzyme. In its most active form the prokaryotic enzyme is associated with nine other proteins to form the **pol III holoenzyme.** The term **holoenzyme** refers to an enzyme that contains several different subunits and retains some activity even when one or more subunits is missing. The smallest aggregate having enzymatic activity is called the **core enzyme.** The activities of the core enzyme and the holoenzyme are usually very different. *E. coli* genes encoding all ten of the polymerase subunits have been identified: one of them, the *dnaE* protein, possesses the major polymerizing activity, but most of the subunits are essential for replication. Pol III shares with pol I a requirement for a template and a primer, but pol III's substrate specificity is much more limited. For instance, neither the core enzyme nor the holoenzyme can act at a nick. The core enzyme is most active in vitro on gapped DNA, where the gap is less than 100 nucleotides long. On the other hand, the holoenzyme is extraordinarily active on long single-stranded DNA templates primed with either a DNA or RNA primer. This reflects the action of the holoenzyme at the replication fork, where, as the parental duplex is unwound, the template is presented essentially as one very long single strand. Pol III cannot carry out strand displacement either, and another system is needed to unwind the helix for a replication fork to be able to proceed (to be discussed in detail in a later section). The pol III enzyme, like pol I, possesses a 3′ → 5′ exonuclease activity that performs the major editing function in DNA replication. This function resides in the *dnaQ* subunit. The activity of this 3′ → 5′ editing exonuclease is stimulated greatly when the *dnaQ* subunit is complexed together with the *dnaE* subunit in either the core enzyme or the holoenzyme. A role for this activity in providing fidelity to the replication process is clearly indicated, since

holoenzyme

core enzyme

Figure 7-14 During nick translation, carried out by pol I, a nucleotide is exonucleolytically removed in the 5′ → 3′ direction for each nucleotide added. The growing strand is shown in blue. The gap in the lower figure is sealed by a separate enzyme-DNA ligase.

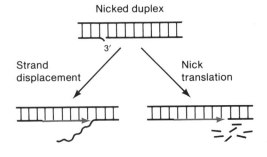

E. coli strains carrying mutations in the *dnaQ* gene have a higher mutation frequency. Unlike pol I, pol III does not possess a $5' \to 3'$ exonuclease activity (see **Table 7-1**).

Although pol III holoenzyme is the major replicating enzyme in *E. coli,* much less is known about it than about pol I because it is a more complex enzyme. Study of pol III is currently an active field of research.

Later in this chapter, when the events at the growing fork catalyzed by the *E. coli* replication system are examined, it will be seen that pol I and pol III holoenzyme are both involved in *E. coli* DNA replication. However, a requirement for two polymerases is not common to all organisms: for instance, *E. coli* phage T4 synthesizes its own polymerase, and this enzyme is capable of carrying out all necessary polymerization functions.

DNA Polymerases Can Work Only in the 5′-P → 3′-OH Direction

All known polymerases (for both DNA and RNA) are capable of chain growth in only the $5' \to 3'$ direction. That is, the growing end of the polymer must have a free 3′-OH group. One might inquire, since the two DNA strands are antiparallel, why a $3' \to 5'$ polymerase isn't also active during DNA replication. After all, if such a $3' \to 5'$ polymerase were available, DNA replication could follow the simplistic scheme illustrated in Figure 7-2.

It is possible, for the following reason, that the polymerization enzymes evolved their $5' \to 3'$ directionality to facilitate editing (of mistakes). If $3' \to 5'$ growth were to occur, the growing strand would also be terminated with a 5′-triphosphate and the 3′-OH group of the incoming nucleotide would react with it. Chemically, this is certainly acceptable. However, since the bonds formed in this case contain only a single phosphate, an exonuclease editing function designed to remove a mismatched base would leave behind a free 5′-monophosphate at the end of the growing strand. In order for chain growth to resume, an enzyme system would be needed to enter the replication fork and convert the monophosphate to a triphosphate. In the interest of speed, and since there is already a great deal going on in the replication fork, it would seem more economical for the cell to require $5' \to 3'$ growth exclusively.

The observation that chain growth proceeds in only one direction introduces, however, what is probably the *greatest complication in the entire replication process,* which will be described next.

TABLE 7-1	COMPARISON OF POL I AND POL III	
Activity	**pol I**	**pol III**
5′P → 3′OH polymerization	+	+
3′OH → 5′P polymerization	−	−
5′P → 3′OH exonuclease	+	−
3′OH → 5′P exonuclease	+	+

Antiparallel DNA Strands and Discontinuous Replication

In the model of replication you were introduced to at the beginning of this chapter (Figure 7-2), both daughter strands are drawn as if replicating continuously. However, no known DNA molecule replicates in this way—instead,

> One of the daughter strands is made in short fragments, which are then joined together.

As just mentioned, polymerase I and polymerase III can add nucleotides only to a 3′-OH group. Examination of the growing fork indicates that if both daughter strands grew in the same overall direction, only one of these strands would have a free 3′-OH group. The other strand would have a free 5′-P group because the two strands of DNA are antiparallel (**Figure 7-15**). One of the following, therefore, must be true:

1. There is another polymerase that can add a nucleotide to the 5′ end; that is, it would catalyze strand growth in the 3′ → 5′ direction, OR,
2. The two strands both grow in the 5′ → 3′ direction, but from opposite ends of the parental molecule. If this were correct, a significant fraction of the unreplicated molecule would have to be single stranded.

Neither of those two possibilities is correct. No DNA polymerase has been discovered that adds nucleotides to the 5′-P terminus, and at most about 0.05% of intracellular DNA is ever in the single-stranded form.

Indeed, both the way replication is accomplished, despite the antiparallel nature of the two strands, and the unidirectional mechanism of action of DNA polymerase are shown in **Figure 7-16**. This mode of synthesis is called **discontinuous synthesis**. The **leading strand** is synthesized continuously, and the **lagging strand** is synthesized discontinuously in the form of short fragments, each growing in the correct 5′ → 3′ direction. In time, the fragments are joined together to form a continuous strand. Critical evidence verifying the discontinuous nature of DNA replication will be reviewed, for the experiments are still considered to be landmarks in the history of molecular biology. These experiments represent a true tour de force, and offer the beginning student an opportunity to gain insight into experimental designs.

discontinuous synthesis
leading strand
lagging strand

So-Called "Okazaki Fragments" Are Joined Together

In 1968, R. Okazaki demonstrated that in *E. coli* some newly synthesized DNA consists of fragments that later are attached to one another to generate long con-

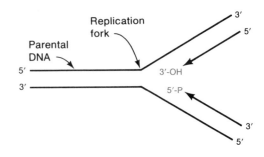

Figure 7-15 The termini (blue) that are present in a replication fork.

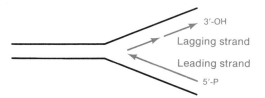

Figure 7-16 A growing replication fork showing the direction of growth of the leading and lagging strands (both in blue).

tinuous strands. The discovery of those fragments supports the discontinuous replication model summarized in Figure 7-16. Okazaki did two experiments to demonstrate discontinuous DNA synthesis. In the first, [³H]dT was added to a growing bacterial culture in order to label the new DNA strands with radioactivity, and 30 seconds later the cells were collected and all of their DNA was isolated. This is called a **pulse-labeling experiment.** The size of the DNA was then measured by observing its sedimentation characteristics in the presence of alkali. In alkali, recall from Chapter 3, double-stranded DNA separates into single strands. Sedimentation reveals the presence of any small DNA fragments that are not joined to a longer strand.

    The type of data obtained (**Figure 7-17**(a)) showed that the most recently made (pulse-labeled) DNA sediments very slowly in comparison with single strands obtained from parent DNA (even though these strands usually break in the course of isolation); from the s value,* it was estimated that the fragments of pulse-labeled DNA range in size from 1,000 to 2,000 nucleotides, whereas the isolated, preexisting parent DNA is usually 20 to 50 times as large. In the second experiment, the bacteria were pulse-labeled for 30 seconds; then, the [³H]dT was replaced with large quantities of nonradioactive dT, and the bacteria were allowed to grow for several minutes. This is called a **pulse-chase experiment.** Flooding the bacterial culture with nonradioactive dT allows one to determine the current size of the radioactive molecules synthesized at an earlier time. That is, the earlier synthesized (radioactive) DNA can be traced. Okazaki observed that the s value (i.e., size) of the radioactive material increased with time of growth in the nonradioactive medium. These experiments are presented and interpreted in Figure 7-17 in terms of the discontinuous replication model shown in Figure 7-16. Apparently the joining of the pulse-labeled fragments to the growing daughter strands caused the label to sediment with the bulk of the longer, continuous strands of DNA.

    These fragments, called **precursor fragments** or **Okazaki fragments,** have the properties predicted by the discontinuous model: they are initially small and then become large as they are attached to previously made DNA. However, the model predicts that only half of the radioactivity should be found in small fragments, whereas the data shown in Figure 7-17(a) indicate that *all* of the newly synthesized DNA consists of fragments. This result should be surprising because there is no reason why the DNA of the 3′-OH-terminated strand should be synthesized discontinuously. In fact, it is not. Occasionally DNA

*The s value, or sedimentation coefficient, is a measure of the speed with which a particle moves under the influence of centrifugal force. For DNA, it is usually related simply to molecular weight.

pulse-labeling experiment

pulse-chase experiment

Okazaki fragments

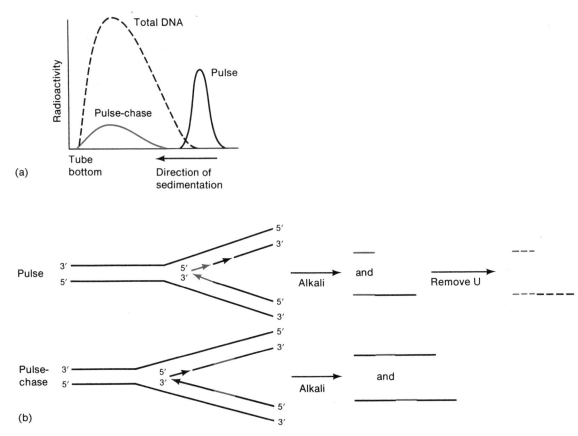

Figure 7-17 (a) The type of data obtained by alkaline sedimentation of pulse-labeled DNA (black) and pulse-chased DNA (blue). Total DNA is the sum of the nonradioactive and radioactive DNA, as might be indicated by optical absorbance. The *s* value of the sedimenting material increases from right to left. (b) The location of radioactive DNA (blue) at the time of pulse labeling and after a chase. The radioactive molecules present in alkali are shown. The fragmentation resulting from removal of uracil (see later in text) accounts for the fact that all pulse-labeled DNA sediments slowly before the chase.

polymerase adds a uracil nucleotide instead of a thymine nucleotide to the end of the growing strand when copying an adenine in the template strand. When this occurs, an enzyme repair system excises the uracil, replacing it with thymine. This repair system does not act at the terminus (in contrast with the proofreading 3′-5′ exonuclease), but instead when the uracil is well within the growing strand. The excision process produces a transient single-strand break, which is sealed by DNA ligase after the correct nucleotide is inserted. Thus, the leading strand is made continuously but, after synthesis, is fragmented at the rare uracil sites.

Additional evidence for the discontinuous-synthesis model comes from high-resolution electron micrographs of replicating DNA molecules. These micrographs show a short, single-stranded region on one side of the replication fork

(**Figure 7-18**). This region results from the fact that synthesis of the discontinuous strand is initiated only periodically; perhaps a particular base sequence or some other signal is required for initiation. In fact, the 3′-OH terminus of the continuously replicating strand is always ahead of the discontinuous strand. This is the origin of the terms **leading strand** and **lagging strand** for the continuously and discontinuously replicating strands, respectively.

A Primer Is Required for Chain Initiation

Like all DNA polymerases, pol III needs a primer. That is, it cannot simply latch onto a template strand and begin polymerizing nucleotides from a new strand. This predicament applies to initiating both the leading strand and the precursor (Okazaki) fragments. Either a ribo- or deoxynucleotide, hydrogen-bonded to the template strand, can serve as a primer. Because all known DNA polymerases can add nucleotides only to the 3′-OH termini such a reactive group must be available on the primer (see Figure 7-12). In every case so far examined, the primer for both leading and lagging strand synthesis is a short RNA oligonucleotide that consists of 1 to 60 bases, the exact number depending on the particular organism. This RNA primer is synthesized by copying a particular base sequence from one DNA strand, and differs from a typical RNA molecule in that, after its synthesis, the *primer remains hydrogen-bonded to the DNA template.*

In bacteria two different enzymes are known to synthesize primer RNA molecules—**RNA polymerase,** which is the same enzyme that is used for synthesis of most RNA molecules, such as messenger RNA (see Chapter 8), and **primase** (the product of the dnaG gene in *E. coli*). Experimentally, these enzymes can be distinguished in vivo by their differential sensitivities to the antibiotic rifampicin—RNA polymerase is inhibited by the antibiotic, but primase is not. In *E. coli*, initiation of **leading-strand** synthesis is rifampicin sensitive, presumably because RNA polymerase is used. Thus, the synthesis of primers is quite different for the leading and the lagging strands. The leading strand only needs to be primed once. Since the leading strand grows continuously, it never has to be reprimed. Thus,

RNA polymerase

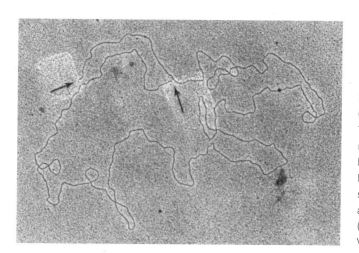

Figure 7-18 A replicating θ molecule of phage λ DNA. The arrows show the two replicating forks. The segment between each pair of thick lines at the arrows is single-stranded DNA; note that it appears thinner and lighter. (Courtesy of Manual Valenzuela.)

the primer synthesized at the *ori* site is the only primer needed for the leading strand. Creating primers for the leading strand is, therefore, not a repetitive process.

However, as we have learned, the **lagging strand** begins as a series of short fragments. The synthesis of each of those Okazaki fragments must begin on a primer. Primers must, therefore, be synthesized repetitively on the lagging strand throughout the elongation process. **Figure 7-19** illustrates the distinction between leading- and lagging-strand primer synthesis.

Like the helicase enzyme, primase is needed to "prepare" the parental DNA for pol III. Those two enzymes are usually linked together. Such a helicase/primase complex is called a **primosome**.

A precursor fragment has the following structure, while it is being synthesized:

PPP-5'————————————————————— 3'-OH
　　　　　　RNA　　　　　　　　　　DNA

Precursor fragments are ultimately joined to yield a continuous strand. The completely synthesized daughter strand contains no ribonucleotides. Assembly of the lagging strand, therefore, requires removal of the primer ribonucleotides, replacement with deoxynucleotides, and then joining. In *E. coli* the first two processes are accomplished by pol I and joining is catalyzed by **DNA ligase.**

DNA ligase

DNA Ligase Joins Precursor Fragments

The assembly of precursor fragments is shown in **Figure 7-20**. Pol III extends the growing strand until the RNA nucleotide of the primer of the previously synthesized precursor fragment is reached. Pol III can go no further since it has no 5' → 3' exonuclease activity (see Table 7-1). It cannot join a 5'-triphosphate at the terminus of a polymer (i.e., on the primer) to a 3'-OH group on the growing strand. Thus, pol III dissociates from the DNA, leaving a nick. *E. coli* DNA ligase cannot seal the nick because a triphosphate is present, even if an additional enzyme could cleave the triphosphate to a monophosphate. DNA ligase would be inactive when one of the nucleotides is in the ribo form. However, pol I works efficiently at a nick as long as there is a 3'-OH terminus. In this case, the enzyme carries out nick translation, probably proceeding into the deoxy section, but there DNA ligase can compete with pol I and seal the nick (**Figure 7-21**).

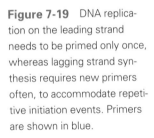

Figure 7-19 DNA replication on the leading strand needs to be primed only once, whereas lagging strand synthesis requires new primers often, to accommodate repetitive initiation events. Primers are shown in blue.

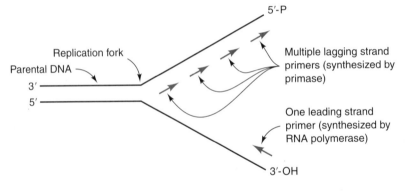

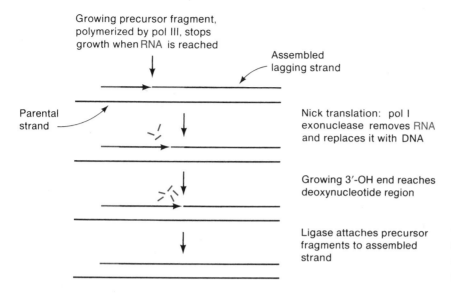

Growing precursor fragment, polymerized by pol III, stops growth when RNA is reached

Assembled lagging strand

Parental strand

Nick translation: pol I exonuclease removes RNA and replaces it with DNA

Growing 3'-OH end reaches deoxynucleotide region

Ligase attaches precursor fragments to assembled strand

Figure 7-20 Sequence of events in assembly of lagging strand fragments. RNA is indicated in blue. The replication fork (not shown) is at the left.

In the usual polymerization reaction, the activation energy for phosphodiester bond formation comes from cleaving the triphosphate. Since DNA ligase has only a monophosphate to work with, it needs another source of energy. It obtains this energy by hydrolyzing either ATP or nicotine adenine dinucleotide (NAD); the energy source depends upon the organism from which the DNA ligase is obtained. The *E. coli* DNA ligase uses NAD.

Thus, the precursor fragment is assimilated into the lagging strand. By this time, the next precursor fragment has reached the RNA primer of the fragment just joined and the sequence begins anew. Another enzyme, RNase H, a riboendonuclease specific for RNA in the form of a RNA-DNA hybrid, can also participate in the removal of primer ribonucleotides.

The Complete DNA Replication System

Let's now begin to understand how all the major components of the replication machinery work in harmony. We'll adopt *E. coli* as our model system since, because of its large number of mutant (regarding DNA replication) genes, it is

KEY CONCEPT

The "Evolution of DNA Replication" Concept

Due to the antiparallel nature of DNA strands, as well as the need for rapid replication, and timely error correction, the discontinuous DNA replication mode has evolved.

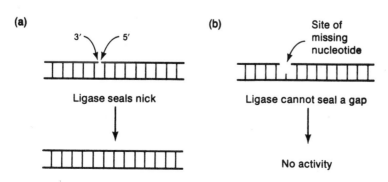

(a)

3'   5'

Ligase seals nick

(b)

Site of missing nucleotide

Ligase cannot seal a gap

No activity

Figure 7-21 The action of DNA ligase. (a) A nick having a 3'-OH and a 5'-P terminus is sealed. (b) If one or more nucleotides are absent, the gap cannot be sealed.

best understood. Similar mechanisms are, however, believed to be active in animal and plant cells.

We will use the diagram in **Figure 7-22** to guide our review. For the sake of focusing on events at the replication fork itself, we will assume that the topoisomerases (e.g., DNA gyrase) described earlier in Figure 7-5 have acted downstream of the replication fork to relieve superhelicity.

Helicase unwinds double-stranded DNA by breaking its hydrogen bonds. Primase, which continuously synthesizes the multiple RNA primers for the lagging strand, complexes with helicase (to form the so-called primosome mentioned earlier). Recall, the single primer for the leading strand is synthesized by a different enzyme, RNA polymerase.

Re-annealing of the separated DNA strands is prevented by the action of ssb proteins. They bind single-stranded DNA and prevent its return to a double-stranded configuration. Due to the opposite polarities of the leading and lagging strands, the lagging strand is looped around the DNA polymerase III enzyme. This

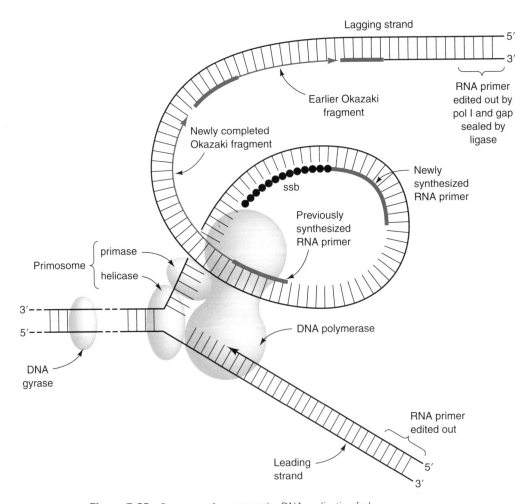

Figure 7-22 Summary of events at the DNA replication fork.

allows one large pol III complex consisting of a pair of subunits to polymerize both leading and lagging strands simultaneously! Replication of both strands is thereby precisely coordinated. The chances of the synthesis of one strand speeding ahead of the other, or the potential for tangling and breaking of single-stranded DNA, are consequently alleviated. Due to the need for a length of template complete with primers for lagging-strand synthesis, synthesis of the lagging strand actually does lag (in time), even though its polymerization occurs simultaneously with that of the leading strand. Thus, the term "lagging" is appropriate for the looped strand.

Clearly, events on the lagging strand are more complicated than on the leading strand, due to the continuous requirement for primers. Once the synthesis of an Okazaki fragment has been completed, it butts up against an earlier synthesized RNA primer/Okazaki fragment combination.

The RNA primers (on both the lagging and leading strands) are edited out, using the nick translation capability of pol I mentioned earlier (see Figure 7-20). The gap between the edited primers (now DNA) and the earlier synthesized Okazaki fragment is sealed by DNA ligase (see Figure 7-21).

The above features pertain mostly to θ replication, described earlier in Figures 7-3 and 7-4. Let us now return briefly to a summary of the covalent extension model of DNA replication.

Replication in the Covalent Extension Mode

This mode employs a nick (single-stranded break in DNA) rather than the use of RNA primers for the initiation process. The nick site provides a free 3′-OH as a polymerization starting point for DNA polymerase. As mentioned previously, many bacteriophages replicate their DNA this way. As DNA polymerase adds nucleotides to that free 3′-OH end and works its way around the circle, the 5′-P end is either displaced (if pol I is employed as a polymerase) or unwound by other enzymes (if pol III is used as the polymerase) (**Figure 7-23**).

That displaced strand serves as the template for lagging-strand synthesis. Eventually, after the circular DNA has been fully replicated once, the newly synthesized leading strand will serve as the template for new lagging-strand synthesis. The inner (non-nicked) strand is never disrupted, hence the alternative names—"rolling circle" or "sigma" (because of its "tail")—for the covalent extension mode of DNA replication. That "inner" circle is employed repeatedly as

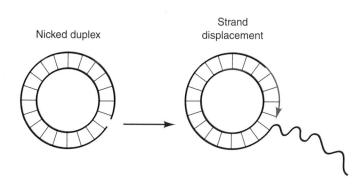

Nicked duplex   Strand displacement

Figure 7-23 Pol I displaces one of the parental strands as it polymerizes nucleotides from the 3′-OH end produced at the nicked site.

a template for leading-strand synthesis, generating a concatemeric length of double-stranded DNA product (see Figure 7-9).

For those phages that employ pol III as the major polymerization enzyme, helicases and ssb proteins operate to unwind the parental strands, instead of the strand displacement process generated by pol I (see Figure 7-23).

Bidirectional Replication Speeds Up DNA Synthesis

Once a D-loop has formed from a *de novo* initiation event (see Figure 7-8), replication will occur in either one direction (**unidirectional**) or in both directions (**bidirectional**) around the loop. During unidirectional replication of a circular DNA, as illustrated in Figure 7-8, one branch point remains at a fixed position, defining the replication origin. In bidirectional replication, both branch points move so that each branch point contains a complete replication fork. If both replication forks move at the same rate, the origin of replication (*ori*) remains close to the midpoint of each branch of the replication loop.

In the illustration of bidirectional replication shown in **Figure 7-24**, the first leftward-moving lagging strand becomes the rightward-moving leading strand, and the second primosome acts to unwind the helix in a clockwise direction and primes the synthesis of the rightward-moving precursor fragments. That is, a primosome (helicase/primase complex) is situated on both sites of the *ori,* so both clockwise and counterclockwise replication occurs around the D-loop. The result of these events is that the DNA molecule will have two replication forks moving in opposite directions around the circle, and the overall rate of duplication of a DNA molecule will be doubled.

Whether a DNA molecule replicates unidirectionally or bidirectionally depends on whether the organism's DNA helicase is capable of moving in opposite

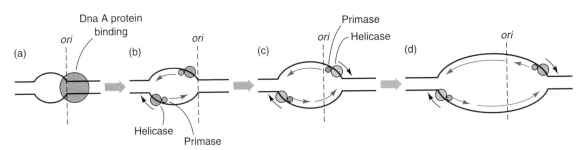

Figure 7-24 The formation of a bidirectionally enlarging replication bubble. (a) DnaA protein binds to the *ori* and opens the double helix, as shown previously in Figure 7-7. (b) Helicase and primase act on both sides of the replication bubble. (c) The leftward-leading strand has progressed far enough that the second rightward precursor (lagging strand) fragment has begun. The first rightward precursor fragment has passed *ori* and has become the rightward-leading strand. (d) The rightward-leading strand has moved far enough that the first leftward (lagging strand) precursor fragment has begun. There are now two complete replication forks. The helicase/primase complex shows how *E. coli* replicates DNA simultaneously in both clockwise and counterclockwise directions. The black arrow illustrates the direction of movement of the helicase (unwinding) protein.

directions away from the *ori*. Some organisms (e.g., some plasmids and a few types of phage) possess helicases that permit solely unidirectional replication. Other organisms (e.g., all eukaryotes, several plasmids, many viruses, and most bacteria) contain helicases that allow for bidirectional replication.

Complex Enough?

Having completed a review of the basic features of DNA replication, let's list the key points:

1. Topoisomerases act to relieve DNA's positive supercoiling.
2. Initiation of synthesis occurs at specific sites (*ori*).
3. Helicase unwinds double-stranded DNA.
4. RNA primers are synthesized so that DNA polymerase can synthesize daughter strands.
5. Lagging-strand precursor fragments are edited and ligated.

By now beginning students of molecular biology can easily understand why many researchers express the opinion that *DNA replication is the most complex process in a living cell!*

Replication of Eukaryotic Chromosomes

The replication of eukaryotic chromosomes presents many problems not found in prokaryotes because of the enormous size of eukaryotic chromosomes and the geometric complexity imposed by the organization of the DNA into nucleosomes. How these problems are handled is described in this section.

The rate of movement of a replication fork in *E. coli* is $\approx 10^5$ base pairs per minute. In eukaryotes, the polymerases are much less active, and the rate ranges from 500 to 5,000 base pairs per minute. Since a typical animal cell contains about 50 times as much DNA as a bacterium, the replication time of an animal cell should be about 1,000 times as great as that of *E. coli,* or about 30 days. However, the duration of the replication cycle is usually several hours, and this is accomplished by the presence of multiple initiation sites. For instance, the DNA of the fruit fly *Drosophila* has about 5,000 initiation sites, each separated by about 30,000 bases, and each site replicates bidirectionally. The number of sites is regulated in a way that is not understood. For example, in the round of replication following fertilization of *Drosophila* eggs, the number of initiation points reaches 50,000, and it takes only 3 minutes to replicate all of the DNA. An example of a fragment of this rapidly replicating *Drosophila* DNA is shown in **Figure 7-25**.

The enormous number of growing forks in eukaryotic cells is reflected in the number of polymerase molecules. In *E. coli* there are between 10 and 20 molecules of pol III holoenzyme. However, a typical animal cell has 20,000 to 60,000 molecules of polymerase α, which is one of the two DNA polymerases thought to be involved in replication.

Replication of double-stranded DNA proceeds through both a polymerization step (nucleotide addition) and a dissociation step (strand separation). The replication of chromatin, which is the form in which DNA exists in eukaryotes, proceeds through an additional dissociation step—namely, dissociation of DNA and

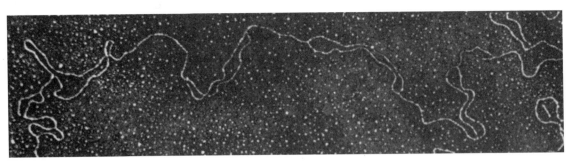

Figure 7-25   Replicating DNA of *Drosophila melanogaster* showing many replicating eyes. The molecular weight of the segment shown is roughly 20×10^6. (Courtesy of David Hogness.)

histone octamers—and a histone-DNA reassociation step (see Chapter 5 for a discussion of the structure of the octamers contained in nucleosomes). In chromatin, DNA is wrapped around a histone octamer to form a nucleosome, and if the DNA were never unwrapped from the histone spool, severe geometric problems would arise at the growing fork. Moreover, after DNA dissociates from the histones, newly formed DNA must rejoin with the nucleosomal octamers so that each daughter molecule can be organized into nucleosomes, just as the parent was.

Examination of replication forks in DNA that has not been deproteinized during isolation indicates that nucleosomes form very rapidly after replication. For example, **Figure 7-26** shows that all portions of a replication eye have the bead-like appearance characteristic of nucleosomes.

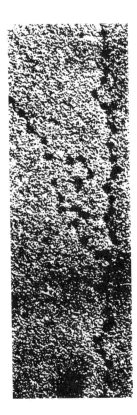

Figure 7-26   A replicating fork showing nucleosomes on both branches. The diameter of each particle is about 110 Å. (Courtesy of Harold Weintraub.)

The synthesis of histones occurs simultaneously with DNA replication—that is, histones are made in the cell as they are needed so that the cell does not contain an appreciable amount of unassociated histone molecules. In light of this, we would like to know whether newly synthesized histones mix with parent histones in the octamers associated with daughter DNA molecules.

Unfortunately, even after many years of investigation, the answer is not clear. In addition, the answer to the other important question about the behavior of histones during replication is unclear—namely, how are the histone octamers resident on the parent duplex distributed on the daughter molecules? (Do they all go to only one daughter molecule, or are they dispersed between both daughter molecules?) Sufficient experimental evidence exists to support either claim. Very recently, histone assembly has been experimentally coupled to replication in vitro. Studies with this type of system should provide definitive answers to more questions.

Enzymology of Human DNA Replication

In prokaryotes, in vitro studies on the replication of infecting phages were the key to unlocking the secrets of the enzymological mechanisms and the identification and purification of the cellular replication proteins. Recently, a similar approach has been used to identify human replication proteins. In particular, these studies have exploited the ability of both adenovirus and the previously mentioned SV40 virus to replicate in human cell cultures.

Adenovirus is a long (about 30,000 nucleotides) linear virus that replicates from the ends. The obvious priming problem (if it replicates from the very last nucleotide, how is a 3'-OH primer end provided?) is solved because the primer is actually a virally encoded protein. The first nucleotide inserted is covalently attached to the primer protein. This protein (called the terminal protein) remains attached to the newly synthesized viral DNA and is encapsulated into the viral particle.

Studies on SV40 replication have revealed two different polymerases, each apparently involved in the synthesis of only one strand. Polymerase α synthesizes part of the lagging strand, while polymerase δ synthesizes the rest of the lagging strand, as well as the entire leading strand. Most eukaryotic α polymerases also have an associated primase activity—thus, its involvement in lagging-strand synthesis, which calls for the repeated generation of primers, seems appropriate. It is interesting to note the functional similarities between development of the eukaryotic and prokaryotic replication forks. One can think of the combination of polymerases α and δ, along with several other proteins that bind to these polymerases, as being equivalent to the prokaryotic polymerase-III holoenzyme. In the next few years, our knowledge of the enzymology of eukaryotic DNA replication should begin to rival that of the prokaryotic system.

A Future Practical Application?

A detailed analysis of the three-dimensional structure of a variety of nucleotide polymerases, including the DNA polymerase described in Figure 7-22 and the reverse transcriptase enzyme that replicates the AIDS virus (human immunodeficiency virus [HIV]—see Chapter 16) has been carried out. Those studies reveal that all polymerases share a common catalytic mechanism. For example, due to similarities in their protein-folding patterns, they all bind their templates and primers in the same way.

Information yielded from such studies is being employed to design anti-HIV drugs. For example, when AZT (3′-azido-2′,3′-dideoxythymidine) is administered to an AIDS patient, it is converted by cellular enzymes to the nucleoside 5′-triphosphate form. That "AZT-triphosphate" is a potent chain-termination inhibitor of reverse transcriptase. It functions much the same way that dideoxyATP functions in the Sanger DNA sequencing reaction (Figure 3-18 in Chapter 3). Thus, it is a potential anti-AIDS drug.

AZT-triphosphate does, however, have toxic side effects. It is believed to inhibit host-cell DNA polymerases in most AIDS patients. Further understanding of the mechanism of action of DNA polymerase, gained through intensive research on the structure of the enzyme, is, therefore, likely to lead to the design of more potent and less toxic drugs that will inhibit reverse transcriptase, but not interfere with human DNA polymerase. Thus, a practical application of basic information on DNA polymerase action will possibly be achieved.

SUMMARY

DNA replication is accomplished by an enzymatic polymerization reaction in which a DNA strand is used as a template to synthesize a copy with a base sequence that is complementary to the template. Replication is semiconservative; each parent single strand is present in one of the double-stranded progeny molecules. The growth of a DNA strand is catalyzed by a DNA polymerase enzyme. Strand growth always moves in the 5′-to-3′ direction. Since double-stranded DNA is antiparallel, only one strand (the leading strand) grows in the direction of movement of the replication fork. The second daughter strand (the lagging strand) is synthesized in the opposite direction as short, precursor fragments that are subsequently joined together. DNA polymerase cannot initiate synthesis, so a primer is always needed. In most replication systems, the primer is a short RNA oligo-nucleotide synthesized either by RNA polymerase or by DNA primase.

In *E. coli,* two enzymes, DNA polymerase I and DNA polymerase III, carry out nucleotide addition. Pol III is responsible for overall chain growth; pol I is used in the completion of precursor fragments and also in repair processes. The primer is removed at later stages of replication by the 5′-3′ exonuclease activity of DNA polymerase I. Pol I and pol III both possess a 3′-5′ exonuclease activity that corrects incorporation errors.

Replication of both linear and circular molecules is usually bidirectional; the replication loops enlarge by movement of replication forks at both ends of the loop. A DNA molecule of a prokaryote is usually initiated at a single site, and has a single replication loop; in contrast, eukaryotic DNA molecules have multiple initiation sites, and, hence, many loops.

DRILL QUESTIONS

1. Is DNA replication conservative or semiconservative?
2. Fill in the blank. A parent strand serves as a _____ for synthesis of a daughter strand.

3. In semiconservative replication, what fraction of the DNA consists of one original parent strand and one daughter strand after 1, 2, and 3 rounds of replication?

4. In which mode of replication does a parent circular DNA molecule yield two daughter circles?

5. In which mode of replication does a parent circle generate a circle with a linear branch?

6. What kind of supercoiling is produced by replicative movement of the growing fork in a circle?

7. What kind of supercoiling is produced by DNA gyrase?

8. Name three enzymatic activities of DNA polymerase I.

9. DNA polymerization occurs by addition of a deoxynucleotide to which chemical group?

10. What two reactions are coupled when nick translation occurs?

PROBLEMS

1. Will a ^{15}N-labeled circle replicating in ^{14}N medium using the rolling circle mode ever achieve the density of $^{14}N^{14}N$ DNA?

2. What are the roles of the various exonuclease activities of the DNA polymerases in DNA replication?

3. How do pol I and pol III differ with respect to their ability to unwind the parent DNA in a replication fork?

4. What is the chemical difference between the groups joined by a DNA polymerase and by DNA ligase?

5. How do organisms allow DNA polymerases to move in the same direction along a template strand, even though double-stranded DNA is antiparallel?

6. What must be done to two precursor fragments before they can be joined together?

CONCEPTUAL QUESTIONS

1. Why isn't there a single mode of replication common to all organisms?

2. What specific constraint to DNA replication occurs at the ends of linear DNA molecules? How does replication occur at the ends of linear DNA molecules?

3. Why has DNA replication been called "the most complex process in a cell"?

in this chapter you will learn

1. The sequence of events that take place in transcription of DNA to RNA, and how those events differ in prokaryotic and eukaryotic systems.

2. About the structures and functions of different classes of RNA.

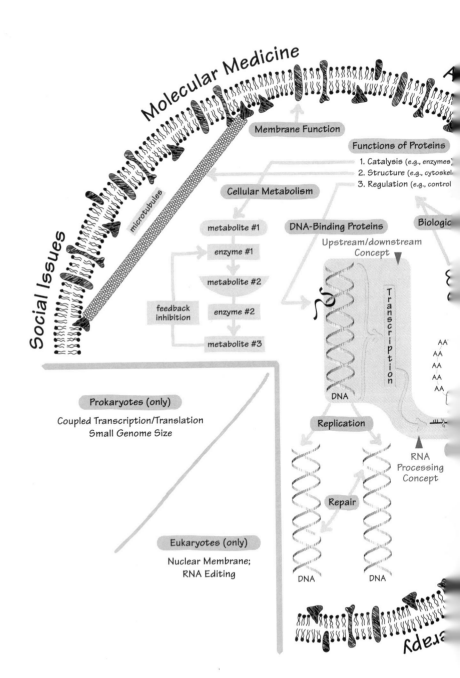

Molecular Medicine

Membrane Function

Functions of Proteins
1. Catalysis (e.g., enzymes)
2. Structure (e.g., cytoskel.
3. Regulation (e.g., control

microtubules

Cellular Metabolism

DNA-Binding Proteins

Biologic

Social Issues

metabolite #1

Upstream/downstream
Concept

enzyme #1

metabolite #2

feedback
inhibition

enzyme #2

Transcription

metabolite #3

AA
AA
AA
AA
AA

DNA

Prokaryotes (only)

Coupled Transcription/Translation
Small Genome Size

Replication

RNA
Processing
Concept

Repair

Eukaryotes (only)

Nuclear Membrane;
RNA Editing

DNA   DNA

erapy

Transcription

Gene expression is accomplished by the transfer of genetic information from DNA to RNA molecules and then from RNA to protein molecules. RNA molecules are synthesized by using the base sequence of one strand of DNA as a template in a polymerization reaction that is catalyzed by enzymes called DNA-dependent RNA polymerases, or simply **RNA polymerases.** The process by which RNA molecules are initiated, elongated, and terminated is called **transcription.**

In this chapter we will quickly realize that, although complementary base pairing is common to both RNA and DNA synthesis, several features of RNA synthesis differ from DNA synthesis. For example, the RNA product is single stranded, unlike the double helix from which it is transcribed. It is also relatively short in length, since only a small portion of the total DNA is represented in any single RNA product.

Initially, we will learn about transcription in prokaryotes. Then we will review the more complicated eukaryotic gene expression system. Because of their relative simplicity, the pioneering experiments on transcription were carried out in prokaryotes. In fact, it was a series of elegant genetic experiments with *E. coli* that led to the discovery of what is now known as messenger RNA (mRNA). Through mRNA, the information stored in DNA is converted to amino acid sequences of the ultimate gene product—protein.

Three aspects of prokaryotic transcription will be considered in this chapter: the enzymology of RNA synthesis, the signals that determine at which locations on a DNA molecule transcription starts and stops, and the types of transcription products and how they are converted to the RNA molecules needed by the cell. Then we will examine eukaryotic transcription, and finish the chapter with a review of a laboratory procedure for studying RNA.

Enzymatic Synthesis of RNA

In this section we describe the basic features of the polymerization of RNA, the identity of the precursors, the nature of the template, the properties of the polymerizing enzyme, and the mechanisms of initiation, elongation, and termination of synthesis of an RNA chain.

The essential chemical characteristics of the synthesis of RNA are the following:

1. The precursors in the synthesis of RNA are the four ribonucleoside 5′-triphosphates (rNTP)—ATP, GTP, CTP, and UTP. On the ribose portion of each NTP, there are two OH groups, one each on the 2′- and 3′-carbon atoms (see Figure 2-5 in Chapter 2).

2. In the polymerization reaction, the 3'-OH terminus of the RNA being synthesized reacts with the 5'-triphosphate of a precursor rNTP; a pyrophosphate is released, and a phosphodiester bond results (**Figure 8-1**). This is the same reaction that occurs in the synthesis of DNA.

3. The sequence of bases in an RNA molecule is determined by the base sequence of the DNA. Each base added to the growing end of the RNA chain is chosen by its ability to base pair with the DNA strand used as a template; thus, the bases, C, T, G, and A in a DNA strand cause G, A, C, and U, respectively, to appear in the newly synthesized RNA molecule.

4. The DNA molecule is double stranded, yet at the time any particular region is transcribed only one strand serves as a template. This comes about because the two DNA strands are separated temporarily to allow one to be used as a template. The meaning of this statement is shown in **Figure 8-2**.

5. The RNA chain grows in the 5' → 3' direction: that is, nucleotides are added only to the 3'-OH end of the growing chain—this is the same as the direction of chain growth in DNA synthesis. Also, the RNA strand and the DNA template strand are antiparallel to one another.

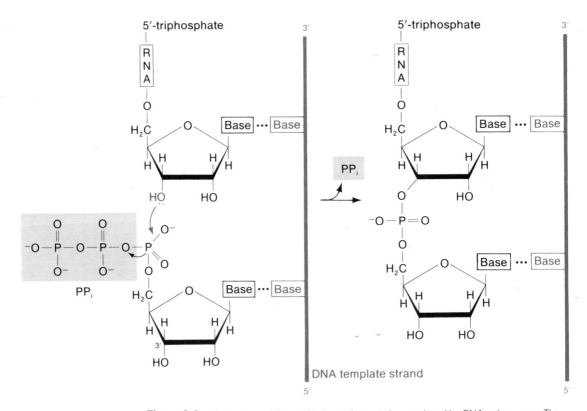

Figure 8-1 Mechanism of the chain-elongation reaction catalyzed by RNA polymerase. The blue arrow joins the reacting groups. The pyrophosphate group (shaded in blue) and the blue hydrogen atom do not appear in the RNA strand. The DNA template and the RNA strands are antiparallel, as in double-stranded DNA.

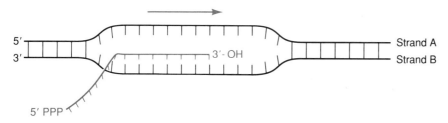

Figure 8-2 An RNA strand (shown in blue) is copied only from strand B of a segment of a DNA molecule. No RNA is copied from strand A in that region of the DNA molecule. However, elsewhere, for example, in a different gene, strand A might be copied; in that case, strand B would not be copied in that region of the DNA. The RNA molecule is antiparallel to the DNA strand being copied and is terminated by a 5'-*tri*phosphate at the nongrowing end. The blue arrow shows the direction of RNA chain growth.

6. RNA polymerases, in contrast to DNA polymerases, are able to initiate chain growth directly—that is, no primer is needed.

7. Only ribonucleoside 5'-triphosphates participate in RNA synthesis and the first base to be laid down in the initiation event retains its triphosphate. Its 3'-OH group is the point of attachment for the subsequent nucleotide. Thus, the 5' end of a growing RNA molecule terminates with a triphosphate (Figure 8-2).

The overall polymerization reaction may be written as

$$n\text{NTP} + \text{XTP} \xrightarrow[\text{Mg}^{2+}]{\text{DNA, RNA-P}} \text{XTP} - (\text{NMP})_n + n\text{PP}_i$$

in which XTP represents the first nucleotide at the 5' terminus of the RNA chain, NMP is a mononucleotide in the RNA chain, RNA-P is RNA polymerase, and PP_i is the pyrophosphate released each time a nucleotide is added to the growing chain. The Mg^{2+} ion is required for all nucleic acid polymerization reactions.

E. coli RNA polymerase consists of five subunits—two identical α subunits and one each of types β, β', and σ. The σ subunit dissociates from the enzyme easily and, in fact, does so shortly after polymerization is initiated. The term **holoenzyme** is used to describe the complete enzyme and **core enzyme** for the enzyme that lacks the σ subunit. We use the name RNA polymerase when the holoenzyme is meant. RNA polymerase is one of the largest enzymes known and can easily be seen by electron microscopy (**Figure 8-3**).

The synthesis of RNA consists of four discrete stages: (1) binding of RNA polymerase to a template at a specific site, (2) initiation, (3) chain elongation, and (4) chain termination and release. A discussion of these stages follows.

Transcription Signals

The first step in transcription is binding of RNA polymerase to a DNA molecule. Binding occurs at particular sites called **promoters,** which are specific sequences of about 40 base pairs at which several interactions occur. The most crucial interactions for positioning of *E. coli* RNA polymerase at a promoter occur at two short-

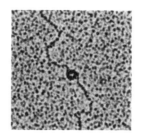

Figure 8-3 *E. coli* RNA polymerase molecule bound to DNA. (× 160,000). (Courtesy of Robley Williams.)

sequence patches on the DNA. These patches are located about 10 and 35 base pairs before the first base, which is copied into RNA. That first base copied is called the start point of transcription, and is assigned as position +1. The two patches, therefore, are at positions −10 and −35, with respect to the start point. The σ subunit of RNA polymerase appears to have two separate α-helical segments with amino acid residues that can make specific contacts with the exposed portions of the base pairs in the major groove of the DNA. The enzyme becomes positioned correctly when these separate α-helical segments contact the base pairs in the two patches simultaneously. Since a DNA molecule makes one helical turn every 10 to 11 base pairs, these contacts would be nearly on the same side of DNA two turns away, or 78 Å along the DNA. **Figure 8-4** shows a diagram of how an extended RNA polymerase could contact those two patches. *E. coli* RNA polymerase is a large protein that covers as much as 70 to 75 base pairs of DNA from position −55 to position +20 when it is bound to a promoter. It can, therefore, easily make the two separate contacts at the same time.

The sequences of many promoters have been identified by researchers. **Figure 8-5** shows portions of the sequences along one strand for a few *E. coli* promoters, with the start points and the −10 and −35 regions shown in red. The strand that is shown has its 5′ end at the left. One striking feature is that no two sequences are identical or even closely similar, except in the two key locations. The sequences shown in blue from −7 to −12 are similar, as are the sequences from about −30 to about −35. The sequences around −10 are considered to be variants of a basic sequence, TATAAT, called the −10 consensus sequence, while the sequence around −35 is a variant of TTGACA, called the −35 consensus sequence. A consensus sequence is a pattern of bases from which actual sequences observed in many different systems differ by usually no more than one or two bases. Although the figure only shows the sequences in one strand, remember that the DNA is double stranded, and that the other complementary strand is part of the recognition site in the −10 and −35 regions.

With some promoters, the sequence at one of the two contact regions differs greatly from the consensus. In this case, RNA polymerase recognizes the sequence very poorly, and the promoter does not function very well. For some

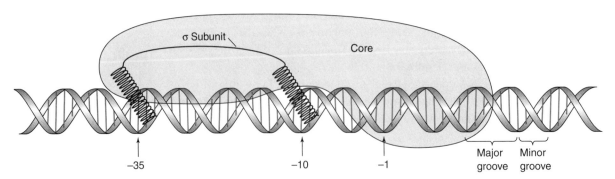

Figure 8-4 A diagram showing RNA polymerase bound to DNA at a promoter. The arrows indicate the position of the transcription start point (+1) and the −10 and −35 regions. The sigma (σ) factor is shown with two cylinders representing α-helical segments that make contact with the DNA in its major groove at the −10 and −35 regions.

```
CCAGGCTTTACACTTTATGCTTCCGGCTCGTATGTTGTGTGGAATTG
CTTTTTGATGCAATTCGCTTTGCTTCTGACTATAAATAGACAGGGTAA
GGCGGTGTTGACATAAATACCACTGGCGGTGATACTGAGCACATCAG
GTGCGTGTTGACTATTTTACCTCTGGCGGTGATAATGGTTGCATGTA
ATTGTTGTTGTTAACTTGTTTATTGCAGCTTATAATGGTTACAAATA
CGTAACACTTTACAGCGGCGCGTCATTTGATATGATGCGCCCCGCTT
```

      –35 Sequence   –10 Sequence   mRNA
                                                             start

Figure 8-5 Base sequences in the nontemplate strand of six different *E. coli* promoters showing the three important regions.

genes, that might be fine because the cell may not need to express that gene very often. For other genes, however, there is often an accessory protein that helps RNA polymerase bind to the promoter. This kind of accessory protein is called a **gene activator protein** and is basically a "positive effector." For example, the λ p_{re} promoter is active only when the λ c11 protein is present. These gene-activator proteins usually bind to specific sequences very near or even within the promoter sequence and they appear to have surfaces to which RNA polymerase can attach, when correctly positioned on the promoter. Thus, a gene-activator protein helps the RNA polymerase bind to a promoter with a poor recognition signal. An important effector protein, which will be discussed further in Chapter 11, is the CAP protein; this protein is needed for the activation of many promoters for genes required for sugar metabolism, and regulation of the binding ability of CAP is a major means of regulating the expression of these genes.

After RNA polymerase binds tightly at the promoter region, it manages to unwind a small part of the DNA, from base pairs +3 to −10, to form the **open-promoter complex.** This unwinding is necessary for pairing of the incoming ribonucleoside triphosphates, the building blocks for synthesis of the RNA. The base composition between the start site through the −10 region is generally rich in A + T, which renders the DNA especially susceptible to unpairing.

Once an open-promoter complex has formed, RNA polymerase is ready to initiate synthesis. RNA polymerase contains two nucleotide binding sites, called the initiation site and the elongation site, respectively. The initiation site prefers to bind purine nucleoside triphosphates, namely ATP and GTP, and one of these is usually the first nucleotide in the chain. Therefore, the first base on the DNA template that is transcribed is usually a thymine or a cytosine. The initiating nucleoside triphosphate binds to the enzyme in the open-promoter complex and forms a hydrogen bond with the complementary DNA base (**Figure 8-6**). The elongation site is then filled with a nucleoside triphosphate that is selected strictly by its ability to form a hydrogen bond with the next base in the DNA strand. The two nucleotides are then joined together by the reaction shown in Figure 8-1. During the joining process, RNA polymerase moves along the DNA to the next base pair. As it moves, it separates the base pair at position +4, and allows the base pair at position −10 to come together again, thus maintaining the same stretch of 13 base pairs unwound, but now centered one base pair further along the DNA. The two RNA nucleotides that were linked together stay paired to their complementary DNA bases, but now a new unpaired base (the one at position +3) is positioned in the elongation site. Again, the elongation site is filled with a nucleoside triphosphate that is selected by its ability to form a hydrogen bond with the DNA base at position 3. This nucleotide

RNA polymerase binds to promoter, slides into place, and forms an open complex. ATP in initiation site binds to T on coding strand.

A NTP is added to the elongation site and is covalently linked to the A.

RNA polymerase moves over to the next DNA base. A NTP enters the elongation site and is covalently linked to the dinucleotide. Then movement of RNA polymerase continues.

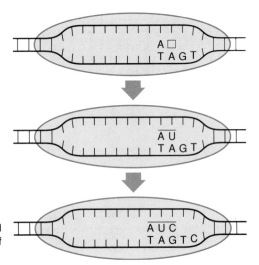

Figure 8-6 A scheme for initiation of RNA synthesis.

is then joined to the two that had been linked previously, and RNA polymerase moves along to the next base pair. Repetition of this cycle of nucleoside triphosphate selection and joining, RNA polymerase motion, DNA base unpairing, and DNA base reannealing gradually elongates the RNA chain.

Note that the selection is made by pairing of the nucleoside triphosphate with only one of the two strands of the DNA. The DNA strands were purposely unwound to allow this selection to occur. The strand that has bases that pair with the incoming RNA nucleotide is the **template strand.** Since the sequence of the RNA is complementary to the template strand, it will have the same sequence as the other strand of DNA.

After several nucleotides (between four and eight) are added to the growing RNA chain, three important changes occur. First, the nucleotide at the 5′ end of the RNA chain (the first one to be incorporated) becomes unpaired from the base on the template strand to allow that base to pair again with its complement on the other DNA strand. Second, RNA polymerase changes its structure and loses the σ subunit. Third, a protein aiding elongation (called NusA protein) binds. The rest of the elongation process is carried out by the core enzyme–NusA complex (**Figure 8-7**). Each new nucleotide that is added at the 3′ end of the growing RNA allows another nucleotide to be released from its binding to the template. Hence, as elongation continues, a section of 4 to 8 nucleotides at the 3′ end of the newly made RNA is paired to the DNA template strand, while the rest of the RNA emerges from RNA polymerase as a single strand.

Termination of RNA synthesis occurs at specific base sequences in the DNA molecule, called **terminators.** Some termination sequences allow RNA polymerase to terminate elongation spontaneously. These sequences are called intrinsic terminators. Other terminators require the action of a protein called Rho; they are called rho-dependent terminators. Intrinsic terminators have three characteristic features (**Figure 8-8**):

1. First, there is an inverted-repeat base sequence containing a central non-repeating segment; that is, the sequence in one DNA strand would read

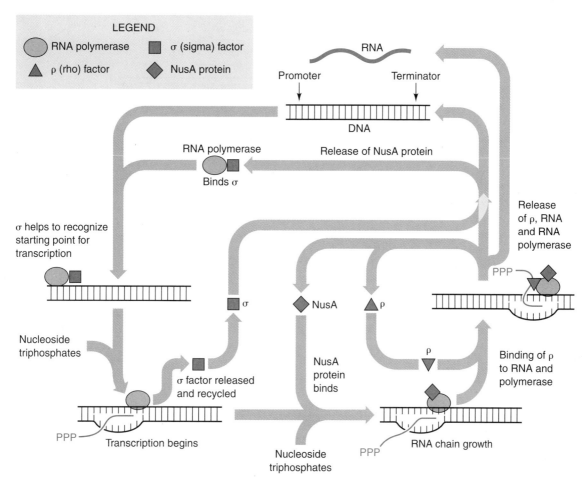

Figure 8-7 Stages in RNA synthesis. Synthesis starts at a promoter on DNA and ends at a terminator. Sigma (σ) factor finds the promoter sequence. NusA protein is an elongation factor, and rho (ρ) is a termination-RNA release factor. (Adapted from Watson et al., *Molecular Biology of the Gene,* Fourth Edition. Menlo Park, CA: Benjamin-Cummings, 1987.)

ABCDEF-XYZ-F′E′D′C′B′A′ in which A and A′, B and B′, and so on, are complementary bases. This sequence is capable of intrastrand base pairing, forming a stem-and-loop configuration in the RNA transcript and, possibly, in the DNA strands.

2. The second region is near the loop end of the putative stem (sometimes totally within the stem) and is a sequence having a high G + C content.

3. A third region is a sequence of A · T pairs (which may begin in the putative stem) that yields in the RNA a sequence of six to eight uracil residues often followed by an adenine.

Rho-dependent terminators lack sequences with a stretch of adenine residues in the template strand. Rho acts by binding to a special sequence on the nascent

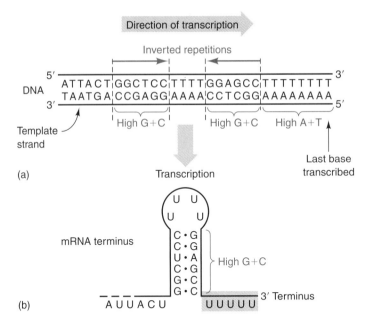

Figure 8-8 Base sequence of (a) the DNA of the *E. coli trp* operon, at which transcription termination occurs, and (b) the 3' terminus of the mRNA molecule. The inverted repeat sequence is indicated by reversed blue arrows. The mRNA molecule is folded to form a stem-and-loop structure thought to exist. The relevant regions are labeled in blue; the terminal sequence of U's in the mRNA is shaded in blue.

RNA and forcibly pulling the RNA away from its contact with the DNA in the transcription-elongation complex with RNA polymerase. Rho derives its force from the hydrolysis of ATP molecules.

The final step in the termination process is dissociation of RNA polymerase from the DNA. This comes about by the rebinding of a σ subunit to the core RNA polymerase after the RNA is released. The released holoenzyme is free to search for a new promoter where it can start synthesis of a new RNA (Figure 8-7).

Classes of RNA Molecules

There are three major classes of RNA molecules—messenger RNA (mRNA), ribosomal RNA (rRNA), and transfer RNA (tRNA). All are synthesized from DNA base sequences. They are all involved in protein synthesis, but each class of RNA has a different function, as will be seen in this and the next chapter. There are also significant differences between the structures and modes of synthesis of the RNA molecules of prokaryotes and eukaryotes, although the basic mechanisms of their functions are nearly the same. The greatest amount of information has been obtained from studies with bacteria and bacterial cell extracts, so that it is here that we begin. Transcription in eukaryotes and the structure and synthesis of eukaryotic RNA molecules are discussed in a later section.

Messenger RNA

The base sequence of a DNA molecule determines the amino acid sequence of every polypeptide chain in a cell, although amino acids have no special affinity for DNA. Thus, instead of a direct pairing between amino acids and DNA, a multistep process is used and the information contained in the DNA is converted to a form in which amino acids can be arranged in an order determined by the DNA base sequence. This process begins with the transcription of the base sequence of one of the DNA strands (the **template strand**) into the base sequence of an RNA molecule, and it is from this molecule, messenger RNA, that the amino acid sequence is obtained by the protein-synthesizing machinery of the cell. As we will see in Chapter 9, the base sequence of the mRNA is read in groups of three bases (a group of three is called a **codon**), from a start codon to a stop point, with each codon corresponding either to one amino acid or a stop signal.

codon

A DNA segment corresponding to one polypeptide chain plus the start-and-stop signals is called a **cistron,** and an mRNA encoding a single polypeptide is called monocistronic mRNA. It is very common for a bacterial mRNA to encode several different polypeptide chains; in this case, it is called a **polycistronic mRNA** molecule.

cistron

polycistronic mRNA

In addition to cistrons and start-and-stop sequences for translation, other regions in mRNA are significant. For example, translation of an mRNA molecule (that is, protein synthesis) seldom starts exactly at one end of the mRNA and proceeds to the other end; instead, initiation of synthesis of the first polypeptide chain of a polycistronic mRNA may begin hundreds of nucleotides from the 5′ terminus of the RNA. The section of nontranslated RNA before the coding regions is called a **leader sequence;** in some cases, the leader contains a regulatory region (called an **attenuator**) that determines the amount of expression of the gene. Untranslated sequences are found at both the 5′ and the 3′ termini:

leader sequence
attenuator

$$\text{P-5}'\underset{\text{protein coding sequence of mRNA}}{\rule{7cm}{0.4pt}}\text{3}'\text{-OH}$$

A polycistronic mRNA molecule may contain intercistronic sequences (**spacers**) (see Figure 9-6 in Chapter 9) hundreds of bases long.

spacers

An important characteristic of prokaryotic mRNA is its short lifetime; usually within a few minutes after an mRNA molecule is synthesized, nuclease degradation occurs. Although this means that continuous synthesis of a particular protein requires ongoing synthesis of the corresponding mRNA molecule, rapid degradation of mRNA is nonetheless advantageous to bacteria, whose environment and needs often fluctuate widely. Synthesis of a necessary protein can be regulated as needed simply by controlling transcription. When a particular protein is needed by a cell, the appropriate mRNA molecule will be made. However, when the protein is no longer needed, inhibition of synthesis of the mRNA is sufficient to prevent wasted synthesis because the previously made mRNA molecules will soon all be degraded.

Stable RNA: Ribosomal RNA and Transfer RNA

During the synthesis of proteins, genetic information is supplied by messenger RNA. RNA also plays other roles in protein synthesis. For example, proteins are synthesized on the surface of an RNA-containing particle called a **ribosome;** these particles consist of three classes of **ribosomal RNA (rRNA),** which are stable mol-

ribosome
ribosomal RNA (rRNA)

transfer RNA (tRNA)

ecules. Also, amino acids do not line up against the mRNA template independently during protein synthesis, but are aligned by means of a set of about 50 adaptor RNA molecules called **transfer RNA (tRNA),** also a stable species. Each tRNA molecule is capable of "reading" three adjacent mRNA bases (a codon) and placing the corresponding amino acid at a site on the ribosome at which a peptide bond is formed with an adjacent amino acid. Neither rRNA nor tRNA is used as a template. The roles of ribosomes and of tRNA molecules will be explained in the following chapter; here we are concerned only with their synthesis, which involves transcription from DNA and some transcriptional modification.

The synthesis of both rRNA and tRNA molecules is initiated at a promoter and completed at a terminator sequence, and in this respect their synthesis is no different from that of mRNA. However, the following three properties of these molecules indicate that neither rRNA nor tRNA molecules are the **primary transcripts** (immediate products of transcription):

1. The molecules are terminated by a 5′-monophosphate rather than the expected triphosphate found at the ends of all primary transcripts.
2. Both rRNA and tRNA molecules are much smaller than the primary transcripts (the transcription units).
3. All tRNA molecules contain bases other than A, G, C, and U, and these "unusual" bases (as they are called) are not present in the original transcript.

All of these molecular changes are made after transcription by a process called **posttranscriptional modification** or, more commonly, **RNA processing.**

Both rRNA and tRNA molecules are excised from large primary transcripts. Often a single transcript contains the sequences for several different molecules—for example, different tRNA molecules, or both tRNA and rRNA molecules. Formation of rRNA molecules is a result of fairly straightforward excision of a single continuous sequence. However, formation of tRNA molecules from primary transcripts requires not only cleavage by several enzymes acting in a particular order, but also chemical modification of various bases. An example of the production of a particular tRNA molecule is given in **Figure 8-9** to illustrate the complexity of the process.

Transcription in Eukaryotes

Most of the basic features of the transcription and the structure of mRNA in eukaryotes are similar to those in bacteria. However, there are five notable differences:

1. Eukaryotic cells contain three classes of nuclear RNA polymerase and these are responsible for the synthesis of different classes of RNA.
2. Many mRNA molecules are very long lived.

cap

3. Both the 5′ and 3′ termini are modified; a complex structure called the **cap** is found at the 5′ end and a long (up to 250 nucleotides) sequence of polyadenylic acid, poly (A), is found at the 3′ end.
4. The mRNA molecule that is used as a template for protein synthesis is usually about one-tenth the size of the primary transcript. During mRNA processing, intervening sequences called **introns** are excised and the fragments are rejoined.

introns

5. All eukaryotic mRNA molecules are monocistronic.

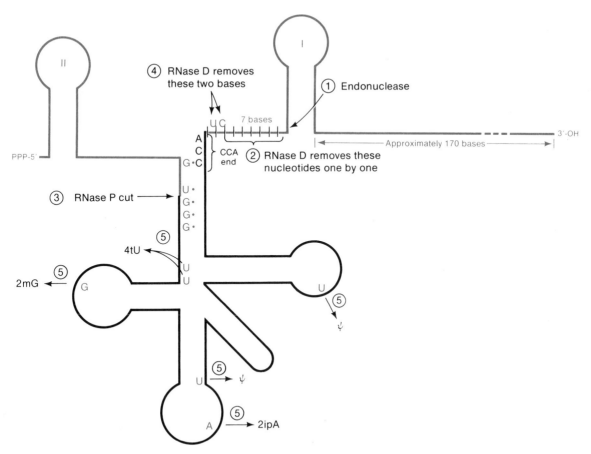

Figure 8-9 The stages in processing of the *E. coli* tRNA^{Tyr} gene transcript. The five stages are given arabic numbers. Step 3 generates the 5′-P end. This step is catalyzed by RNase P, which is a ribozyme, an unusual RNA enzyme (see Chapter 3). Step 4 generates the 3′-OH end (the CCA end). In step 5, six bases, all in or near the loops of the tRNA molecule, are modified to form pseudouridine (Ψ), 2-isopentenyladenosine (2ipA), 2-*o*-methylguanosine (2mG), and 4-thiouridine (4tU). The continuous sequence that forms the final tRNA molecule is given in black.

These points are illustrated in **Figure 8-10**, which shows a schematic diagram of a typical eukaryotic mRNA molecule and how it is produced. There is clearly more to the production of mRNA in eukaryotes than just transcription of a DNA!

Initiation of transcription in eukaryotes is not well understood. A sequence, TATAAAA, analogous to the −10 consensus sequence in prokaryotes, but centered at −29, is a part of many promoters. Other sequences called upstream activation sites and enhancers, which will be discussed in Chapter 12, affect the efficiency of initiation. These are usually sites where certain proteins bind. These proteins, called *transcription factors,* generally increase the transcription from promoters that are within the vicinity of the sequences to which they bind. In some

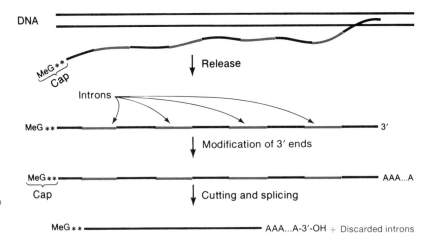

Figure 8-10 Schematic drawing showing production of eukaryotic mRNA. The primary transcript is capped before it is released. Then, its 3′-OH end is modified, and finally the intervening regions are excised. MeG denotes 7-methylguanosine and the two asterisks indicate the two nucleotides whose riboses are methylated.

cases, promoters as far as 2,000 and 3,000 bases away from an enhancer site (along the DNA) (see Chapter 12) are activated.

In eukaryotic cells, DNA is bound tightly to histone proteins in a structure called chromatin. The structure of chromatin (see Figure 5-6) makes transcription more complicated in eukaryotes than in prokaryotes, and evidence exists suggesting that chromatin structure is temporarily reconfigured in regions being transcribed, although how it is altered is far from clear. There is also a mechanism for terminating transcription in eukaryotic cells, but very little is known about the general properties of the sequences or the factors involved.

Three Types of RNA Polymerases Function in Eukaryotes

The three major classes of eukaryotic RNA polymerases are denoted I, II, and III; they can be distinguished by the ions required for their activity, their optimal ionic strength, and their sensitivity to various inhibitory compounds. All are found in the eukaryotic nucleus. RNA polymerases are also found in mitochondria and chloroplasts. The locations and products of the nuclear RNA polymerases are listed in **Table 8-1**. Note that RNA polymerase II is the enzyme responsible for all mRNA synthesis.

The biochemical reaction catalyzed by the eukaryotic RNA polymerases is the same as that catalyzed by *E. coli* RNA polymerase.

The 5′ terminus of a eukaryotic mRNA molecule carries a methylated guanosine derivative, 7-methylguanosine (7-MeG), in an unusual 5′-5′ linkage to the

TABLE 8-1	LOCATIONS AND PRODUCTS OF NUCLEAR RNA POLYMERASES		
	Class I	Class II	Class III
Location	Nucleolus	Nucleoplasm	Nucleoplasm
Product	rRNA	mRNA	tRNA, 5S RNA

5′-terminal nucleotide of the primary transcript (see Figure 9-17 in Chapter 9). Occasionally, the sugars of the adjacent nucleotides are also methylated. The unit

(7-MeG)-5′-PPP-5′-(G or A, with possibly methylated ribose)-3′-P-

in which P and PPP refer to mono- and triphosphate groups, respectively, is called a **cap** (its chemical structure is presented in the next chapter).

Capping occurs shortly after initiation of synthesis of the mRNA, possibly before RNA polymerase II leaves the initiation site, and precedes all excision and splicing events. The biological significance of capping has not yet been unambiguously established, but it is believed that it is required for efficient protein synthesis. Capping may function to protect the mRNA from degradation by nucleases and to provide a feature for recognition by the protein-synthesizing machinery.

Most, but not all, animal mRNA molecules are terminated at the 3′ end with a poly(A) tract, which is added to the primary mRNA by a nuclear enzyme, poly(A) polymerase. The adenylate residues are not added to the 3′ terminus of the primary transcript. Transcription normally passes the site of addition of poly(A), so that an endonucleolytic cleavage must occur before the poly(A) is added. A base sequence—AAUAAA—10 to 25 bases upstream from the poly(A) site, is a component of the system for recognizing the site of poly(A) addition. Interestingly, some primary transcripts contain two or more sites at which poly(A) can be added (see Figure 12-5(b) in Chapter 12). The differentially terminated mRNA molecules usually play different roles in the life cycle of the particular organism. The length of the poly(A) segment can be from 50 to 250 nucleotides. The significance of the poly(A) terminus is unknown at present, but it is believed that it increases the stability of mRNA; a possible role in enhancement of mRNA translation has also been proposed. Some cellular mRNA molecules lack poly(A), so its presence is not obligatory for successful translation.

Eukaryotic Primary Transcripts Contain Intervening Sequences (Introns)

Most of the primary transcripts of higher eukaryotes contain untranslated intervening sequences (**introns**) that interrupt the coding sequence and are excised in the conversion of the primary transcript to mRNA (Figure 8-10 and **Figure 8-11**). The amount of discarded RNA ranges from 50% to more than 90% of the primary transcript. The remaining segments (**exons**) are joined together to form the finished mRNA molecules. The excision of the introns and the formation of the final mRNA molecule by joining of the exons is called **RNA splicing.**

exons

RNA splicing

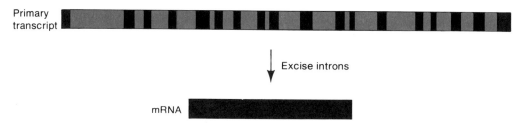

Primary transcript

Excise introns

mRNA

Figure 8-11 A diagram of the conalbumin primary transcript and the processed mRNA. The 17 introns, which are excised from the primary transcript, are shown in blue.

TABLE 8-2	TRANSLATED EUKARYOTIC GENES IN WHICH INTRONS HAVE BEEN DEMONSTRATED

Gene	Number of Introns
α-Globin	2
Immunoglobulin L chain	2
Immunoglobulin H chain	4
Yeast mitochondria cytochrome *b*	6
Ovomucoid	6
Ovalbumin	7
Ovotransferrin	16
Conalbumin	17
α-Collagen	52

Note: Genes for histones and interferon are among the few protein-coding genes in higher organisms that lack introns. Many human genes are comprised of more than 95% intron sequence.

The number of introns per gene varies considerably (**Table 8-2**) and is not the same in all organisms for a given protein. Furthermore, within a particular gene, introns have many different sizes and are usually larger than exons.

Splicing occurs in the nucleus and, in some cases, starts on an RNA that is in the process of being elongated by RNA polymerase (a nascent RNA). Usually, however, splicing is carried out after transcription and poly(A) addition are completed. The existence of splicing explains how the nucleus can contain an enormous number of different RNA molecules whose size distribution is very great. The precursor and partially processed molecules are called **heterogeneous nuclear RNA (HnRNA)** because of the great diversity in size of the molecules. After processing is completed, the mature mRNA is transported to the cytoplasm to be translated.

Removing an intron without altering the coding sequence of the resulting mRNA molecule requires great fidelity in the cleavage process. For example, a cutting error that displaces the cutting site by a single base would completely destroy the reading frame for protein synthesis of the bases in the mRNA. Fidelity is provided by the base sequence itself. In all genes observed in which splicing occurs, the splice sites of the primary transcript are marked by a sequence resembling

$$5'—{}^{A}_{C}AGG\underset{\downarrow}{U}{}^{A}_{G}AGU\;.\;.\;.\overset{intron}{.}\;.\;.\;.\;(Py)_6XCA\underset{\downarrow}{G}G{}^{G}_{U}\;-3'$$

in which Py is any pyrimidine, X is any base, and the arrows are the splice points. The underlined bases are the same for all observed introns (that is, they are conserved), whereas the remainder is a consensus sequence in which there is some variation from one splice site to the next.

In the nucleus, the precursors to mRNA molecules (HnRNAs) are complexed with specific proteins, forming a **ribonucleoprotein particle (RNP).** In the course of the splicing reaction, primary transcripts are assembled into particles called HnRNP.

The process of RNA splicing occurs on a separate structure called a **spliceosome.** Spliceosomes actually form on the splice sites through interactions with a

heterogeneous nuclear RNA

number of small nuclear ribonucleoproteins called **snRNPs.** A key component of an snRNP is at least one small nuclear RNA (snRNA). Each snRNP has a separate function. The one with the 165-nucleotide U1 snRNA binds to the consensus sequence at the 5′ end of an intron, while that with the U5 snRNA aligns the two exon sequences that are to be linked together. These snRNPs bring the two junctions into close proximity at the spliceosome, which has enzymatic activities that perform the intricate process of cutting out the intron sequences and joining the two exon segments.

The existence of introns seems to be fairly wasteful, especially in view of the relative amounts of intron and exon RNA. In a few cases, the presence of introns increases the coding capacity of a particular stretch of genes in that alternative splicing patterns yield different mRNA molecules (discussed in Chapter 12). This is common in DNA viruses, but not particularly frequent in cellular DNA, although several examples are known.

Several theories have been proposed to explain why genes from eukaryotic cells are interrupted by introns. Since introns are almost never found in the genes from bacteria, which are considered to be simpler forms of life, it is reasonable to suppose that gene interruptions were added during the course of evolution. Various types of recombination allow segments of DNA to be exchanged between different places on a chromosome. Thus, an interruption of a gene could appear by this type of recombination, and as long as the cell had acquired the means of splicing RNA, the right kinds of interruptions would yield a functional gene.

An alternative idea is that interrupted genes were present even in the earliest organisms, and during the course of evolution were lost by the bacteria in order to simplify their genomes and allow them to duplicate as rapidly as possible. This alternative provides a plausible explanation of how genes for complex proteins might have evolved. Each exon could represent the coding sequence of an ancestral small protein and rearrangements that occurred during the course of evolution could have brought the sequences together from various parts of the genome to form new genes. The use of an RNA splicing mechanism that could work to remove all introns would allow a gene to be assembled and tried in new combinations without requiring precise recombination events to fuse parts and form a gene lacking interruptions.

Examination of the structure of proteins has revealed that many consist of several independently folded regions, each separated by a short polypeptide segment. Each individually folded region is called a domain. A protein with two very distinct domains is shown in **Figure 8-12**. This particular protein, a part of an antibody molecule, is encoded by a gene consisting of two large and two small exons. The two large exons code for the two separated domains, suggesting that each domain is derived from an ancestral protein.

Means of Studying Intracellular RNA

Several procedures are used to study RNA metabolism in vivo. In most of the techniques, radioactive RNA is prepared by adding ^{3}H-labeled or ^{14}C-labeled uridine to the growth medium. However, this also produces radioactive DNA, because uridine can be metabolized to cytosine and thymine; this is significant when studying RNA

KEY CONCEPT

The "RNA Processing" Concept

Eukaryotic mRNAs are extensively processed ("cut and shut") prior to functioning as template for protein synthesis.

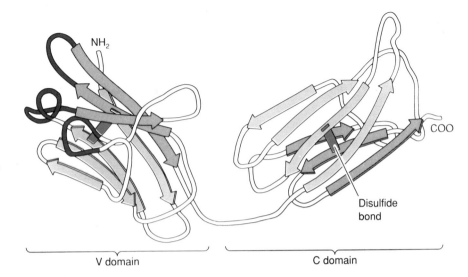

Figure 8-12 A diagram of a subunit of an antibody molecule, a protein with two domains, one black and one blue. (Adapted from Watson et al., *Molecular Biology of the Gene*, Fourth Edition. Menlo Park, CA: Benjamin-Cummings, 1987.)

metabolism because, when RNA is isolated from cells, it is usually contaminated with DNA. The contaminating DNA can be removed by treatment of the sample with pancreatic DNase, for this enzyme degrades the DNA to mononucleotides and small oligonucleotides, which can be easily separated from the RNA.

Most investigations of RNA in vivo are concerned with either the presence, synthesis, or degradation of specific mRNA species—for example, the mRNA from a particular gene. In order to carry out such analyses, it is necessary to distinguish the particular RNA molecule from all other RNA molecules. This is usually done by **DNA-RNA hybridization.** With this technique, DNA with a sequence complementary to the RNA molecule of interest is permanently fixed to a nitrocellulose filter, as described earlier in Chapter 3 (Figure 3-11). The filter is subsequently incubated with an extract containing radioactive RNA, using conditions leading to base pairing between the RNA and DNA strands, and then washed repeatedly. RNA that is not present in DNA-RNA hybrids is removed; the amount of radioactivity remaining on the filter is a measure of the amount of the particular RNA present in the extract.

A variety of procedures are used to obtain the DNA used in a hybridization experiment. In the most useful experiments, a DNA segment containing a sequence complementary to the RNA of interest is inserted (by genetic engineering techniques described in Chapter 15) into a piece of foreign DNA that has no sequences in common with the RNA and that is easily purified. In this case, the DNA sequence of interest is said to be **cloned.** Typical foreign DNA molecules are plasmids, phage, and viral DNA. A DNA molecule carrying a cloned DNA sequence to be used in a hybridization experiment is called a **probe.** Similarly, an RNA molecule can be used in hybridization experiments (see below) and, in this instance, is also referred to as a "probe."

The technique most commonly used is the **Southern transfer,** or **Southern blotting,** procedure (named after E. Southern, its developer), a method in which

DNA-RNA hybridization

cloned

Southern transfer
Southern blotting

hybridization is performed with a large number of distinct DNA segments simultaneously. In this procedure, the total DNA of an organism is broken into discrete fragments by restriction enzymes (Chapter 15), which make cuts in unique base sequences. Fragments are separated by gel electrophoresis (as in the di-deoxy base-sequencing technique, Chapter 3) DNA, and hybridization is carried out on all of the fragments represented in the gel. Various techniques enable the positions of particular DNA fragments to be identified. The location of hybridized radioactive RNA can then be matched with these fragments. The technique is performed as follows (**Figure 8-13**).

DNA is enzymatically fragmented and then electrophoresed through an agarose gel. Following electrophoresis, the gel is soaked in a denaturing solution (usually NaOH) so that all the DNA in the gel is converted to single-stranded DNA, which is a prerequisite for the hybridization to be performed later. After the gel (which is typically in the form of a long flat sheet) is denatured, it is placed on a piece of nitrocellulose paper. The denatured DNA binds tightly to the nitrocellulose paper. DNA molecules do not diffuse very much, so if the gel and the nitrocellulose filter are in firm contact, the positions of the DNA molecule on the filter will be identical to their positions in the gel. The nitrocellulose filter is dried in vacuum to insure that the DNA remains on the filter during the hybridization step. The dried filter is then moistened with a very small volume of a solution of ^{32}P-labeled RNA "probe," placed in a tight-fitting plastic bag to prevent drying, and held at a temperature suitable for renaturation (usually 65°C) for 16–24 hours. The filter is then removed, washed to remove unbound radioactive molecules, dried, and autoradiographed with x-ray film. The blackened position on the film indicates the locations of the DNA molecules whose DNA base sequences are complementary to the sequences of the added radioactive probe. Since the genes contained in each fragment on the gel usually are known, specific mRNA molecules in a mixture of RNA probes can be identified as corresponding to specific genes. The degree of blackening on the film is easily measured quantitatively and is usually proportional to the amount of RNA probe that has hybridized. Therefore, the amount of mRNA transcribed from each region of a DNA molecule can be measured with this method. Note that many different mRNA molecules can be studied simultaneously using this technique. The Southern transfer method also has many other uses, particularly in DNA–DNA hybridization analysis.

A Future Practical Application?

As an alternative to conventional drug therapy as a method for combating viral infections, some researchers are investigating the use of *antisense oligonucleotides* to disrupt the metabolism of the virus. An antisense oligonucleotide is a nucleotide sequence that is complementary to either a single-stranded protein-coding (sense) DNA segment or an mRNA protein-synthesis template.

In a living cell, an antisense oligonucleotide that has a base sequence complementary to a gene will bind directly to the target gene and thereby prevent RNA

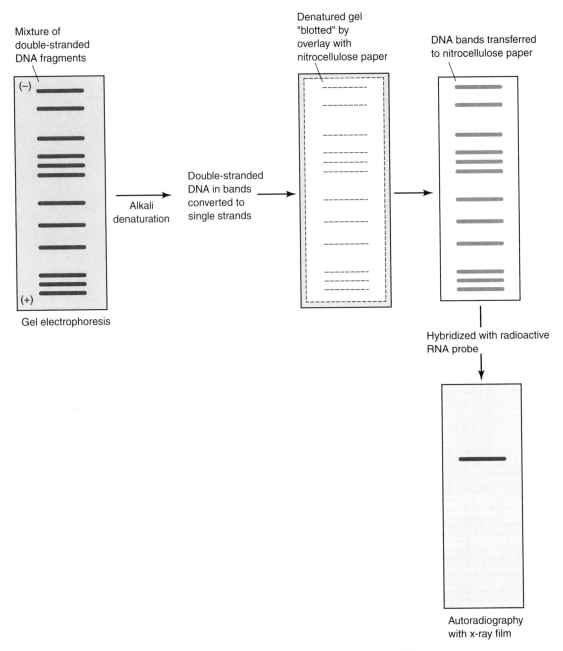

Figure 8-13 The Southern transfer technique is used to detect, in a mixture of RNAs, mRNAs that hybridize to specific gene fragments (bands in the gel). DNA bands from a gel are blotted onto nitrocellulose paper for subsequent hybridization to a radioactive RNA probe. Similarly, RNAs can be separated on a gel, blotted onto nitrocellulose paper, and hybridized with a radioactive DNA probe (so-called Northern transfer).

polymerase from transcribing the gene. Similarly, an oligonucleotide that hydrogen bonds to a mRNA segment can disrupt its translation.

Short oligonucleotides, consisting of as few as a dozen nucleotides, have so far been demonstrated to be effective in disrupting either transcription or translation in test systems that consist of virus-infected cultured mammalian cells.

A key advantage of this method is that it lends itself to the so-called "rational" approach to drug design. That is, once candidate target nucleic acid sequences are known, researchers can design appropriate oligonucleotides based on the simple Watson-Crick base-pairing rules. Another advantage of this method is its specificity. The oligonucleotides should bind only to their target genes and, therefore, not interfere with other essential metabolic functions. Finally, the relative ease with which oligonucleotides can be synthesized (see Figure 3-18 in Chapter 3) provides for quick experimentation.

Disadvantages to this pioneering technology do, however, exist. A major one concerns "drug delivery." Difficulties have been encountered with administering oligonucleotides to mammals. Frequently, these agents either do not penetrate target cells, or, once within a cell, they are degraded by cellular RNases. Nevertheless, novel delivery strategies, which involve using lipid vesicles or carrier proteins, and further refinements in the design of RNase-resistant oligonucleotide structures, are expected to alleviate those problems.

SUMMARY

RNA polymerase initiates transcription by first binding to a promoter sequence on DNA. The bacterial enzyme does this by contact between the σ subunit of the holoenzyme and sequences at 10 and 35 base pairs upstream from the start point. Eukaryotic RNA polymerases rely on recognition of the promoter by separate initiation factors. RNA synthesis is initiated by a reaction between two nucleoside triphosphates paired to residues +1 and +2 of the template strand. Chains are elongated by addition of nucleotides at the 3′ end of the growing RNA at a rate of about 50 per second. Termination occurs spontaneously at certain DNA sequences, called intrinsic terminators, or by action of a protein factor at a "factor-dependent" terminator sequence. In *E. coli*, that factor is called Rho, which acts at rho-dependent terminators. The three major classes of RNA are messenger RNA (mRNA), ribosomal RNA (rRNA), and transfer RNA (tRNA). rRNA and tRNA are derived from primary transcripts, synthesized as previously described, by various processing reactions that include cutting, trimming, and nucleotide modification. In prokaryotes, mRNAs are usually primary transcripts, while in eukaryotes, mRNA are processed from precursor transcripts by addition of 5′ caps and poly(A) tails, and usually by splicing together exon sequences.

DRILL QUESTIONS

1. a. From what substrates is RNA made?
   b. On what template?
   c. With what enzyme?
   d. Is a primer required?
2. Describe the differences, if any, between the chemical reactions catalyzed by DNA polymerase and RNA polymerase.

3. What chemical groups are present at the origin and terminus of a molecule of mRNA that has just been synthesized?

4. a. What is mRNA?
   b. How does mRNA sometimes differ from a primary transcript?
   c. Define cistron and polycistronic mRNA.
   d. Which parts of a mRNA molecule are not translated?

5. How many subunits are in *E. coli* RNA polymerase, and which one is responsible for correctly positioning the enzyme on a promoter?

6. a. What two regions are common to most prokaryotic promoters?
   b. What consensus sequence is present in a large number of eukaryotic promoters?

7. Answer these questions about eukaryotic RNA.
   a. What is a cap?
   b. At which end of the mRNA is the poly(A)?
   c. Are there eukaryotic mRNA molecules that do not contain either feature?

8. a. What are intervening sequences or introns?
   b. What is meant by mRNA "splicing"?

PROBLEMS

1. The −10 region of an *E. coli* promoter is an example of what is termed a consensus sequence—namely, a base sequence from which other sequences having similar functions can be obtained by changing only one or two bases. Answer these questions about consensus sequences.
   a. What is the evolutionary significance of such a sequence?
   b. What is the biochemical significance of a conserved base, such as the conserved T in the −10 region?
   c. What biochemical differences might you expect between sequences that differ only slightly in the nonconserved bases?

2. An RNA molecule that has a 3′-OH terminus and a 5′-P terminus is isolated. What information does this fact provide?

3. A chromatographic column in which oligo-dT is linked to an inert substance is useful in separating eukaryotic mRNA from other RNA molecules. On what principle does this column operate?

4. A particular sequence containing six base pairs is located in ten different organisms. The observed sequences are ACGCAC, ATACAC, GTGCAC, ACGCAC, ATACAC, ATGTAT, ATGCGC, ACGCAT, GTGCAT, and ATGCGC. What is the consensus sequence?

5. Write down the two RNA sequences that could conceivably result from complete transcription of a DNA molecule that has the following sequence in one of the strands—5′-AGCTGCAATG-3′. Indicate the 5′ and 3′ end of each transcript.

6. Genetic engineering techniques allow the movement of genes from one organism to another. When eukaryotic genes are placed into *E. coli*, it has been observed that these genes are not always transcribed, while transcription occurs much more frequently when eukaryotic genes are transferred to yeast. Why?

CONCEPTUAL QUESTIONS

1. How might the expression of genes that are developmentally controlled be regulated at the level of their mRNA?
2. What could be the consequence(s) of mutations in various regions of a promoter?
3. How might an organism compensate for severe promoter mutations in order to survive?

in this chapter you will learn

1. The nature of codons, anticodons, and the "universal" genetic code.

2. The mechanisms by which polypeptides are synthesized.

3. The structure of transfer RNA and the function of aminoacyl tRNA synthetases.

4. The general structure and function of ribosomes.

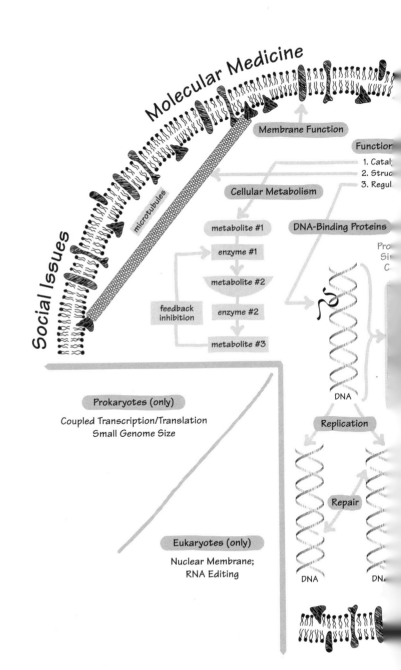

Translation

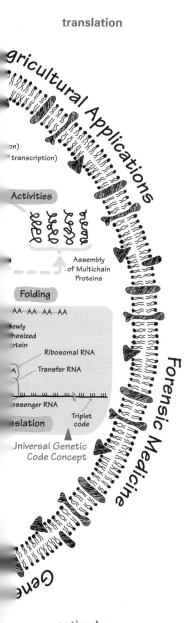

The synthesis of every protein molecule in a cell is directed by intracellular DNA. There are two aspects to understanding how this is accomplished—the **information** or **coding problem** and the **chemical problem.** The information problem is the mechanism by which a base sequence in a DNA molecule is translated into an amino acid sequence of the polypeptide chain. The chemical problem includes the actual process of synthesizing a protein: the means of initiating synthesis; linking together the amino acids in the correct order; terminating the chain; releasing the finished chain from the synthetic apparatus; folding the chain; and, often, postsynthetic modification of the newly synthesized chain. The overall process is called **translation.**

This chapter presents an outline of the process in order to introduce the terminology and present the major features of the coding and decoding systems, and the mechanism for polypeptide synthesis.

We will learn that once the necessary information is coded (in a triplet format) into mRNA, other types of RNAs, as well as several enzymes, join up to form the macromolecular complex that carries out protein synthesis. Eventually, we will begin to appreciate that many features of protein synthesis are much more complicated than nucleic acid synthesis. That complexity derives from the fact that amino acids do not have a direct affinity for the nucleotide bases of mRNA. Special molecules are, therefore, required to hold the mRNA in position while amino acids are lined up and covalently linked to one another to form a protein.

Outline of Translation

Protein synthesis occurs on intracellular particles called **ribosomes.** In prokaryotes, these particles consist of three RNA molecules and about 55 different protein molecules. Those proteins include the enzymes needed to form a peptide bond between amino acids, a site for binding the mRNA, and sites for bringing in and aligning the amino acids in preparation for assembly into the finished polypeptide chain. Amino acids themselves are unable to interact with the ribosome and cannot recognize bases in the rRNA molecule. Thus, there exists a collection of carrier molecules mentioned in the previous chapter, **transfer RNAs (tRNAs).** These molecules contain a site for amino acid attachment, and a region called the **anticodon** that recognizes the appropriate base sequence (the **codon**) in the mRNA. Proper selection of the amino acids for assembly is determined by the positioning of the tRNA molecules, which in turn is determined by hydrogen-bonding between the anticodon of each tRNA molecule and the corresponding codon of the mRNA.

A schematic diagram showing the events that occur on the ribosome is given in **Figure 9-1**. The scheme shown applies equally to prokaryotes and eukaryotes, with a single exception. In eukaryotes, since transcription occurs in the nucleus and protein synthesis occurs in the cytoplasm, the mRNA is not attached to the DNA during protein synthesis, in contrast with what is shown in the figure.

Clearly, there need to be many different tRNA molecules, because each amino acid must be brought in to the ribosome in a way that insures that it corresponds to the base-sequence of the mRNA. There are specific tRNA molecules that correspond to each amino acid. Furthermore, the linkage of each amino acid to its tRNA molecule is catalyzed by a specific enzyme, which insures that the appropriate amino acid will be attached to the correct tRNA molecule.

In earlier chapters, several examples of directional synthesis of macromolecules were seen—for example, DNA and RNA. This is also the case for polypeptide synthesis, which begins at the amino terminus. Furthermore, translation of an mRNA molecule occurs in only one direction namely 5′ to 3′. **Figure 9-2** summarizes the polarity of synthesis of mRNA and protein with respect to the DNA coding strand. These directionalities are also shown in Figure 9-1.

The Genetic Code

genetic code

The **genetic code** is the collection of base sequences (codons) that corresponds to each amino acid and to translation signals.

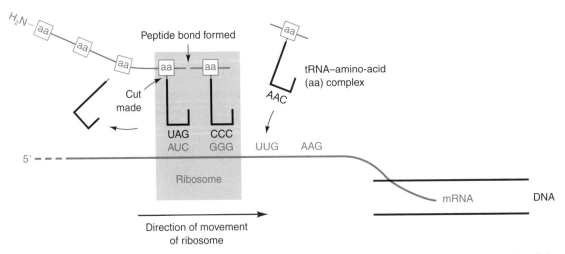

Figure 9-1 A diagram showing how a protein molecule is synthesized. Note that the relative directions of polypeptide and RNA synthesis are such that polypeptide synthesis can occur before the mRNA is completed. This is, in fact, the case in prokaryotes.

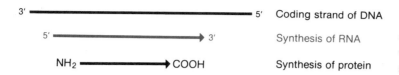

Figure 9-2 Directions of synthesis of RNA and protein with respect to the coding strand of DNA.

Since there are 20 different amino acids that occur in proteins, there must be more than 20 codons to include signals for starting and stopping the synthesis of particular protein molecules. If all codons have the same number of bases, then each codon must contain at least three bases. The argument for this conclusion is the following: a single base cannot be a codon because there are 20 amino acids and only four bases. Pairs of bases also cannot serve as codons because there are only 4^2 or 16 possible pairs of the four bases. Triplets of bases are possible because there are 4^3 or 64 triplets, which is more than adequate. In fact, the genetic code is triplet code, and all 64 possible codons carry information of some sort. Furthermore, in translating mRNA molecules, the codons do not overlap, but are read sequentially (**Figure 9-3**).

The general properties of the code—for example, that each codon contains three bases, and that codons do not overlap—were deduced from genetic experiments. The sequence of each codon was determined from in vitro protein-synthesizing experiments in which synthetic mRNA molecules of known sequence were used; for example, when polyuridylic acid was used as an mRNA, the peptide polyphenylalanine was made, indicating that UUU is the codon for phenylalanine. Use of poly(U), terminated with a 3′ guanine, yielded polyphenylalanine with a carboxyl-terminal leucine, indicating that UUG is a leucine codon. These and other experiments identified all of the codons. The code is shown in **Table 9-1**. Keep in mind that Table 9-1 lists mRNA bases, which comprise the various codons. Those mRNA codons will be read by tRNAs, each of which contains an anticodon base sequence.

The following features of the code should be observed:

1. Most amino acids have more than one codon. In fact, only methionine and tryptophan have a single codon. Furthermore, multiple codons corresponding to a single amino acid usually differ only by the third base. For example, GGU, GGC, GTA, and GGG all code for glycine. Thus, the code is said to be redundant (the term "degenerate" is also used).

2. Three codons signal termination of polypeptide synthesis—the stop codons UAA, UAG, and UGA.

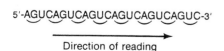

5′-AGUCAGUCAGUCAGUCAGUCAGUC-3′

Direction of reading

Figure 9-3 Bases in mRNA are read sequentially in the 5′ to 3′ direction, in groups of three.

TABLE 9-1	THE "UNIVERSAL" GENETIC CODE				
First Position (5′ end)	**Second Position**				**Third Position (3′ end)**
	U	C	A	G	
U	Phe	Ser	Tyr	Cys	U
	Phe	Ser	Tyr	Cys	C
	Leu	Ser	Stop	Stop	A
	Leu	Ser	Stop	Trp	G
C	Leu	Pro	His	Arg	U
	Leu	Pro	His	Arg	C
	Leu	Pro	Gln	Arg	A
	Leu	Pro	Gln	Arg	G
A	Ile	Thr	Asn	Ser	U
	Ile	Thr	Asn	Ser	C
	Ile	Thr	Lys	Arg	A
	Met	Thr	Lys	Arg	G
G	Val	Ala	Asp	Gly	U
	Val	Ala	Asp	Gly	C
	Val	Ala	Glu	Gly	A
	Val	Ala	Glu	Gly	G

Note: The boxed codons are used for initiation. GUG is very rare.

3. One codon signals initiation of polypeptide synthesis—the start codon AUG, which codes for methionine. An important question is how a particular AUG sequence is designated as a start codon, whereas others located in a coding sequence act only as an internal codon for methionine. In prokaryotes, special base-sequences serve this function; a different mechanism is used for eukaryotes. In some organisms, GUG is also used as a start codon for some proteins.

To date, the same codon-amino acid relations seem to exist for all organisms—viruses, prokaryotes, Archaebacteria, and eukaryotes—and the code is said to be universal. A few exceptions exist: variations in the genetic code have been observed in certain mitochondria, in the nucleus of certain yeasts and ciliates, and elsewhere. Furthermore, the particular exceptions appear to be species specific. The evolutionary significance of these differences is a widely discussed topic.

Transfer RNA and the Aminoacyl Synthetases

aminoacyl-tRNA synthetases

The decoding operation by which the base sequence within an mRNA molecule becomes translated to an amino acid sequence of a protein is accomplished by the tRNA molecules and a set of enzymes called the **aminoacyl-tRNA synthetases.** These synthetase enzymes covalently attach the amino acid to the tRNA. Their action will be described shortly.

The tRNA molecules are small, single-stranded nucleic acids ranging in size from 73 to 93 nucleotides. Like all RNA molecules, they have a 3′-OH terminus, but the opposite end terminates with a 5′-monophosphate rather than a 5′-triphosphate, because tRNA molecules are cut from a large primary transcript (see Figure 8-9 in Chapter 8). Due to pairing of complementary base sequences, short double-stranded regions form, causing the molecule to fold into a structure in which open loops are connected to one another by double-stranded stems (**Figure 9-4**). In two dimensions, a tRNA molecule is drawn as a planar cloverleaf. Its three-dimensional structure is more complex, as is shown in **Figure 9-5**. Panel (a) shows a skeletal model of the yeast tRNA molecule that carries phenylalanine, and panel (b) shows an interpretive drawing.

Three regions of each tRNA molecule are used in the decoding operation. One of these regions is the **anticodon,** a sequence of three bases that can form base pairs with a codon sequence in the mRNA. No normal tRNA molecule has an anticodon complementary to any of the stop codons UAG, UAA, or UGA, which is why these codons are recognized as stop signals. A second site is the **amino acid attachment site.** The amino acid corresponding to the particular mRNA codon that base pairs with the tRNA anticodon is covalently linked to this terminus. These bound amino acids are joined together during polypeptide synthesis. A specific aminoacyl tRNA synthetase matches the amino acid with the anticodon; to do so, the enzyme must be able to distinguish one tRNA molecule from another. The necessary distinction

KEY CONCEPT

The "Universal Genetic Code" Concept

The triplet code is universal (i.e., the same code is used by all organisms). It also has redundancies (i.e., multiple codons exist for most amino acids).

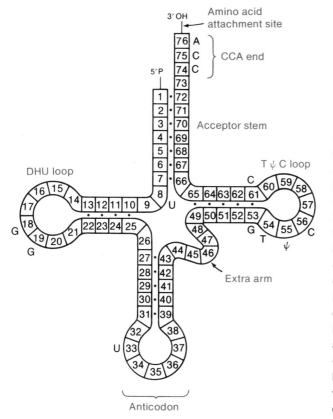

Figure 9-4 The consensus tRNA cloverleaf structure with its bases numbered. The hydrogen-bonded regions are indicated by dots between the bases. A few bases present in almost all tRNA molecules are shown. tRNA contains unusual bases: one of these, dihydrouracil (DHU), is in the DHU loop, and another, pseudouridine (Ψ), is in the T Ψ C loop. The extra arm varies in length in different tRNAs.

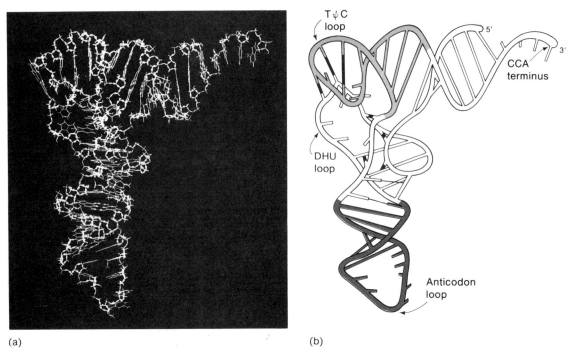

Figure 9-5 (a) Photograph of a skeletal model of yeast tRNA^Phe. (b) Schematic diagram of the three-dimensional structure of yeast tRNA^Phe. (Courtesy of Dr. Sung-Hou Kim.)

is provided by an as yet ill-defined region encompassing many parts of the tRNA molecule called the **recognition region.**

The different tRNA molecules and synthetases are designated by stating the name of the amino acid that can be linked to a particular tRNA molecule by a specific synthetase; for example, leucyl-tRNA synthetase attaches leucine to tRNA^Leu. When an amino acid has become attached to a tRNA molecule, the tRNA is said to be **acylated** or **charged.** An acylated tRNA molecule is designated in several ways. For example, if the amino acid is glycine, the acylated tRNA would be written glycyl-tRNA or Gly-tRNA. The term **uncharged tRNA** refers to a tRNA molecule lacking an amino acid, and **mischarged tRNA** to one acylated with an incorrect amino acid.

At least one (and usually only one) aminoacyl synthetase exists for each amino acid. For a few of the amino acids specified by more than one codon, more than one synthetase exists.

Accurate protein synthesis (the placement of the correct amino acid at the appropriate position in a polypeptide chain) requires:

1. attachment of the correct amino acid to a tRNA molecule by the synthetase;
2. accuracy in codon-anticodon binding.

Several important experiments showing that codon-anticodon binding is based only on base-pair recognition and that the identity of the amino acid attached to the tRNA molecule does not influence this recognition will be described below.

Experimental Mischarging of a tRNA Molecule

In this experiment, tRNA^{Cys} was charged with radioactive cysteine. The cysteine was then chemically converted to radioactive alanine, yielding alanyl-tRNA^{Cys}. This mischarged tRNA molecule was then used in an in vitro system containing hemoglobin mRNA and capable of synthesizing hemoglobin, the complete amino acid sequence of which was known. Hemoglobin containing radioactive alanine was made, but the radioactive alanine was present at the sites normally occupied by cysteine rather than at normal alanine positions. This experiment confirmed the hypothesis that tRNA is a carrier molecule, and that the amino acid recognition region and the anticodon are distinct regions.

Experimental Alteration of an Anticodon

The anticodon in the tRNA molecule is responsible for inserting the amino acid into the correct position. This was demonstrated by chemical alterations of a single base of the anticodon that was sufficient to change the specificity of the tRNA. Glycine tRNA with the anticodon UCC was chemically altered to UCU, which pairs with the codeword for arginine: AGA. The altered glycine tRNA, when charged with radioactive glycine, inserted the glycine into the arginine positions in the newly synthesized polypeptide, directed by a synthetic mRNA (AGA)n. The native glycine (unaltered) tRNA was not active with this message. Native arginine tRNA was also capable of inserting its labeled arginine into polypeptide directed by (AGA)n synthetic mRNA.

Experimental Analysis of Error Frequency

Several amino acids are structurally similar and it is to be expected that synthetases might make occasional mistakes. Valine and isoleucine constitute such a possibly ambiguous pair of amino acids and, in fact, isoleucyl-tRNA synthetase activates valine with ATP to form valyl-AMP, which remains bound to the synthetase, at a frequency of about one per 225 activation events. This would mean that 1/225 of all isoleucine positions in proteins could contain a valine. For a typical protein containing 500 amino acids, of which 25 were isoleucines, about one copy of the protein in nine could be altered. Since there are at least 10 known examples of misacylation, there could be an error in almost every protein molecule, although this does not occur. An editing mechanism exists.

The editing mechanism that corrects the valine-isoleucine error is a hydrolytic step in which valyl-AMP is cleaved and removed from the enzyme. The hydrolysis is carried out by the isoleucyl-tRNA synthetase itself, for the synthetases possess a second active site for editing mischarged tRNAs. Interestingly, the signal that activates the hydrolytic function is the attempted binding of valine to tRNA^{Ile}. The number of times this editing system fails and valyl-tRNA^{Ile} is formed is about 1 in 800. Thus, the overall error frequency—that is, the fraction of isoleucine sites occupied by valine—is $(1/225)(1/800) = 1/180{,}000$. If all possible amino acid misacylations occur at this frequency, only about 0.17% of the proteins would be defective. A similar case, in which methionyl-tRNA synthetase forms threonyl-AMP and homocysteyl-AMP, has also been analyzed.

The Wobble Phenomenon

The pattern of redundancy in the code suggests that something is missing in the explanation of codon-anticodon binding; the most striking aspect of the redundancy is that, with only a few exceptions, the identity of the third codon base appears to be unimportant. That is, XYA, XYB, XYC, and XYD usually correspond to the same amino acid.

wobble hypothesis

In 1965, Francis Crick made a proposal, known as the **wobble hypothesis,** that explains how some tRNA molecules respond to several codons and also provides insight into the pattern of redundancy in the code. Up to that time it was generally assumed that no base pair other than G · C, A · T, or A · U would be found in a nucleic acid. This is true of DNA because the regular helical structure of double-stranded DNA imposes two steric constraints:

1. Two purines cannot pair with one another because there is not enough space for a planar purine-purine pair.
2. Two pyrimidines cannot pair because they cannot reach one another.

Crick proposed that, since the anticodon is located within a single-stranded RNA loop, the codon–anticodon interaction might not be constrained to the same degree as is the DNA double helix. By model-building he showed that the steric requirements were less stringent at the third position of the codon. By allowing a little play in the structure (this play is called "wobble"), Crick demonstrated that other base pairs can exist between codon and anticodon. He required, first, that the first two base pairs be of the standard type in order to maximize stability, and second, that the third base pair not produce as much distortion as a purine-purine pair might cause. He included inosine (to replace G) in his model because it was known to be in the anticodons of several tRNA molecules, and he proposed that the base pairs listed in **Table 9-2** were possible in the third position of the codon.

Analysis of the table reveals the four pairings we would expect, plus five others. Three of these (A-I, U-I, and C-I) arise from the ability of inosine (a purine) to base pair with both the RNA pyrimidines (C and U) and with the purine adenine. Additionally, under wobble conditions, guanine and uracil can be expected to pair (a purine-pyrimidine pair), which provides the other two unexpected pairs (G-U and U-G). This information explains how a single tRNA molecule can theoretically respond to several codons.

There are two major species of the alanine tRNA of yeast. One of these responds to the codons GCU, GCC, and GCA. Its anticodon is IGC, which is consistent with the entries in Table 9-2, and shows that only inosine can pair with U, C, and A. (Remember the convention for naming the codon and the anticodon— *always with the 5′ end at the left.* Thus, the codon 5′-GCU-3′ is matched by the anticodon 5′-IGC-3′.) Similarly, yeast $tRNA_{II}^{Ala}$ responds only to GCG; there are two possible anticodons (CGC and UGC) because both C and U can bond to G. If the anticodon were UGC, $tRNA_{II}^{Ala}$ would respond to both GCG and GCA, which is not the case. Thus, the anticodon cannot be UGC. The anticodon must, therefore, be CGC, as indeed it is.

The most striking achievement of the wobble hypothesis is that it explains the arrangement of all the synonyms in the code.

TABLE 9-2	ALLOWED PAIRINGS ACCORDING TO THE WOBBLE HYPOTHESIS
Third Position (3′ OH) Codon Base (in mRNA)	First Position (5′p end) Anticodon Base (in tRNA)
A	U, I
G	C, U
U	G, I, A
C	G, I

Polycistronic mRNA

Many prokaryotic mRNA molecules are polycistronic; they contain sequences specifying the synthesis of several proteins. A polycistronic mRNA molecule must possess a series of start and stop codons. If an mRNA molecule encodes three proteins, the minimal requirement would be the sequence:

Start, protein 1, stop—start, protein 2, stop—start, protein 3, stop.

Actually, an actual mRNA molecule is probably not so simple—the so-called "leader sequence," which precedes the first start signal, may be several hundred bases long, and there is usually a sequence called a **spacer** of from 5 to 20 bases between one stop codon and the next start codon. The structure of a tricistronic mRNA more typically resembles that shown in **Figure 9-6**.

spacer

Overlapping Genes

In all that has been said so far about coding and signal recognition, an implicit assumption has been that the mRNA molecule is scanned for start signals to establish the **reading frame** (the actual nucleotide sequence translated into a protein). It is also assumed that reading then proceeds in a single direction within the frame. The idea that several reading frames might exist in a single segment was not considered for many years, primarily because a mutation in a gene that overlapped another gene would then have to often produce a mutation in the second gene, and mutations affecting two genes had not been observed. The notion of overlapping reading frames was also rejected on the grounds that severe constraints would be

reading frame

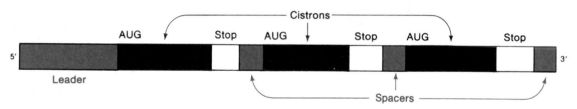

Figure 9-6 Arrangement of cistrons (black) and untranslated regions (blue) in a typical polycistronic mRNA molecule.

placed on the amino acid sequences of two proteins translated from the same portion of mRNA. However, because the code is highly redundant, the constraints are actually not so rigid.

If multiple reading frames were present in an organism, a single DNA segment could be utilized to maximum efficiency. A disadvantage, however, is that evolution might be slowed. Single-base-change mutations would be deleterious more often if multiple reading frames were present than if there were a unique reading frame. Nonetheless, some organisms—namely, small viruses and the smallest phages—have evolved with overlapping reading frames.

The *E. coli* phage φX174 contains a single strand of DNA consisting of 5,386 nucleotides whose base sequence is known. If a single reading frame were used, at most 1,795 amino acids could be encoded in the sequence and, if we take 110 as the molecular weight of an "average amino acid, at most 197,000 molecular weight units of protein could be made. However, the phage makes 11 proteins and the total molecular weight of these proteins is 262,000. This paradox was resolved when it was shown that translation occurs in several **reading frames** from three mRNA molecules (**Figure 9-7**). For example, the sequence for protein B is contained within the sequence for protein A, but translated in a different reading frame. Similarly, the protein E sequence is totally within the sequence for protein D. Protein K is initiated near the end of gene A, includes the base sequence of B, and terminates in gene *C;* synthesis is not in phase with either gene *A* or gene *C*. Of note is protein A′ (also called A*), which is formed by reinitiation within gene *A* and in the same reading frame, so that it terminates at the stop codon of gene *A*. Thus, the amino acid sequence of A′ is identical to a segment of protein A. In total, five different proteins obtain some or all of their primary structure from shared base sequences in φX174. This phenomenon, known as **overlapping genes,** has been observed in the related phage G4 and in the small animal virus SV40.

It should be realized that the single structural feature responsible for gene overlap is the location of each AUG initiation sequence.

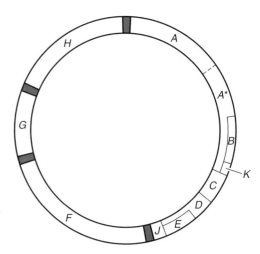

Figure 9-7 Genetic map of φX174 showing the overlapping genes. Spacers are darkened.

Polypeptide Synthesis

Preceding sections discussed how the information in an mRNA molecule is converted to an amino acid sequence having specific start and stop points. This section examines the chemical attachment of the amino acids to one another.

Polypeptide synthesis can be divided into three stages: (1) **initiation,** (2) **elongation,** and (3) **termination.** The main features of the initiation stage are the binding of mRNA to the ribosome, selection of the initiation codon, and the binding of acylated tRNA bearing the first amino acid. In the elongation stage, there are two processes: joining together two amino acids by peptide bond formation, and moving the mRNA and the ribosome with respect to one another so that the codons can be translated successively. In the termination stage, the completed protein is dissociated from the synthesis machinery, and the ribosomes are released to begin another cycle of synthesis.

We begin our discussion with a description of the structure of the ribosome.

Ribosomes

A ribosome is a multicomponent particle containing several enzymes needed for protein synthesis. It also brings together a single mRNA molecule and charged tRNA molecules in the proper position and orientation to allow the base sequence of the mRNA molecule to be translated into an amino acid sequence (**Figure 9-8**(a)). The properties of the *E. coli* ribosome are understood best, and it serves as a useful model for discussion of all ribosomes.

All ribosomes contain two subunits (Figure 9-8(b)). For historical reasons, the intact ribosome and the subunits have been given numbers that describe how fast they sediment when centrifuged. For *E. coli* (and for all prokaryotes) the intact particle is

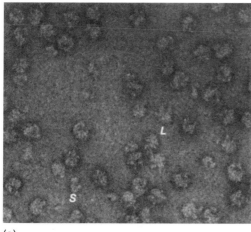
(a)

(b)

Figure 9-8   Ribosomes. (a) An electron micrograph of *E. coli* 70S ribosomes. A few ribosomal subunits are also in the field; S denotes a 30S particle and L denotes a 50S particle. (b) A three-dimensional model of the 70S ribosome. The small subunit is light, and the large subunit is dark. (Courtesy of James Lake.)

called a **70S ribosome** (S is a measure of the sedimentation rate, described earlier for Figure 7-17) and the subunits, which are unequal in size and composition, are termed **30S** and **50S.** A 70S ribosome consists of one 30S subunit, and one 50S subunit.

Both the 30S and the 50S particles can be dissociated into RNA (called **rRNA,** for ribosomal RNA) and protein molecules under appropriate conditions (**Figure 9-9**). Each 30S subunit contains one 16S rRNA molecule and 21 different proteins; a 50S subunit contains two RNA molecules (one 5S rRNA molecule and one 23S rRNA molecule) and 32 different proteins. In each particle, usually only one copy of each protein molecule is present, although a few are duplicated or modified. Like tRNA molecules, rRNA molecules are cut from large primary transcripts. In many organisms, some tRNA molecules are also cut from the large transcripts containing rRNA.

The basic features of eukaryotic ribosomes are similar to those of bacterial ribosomes, but all eukaryotic ribosomes are somewhat larger than those of prokaryotes. They contain a greater number of proteins (about 80) and additional RNA molecules (four in all). The biological significance of the differences between prokaryotic and eukaryotic ribosomes is unknown. A typical eukaryotic ribosome (an **80S ribosome**) consists of two subunits, **40S** and **60S.** These sizes may vary by as much as ±10% from one organism to the next, in contrast to bacterial ribosomes, which have sizes that are nearly the same for all bacterial species examined. The best-studied eukaryotic ribosome is that of the rat liver. The 40S and 60S subunits can also be dissociated. A 40S subunit, which is analogous to the 30S subunit of prokaryotes, consists of one 18S rRNA molecule and about 30 proteins, and the 60S subunit contains three rRNA molecules (one 5S, one 5.8S, and one 28S rRNA molecule) and about 50 proteins. The 5.8S, 18S, and 28S rRNA molecules of eukaryotes correspond functionally to the 5S, 16S, and 23S molecules of bacterial ribosomes. The bacterial counterpart to the eukaryotic 5S rRNA is very likely present as part of the 23S rRNA sequence. The eukaryotic 5S ribosomal RNA is synthesized in the nucleus while the 28S, 18S, and 5.8S rRNAs are

Figure 9-9 Dissociation of a prokaryotic ribosome. The configuration of two overlapping circles will be used throughout this chapter for the sake of simplicity. The correct configuration is shown in Figure 9-8(b).

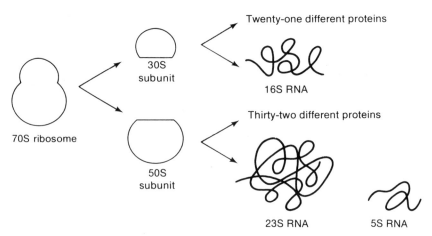

transcribed in the nucleolus as a 45S precursor and then processed to the three rRNA species.

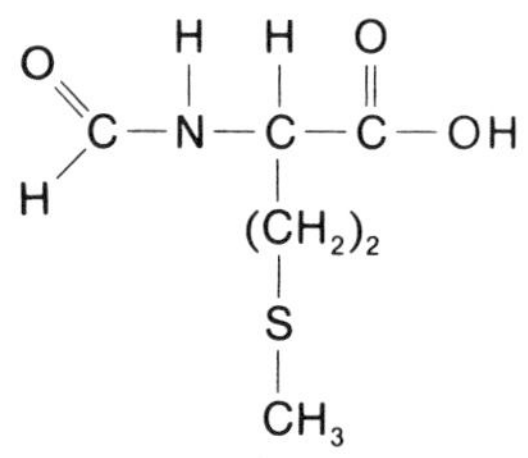

Figure 9-10 Chemical structure of *N*-formylmethionine. If the HC = O group at the left were an H, the molecule would be methionine.

Stages of Polypeptide Synthesis in Prokaryotes

Polypeptide synthesis in prokaryotes and eukaryotes follows the same overall mechanism, although there are differences in detail, the most important being the mechanism of initiation. The prokaryotic system is best understood and will serve as a model for our discussion.

An important feature of initiation of polypeptide synthesis is the use of a specific initiating tRNA molecule. In prokaryotes the methionine in the tRNA^{Met} is formylated to yield *N*-formylmethionine tRNA; the tRNA is often designated tRNA^{fMet} (**Figure 9-10**). Both tRNA^{fMet} and tRNA^{Met} recognize the codon AUG, but only tRNA^{fMet} is used for initiation. The tRNA^{fMet} molecule is first acylated with methionine, and an enzyme (found only in prokaryotes) adds a formyl group to the amino group of the methionine. In eukaryotes there is also a specific tRNA^{Met} for initiation. The initiating tRNA molecule is also charged with methionine, but formylation does not occur. The use of these initiator tRNA molecules means that *while being synthesized,* all prokaryotic proteins have *N*-formylmethionine at the amino terminus, and all eukaryotic proteins have methionine at the amino terminus. However, these amino acids are frequently removed later (this is called **processing**), and all of the amino acids have been observed at the amino termini of one type or another completed protein molecule isolated from cells.

Polypeptide synthesis in bacteria begins by the association of one 30S subunit (not the entire 70S ribosome), an mRNA molecule, fMet-tRNA, three proteins known as **initiation factors,** and guanosine 5′-triphosphate (GTP). These molecules constitute the **30S preinitiation complex** (**Figure 9-11**). Since polypeptide synthesis begins at an AUG start codon and AUG codons are found within coding sequences (that is, methionine occurs within a polypeptide chain), some signal must be present in the base sequence of the mRNA molecule to identify a particular AUG codon as a start signal. The means of selecting the correct AUG sequence differs in prokaryotes and eukaryotes. In prokaryotic mRNA molecules, a particular base sequence (AGGAGGU—called the **ribosome binding site** or sometimes the **Shine-Dalgarno sequence**) exists near the AUG codon used for initiation. It forms base pairs with a complementary sequence near the 3′ terminus of the 16S rRNA molecule of the ribosome (**Figure 9-12**). In eukaryotic mRNA molecules, the 5′ terminus binds to the ribosome with the help of certain proteins that recognize the 5′ cap (see Figure 9-17). Then the mRNA molecule slides along the ribosome until the first AUG codon in frame, *nearest the 5′ terminus,* is in contact with the ribosome. Initiation of translation in eukaryotes is, thus, said to employ a **scanning mode** for locating the initiator AUG codon. The consequence of this mechanism for initiation in eukaryotes will be explained at the end of this section.

initiation factors

30S preinitiation complex

Shine-Dalgarno sequence

scanning mode

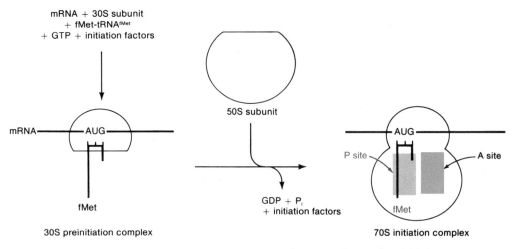

Figure 9-11 Early steps in protein synthesis in prokaryotes; formation of the 30S preinitiation complex and of the 70S initiation complex.

Following formation of the 30S preinitiation complex, a 50S subunit joins with this complex to form a **70S initiation complex** (Figure 9-11).

The 50S subunit contains two tRNA binding sites. These sites are called the **A (aminoacyl) site** and the **P (peptidyl) site.** When joined with the 30S preinitiation complex, the position of the 50S subunit in the 70S initiation complex is such that the fMet-tRNA^{fMet}, which was previously bound to the 30S preinitiation complex, occupies the P site of the 50S subunit. Placement of fMet-tRNA^{fMet} in the P site fixes the position of the fMet-tRNA anticodon so that it pairs with the AUG initiator codon in the mRNA. *Thus, the **reading frame** (nucleotide sequence actually translated into a protein) is unambiguously defined upon completion of the 70S initiation complex.*

Once the P site is filled, the A site of the 70S initiation complex becomes available to any tRNA molecule whose anticodon can pair with the codon adjacent to the initiation codon. An elongation protein (**EF-Tu**) facilitates delivery of charged tRNA to the A site. After occupation of the A site, a peptide bond is formed between *N*-formylmethionine and the adjacent amino acid by an enzyme complex called **peptidyl transferase.** As the bond is formed, the *N*-formylmethionine is cleaved from the fMet-tRNA in the P site, and a peptide bond is formed between

aminoacyl (A) site

peptidyl (P) site

peptidyl transferase

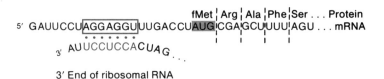

Figure 9-12 Initiation of translation in prokaryotes. Base pairing between the Shine-Dalgarno sequence (in box) in the mRNA and the complementary region (blue) near the 3′ terminus of 16S rRNA. The AUG start codon is shaded.

the initial methionine and the amino acid on the tRNA in the A site. Thus, a dipeptide is formed on the A site (**Figure 9-13**).

After the peptide bond forms, an uncharged tRNA molecule occupies the P site and a dipeptidyl-tRNA occupies the A site. At this point, two movements occur:

1. The tRNA^{fMet} in the P site, now no longer linked to an amino acid, leaves this site.
2. The A site peptidyl-tRNA remains hydrogen-bonded to the mRNA (in the A site codon). This tRNA-mRNA unit moves relative to the ribosome, so that the codon/anticodon bond is now in the ribosome's P-site.

The movement of peptidyl-tRNA from the A site to the P site, and the movement of mRNA in relation to the ribosome, is termed **translocation** (see Figure 9-13). Step (2) requires the presence of an elongation protein **EF-G** and GTP. After mRNA movement has occurred, the A site is again available to accept a charged tRNA molecule having a correct anticodon.

When a chain termination codon (either UAA, UAG, or UGA) is reached, no acylated tRNA exists that can fill the A site, so chain elongation stops. How-

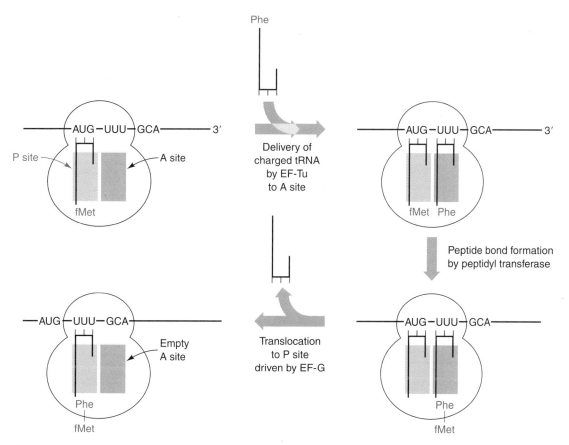

Figure 9-13 Elongation phase of protein synthesis: binding of charged tRNA, peptide bond formation, and translocation.

ever, the polypeptide chain is still attached to the tRNA occupying the P site. Release of the protein is accomplished by proteins called **release factors.** In the presence of release factors, peptidyl transferase separates the polypeptide from the tRNA; then the polypeptide chain, which has been held on the ribosome solely by the interaction with the tRNA in the P site, is released from the ribosome. The 70S ribosome then dissociates into its 30S and 50S subunits, completing the cycle.

If the mRNA molecule is polycistronic and the AUG codon initiating the second polypeptide is not too far from the stop codon of the first, the 70S ribosome will not always dissociate completely, but may re-form an initiation complex with the second AUG codon. The probability of such an event decreases with increasing separation of the stop codon and the next AUG codon. In some genetic systems, the separation is sufficiently great that more protein molecules are always translated from the first gene than from subsequent genes. This is a mechanism for maintaining particular ratios of gene products (discussed later in Chapter 11). Mutations sometimes arise that convert a sense codon (amino-acid specifying to a stop codon—for example, the mRNA molecule has a UAG codon at the site of a UAC codon. Such mutations, which will be discussed further in the next chapter, cause premature termination of a polypeptide. If the mutation is sufficiently far upstream from the normal stop codon of the gene containing the mutation, the distance to the next AUG codon may be so great that separation of the translating 70S ribosome and the mRNA molecule is almost inevitable. In this case, genes downstream from the mutation will rarely be translated. Such mutations are the most common type of polar mutation.

In eukaryotes immediate reinitiation of polypeptide synthesis following an encounter of a ribosome with a stop codon *does not occur.* Also, as pointed out earlier in this section, polypeptide synthesis in eukaryotes is initiated when a ribosome binds to the 5′ terminus of an mRNA molecule and slides along the first AUG codon. There is no mechanism for initiating polypeptide synthesis at any AUG other than the first one encountered; *eukaryotic mRNA is always monocistronic* (**Figure 9-14**). However, a primary transcript can contain coding

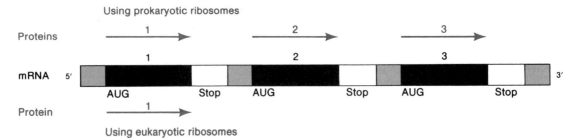

Figure 9-14 Difference in the products translated from a tricistronic mRNA molecule by the ribosomes of prokaryotes and eukaryotes. The prokaryotic ribosome translates all of the cistrons, but the eukaryotic ribosome translates only one cistron—the one nearest the 5′ terminus of the mRNA. Translated sequences are in black, stop codons are white, and the leader and spacers are shaded.

sequences for more than one polypeptide chain and, in fact, this is a frequent arrangement in animal viruses. In these cases, differential splicing generates several different mRNA molecules from the transcript (see Chapter 12). For example, if the AUG nearest the 5' terminus is excised, the second AUG will become available. Another mechanism for producing several proteins from a single transcript is protein processing. In this case, a single giant polypeptide chain, called a **polyprotein,** is made and then cleaved into several component polypeptide chains, each constituting a distinct protein.

Complex Translation Units

The unit of translation is almost never simply a ribosome traversing an mRNA molecule, but is a more complex structure, of which there are several forms. Some of these structures are described in this section.

After about 25 amino acids have been joined in a polypeptide chain, the AUG initiation site of the encoding mRNA molecule is completely free of the ribosome. A second initiation complex then forms. The overall configuration is two 70S ribosomes moving along the mRNA at the same speed. When the second ribosome has moved along a distance similar to that traversed by the first, a third ribosome is able to attach. This process, movement and reinitiation, continues until the mRNA is covered with ribosomes at a density of about one 70S particle per 80 nucleotides. This large translation unit is called a **polyribosome,** or simply a **polysome.** This is the usual form of the translation unit in all cells (both prokaryotic and eukaryotic). An electron micrograph of a polysome and an interpretive drawing are shown in **Figure 9-15**.

polysome

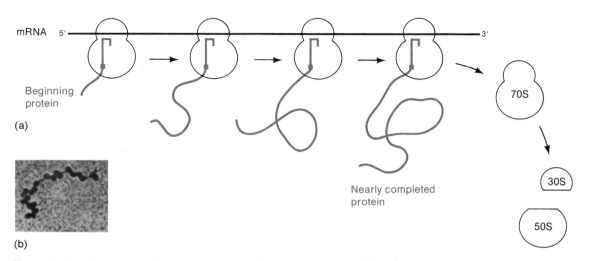

Figure 9-15 Polysomes. (a) Diagram showing relative movement of the 70S ribosome and the mRNA, and growth of the protein chain. (b) Electron micrograph of an *E. coli* polysome. (Courtesy of Barbara Hamkalo.)

The use of polysomes has a particular advantage to a cell; it increases the overall rate of protein synthesis compared to the rate that would occur if there were no polysomes.

An mRNA molecule being synthesized has a free 5′ terminus; since translation occurs in the 5′ → 3′ direction, each cistron contained in the mRNA is synthesized in a direction that is appropriate for immediate translation. The ribosome binding site is transcribed first, followed in order by the AUG codon, the region encoding the amino acid sequence, and finally the stop codon. Therefore, in bacteria in which no nuclear membrane separates the DNA and the ribosome, there is no obvious reason why the 70S initiation complex should not form before the mRNA is released from the DNA! With prokaryotes this does indeed occur; this process is called **coupled transcription-translation.** This coupled activity does not occur in eukaryotes. Recall that the eukaryotic mRNA is first synthesized and processed in the nucleus and then later transported through the nuclear membrane to the cytoplasm where the ribosomes are located.

Coupled transcription-translation speeds up protein synthesis in the sense that translation does not have to await release of the mRNA from the DNA. Translation can also be started before the mRNA is degraded by nucleases. **Figure 9-16**(a)

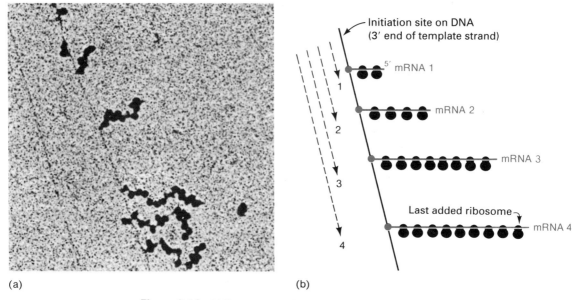

(a)   (b)

Figure 9-16   (a) Transcription of a section of the DNA of *E. coli* and translation of the nascent mRNA. Only part of the chromosome is being transcribed. The dark spots are ribosomes, which coat the mRNA. (From O. L. Miller, Barbara A. Hamkalo, and C. A. Thomas, *Science;* 1977: 169, 392.) (b) An interpretation of the electron micrograph of part (a). The mRNA is in blue and is coated with black ribosomes. The large blue spots are the RNA polymerase molecules; they are actually too small to be seen in the photo. The dashed arrows show the distances of each RNA polymerase from the transcription initiation site. Arrows 1, 2, and 3 have the same length as mRNA 1, 2, 3; mRNA 4 is shorter than arrow 4, presumably because its 5′ end has been partially digested by an RNase.

shows an electron micrograph of a DNA molecule to which are attached a number of mRNA molecules, each associated with ribosomes. The micrograph is interpreted in panel (b). Note that the lengths of the polysomes increase with distance from the transcription initiation site because the mRNA is farther from that site and, hence, longer.

Eukaryotic mRNA differs from prokaryotic mRNA in two major features. It is modified both by the addition of a cap at the 5′ end of the message and by the addition of about 200 adenlynic residues at the 3′ end, which are termed **poly A,** as described in the previous chapter. These modifications do not take place in mitochondrial and chloroplast mRNA. The role of the poly A in translation is not yet clear. The cap adds a 7-methylguanosine residue linked to a triphosphate at the 5′ end of the mRNA. Thus, this is a 5′ to 5′ linkage instead of the usual 5′ to 3′ linkage (**Figure 9-17**). The cap may protect the mRNA from exonucleases, but its main role is to promote and stimulate efficient translation. The 5′ untranslated region at the cap end of the eukaryotic mRNA is folded into a secondary structure held together by base pairing. This secondary structure inhibits efficient translation. Specific proteins, called cap-binding proteins, bind to the cap and unwind the secondary structure of the 5′ end of the mRNA. This facilitates efficient translation. These proteins are also involved in guiding the 40S eukaryotic ribosomal subunit to the cap, and, thus, to the 5′ end of the mRNA.

KEY CONCEPT

The "Prokaryotic Simplicity" Concept

RNA synthesis and protein synthesis are coupled in prokaryotes, separate (due to the nuclear membrane) in eukaryotes.

poly A

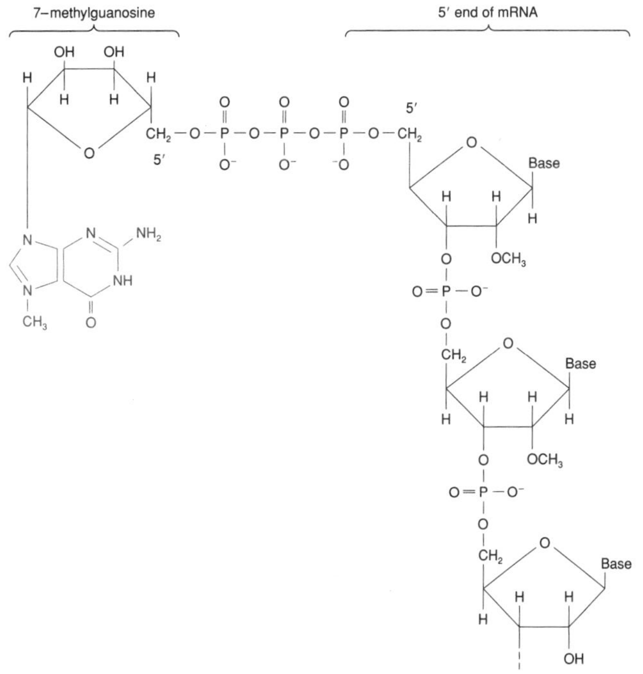

Figure 9-17 Structure of the cap at the 5′ end of eukaryotic mRNA. Note the unusual 5′ to 5′ linkage. All caps contain 7-methylguanylate (shown in blue) attached by a triphosphate linkage to the ribose of the 5′ end of the mRNA. This cap was encountered earlier (Figure 8-10). The 2′ hydroxyl group of the first mRNA ribose is also methylated, while the second sugar is not always methylated.

Antibiotics

antibiotics

Many antibacterial agents (**antibiotics**) have been isolated from fungi. Most of these are inhibitors of protein synthesis. For example: streptomycin and neomycin bind to a particular protein in the 30S particle and thereby prevent binding of tRNA^{fMet} to the P site; the tetracyclines inhibit binding of charged tRNA; lincomycin and chloramphenicol inhibit the peptidyl transferase; and puromycin causes premature chain termination; erythromycin binds to a free 50S particle and prevents formation of the 70S ribosome. Indeed, the experimentation that was carried out to unravel the mechanism of action of those antibiotics led, in many instances, to further understanding of several of the details of protein synthesis described in Figures 9-11 and 9-13!

A particular antibiotic has clinical value only if it acts on bacteria and not on animal cells; the clinically useful antibiotics usually either fail to pass through the cell membrane of animal cells, or do not bind to eukaryote ribosomes because of some unknown feature of their structure.

Some disease-causing bacteria exert their pathogenic effect because they excrete inhibitors of mammalian protein synthesis. The agent causing diphtheria is an example; it binds to a factor necessary for movement of mammalian ribosomes along the mRNA.

KEY CONCEPT

The "Antibiotic Action" Concept

The mechanism of action of most antibiotics is understood in terms of the antibiotic interfering with a specific step in protein synthesis, or a step in the synthesis of the cell wall, or other indispensable component of the bacterial cell.

A Future Practical Application?

Twenty amino acids are normally incorporated into proteins synthesized by living cells. Researchers are now, however, experimenting with the synthesis of proteins containing novel amino acids. Some of these "unnatural" amino acids, such as fluorotyrosine (a substituted form of naturally occurring tyrosine), are expected to make the properties of the proteins containing them more desirable for commercial purposes. For example, the folding properties of an enzyme used in a laundry detergent might be altered so as to enhance the stability of the protein in hot water.

The incorporation of novel amino acids requires, of course, artificial expansion of the genetic code. First, new bases, representing structural analogs of the naturally occurring ones (C, U, G, A) have been synthesized directly into mRNAs. Second, tRNAs containing a base in the anticodon loop that pairs with the novel base in the synthetic mRNA were prepared. Third, those new tRNAs were charged with the novel amino acids (e.g., iodotyrosine). Finally, those components were formulated into an in vitro translation system. The new base pairings were observed to function as efficiently as the standard ones!

New proteins have thereby been synthesized. The horizons for the development of proteins having either research value or industrial applications have been expanded. Although presently the process is probably too expensive for preparing vast quantities of specific proteins, even the small quantities of the novel proteins produced in this fashion can be employed by researchers to learn how to enhance the worth of nature's own proteins to commercial ventures.

SUMMARY

Translation is the production of proteins from the information encoded in DNA. That information (the genetic code) consists of the 64 three-base codons standing for translational starts, stops, and amino acids. For many amino acids, more than one codon exists, differing only in the third position. Charged tRNA molecules have an amino acid attached to the ends by aminoacyl tRNA synthetases, and their anti-codon base pairs were mRNA codons. One tRNA reads several codons since the first base of the anti-codon (wobble position) pairs with the third base of several codons. A reading frame is a series of sequential, nonoverlapping codons with one start and stop codon. Translation occurs on ribosomes and consists of three stages: (1) Initiation, which brings a small ribosomal subunit, initiating tRNA and mRNA together. The start codon is chosen by the Shine-Dalgarno sequence in prokaryotes that use formyl-methionine to initiate eukaryotes use the first AUG and methionine. The large subunit, containing the P (peptidyl) and A (aminoacyl) tRNA sites, then joins; (2) Elongation, which involves the binding of charged tRNAs, peptide bond formation by peptidyl transferase, and the translocation of the mRNA by one codon. GTP and initiation and elongation factors are required; (3) Termination, which occurs when a stop codon is reached. Release factors separate the polypeptide from the tRNA in the P sites. Following termination, the ribosome may dissociate, or in prokaryotes, may begin translation again at a new 3' AUG. Prokaryotic mRNAs, thus, can be polycistronic. Eukaryotic mRNAs are monocistronic, although one mRNA can produce several proteins by differential mRNA processing or cleavage of a polyprotein. In prokaryotes, transcription and translation are coupled, while in eukaryotes, they are separate.

DRILL QUESTIONS

1. Which of the following statements are true of tRNA molecules? If false, explain.
   (a) They are needed because amino acids cannot stick to mRNA.
   (b) They are much smaller than mRNA molecules.
   (c) They are synthesized without the need for intermediary mRNA.
   (d) They bind amino acids without the need for any enzyme.
   (e) They occasionally recognize a stop codon if the Rho factor is present.

2. Which of the following properties are part of the normal functioning of aminoacyl tRNA synthetases?
   ·(a) Recognition of the codon.
   (b) Recognition of the anticodon of a tRNA molecule.
   (c) Recognition of the amino acid recognition region of a tRNA molecule.
   (d) Ability to distinguish one amino acid from another.
   (e) Ability to remove an incorrectly coupled amino acid from a tRNA molecule.

3. What are the s values of prokaryotic and eukaryotic ribosomes, their subunits, and their RNA molecules?

4. List three differences between prokaryotic and eukaryotic translation (ignore differences in ribosome structure).

5. Which of the following are steps in protein synthesis in prokaryotes? If false, explain.

(a) Binding of tRNA to a 30S particle.
(b) Binding of tRNA to a 70S ribosome.
(c) Coupling of an amino acid to ribosome by an aminoacyl synthetase.
(d) Separation of the 70S ribosome to form 30S and 50S particles.

6. What is the reading frame of an mRNA? What additional features would you expect an mRNA to have?

7. Translation has evolved in a particular polarity with respect to the mRNA molecule. What is this polarity, and what would be the disadvantages of having the reverse polarity?

8. Which of the following is the normal cause of chain termination?
   (a) The tRNA corresponding to a chain-termination triplet cannot bind an amino acid.
   (b) There is no tRNA with an anticodon corresponding to a chain-termination triplet.
   (c) Messenger RNA synthesis stops at a chain termination triplet.

9. Which processes in protein synthesis require hydrolysis of GTP?

PROBLEMS

1. A DNA molecule has the structure

   TACGGGAATTAGAGTCGCAGGATC
   ATGCCCTTAATCTCAGCGTCCTAG

   The upper strand is the template strand, and is transcribed from left to right. What is the amino acid sequence of the protein encoded in this DNA molecule?

2. Which amino acids can replace arginine by a single base pair change?

3. The amber codon UAG does not correspond to any amino acid. Some strains carry suppressors that are tRNA molecules mutated in the anticodon, enabling an amino acid to be placed at a UAG site. Assuming that the anticodon of the suppressor differs by only one base from the original anticodon, which amino acids could be inserted at a UAG site? At a UAA site?

4. There are several arginine codons. Suppose you had a protein that contained only three arginines (Arg-1, Arg-2, Arg-3). In a particular mutant, Arg-1 is replaced by glycine. In another, Arg-2 is replaced by methionine. In still another mutant, Arg-3 is replaced by isoleucine. Suppose several hundred other mutants at various sites are isolated. Which other amino acids would you expect to find replacing Arg-1, Arg-2, and Arg-3, assuming only single-base changes?

5. Suppose that you are making use of the alternating copolymer GUGU-GUGUGU . . . as an mRNA in an in vitro protein-synthesizing system. Assuming that an AUG start codon is not needed in the in vitro system, what peptides are made by this mRNA?

6. Assume that the anticodon of a tRNA is 5′-IGU-3′.
   (a) Which amino acid would this tRNA insert during translation?
   (b) Which codons would be read by this tRNA?
   (c) Assuming that only two tRNAs are needed to read this family of codons, what would probably be the anticodon of the other tRNA?

CONCEPTUAL QUESTIONS

1. How might various antibiotics be used to study the process of translation?
2. Consider the advantages and disadvantages of overlapping reading frames.
3. What do you think is the significance of the universality of the genetic code? What does this say about the biological evolution of the system and its ability to change?

CHAPTER

10

in this chapter you will learn

1. About different types of mutations
2. How mutations are produced
3. How damage to DNA is repaired

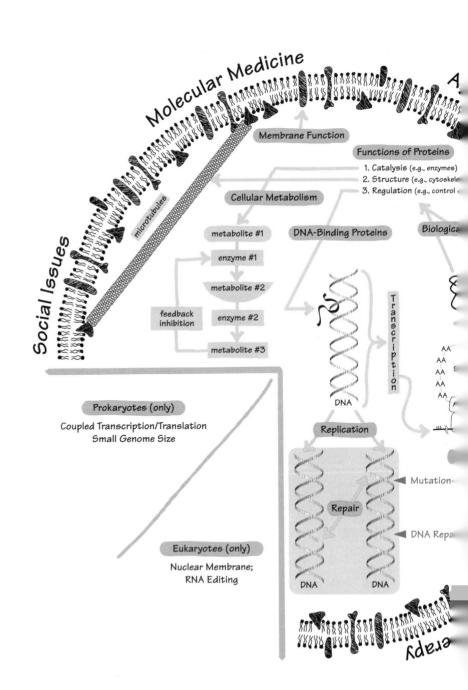

Molecular Medicine

Social Issues

microtubules

Membrane Function

Functions of Proteins

1. Catalysis (e.g., enzymes)
2. Structure (e.g., cytoskele
3. Regulation (e.g., control

Cellular Metabolism

metabolite #1

enzyme #1

metabolite #2

feedback inhibition

enzyme #2

metabolite #3

DNA-Binding Proteins

Biologica

Transcription

AA
AA
AA
AA
AA

DNA

Prokaryotes (only)

Coupled Transcription/Translation
Small Genome Size

Replication

Mutation

Repair

DNA Repa

Eukaryotes (only)

Nuclear Membrane;
RNA Editing

DNA   DNA

erapy

Mutations, Mutagenesis, and DNA Repair

There is no single molecule whose integrity is as vital to the cell as DNA. Indeed, survival of the species depends upon maintaining the nucleotide sequence intact. The genetic material is, however, subject to constant challenge. During DNA replication, RNA transcription, and even while in an inactive, resting state, damage can occur. Copying errors, breaks, and damage to nucleotide bases, if not corrected, lead to permanent change—mutation. Most mutations are potentially dangerous, since they lead to either blocks in DNA replication or the production of defective proteins. Thus, in the course of hundreds of millions of years there have evolved efficient systems for correcting occasional replication errors and for eliminating damage to DNA caused by environmental agents and intracellular chemicals.

We will begin our study of mutations and DNA repair by defining what it means to be a "mutant" and briefly surveying the various types of mutations that occur in living organisms. Next, we will consider some of the ways in which both spontaneous and induced mutation may arise, and examine the mechanisms of action of specific mutagenic agents. Finally, we will review some of the systems that have evolved to maintain the integrity of the DNA and to repair DNA damage once it does occur.

Types of Mutations

Mutant refers to an organism or a gene that is different from the normal or wild type. His⁻ yeast or white-eyed *Drosophila* are examples of mutants. Sometimes, the "normal" form of an organism found in nature **lacks** the ability to carry out a certain biochemical reaction (for example, *E. coli* isolated from nature is unable to metabolize lactose, or has a *lac⁻* phenotype). In many such instances, the "+" form is referred to as the "wild type," and any "−" form is referred to as a mutant.

Mutation refers to any heritable change in the base sequence of a DNA molecule. The most common type of change is a substitution, addition, or deletion of one or more bases (**Table 10-1**).

A **mutagen** is a physical or chemical agent that causes mutations to occur (or increases the frequency of their occurrence).

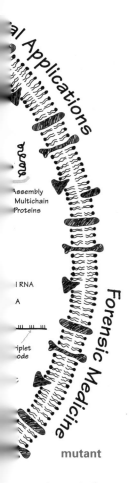

al Applications

Assembly
Multichain
Proteins

I RNA

A

riplet
ode

Forensic Medicine

mutant

mutation

mutagen

TABLE 10-1	EXAMPLES OF TYPES OF POINT MUTATIONS	
Type of Mutation	**Result at Molecular Level**	**Example**
A. Base Substitution Mutations:		
Changes in DNA		
1. Transition:	One purine replaced by a different purine; *or* one pyrimidine replaced by a different pyrimidine.	A-T → G-C
2. Transversion:	A purine replaced by a pyrimidine or vice versa.	A-T → T-A
Changes in protein		
1. Silent mutation:	Altered codon codes for same amino acid.	GAG → GAA Glu Glu
2. Neutral mutation:	Altered codon codes for a different but functionally similar amino acid. (Protein *may* be functional.)	GAG → GAU Glu Asp
3. Missense mutation:	Altered codon codes for a different, dissimilar amino acid. (Protein often nonfunctional.)	GAG → AAG Glu Lys
4. Nonsense mutation (= Chain termination mutation):	New codon is a termination codon. (Protein synthesis stops. Protein is nonfunctional.)	GAG → UAG Glu Stop

B. *Frameshift Mutations:* Addition or deletion of one or more base pairs will result in a shift in the reading frame of the resulting mRNA molecule, and lead to production of a nonfunctional protein.

1. Wild type base sequence: ATG ACC AGG TC

2. Base addition: ATG AC<u>A</u> CAG GTC
                       *|_____|

3. Base deletion: ATG ACA GGT C
                  Missing C ↑
                       *|_____|

*Horizontal brackets indicate the affected segment.

mutagenesis

Mutagenesis is the process of producing a mutation. If the mutation occurs in nature without the addition of a mutagen, it is referred to as a **spontaneous mutation,** and the resulting mutants are **spontaneous mutants.** If it is caused by a mutagen, the process is **induced mutagenesis.** In this chapter, we will concentrate on mutations occurring in living organisms. Later, in Chapter 16, we will discuss **site-specific mutagenesis,** a form of induced mutagenesis in which genetic engineering techniques are used to construct mutant DNA molecules containing mutations at specific, preselected locations.

Mutations are classified in several ways. One distinction is based upon the number of bases changed in the DNA molecule. A **point mutation,** for example, involves a single changed base pair, while a **multiple mutation** involves alterations in two or more base pairs. A point mutation may be a **base substitution,** a

base addition, or a **base deletion** (Table 10-1), but the term is usually reserved for base substitutions.

A second distinction is based upon the consequence of the change in terms of the amino acid sequence affected. As described in Table 10-1, a single-base substitution may cause little or no change in the amino acid sequence of the protein or in the ability of the protein to function normally (**silent mutations, neutral mutations),** or it may lead to the production of a nonfunctional and/or truncated polypeptide (**missense mutations, nonsense mutations).**

If the substitution produces a protein that is active at one temperature (typically at 30°C or lower) and inactive at a higher temperature (usually 40°–42°C), the mutation is called a **temperature-sensitive** or **ts mutant.** If the mutation generates a stop codon, causing protein synthesis to cease, the mutation is called a **chain-termination** or **nonsense mutation.** Temperature-sensitive and chain-termination mutations are considered to be **conditional mutations** because they exhibit the mutant phenotype only under certain conditions. Such mutations are extremely useful to the molecular biologist, since they can facilitate the study of essential processes, such as DNA replication, without which a mutant organism could not survive.

With microorganisms, the phenotype is capitalized (Lac$^+$ or Lac$^-$), while the genotype is written in lower-case italics (*lac$^+$* or *lac$^-$*). This convention does not apply to higher organisms. The + or − notation also indicates whether the bacterium in question is able (+) or unable (−) to synthesize or utilize a specific substance. (His$^+$ bacteria can synthesize histidine, while His$^-$ bacteria cannot.) Another commonly used notation indicates whether a particular bacterial strain is resistant (Amp-r) or sensitive (Amp-s) to a particular antibiotic, ampicillin in this instance. The corresponding genotypes of organisms having these phenotypes would be designated *amp-r* and *amp-s,* respectively.

missense mutations

nonsense mutations

temperature-sensitive

conditional mutations

Biochemical Basis of Mutants

A mutant may be defined as an organism in which either the base sequence of the DNA or the phenotype has been changed. These definitions are often the same, since a single base change in a DNA molecule can lead to an alteration in the amino acid sequence of a protein. (Exceptions to this general rule occur in the cases of **silent mutations** and **neutral mutations,** which will be discussed soon.) The chemical and physical properties of each protein are determined by its amino acid sequence, so that a single amino acid change is capable of inactivating a protein.

From the discussion of protein structure in Chapter 4, it is easy to understand how an amino acid substitution can change the structure and biological activity of a protein. For instance, consider a hypothetical protein whose three-dimensional structure is determined entirely by an interaction between one positively charged amino acid (for example, lysine) and one negatively charged amino acid (glutamic acid). A substitution of methionine, which is uncharged, for lysine would clearly destroy the three-dimensional structure, as would the substitution of histidine, which is positively charged, for glutamic acid. Similarly, a protein might be stabilized by a hydrophobic cluster, in which case substitution of glutamine (polar) for leucine (nonpolar) would also be disruptive.

A base substitution does not always yield a mutant phenotype. Because of the redundancy in the code, some base changes do not alter the amino acid sequence **(silent mutations),** and some amino acid changes do not significantly affect the structure of the protein **(neutral mutations).** (See Table 10-1.)

The shapes of proteins are determined by such a variety of interactions that sometimes an amino acid substitution is only partially disruptive. For instance, an isoleucine might substitute successfully for leucine (and be neutral), but replacement with a more bulky amino acid such as phenylalanine might cause subtle stereochemical changes, although the hydrophobic cluster is preserved. This could be manifested as a reduction, rather than a loss, of activity of an enzyme. A bacterium carrying such a mutation in the enzyme that synthesizes adenine might grow very slowly (but it would grow), unless adenine were provided in the growth medium. Such a mutation is called a **leaky mutation;** these mutations are not particularly useful for most genetic studies. However, several hereditary disorders in humans are due to such leaky mutations. For example, in certain individuals, the gene that codes for the essential enzyme, glucose-6-phosphate dehydrogenase, contains a point mutation, resulting in the production of a defective form of the enzyme having greatly reduced catalytic activity. Individuals who possess this mutant gene develop severe hemolytic anemia when exposed to a variety of common substances, including sulfonamide antibiotics, antimalarial drugs, mothballs, and even certain types of dried beans (fava beans).

Generally speaking, the following types of amino acid substitutions are expressed as nonleaky mutations:

Polar amino acid ↔ Nonpolar amino acid

Change of sign of amino acid (+ ↔ −)

Small side chain ↔ bulky side chain

Sulfhydryl ↔ any other side chain

Hydrogen-bonding amino acid ↔ Non-hydrogen-bonding amino acid

Proline (changes the shape of the polypeptide backbone) ↔ any other amino acid

Any change in the substrate-binding site

So far, only amino acid substitutions have been discussed. Certain other types of mutations tend to eliminate the activity of the protein totally. These include **base additions** and **base deletions (= frameshift mutations),** in which all of the amino acids starting from the mutant site are different, and **chain termination mutations (= nonsense mutations),** in which a protein chain is terminated prematurely. Both will be discussed in greater detail later in this chapter.

Mutagenesis

The production of a mutant requires that a change occur in the DNA base sequence. There are a number of distinct mechanisms for altering the structure of

DNA. These include base substitutions, additions, or deletions during replication, base changes resulting from the inherent chemical instability of the bases or of the *N*-glycosylic bond, and alterations resulting from the action of other chemicals and environmental agents. These mechanisms are summarized in **Table 10-2**.

TABLE 10-2   COMMON DEFECTS IN DNA AND THEIR ORIGINS

Type of Defect	How Does This Type of Change Arise?	
1. Incorrect base in one strand cannot hydrogen bond with corresponding base in the opposite strand	Normal base tautomerizes (i.e., isomerizes in such a way that it is capable of an alternative form of hydrogen bonding); base substitution occurs during subsequent DNA replication 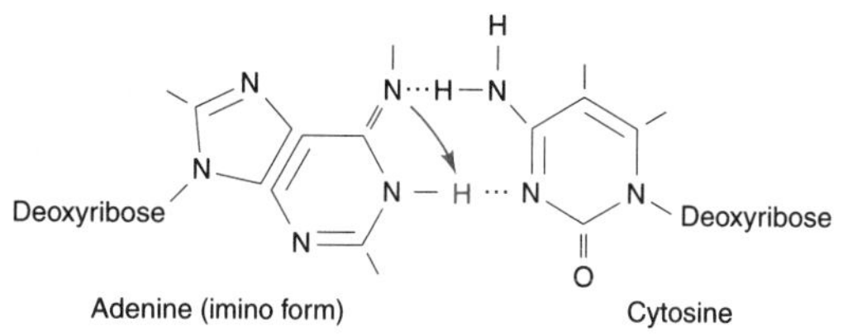 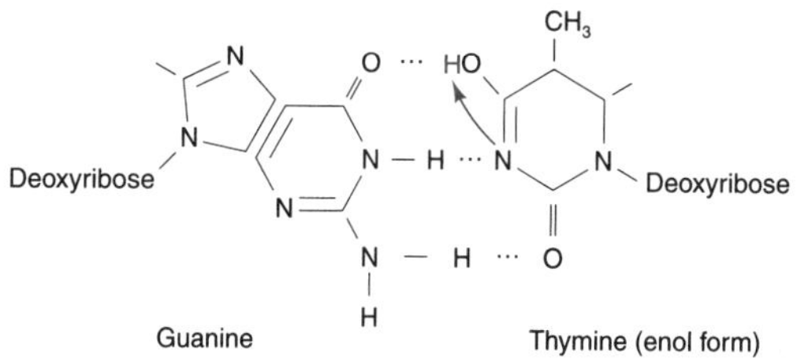	 Adenine (imino form) Cytosine Guanine Thymine (enol form)
2. Missing bases	Depurination: *N*-glycosylic bond joining purine base to deoxyribose is spontaneously broken without breaking DNA backbone 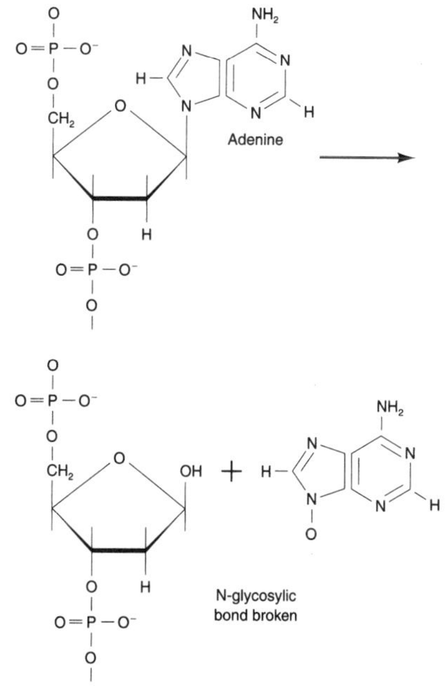	 Adenine N-glycosylic bond broken

(continued on next page)

TABLE 10-2	COMMON DEFECTS IN DNA AND THEIR ORIGINS	
Type of Defect	**How Does This Type of Change Arise?**	
3. Altered bases	Alkylating agents add methyl or ethyl groups to existing bases	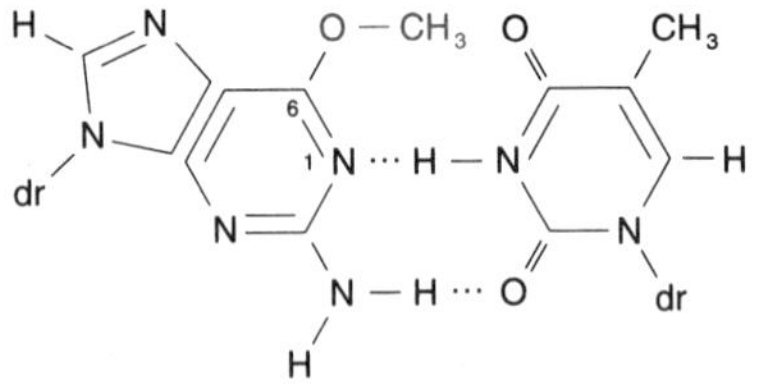 O⁶-Methylguanine, Thymine structures
4. Addition or deletion of one or more bases	May occur spontaneously, or be induced by chemical mutagens (intercalating agents) or biological agents (transposable elements)	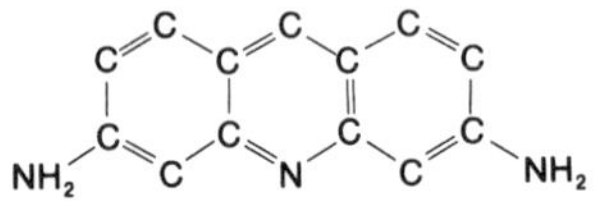 Proflavine, 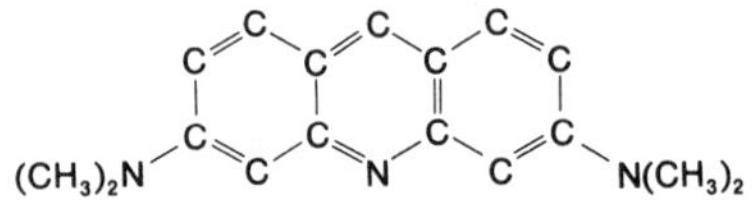 Acridine orange structures
5. Single-strand breaks	Phosphodiester bond is broken as a result of exposure to chemical agents or ionizing radiation	Peroxides, metal ions such as Fe^{2+} and Cu^{2+}, x-rays
6. Double-strand breaks	Phosphodiester bonds on opposite DNA strands are broken as a result of exposure to high doses of chemical agents or ionizing radiation	Peroxides, metal ions such as Fe^{2+} and Cu^{2+}, x-rays
7. Cross-linking of complementary DNA strands	Certain antibiotics (mitomycin-C) or reagents (nitrite ions) form covalent bonds between two bases on complementary DNA strands, preventing strand separation during DNA replication	Mitomycin-C, psoralen, *cis*-platinum

Induced Mutations

During its lifetime, an organism may be exposed to a variety of physical, chemical, and biological agents that are capable of causing damage to its genetic material. Each agent has its own characteristic mechanism of action, and each has a tendency to cause specific types of damage to the DNA molecule. With the widespread use of site-specific mutagenesis, as discussed in Chapter 16, chemical mutagens are no longer as widely used experimentally as they were in the past. Nevertheless, they have played a key role in helping us to gain an understanding of mutations and mutagenesis. We will briefly consider some of these classical mutagenic agents. An overview of the mode of action of a number of chemical mutagens is presented in **Table 10-3**.

TABLE 10-3	SOME COMMONLY USED CHEMICAL MUTAGENS AND THEIR MECHANISMS OF ACTION		
Type of Mutagenic Agent	**Example of This Type of Agent**	**Mode of Action of This Agent**	**Diagram of Mode of Action**
1. Base analogue: Resembles a normal base found in DNA and is incorporated into DNA; later undergoes a tautomeric shift and mispairs, causing a transition mutation.	5-Bromouracil (base analogue of T) Structure: 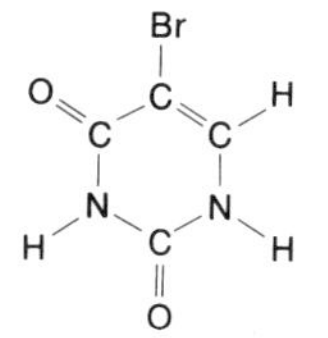 5-Bromouracil (keto form)	Normally pairs with A, but can undergo a tautomeric shift and pair with G. This results in incorporation of C into the daughter DNA strands during subsequent rounds of DNA replication.	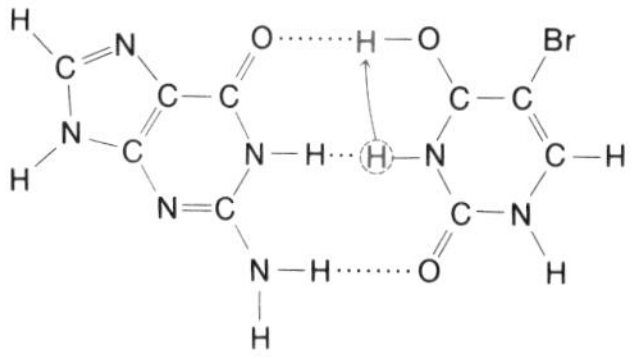 Guanine 5-Bromouracil (enol form) $AT \rightarrow GC$
2. Nitrous acid	Structure: 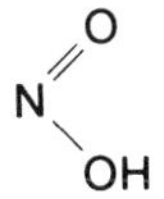 **Nitrous acid**	Converts amino groups to keto groups by oxidative deamination. $C \rightarrow$ uracil (U) (pairs with A) $A \rightarrow$ hypoxanthine (H) (pairs with C) $G \rightarrow$ xanthine (X) (pairs with C)	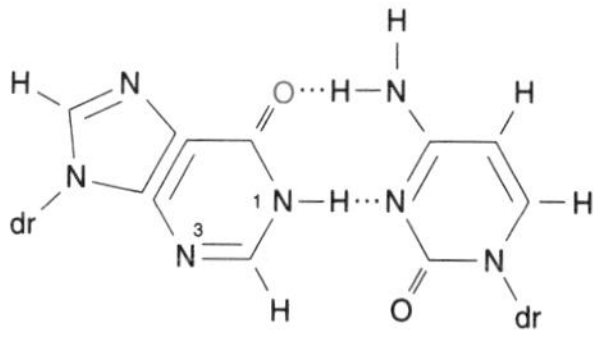 Hypoxanthine Cytosine $AT \rightarrow GC$
3. Hydroxylamine	Structure: 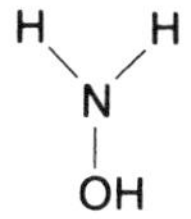 **Hydroxylamine**	Reacts with C and converts it to a modified base that pairs only with A.	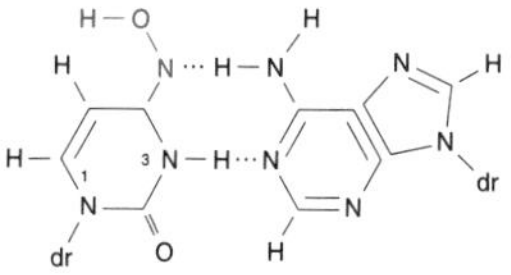 Hydroxylaminocytosine Adenine $CG \rightarrow TA$
4. Alkylating agents	EMS (ethylmethane sulfonate) and MMS (methylmethane sulfonate) Structure: 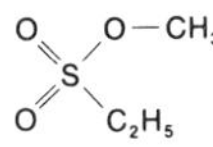 **Ethylmethane sulfonate**	Add alkyl groups (ethyl or methyl) to the hydrogen-bonding oxygen of G and T, producing O-6 alkylguanine (pairs with T) and O-4-alkylthymine (pairs with G).	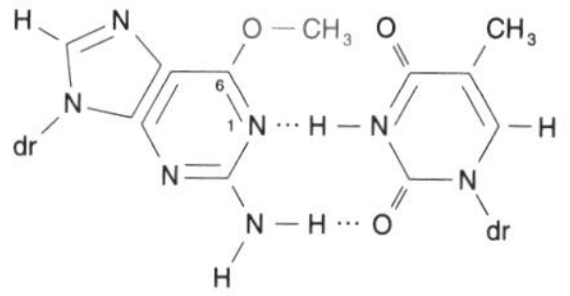 O^6-Methylguanine Thymine $GC \rightarrow AT$
5. Intercalating agents: Planar, three-ringed molecules whose dimensions are roughly the same as those of a purine-pyrimidine base pair.	Proflavine, acridine orange Structures: 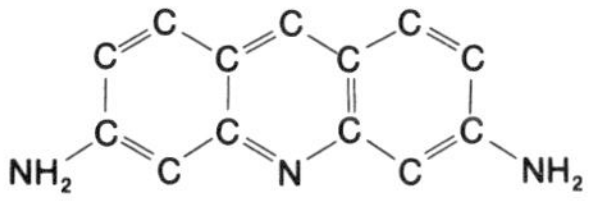 **Proflavine** 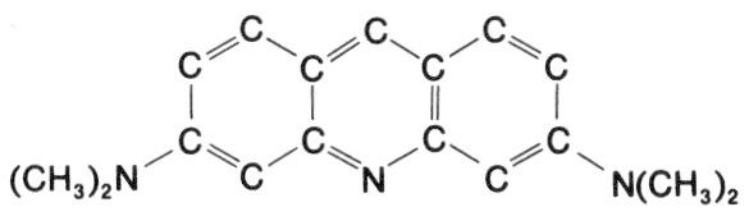 **Acridine orange**	Insert between two adjacent base pairs in a DNA molecule, causing insertions or deletions.	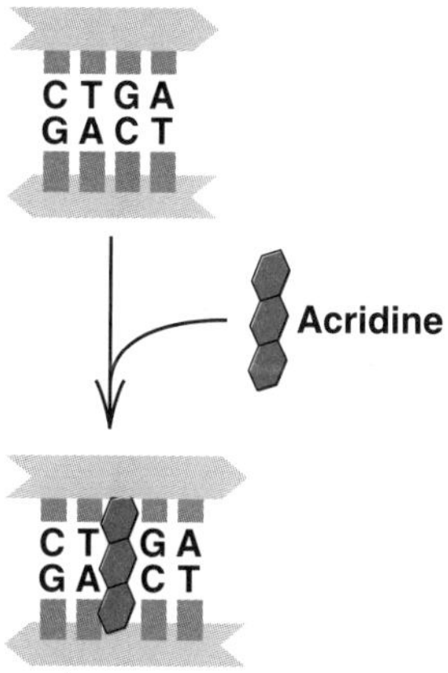 C T G A / G A C T → Acridine C T G A / G A C T

Ultraviolet Irradiation

Ultraviolet light is a fairly potent mutagen. The main DNA lesions are two chemically different types of covalent dimers: cyclobutane-pyrimidine dimers and (604) pyrimidine dimers (**Figure 10-1**). The latter only account for 20% of the total dimers, but are correspondingly more mutagenic. In *E. coli* the number of mutants induced by ultraviolet (UV) light can be reduced by exposure to visible light **(photoreactivation),** which implicates cyclobutane pyrimidine dimers since (6-4) dimers cannot be photoreactivated. Bacterial mutants that lack the ability to carry out SOS repair (*lexA⁻* and *recA⁻* mutants) are not mutagenized by ultraviolet light. These and other experimental results have made it clear that mutagenesis by UV light is almost exclusively a result of error-prone SOS repair. Error-prone repair of pyrimidine dimers leads to the production of both transitions and transversions. Photoreactivation, SOS repair, and two other repair systems employed by cells to deal with UV damage are discussed later in this chapter.

Mutagenesis by Insertion of Long Segments of DNA (Transposable Elements)

E. coli, and many other organisms as well, contain mobile DNA segments that are hundreds to thousands of base-pairs long (called **transposable elements**). When a transposable element replicates, one replica often remains in the original insertion site, and the other replica is inserted in another region of the chromosome in a complex process discussed in Chapter 14. This process of insertion of a replica at a second site is called **transposition.** When transposition occurs, the sequence frequently inserts itself into a bacterial gene, thereby mutating that gene.

Some transposable elements contain sequences for termination of transcription. If such an element inserts between two genes that are transcribed as a polycistronic mRNA, or between the promoter and the first gene, all genes downstream

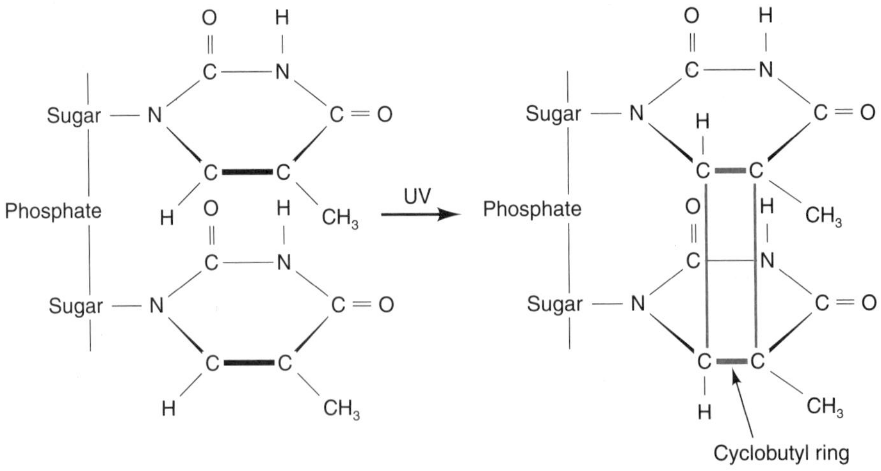

photoreactivation

transposition

Figure 10-1 Structure of a cyclobutylthymine dimer. Following ultraviolet (UV) irradiation, adjacent thymine residues in a DNA strand are joined by formation of the bond shown in blue. Although not drawn to scale, these bonds are considerably shorter than the spacing between the planes of adjacent thymines, so that the double-stranded structure becomes distorted. The shape of the thymine ring also changes as the C=C double bond of each thymine is converted to a C—C single bond in each cyclobutyl ring.

from the insertion site will not be transcribed. These mutations belong to a class called **polar mutations.**

Mutator Genes

In *E. coli* there are genes that, when present in a mutant state, cause mutations to appear very frequently in other genes throughout the genetic map. These genes are called **mutator genes.** This is a misnomer, because the function of these genes is probably to keep the mutation frequency low; only when the product of the mutator gene is itself defective will there be widespread production of mutations.

Of the many mutator genes that have been observed, four types that yield strong phenotypic effects are:

1. A mutant DNA polymerase that reduces or eliminates the $3' \rightarrow 5'$ exonuclease activity of the editing function.
2. A mutant methylating enzyme (the *dam* enzyme) responsible for methylation of the sequences that the mismatch repair system uses to discriminate parental strands from daughter strands.
3. A mutant enzyme that cannot carry out the excision step of mismatch repair.
4. Mutations in the regulatory circuits that maintain the error-prone SOS repair system in an "off" state.

Reversion

So far we have discussed changes from the wild type to the mutant state. The reverse process, in which the wild-type phenotype is regained, also occurs; this process is called **back mutation, reverse mutation,** or, most commonly, **reversion.** It may occur spontaneously, or may be induced by a variety of mutagenic agents. There are several different ways in which reversion may occur. In some cases, the back mutation occurs at the **same site** as the original mutation and restores the wild-type base sequences. This represents a **"true reversion."** In other instances, the wild-type phenotype is restored as a result of a second mutation occurring at a *different* site. Such an event may be thought of as a **"pseudoreversion,"** and mutations of this latter type are described as **second-site** or **suppressor mutations.** This second site may be situated elsewhere within the *same* gene (**intragenic suppression**) or within a *different* gene (**intergenic suppression**). We will first consider the intragenic type of revertants.

Intragenic Revertants

Let's consider a hypothetical protein containing 97 amino acids whose structure is determined entirely by an ionic interaction between a positively charged (+) amino acid at position 18 and a negative one (−) at position 64 (**Figure 10-2**). If the (+) amino acid is replaced by a (−) amino acid, the protein is clearly inactive. Three kinds of reversion or suppression events would restore activity (Figure 10-2(a)):

1. The original (+) amino acid could be put back, either as a result of restoration of the *identical* base sequence or because the genetic code is redundant, by substitution of a base sequence that codes for the same amino acid.
2. A different (+) amino acid could be put in position 18.

mutator genes

KEY CONCEPT

The "Mutation → Variation" Concept

Mutations, which represent changes in the nucleotide sequence of genes, generate the wide-ranging diversity that characterizes the biological kingdom.

reversion

suppressor mutations

intragenic suppression
intergenic suppression

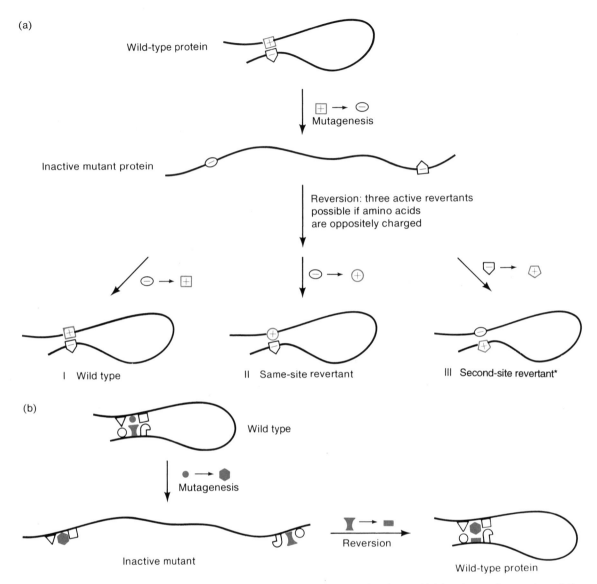

Figure 10-2 Several mechanisms of reversion. In panel (a) the charge of one amino acid is changed, and the protein loses activity. The activity is returned by (I) restoring the original amino acid, (II) replacing the (−) amino acid by another (+) amino acid, or (III) reversing the charge of the original (−) amino acid. In each case the attraction of opposite charges is restored. In panel (b) the structure of the protein is determined by interactions between six hydrophobic amino acids. Activity is lost when the small circular amino acid is replaced by the bulky hexagonal one, and is restored when space is made by replacing the convex amino acid with the small rectangle.

3. The (−) amino acid at position 64 could be replaced by a (+) amino acid; this second mutation would restore the activity of the protein. A possibility that would not generally work, but which might work in a specific case, is to insert a (+) amino acid at position 17 or 19.

Figure 10-2(b) shows another more complicated example of an intragenic reversion. In this case, the structure of a protein is maintained by a hydrophobic interaction. The replacement of an amino acid with a small side chain by a bulky pheny-lalanine changes the shape of that region of the protein. A second amino acid sub-stitution providing space for the phenylalanine might restore the protein structure.

The analysis of second-site amino acid substitutions has been an important aid in determining the three-dimensional structure of proteins because the fol-lowing rule is always obeyed: If a substitution of amino acid A by amino acid X, which creates a mutant, is compensated for by a substitution of amino acid B by amino acid Y, then A and B are either three-dimensional neighbors, or both are contained in two interacting regions.

Revertants of frameshift mutations usually occur at a second site. It is, of course, possible that a particular added base could be removed or a particular deleted base could be replaced by a spontaneous event, but this would not occur very often. Second-site reversion of a frameshift mutation has two requirements, illustrated in **Figure 10-3**:

1. The reverting event must be very near the original site of mutation so that very few amino acids are altered between the two sites.
2. The segment of the polypeptide chain in which both changes occur must be able to withstand substantial alteration.

Intergenic Suppression

Intergenic suppression refers to a mutational change that eliminates or suppresses the mutant phenotype. One type, which occurs when two proteins interact, is a mutation in the binding site of a protein A that prevents the protein from inter-acting with protein B. A second mutation, in the binding site of protein B, alters this binding site so that the mutant protein B can bind to the mutant protein A. As a result, the interaction between the two proteins is restored. The occurrence of intergenic suppression of this kind is an important indicator of the interaction be-tween two proteins and is a splendid example of a genetic result giving informa-tion about molecular structure.

A second type of intergenic suppression has the remarkable property that the second-site mutation not only eliminates the effect of the original mutation, but

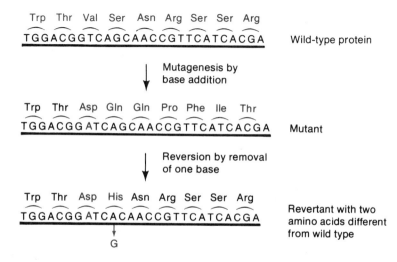

Figure 10-3 Reversion by base deletion from an acridine-induced, base-addition mutant.

also suppresses mutations in many other genes as well. This type of suppression, which is produced by mutations in certain tRNA molecules and aminoacyl tRNA synthetases, can be seen most clearly when suppression of chain-termination mutations is examined.

Chain-termination mutations (also called **nonsense mutations**) are common, for they can arise in many ways. For example, a single-base change in any of the codons AAG, CAG, GAG, UCG, UUG, UGG, UAC, and UAU can give rise to the chain-termination codon UAG. If such a mutation occurs within a gene, a mutant protein with little or no function will result because no normal tRNA molecule exists whose anticodon is complementary to UAG. Only a fragment of the wild-type protein is produced, and this usually fails to function unless the mutation is very near the carboxyl terminus of the protein.

In certain bacterial strains, the presence of a chain-termination mutation is not sufficient to stop polypeptide synthesis. For example, a phage may have acquired a UAG codon in a gene encoding a critical protein. When the phage infects most host bacteria, no phage progeny will be produced. In a particular bacterial strain, however, the mutant phage may grow normally, indicating that the mutation is made silent by some element in the bacterium. Such a bacterium **suppressor** is said to be able to **suppress** the mutation and to contain a **suppressor.** A bacterium able to suppress a particular type of chain-termination mutation—a UAG codon in the example—is usually able to do so with a large number of mutations in that class, whether the mutation is in a phage or in the bacterium itself. In general, other types of chain-termination mutations, for example, UGA, will not be suppressed. The explanation for this phenomenon is that the bacterium (called a **suppressor mutant** **suppressor mutant**) contains an altered tRNA molecule that can respond to a particular stop codon. For UAG, the altered tRNA might contain the anticodon **suppressor tRNA** CUA, which can pair with that codon. Such a tRNA molecule is called a **sup-** **nonsense suppressor** **pressor tRNA** (or a **nonsense suppressor tRNA**), and the mutation on which it **tRNA** can act is said to be suppressor sensitive.

What has been mutated in the production of a suppressor tRNA? Clearly, it must be a normal tRNA gene. Therefore, in the example just given, a tRNA^Lys molecule whose anticodon is CUU has been altered to have an anticodon CUA, which can hydrogen bond to the codon UAG.

Inasmuch as a single-base change is sufficient to alter the complementarity of an anticodon and a codon, there are (at most) eight tRNA molecules having a complementary anticodon that, with a single changed base, will also suppress a UAG codon. Thus, the following amino acids (whose codons are also indicated) can be put in the site of a chain-termination codon: Lys (AAG), Gln (CAG), Glu (GAG), Ser (UCG), Trp (UGG), Leu (UUG), and Tyr (UAC and UAU). Note that these are the same amino acid codons that can be altered by mutation to form a UAG site. Suppressors also exist for chain-termination mutants of the UAA and UGA types. These too are mutant tRNA molecules whose anticodons are altered by a single-base change.

In conventional notation, suppressors are given the genetic symbol *sup,* followed by a number (or occasionally a letter) that distinguishes one suppressor from another. A cell lacking a suppressor is designated *sup0* and *sup⁻*.

Several features of nonsense suppressors should be recognized:

1. Not every UAG suppressor can restore a functional protein by suppressing each UAG chain-termination codon. Thus, a UAG codon produced by mutating the leucine UUG codon might be suppressed by a suppressor tRNA that inserts tyrosine, serine, or tryptophan, but might not be able to tolerate a substitution by the electrically charged amino acids lysine, glutamine, or glutamic acid.

2. Suppression may be incomplete in that the activity of the suppressed mutant protein may not be as great as that of the wild-type protein, and the stop codon may not always be read as a sense codon.

3. A cell can survive the presence of a suppressor only if the cell also contains two or more copies of the tRNA gene. Clearly, if a tRNA^{Ser} molecule that reads the UCG codon is mutated, then the UCG codon can no longer be read as a sense codon. This will lead to chain termination wherever UCG occurs, and a cell harboring such a mutant tRNA molecule will fail to complete virtually every protein made by the cell. Therefore, in any living cell containing a suppressor tRNA, there must always be an additional copy of a wild-type tRNA that can function in normal translation.

If a cell contains a UAG suppressor, the proteins terminated by a single UAG codon will not be terminated, and the existence of a suppressor tRNA should be lethal. There are two ways that this problem can be avoided:

1. Protein factors active in termination (see Chapter 9) respond to chain-termination codons even though a tRNA molecule that recognizes the codon is present; i.e., suppression is weak.

2. Normal chain termination often uses pairs of distinct termination codons such as the sequence UAG-UAA. The existence of a UAG suppressor, thus, would not prevent termination of a double-terminated protein.

Suppression of missense mutations also occurs. For example, a protein in which valine (nonpolar) has been mutated to aspartic acid (polar), resulting in loss of activity, is restored to the wild-type phenotype by a missense suppressor that substitutes alanine (nonpolar) for aspartic acid. Such a mutation can occur in three ways: (1) a mutant tRNA molecule may recognize two codons, possibly by a change in the anticodon loop; (2) a mutant tRNA can be recognized by a noncognate aminoacyl tRNA synthetase and be misacylated; and (3) a mutant synthetase can charge a noncognate tRNA molecule. Examples of each type of suppression are known. Suppression of missense mutations is necessarily inefficient. If a suppressor that substitutes alanine for aspartic acid worked with, say, 20% efficiency, then, in virtually every protein molecule synthesized by the cell, at least one aspartic acid would be replaced and the cell could not possibly survive. The usual frequency of missense suppression is about 1%.

Reversion As a Means of Detecting Mutagens and Carcinogens

In view of the increased number of chemicals present as environmental contaminants, and because many cancer-causing agents (**carcinogens**) are also mutagens, tests for the mutagenicity of these substances have become important. One sim-

carcinogens

ple method for screening large numbers of substances for mutagenicity is a reversion test using nutritional mutants of bacteria. In the simplest reversion test, known numbers of a mutant bacterium are plated on growth medium containing a potential mutagen, and the number of revertant colonies that arise is counted. If the substance is a mutagen, the number of colonies will be greater than that obtained in the absence of the mutagen. However, such simple tests fail to demonstrate the mutagenicity of many carcinogens. These substances are not directly mutagenic (or carcinogenic), but are converted to actively mutagenic compounds by enzymatic reactions that occur in the livers of mammals, and that have no counterpart in bacteria. The normal function of these enzymes is to protect the organism from various noxious substances that occur naturally by enzymatically converting them to nontoxic substances. However, when the enzymes encounter certain man-made and natural compounds, they convert these substances (which may not themselves be directly harmful) to mutagens or carcinogens. The enzymes are contained in the microsomal fraction of liver cells. Addition of the microsomal fraction to the bacterial growth medium allows these substances to undergo enzymatic "activation," and enables researchers to determine their mutagenicity under conditions more closely resembling those found in the liver. This

Ames test   is the basis of the **Ames test** for potential carcinogens.

In the Ames test, histidine-requiring (His^-) mutants of the bacterium *Salmonella typhimurium,* which contain either a base substitution or a frameshift mutation, are used to test for reversion to His^+. The frequency of spontaneous His^+ revertants is low in this mutant, but mutants are readily produced in one or both of these mutants by most known mutagens. Solid medium is prepared containing a very small amount of histidine, sufficient to initiate the growth of individual cells, but not enough to support colony formation. Minimal histidine is necessary because many mutagenic agents act only on cells that are actively involved in DNA replication, and two rounds of DNA replication and cell division are often required for the new mutant phenotype to be expressed. A small amount of rat liver extract and about 10^8 His^- mutants are spread on each plate. Then the substance to be tested is applied to one set of plates (Group A = experimental group), while distilled water is applied to the plates in the control group (Group B), which lacks the carcinogen. The number of colonies appearing on the plates in Group B (control) is usually about 5 to 10 (these are the spontaneous revertants). When a known mutagen is present, more colonies will be present on the plates in Group A. The number of colonies on the Group A plates depends upon the concentration of the substance being tested and, for a known mutagen, correlates roughly with its known effectiveness as a mutagen (and, hence, as a potential carcinogen).

The Ames test has not yet been used with thousands of substances and mixtures (industrial chemical, food additives, pesticides, hair dyes, and the like) and numerous unsuspected substances have been found to stimulate reversion in this test. This does not mean that the substance is definitely a carcinogen, but only that it has a high probability of being one. As a result of these tests, many industries have reformulated their products to render them nonmutagenic. Ultimate proof of carcinogenicity is determined by testing for tumor formation in laboratory animals. The Ames test and several other microbiological tests are used to reduce the number of substances that need to be tested in animals, since, to date, only a few percent of the more than 300 substances known to be carcinogens from animal experiments have failed to increase the reversion frequency in the Ames test.

DNA Repair Mechanisms

Maintaining the integrity of the genetic information is essential to the survival of the cell. Without a doubt, this is the reason that living organisms have evolved such a diverse repertoire of DNA repair mechanisms. The process begins during DNA replication, and continues, in various forms, throughout the postreplication period. Let us now begin to consider the intricacies of DNA repair.

Spontaneous Mutations and Their Repair

Spontaneous mutations are those that arise naturally, not through the action of a mutagenic agent. They may arise through (1) errors in DNA replication or (2) spontaneous alteration of a nucleotide within an existing DNA molecule.

As has already been described in Chapter 7, DNA polymerases occasionally catalyze the incorporation of an incorrect base that cannot form a hydrogen bond with the template base in the parent strand. Such errors are usually corrected by the editing (or proofreading) function of these enzymes. In the event that an occasional misincorporated base is overlooked during proofreading, a second correction system, called **mismatch repair,** exists to deal with such mispaired bases (**Figure 10-4**). In mismatch repair, a pair of non–hydrogen-bonded bases is recognized as incorrect, and a polynucleotide segment is excised from one strand, thereby removing one member of the unmatched pair.

If it is to eliminate errors, the mismatch repair system must be able to distinguish the correct base in the parental strand from the incorrect base in the daughter strand. In *E. coli,* the critical information is provided by the presence or absence of **adenine methylation** in the base sequence G-A-T-C. These bases carry methyl groups not found elsewhere in the DNA. Methylation of the adenines in this sequence is associated with DNA replication. However, it does not take place in the replication fork, but is somewhat delayed. The effect is that, while parental strands are fully methylated, the newly synthesized daughter strands near the replication fork are not yet methylated. The mismatch repair system recognizes the degree of methylation of each strand, and when a mismatch is found, it preferentially excises nucleotides from the *undermethylated* strand—that is, from the daughter strand (see Figure 10-4). Thus, the parent strand is always the template, enabling the repair system to correct misincorporation errors. However, despite the existence of proofreading and mismatch repair, errors occasionally escape detection and give rise to mutations.

Spontaneous mutations may also arise as a consequence of tautomeric shifts in the nucleotide bases. Each of the four nitrogen bases is capable of existing in a rare, alternate isomeric form (**tautomer,**) having different hydrogen-bonding properties (Table 10-2). The rare imino forms of adenine (**A***) and cytosine (**C***) base pair with cytosine and adenine, respectively, while the rare enol forms of thymine (**T***) and guanine (**G***) hydrogen bond with guanine and thymine, respectively. A base might be incorporated in its normal form, then undergo a tautomeric shift after incorporation. During a subsequent round of DNA replication, the presence of one of these rare tautomers in the template could result in the incorporation of an incorrect base into the daughter DNA molecule. (A similar process occurs when mutagenic agents called **base analogues** undergo tautomerization, as mentioned earlier in Table 10-3.)

mismatch repair

base analogues

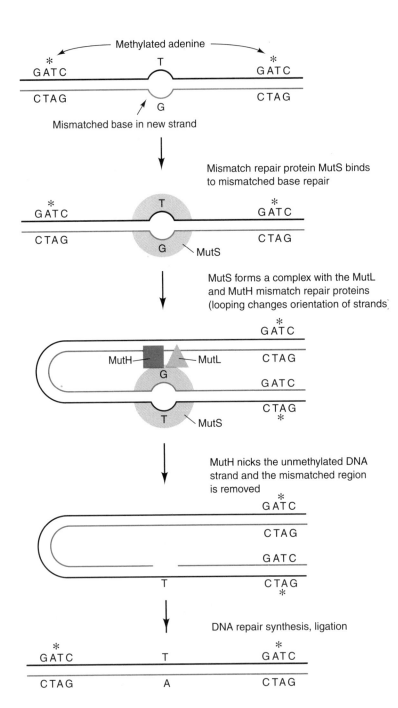

Figure 10-4 Possible mechanism of mismatch repair in *E. coli*. Mismatch repair enzyme recognizes mismatched base pair and undermethylated daughter DNA strand. Mismatched region is excised and repaired.

Depurination refers to the breakage of the *N*-glycosylic bond between a base (in this case a purine) and the deoxyribose of the nucleotide, with the subsequent loss of the nitrogen base. Unless such damage is repaired prior to DNA replication, there is a great likelihood that a mutation will arise, since the **apurinic site** lacks the "information" necessary to specify the insertion of the correct complementary base into the new DNA strand. If the DNA polymerase should happen to reach the apurinic site before repair has taken place, replication might come to a halt, and/or an incorrect base might be inserted into the new DNA strand opposite the apurinic site. Once again, specific repair mechanisms involving enzymes called **AP endonucleases (AP = apurinic)** exist to deal with such damage, and will be discussed below.

Deamination is another process that can give rise to mutations. It has been estimated that, in an average human cell, cytosine deamination occurs at a rate of 100 cytosines per genome each day. When cytosine loses an amino group, it becomes uracil. After one round of replication, this would lead to replacement of the original G-C pair with an A-U pair, which would then become an A-T pair after another round of replication. Since this would be mutagenic, cells have evolved a mechanism for replacing the unwanted U by a C. The first step in their repair cycle (Figure 10-5) is removal of the uracil by the enzyme **uracil *N*-glycosylase.** This enzyme cleaves the *N*-glycosylic bond and leaves the deoxyribose in the backbone. A second enzyme, **AP endonuclease** (which acts on apurinic or "baseless" sites in general), makes a single cut in the DNA backbone on the 5′ side of the damaged site, and the sugar phosphate residue is removed by a **phosphodiesterase.** DNA polymerase then fills the gap with the correct nucleotide and ligase seals the gap. (This sequence, endonuclease-exo-nuclease-polymerase, is an example of a general repair mechanism called excision repair, described later in the chapter.) A specific glycosylase is also available to remove hypoxanthine, the consequence of adenine deamination.

The latter part of this repair mechanism, namely that beginning with AP endonuclease (shown in Figure 10-5), is also used in the repair of missing bases arising by depurination.

Deamination

uracil *N*-glycosylase

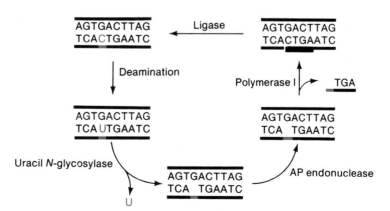

Figure 10-5 Scheme for repair of cytosine deamination. The same mechanism could remove a uracil that is accidentally incorporated.

Mutational Hot Spots and
5′-Methylcytosine Deamination

If a hundred mutations in a single gene are mapped, they are, for the most part, distributed roughly equally over the mutated sites. However, a few sites are represented by as many as 100 times the typical number of mutations; these sites are called **hot spots.**

About 4% of the cytosines in a typical DNA molecule are in the methylated form 5′-methylcytosine (MeC). The role of methylcytosine (and other methylated bases) is not clearly understood in many cases. Earlier in this chapter, we described the role of adenine methylation in mismatch repair, and we shall examine the additional role of cytosine methylation in regulation of gene expression in vertebrates in Chapter 12. At any rate, methylated cytosines are not inherently harmful and do not change the hydrogen bonding properties of the base. MeC pairs with guanine just as cytosine does.

MeC is also subject to alteration by spontaneous deamination, and this leads to a curious situation. When MeC is deaminated, it becomes 5-methyluracil, which is another name for the normal base thymine. Therefore, the G-MeC base pair becomes a G-T pair, which in subsequent replication yields an A-T pair. Note that the main mismatch repair system is certainly able to convert the G-T pair back to a correct G-C pair, but *only* if it is able to distinguish between the parental and daughter strand on the basis of degree of adenine methylation. Unfortunately, spontaneous deamination can also occur in nonreplicating DNA (e.g., in a resting cell or in a phage), where both strands have been equally methylated by the adenine-methylating system. The mismatch repair system, thus, receives no signal, indicating that the G-C pair is the correct one, and could just as well convert the G-T pair to an A-T pair. The mutation frequency consequently can be quite high at a MeC site. A given G-MeC → A-T transition will, of course, only produce a mutation if the change causes an amino acid substitution that affects the activity of the gene product. Such MeC sites do not occur very often, so hot spots should not be particularly frequent. Direct determination of the base sequence of several genes and of hot spot mutants has shown that, indeed, MeC accounts for most of the hot spots for spontaneous mutagenesis.

A repair system does exist that is capable of correcting the results of MeC deamination. A specific mismatch repair enzyme recognizes mismatched G-T base pair and *always* removes the mispaired thymine, rather than the guanine, so that the correct (G-C) base-pairing sequence can be restored. The new cytosine can be remethylated at a later time.

Repair by Direct Reversal

The simplest mechanism for repair of altered bases is to restore them to normal without major surgery on the DNA molecule. In the case of pyrimidine dimers formed in DNA by ultraviolet light (see Figure 10-1), this reversal can be accomplished very simply by an enzyme that recognizes and binds to these dimers. This enzyme, called **photolyase,** is activated by visible light and cleaves the dimers to yield intact pyrimidines. Photolyase is found in many types of cells, but operates only on pyrimidine dimers and only when activated by light (300–600 nm).

Another example of direct reversal is the methyltransferase (an alkyltransferase) that recognizes 0^6 methyl guanine in DNA and removes the offending methyl group by attaching it instead to its own cysteines. The exchange reaction restores the normal guanine in the DNA, but inactivates the enzyme. The importance of this particular repair scheme, which is used to repair damage caused by alkylating agents such as MMS, is underscored by the fact that an entire protein molecule is expended for each 0^6 methyl guanine repaired.

Excision Repair

The most ubiquitous repair scheme, and one that can deal with a large variety of structural defects in DNA, is called excision repair. This multistep enzymatic process is best understood in *E. coli* (in which it was originally discovered). It is a mechanism by which pyrimidine dimers can be repaired in the DNA without photolyase. The essential steps of this process, as it occurs in *E. coli,* are illustrated in **Figure 10-6**. Variations on this basic "cut and patch" theme occur in other organisms. In the first step, **incision,** a repair endonuclease recognizes the distortion produced by a thymine dimer (or other bulky lesions) and makes two cuts in the sugar-phosphate backbone, 12 to 13 nucleotides apart, on each side of the dimer (one cut is located eight nucleotides to the 5′ side, the other four to five nucleotides to the 3′ side). In *E. coli,* the product of three genes (*uvrA, uvrB,* and *uvrC*) act in concert to produce the incisions. In other systems, a single incision is adequate to initiate the repair sequence. At the incision site illustrated in Figure 10-14, there is a 5′-P group on the side of the cut containing the dimer, and a 3′-OH group on the other side. The 3′-OH group is recognized by polymerase I, which can then synthesize a new strand while simultaneously displacing the DNA segment carrying the thymine dimer. In the final step of the process, ligase joins the newly synthesized segment to the original DNA strand.

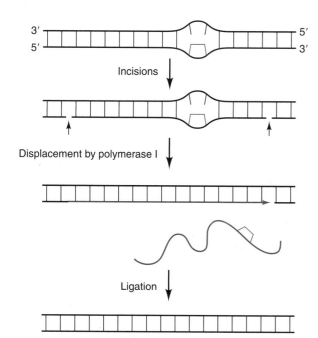

Figure 10-6 Scheme for excision repair of a thymine dimer by the "cut and patch" mechanism. The thymine dimer and the repair patch are both shown in blue.

The excision-repair system has been shown to preferentially repair pyrimidine dimers that are located in certain essential regions of the genome, notably within expressed genes. In genes that are being actively transcribed, the repair is somehow targeted selectively to the strand that is being used as a template for mRNA synthesis. This special treatment makes sense because a pyrimidine dimer poses an absolute block to the process of transcription and could kill a cell quite directly if it prevented expression of an essential gene.

In mammalian cells, recall that chromatin structure is much more complex than it is in bacteria such as *E. coli*. Thus, the process of excision repair involves many more enzymes. At least a dozen different genes are involved in recognizing and excising the damaged DNA segment. Individuals who suffer from a rare genetic disorder known as **xeroderma pigmentosum (XP)** are extremely sensitive to sunlight and have a tendency to develop skin cancer after very short exposures to the sun. Recent evidence has shown that xeroderma pigmentosum can result from a mutation in any one of seven different genes involved in various steps in the human excision-repair process. Cells from XP patients are sensitive to UV light and have a reduced ability to remove thymine dimers from their DNA. Another known genetic disease, Cockayne's syndrome, is also characterized by sensitivity to UV light, but without the tendency to develop skin cancer. Its victims do suffer from a variety of other problems, including neurological and developmental abnormalities. It now appears that some of the genes involved in the excision-repair pathway may play a dual role. In fact, one of the proteins involved in the unwinding of the damaged segment of DNA (the *XPB* gene product, which functions as a helicase) also functions as a component of a crucial transcription factor (TFIIH), which is involved in regulating the activity of all protein-encoding genes. This could explain the observation that the defect in Cockayne's syndrome affects the preferential repair of expressed genes, as previously discussed.

Recombinational Repair

The excision-repair systems just described are responsible for the removal of many thymine dimers and other lesions. However, before sufficient time has elapsed for their repair, many dimers have interfered with various cellular processes. The deleterious effects of some of the remaining dimers are eliminated by recombinational repair, which is carried out by the system responsible for genetic recombination.

In order to discuss the mechanism of recombinational repair, it is necessary to know the effect of a thymine dimer on DNA replication. When polymerase III reaches a thymine dimer, the replication fork fails to advance. A thymine dimer is still capable of forming hydrogen bonds with two adenines because the chemical change in dimerization does not alter the groups that engage in hydrogen bonding. However, the dimer introduces a distortion into the helix, and when an adenine is added to the growing chain, polymerase III reacts to the distorted region as if a mispaired base had been added; the editing function then removes the adenine. The cycle begins again—an adenine is added and then it is removed; the net result is that the polymerase is stalled at the site of the dimer. (The same effect would occur if, instead of a dimer, radiation or chemical damage resulted in the formation of a base with which no nucleoside triphosphate could base pair.) Evidence that such a phenomenon occurs after ultraviolet irradiation is the existence of a UV-light-induced idling process—that is, rapid cleavage of deoxynucleoside triphosphates to monophosphates without any net DNA synthesis (i.e., without advance of the replication fork). A cell in which DNA synthesis is permanently stalled cannot complete a round of replication, and does not divide.

There are basically two different ways in which DNA synthesis can get going again—**postdimer initiation** and **transdimer synthesis.**

One way a cell could deal with a thymine dimer is to pass it by and reinitiate chain growth beyond the block, perhaps at the starting point for the next Okazaki fragment (**Figure 10-7**). The result of this process is that the daughter strands would have large gaps, one for each unexcised thymine dimer. There is no way to produce viable daughter cells by continued replication alone, because the strands with the thymine dimers would continue to turn out gapped daughter strands, and the first set of gapped daughter strands would be fragmented when the growing fork entered a gap. However, by a recombination mechanism known as **sister-strand exchange,** proper double-stranded DNA molecules can be produced.

The essential idea in sister-strand exchange is that a single-stranded segment free of any defects is excised from a "good" strand on the homologous DNA segment and inserted into the gap created by the bypassing of the thymine dimer during replication (**Figure 10-8**). This recombinational event requires the RecA protein. The gap created in the donor molecule by excision of the "patch" is then repaired by polymerase I and ligase. If this exchange and gap filling are done for each thymine dimer, two complete daughter strands can be formed, and each can serve in the *next round* of replication as a template for synthesis of normal DNA molecules. Note that the system fails if two thymine dimers in opposite strands are very near to one another because then no undamaged sister-strand segments are available to be excised. Many of the molecular details of recombinational repair are not yet known, so the model shown in Figure 10-8 must be considered to be a working hypothesis reflecting our present state of knowledge.

Recombinational repair is an important mechanism because it eliminates the necessity for delaying replication for the many hours that would be needed for excision repair to remove all thymine dimers. It may also be the case that some kinds of damage cannot be removed by excision repair—for example, alterations that do not cause helix distortion, but do stop DNA synthesis.

Since recombinational repair occurs after DNA replication, in contrast with excision repair, it is often called **postreplicational repair.** It should be noted, however, that what is repaired is the gap in the daughter strand, rather than the thymine dimer.

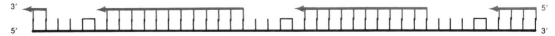

Figure 10-7 Blockage of replication by thymine dimers (represented by joined lines) followed by restarts several bases beyond the dimer. The black region is a segment of ultraviolet-light-irradiated parent DNA. The blue region represents synthesis of a daughter molecule from right to left. The daughter strand contains gaps.

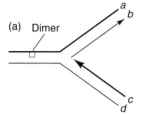

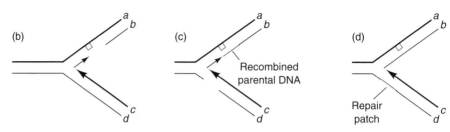

Figure 10-8 Recombinational bypass. (a) In strand *a,* a molecule containing a thymine dimer (blue box) is being replicated. (b) Replication is blocked at the site of the dimer, leaving a gap because an Okazaki fragment cannot be completed. (c) That gap is filled by a sister–strand recombination event in which a segment of the parent strand *d* is utilized. That now leaves a gap in strand *d.* (d) The daughter strand *c* can serve as a template for repair DNA synthesis to fill the gap in strand *d.* DNA synthesized after irradiation is shown in blue. Heavy and thin lines are used only to identify strands of the same polarity.

Despite the evidence that thymine dimers and other helix-distorting lesions block the progression of DNA polymerase, it is clear that most types of cells, including mammalian cells, can tolerate large numbers of persisting lesions without resorting to recombinational strand exchange. Somehow replication is able to overcome the lesions, albeit with an expected high error rate at those noncoding sites.

The SOS Response

SOS response system

While most DNA repair mechanisms are constitutive (i.e., active all the time), a few are activated in response to some signal, such as a blocked replication fork. Most notable is the **SOS response system,** characterized in *E. coli* as a complex regulatory scheme in which the products of two genes, *recA* and *lexA,* govern the expression of a number of other genes involved in DNA repair (**Figure 10-9**). Some of these genes (*uvrA* and *uvrB*) are involved in excision repair, while others like *umuC* and *umuD* are required to help replication bypass the offending lesion. In order to continue DNA replication across a region containing a thymine dimer, the editing function of polymerase III must be relaxed, otherwise the helical distortion caused by the dimer would trigger pol III's $3' \rightarrow 5'$ proofreading activity, and the replication fork would become stalled. Relaxation of proofreading is not without its own inherent hazards. It leads to **error-prone DNA replication,** and results in a higher-than-normal level of mutagenesis. Because this is detrimental to the cell, expression of the genes involved in the SOS response must be carefully regulated in order to maintain the fidelity of DNA replication under normal circumstances. In the presence of certain types of potentially lethal DNA damage, the system is switched "on" as a last-ditch effort to allow the cell to survive (hence the name "SOS" response). The enhanced mutagenesis that results may be seen as a benefit in that it may result in progeny that, through mutation, are better adapted to live in the noxious environment that led to the induction of the SOS system.

The genes of the SOS response are ordinarily maintained in an "off" state. This is accomplished by the *lexA* gene product, the LexA repressor, which binds to an operator sequence near each gene and prevents its transcription. The LexA

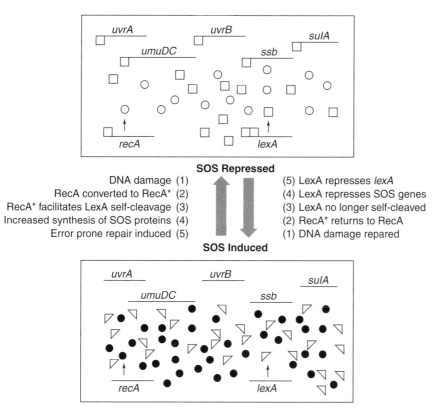

Figure 10-9 The SOS response. In uninduced cells, the RecA protein [○] is not activated and, thus, does not facilitate self-proteolysis of the LexA protein [□]. The LexA protein functions as a repressor, turning off transcription of many different genes, including the *recA* gene and the *lexA* gene itself (i.e., it is autoregulatory). DNA-damaging agents induce the SOS response by activating RecA protein to RecA*[●], which facilitates the self-cleavage of LexA protein [▽] and several other proteins, including the λ repressor. After LexA protein is cleaved, it cannot function as a repressor, resulting in transcription of all of the genes regulated by the LexA protein. As the SOS genes repair the DNA damage, RecA* returns to RecA, LexA is no longer cleaved, and accumulation of LexA represses the SOS genes.

repressor even represses its own expression, so that only small amounts of this protein are synthesized routinely. (Regulation of gene expression in prokaryotes will be discussed in greater detail in Chapter 11.) The RecA protein plays two different, complementary roles in the SOS response. First, it inhibits the editing function of DNA polymerase III. It binds to the distorted region of the DNA molecule containing the dimer; when pol III encounters this region, the RecA protein interacts with the polymerase subunit involved in proofreading. This results in inhibition of pol III's editing function and allows the replication to proceed.

Binding of the RecA protein to the single-stranded DNA also has a second effect: it causes a conformational change in the RecA protein (→ RecA*). This activated form of the RecA* protein interacts with the LexA repressor protein, stimulating the LexA repressor to inactivate itself by proteolytic cleavage. With the repressor inactivated, all of the genes involved in the SOS response can be expressed, and the enzymes required for DNA repair are synthesized. (Although this also leads to high levels of expression of the *lexA* gene, the repressor will continue to be cleaved as long as the activated RecA* is present.) Eventually, DNA repair will be completed and the RecA protein will once again lose its proteolytic activity. This will allow LexA repressor levels to increase, shutting off the expression of the genes involved in the SOS response, and allowing pol III to resume normal proofreading.

The first evidence for the SOS system was obtained 25 years before the mechanism began to be understood. The experiment, illustrated in **Figure 10-10**, involved an analysis of the survival of UV-irradiated bacteriophage λ plated on *E. coli*. The survival of the UV-irradiated phage was markedly enhanced if the bacterial host had also been irradiated with a low UV dose. However, in that case, a much higher percentage of mutants were obtained among the surviving phages. This phenomenon (termed **UV reactivation**) is now understood in terms of the SOS system. The low UV dose to the bacteria activates the SOS error-prone replication of the damaged phage. (As you recall, bacteriophages are dependent upon *host enzymes* for bacteriophage DNA replication.)

There are examples of inducible cellular responses to other types of environmental stress, such as heat shock, oxidative DNA damage, or agents that produce 0^6 methyl guanine. In the latter case, there is an *adaptive response* in *E. coli* that increases the level of the methyltransferase enzyme 100-fold when the cells are "conditioned" by growth in the presence of agents that produce this lesion.

KEY CONCEPT

The "DNA Repair" Concept

Despite the inherent chemical stability of double-stranded DNA, from time to time it sustains damage, and, hence, its continual repair is necessary for cells to function properly.

Future Practical Applications?

Mutations in one particular gene, the p53 gene (a 53,000 dalton molecular weight protein) are known to be responsible for more than 50% of human cancers, including breast, prostate, brain, lung, liver, pancreatic, colorectal and adrenal cancers, soft tissue sarcomas, osteosarcoma, lymphoma, and melanoma. The protein product of the normal, nonmutant p53 gene acts as a transcription factor and tumor suppressor, and is involved in regulation of the cell cycle, DNA repair, and programmed cell death. When DNA damage occurs, expression of p53 increases, leading to the arrest of the cell cycle. This gives the cell time to repair its DNA before it is replicated and passed along to a new generation of daughter cells. In cells in which extensive DNA damage has occurred, the p53 gene triggers the cell to undergo apoptosis, or programmed cell death.

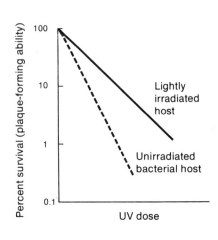

Figure 10-10 Ultraviolet (UV) reactivation of UV-light-irradiated phage λ. The dashed line shows the survival curve (for plaque-forming ability) obtained when λ phage irradiated with various doses of UV light are plated on unirradiated bacteria. The solid line represents survival of plaque-forming ability, when UV-light irradiated λ are plated on lightly irradiated bacteria.

If the p53 gene is mutated, it is no longer able to arrest the cell cycle in response to DNA damage. Mutations begin to accumulate in a variety of other genes, the end result being development of certain forms of cancer. As our understanding of the role of the p53 gene has increased, it has become possible to use this knowledge to predict the likelihood that patients will develop cancer or that they will respond well to chemotherapy. We can even design a gene therapy protocol that may be effective against their particular form of cancer. Overexpression of the p53 gene is associated with an increased risk of developing cancer, and screening for levels of p53 expression may prove to be a valuable diagnostic tool in individuals who are at high risk of developing cancer.

Once patients have been diagnosed with a malignancy, analysis of their levels of p53 expression may enable physicians to determine whether traditional chemotherapy is likely to be an effective treatment. Conventional chemotherapy is largely dependent upon the ability of the anticancer drug to induce tumor regression by stimulating the tumor cells to undergo apoptosis. In patients possessing a mutant p53 gene and overexpressing the p53 protein, tumors are likely to be resistant to treatment involving conventional chemotherapy drugs.

Now that the p53 gene has been characterized, clinical trials involving its use in gene therapy have been initiated. Liver, head and neck cancer, lung cancer, and even oral cancers have been targeted in recent human gene therapy trials involving recombinant adenoviral vectors carrying the human p53 allele. Preliminary results are encouraging. In the case of gene therapy directed against head and neck cancer, for example, tumors shrank in 25 out of 30 patients. Researchers are optimistic that p53 gene therapy may offer new hope of successful treatment of a variety of cancers, particularly when used in conjunction with chemotherapy.

SUMMARY

Mutagenesis is the process by which mutations (changes in base sequence of a DNA molecule) are created. An organism carrying a mutation is called a mutant. A base substitution is a change in a single nucleotide, and can be a missense mutation (causing the change of a single amino acid) or a nonsense mutation (causing protein-synthesis termination). Base substitutions may be either transitions, e.g., Py-Pu to Py-Pu, or transversions, e.g., Py-Pu to Pu-Py. A frameshift mutation is an addition or loss of a nucleotide causing an alteration of the reading frame. A silent mutation causes no change in the phenotype. The phenotype of a conditional mutant is seen only under certain conditions. Mutations may be spontaneous (due to uncorrected replication errors or natural changes), or induced by a chemical or physical mutagen. Base-analogue mutagens are incorporated into DNA via normal base-pairing interactions during DNA replication, but in future rounds of replication they form incorrect base pairs, leading to a mutation. Chemical mutagens alter existing bases in DNA, changing their base-pairing properties or leading to error-prone DNA replication. Ultraviolet radiation can also induce error-prone replication. Intercalating agents insert between adjacent base pairs in DNA and cause frameshift mutations during DNA replication. Mutations can also be caused by transposable elements that insert into genes, or by mutator genes that, if mutated, lead to mutation in other genes. Mutational hot spots are highly mutable regions of DNA. Reversion or suppression is the process by which a mutant phenotype is returned to wild type, and may occur by a change at another site in the same gene (intragenic)

(*continued*)

SUMMARY (*Continued*)

or in a different gene (intergenic). The Ames test uses reversion to test for the possible mutagenic effects of chemical compounds.

Processes exist to repair the various DNA alterations that may occur in living organisms. Mismatch repair is used to replace an incorrect base that was not corrected by proofreading. This repair system recognizes a pair of non-hydrogen-bonded bases, excises the one that is incorrect, and replaces it with the proper base. A pathway involving the enzyme uracil-*N*-glycosylase is responsible for repair of cytosine deamination. Four mechanisms exist for the repair of thymine dimers. Photolyase is an enzyme that

cleaves the cyclobutyl ring of the dimer when activated by visible light. The excision repair system makes a cut in the backbone on either side of the dimer, allowing the displacement and subsequent replacement of the portion of the strand containing the dimer. Recombinational repair involves postdimer reinitiation of DNA synthesis and sister-strand exchange, and prevents long delays of replication. The SOS response system also prevents delay of replication; this system allows for transdimer synthesis—at the expense of a higher rate of base misincorporation by the replication enzymes.

DRILL QUESTIONS

1. In what way does 5-bromouracil (BU) function as a mutagen? What type of mutations would occur from its use?

2. Several hundred independent missense mutants, altered in the A protein of tryptophan synthetase, have been collected. Originally, it was hoped that at least one mutant for each of the 186 amino acid positions in the protein would be found. However, fewer than 30 of the positions were represented with one or more mutants. Suggest some possibilities to explain why this set of missense mutants was so limited.

3. Since nonsense suppressors are mutant tRNA molecules, how does the cell survive loss of a needed tRNA by such a mutation?

4. Suppose a hypothetical enzyme contains 156 amino acids. (This number has no significance in the problem.) Assume amino acid 28, which is glutamic acid, is replaced in a mutant by asparagine and, as a result, all enzymic activity is lost. Suppose in this mutant protein, amino acid 76, which is asparagine, is replaced by glutamic acid and full activity of the enzyme is restored. What can you say about amino acids 28 and 76 in the normal protein?

5. Two DNA alterations occur so frequently that they are considered to be the weak points of a DNA molecule. What are these two changes?

6. Uvr⁺ bacteria possess the excision-repair system. The ability of UV-irradiated T4 phage to form plaques is the same on both Uvr⁺ and Uvr⁻ bacteria. How might you explain this fact?

7. Which of the enzymes listed below is involved in repair of *both* thymine dimers and deaminate cytosine in *E. coli*?
   (a) DNA polymerase III
   (b) Photolyase
   (c) Uracil *N*-glycosylase
   (d) DNA polymerase I
   (e) AP endonuclease

PROBLEMS

1. Which of the following amino acid substitutions would probably yield a mutant phenotype? Explain your reasoning.
   (a) Pro to His
   (b) Lys to Arg
   (c) Ile to Thr
   (d) Ile to Val
   (e) Ala to Gly
   (f) Phe to Leu
   (g) Try to His
   (h) Arg to Ser

2. Consider a bacterial gene containing 1,000 base pairs. As a result of treatment of a bacterial culture with a mutagen, mutations in this gene are recovered at a frequency of one mutant per 10^5 cells. One of these mutants is grown, and a pure culture of this mutant is obtained. This culture is then treated with the same mutagen and revertants are found at a frequency of one per 10^5 cells. Would you expect the gene product obtained from the revertant to have the same amino acid sequence as the wild-type cell? Explain.

3. A particular mutant shows absolutely no activity. Despite an exhausting search for revertants by your colleagues, using such mutagens as 5-bromouracil, nitrous acid, and UV light, none is found. What sort of mutation do you think exists in the original mutant? Why? What mutagen would you use to try to obtain a revertant?

4. A bacterial repair system called X removes thymine dimers. You have in your bacterial collection the wild-type (X^+) and an X^- mutant. Phage, when UV-irradiated and then plated, gives a larger number of plaques on X^+ than on X^- bacteria. It has been proposed on the basis of survival curve analysis that the X enzyme is inducible. To test this proposal, UV-irradiated phage are adsorbed to both X^+ and X^- bacteria in the presence of the antibiotic chloramphenicol (which inhibits protein synthesis). No thymine dimers are removed in the X^+ cell and 50% are removed in the X^- cell. The same results were obtained in the absence of chloramphenicol.
   (a) Is X an inducible system?
   (b) Suppose 5% of the thymine dimers had been removed in the presence of chloramphenicol and 50% in its absence; how would your conclusion be changed?

CONCEPTUAL QUESTIONS

1. Consider the effects of life on Earth from the breakdown of the ozone layer (which would allow more UV light to penetrate to the Earth's surface). What might be the eventual outcome of that situation?

2. Would different types of organisms—for example, bacteria, fruit flies, frogs, and humans—be expected to sustain different types of DNA damage?

3. How might it be established whether DNA damage-repair mechanisms evolved first to repair replication errors, and only secondarily to repair damage from environmental influences, or vice versa?

Coordination of Macromolecular Function in Cells

in this chapter you will learn

1. About the need for regulation of gene expression, and the differences in positive and negative regulation of transcription.

2. How mutations led to the understanding of how certain operons are regulated.

3. How regulation of several specific operons is accomplished.

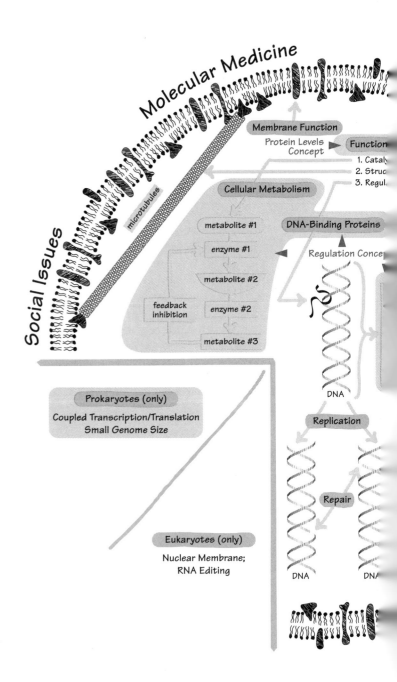

Molecular Medicine

Social Issues

microtubules

Membrane Function

Protein Levels Concept

Function
1. Cataly
2. Struc
3. Regul.

Cellular Metabolism

metabolite #1

enzyme #1

metabolite #2

feedback inhibition

enzyme #2

metabolite #3

DNA-Binding Proteins

Regulation Concep

Regulation Conce

DNA

Replication

Repair

DNA

DNA

DN.

Prokaryotes (only)

Coupled Transcription/Translation Small Genome Size

Eukaryotes (only)

Nuclear Membrane; RNA Editing

Regulation of Gene Activity in Prokaryotes

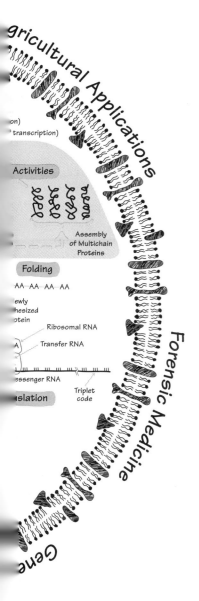

I n earlier chapters, we have learned that chromosomal DNA is organized into genes, and how the encoded one-dimensional information is first transcribed into RNA and then translated into protein as three-dimensional information. This three-dimensional information is responsible for all biological processes, sential or auxiliary. However, the behavior of an organism depends not only on the nature of the proteins that are expressed but also on the extent to which they are expressed and the environmental conditions under which this expression occurs. Of course, some housekeeper proteins—proteins that build cell architecture, for example—must be made all the time. If all proteins were made all the time, there would not be any variation in the behavior patterns of cells. Remember that organisms consume large amounts of energy in the processes of transcription and translation. The ability to synthesize materials only as needed, therefore, would make sense for economy and adaptation. There are devices to insure that proteins are synthesized in exactly the amounts they are needed and only when they are needed. In general, the synthesis of particular gene products is controlled by mechanisms that are collectively called "gene regulation." Regulation of gene expression not only exists, but makes cellular adaptation, variation, differentiation, and development possible.

The regulatory systems of prokaryotes and eukaryotes are somewhat different. Prokaryotes are generally free-living unicellular organisms that grow and divide indefinitely as long as environmental conditions are suitable and the supply of nutrients is adequate. Thus, their regulatory systems are geared to provide the maximum growth rate in a particular environment, except when such growth would be detrimental. This strategy also seems to apply to the free-living unicells such as yeast, algae, and protozoa.

The requirements of tissue-forming eukaryotes are different from those of prokaryotes. In a developing organism—for example, in an embryo—a cell must not only grow and produce many progeny cells, but must also undergo considerable change in morphology and biochemistry, and then maintain that changed state. Furthermore, during the growth and cell division phases of the organism, these cells are challenged less by the environment than are bacteria, since the composition of their growth media does not change drastically with time. Some examples of such media are blood, lymph, other body fluids, or, in the case of marine animals, sea water. Finally, in an adult organism, growth and cell division in most cell types has stopped, and each cell needs only to maintain itself and its properties. Many other examples could be given; the main point is that because a typical eukaryotic cell faces different contingencies than a bacterium, the regulatory mechanisms of eukaryotes and prokaryotes are not the same. In this chapter, we consider regulation in prokaryotes. Eukaryotes are examined in Chapter 12.

Principles of Regulation

KEY CONCEPT

The "Regulation" Concept

A multitude of regulatory mechanisms function in cells to ensure that the supply of various macromolecules matches up with the cell's ever-changing needs.

Since the best understood regulatory mechanisms are those used by bacteria and bacteriophages, these are the bases of all important present-day concepts of such regulation. Obviously, gene expression can be regulated by controlling the synthesis of mRNA or protein. Mechanisms for both exist in prokaryotes. In *E. coli,* the maximum number of copies of a protein present depends upon its nature and need. For example, a cell may contain only 20 copies of *lac* repressor, a protein described later, but possesses over 100,000 copies of EF-Tu, a protein factor needed for polypeptide synthesis. Regulatory signals exist to determine the maximum amount of each protein by controlling the rate of transcription or translation. *E. coli* cells have enzymes that give them the capability of using compounds other than glucose for food, protecting themselves from toxic environments, synthesizing useful compounds only when needed, etc. However, the enzymes that participate in these adaptive processes are normally absent; they are made only when the need arises. Then on-off regulatory activities are obtained by allowing synthesis (or translation) of a particular mRNA when the gene product is needed, and prohibiting synthesis when the product is not needed.

In bacteria, few examples are known in which a system is completely switched off. In the off state, a basal level of gene expression almost always remains, often consisting of only one or two transcriptional events per cell generation; hence, very little synthesis of the gene product occurs. For convenience, when discussing transcription, the term "off" will be used; but it should be kept in mind that usually levels are very low rather than absent. In only one known case in bacteria (namely, in spores) is expression of most genes totally turned off. In eukaryotes, however, complete turning off of a gene is quite prevalent.

In bacterial systems, when several enzymes act in sequence in a single metabolic pathway, usually either all or none of these enzymes are produced. This phenomenon, **coordinate regulation,** results from control of the synthesis of a single polycistronic mRNA molecule that encodes all of the gene products. That is, genes in prokaryotes are often clustered on the chromosome, and thus, coordinate regulation is possible. In eukaryotes, a different situation pertains: Genes are frequently organized separately, and dispersed among different chromosomes, and mRNA is usually monocistronic, as discussed in Chapter 8.

Transcriptional Regulation

Several mechanisms for regulation of transcription are common; the particular one used often depends on whether the enzymes being regulated act in degradative or synthetic metabolic pathways. That is, does the action of the enzymes in question break down a substance into a more useful compound (degradative), or is the desired molecule being "built" (synthetic)? In a multistep degradative system, the availability of the molecule to be degraded frequently determines whether or not the enzymes involved in the pathway will be synthesized. In contrast, in a biosynthetic pathway the final product is often the regulatory molecule. The molecular mechanisms for each of the two regulatory patterns vary widely, but usually fall into one of two major categories—**negative** or **positive regulation** (**Figure 11-1**).

In negatively regulated systems, a specific protein (called a **repressor protein**) that inhibits transcription of a specific gene (or genes) may be present in the cell. In some cases, the repressor alone acts to prevent transcription, and a molecule (called an **inducer**), which is an antagonist of the repressor, is needed to allow transcription. In other instances of negative regulation, the repressor on its own does not inhibit transcription—it does so only when complexed with a specific signal molecule. In a positively regulated system, a protein called an **activator** works to increase the frequency of transcription of an operon. To be effective inducers, some activator molecules must first associate with a signal molecule. Keep in mind that the terms positive or negative, when applied to regulation, refer to the action of a protein that can bind to the gene's regulatory region. If, when bound, this protein *allows* transcription, the system is said to be positively regulated. On the other hand, if the bound protein *prevents* transcription, the system is considered to be negatively regulated.

Repressors and activators are not mutually exclusive in their action and nature. Some systems are both positively and negatively regulated; such a system can respond with a high degree of sensitivity to different conditions in the cell. In addition, systems exist in which the same regulatory protein acts as a repressor under one set of circumstances, and as an activator under another.

A degradative system may be regulated either positively or negatively. In a biosynthetic pathway, the final product usually negatively regulates its own synthesis; in the simplest type of negative regulation, the absence of the product results in an increase in its synthesis, and the presence of the product decreases its synthesis.

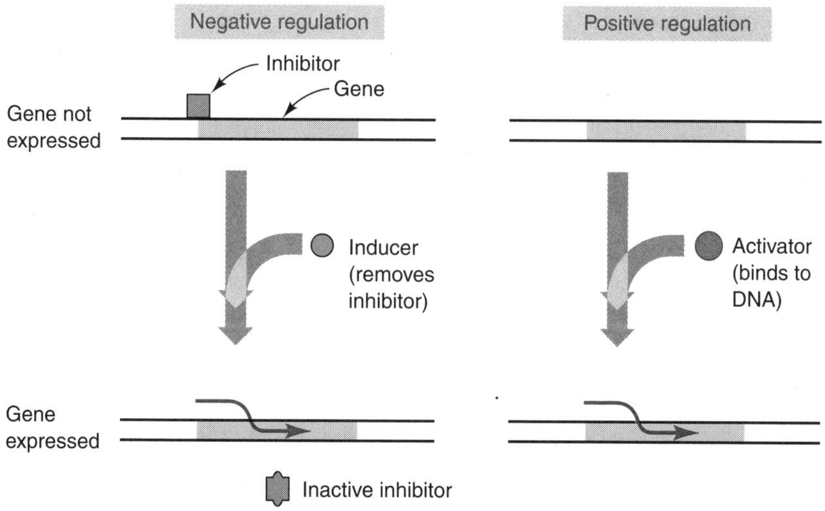

negative regulation
positive regulation
repressor protein

inducer

activator

Figure 11-1 The distinction between negative and positive regulation. In negative regulation, an inhibitor, bound to the DNA, must be removed before transcription can occur. In positive regulation, an activator molecule must bind to the DNA. A system may also be regulated both positively and negatively; then, the system is "on" when the positive regulator is bound to the DNA and the negative regulator is not bound to the DNA.

In both prokaryotic and eukaryotic systems, enzyme activity rather than enzyme synthesis may also be regulated. That is, the enzyme may be present, but its activity is turned on or off. When this occurs, the product of the enzymatic reaction (or in the case of a biosynthetic pathway, the final product of a sequence of reactions) usually inhibits the enzyme activity. This mode of regulation is called **feedback inhibition.** Small molecules other than reaction products are also frequently used either to activate or inhibit a particular enzyme. Such a molecule is termed an allosteric effector molecule, and will be described later in this chapter.

feedback inhibition

The *E. coli* Lactose System and the Operon Model

In *E. coli*, two proteins are necessary for the metabolism of lactose—the enzyme **β-galactosidase** (which cleaves lactose to yield galactose and glucose), and a transport protein called **lactose permease** (required for the entry of lactose into a cell). The existence of two different proteins in the lactose-utilization system was first shown by a combination of genetic experiments and biochemical analysis.

β-galactosidase
lactose permease

In order to carry out genetic experiments involving bacteria, it is often essential to introduce a second copy of the gene of interest into the bacterial cell. One of the easiest ways to accomplish this is by means of a **plasmid,** a small, circular double-stranded DNA molecule capable of independent replication within the cell. Some plasmids, such as the **F plasmids** (F plasmids are discussed in detail in Chapter 14), can also become integrated into the bacterial chromosome and replicate along with the bacterial DNA. As well, many bacterial cells are capable of transferring a copy of a plasmid to another bacterium that lacks a plasmid. Occasionally, the plasmid that is transferred had earlier acquired additional genes originally found on a bacterial chromosome.

F plasmids

Those genetic features have been exploited by microbial geneticists to understand many aspects of gene function. For example, if a bacterium containing an F′ plasmid transfers a copy of its plasmid to a new bacterium, the recipient will then possess two copies of certain bacterial genes—the original copy present on its chromosome, plus a second copy carried by the F′ plasmid. Such cells are referred to as partial diploids (**Figure 11-2**), and have been very informative, as will be detailed shortly.

First, hundreds of mutants unable to use lactose as a carbon source (Lac⁻ mutants) were isolated. Some of the mutations were in the *E. coli* chromosome itself, and others were in a second copy of the lactose utilization genes carried on the F′*lac* plasmid. A bacterial cell lacking an F plasmid was "mated" with one containing an F plasmid. The partial diploid cell that resulted contained two copies of the lactose utilization genes—one on the bacterial chromosome and the other

Figure 11-2 Bacterial cell containing two copies of the *lac* operon genes: one on the chromosome and one on the F′ plasmid. In Table 11-1, the plasmid genes are printed in blue.

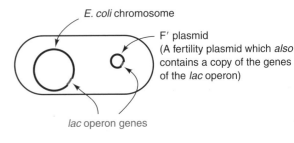

on the plasmid. For example, if the bacterial plasmid was *lac*⁺ (i.e., carried the genes necessary for lactose metabolism) and the bacterial chromosome was *lac*⁻ (unable to use lactose as a carbon source), we would write the genotype of the partial diploid cell as F′*lac*⁺/*lac*⁻. (By convention, the genotype of the plasmid is to the left of the diagonal line, and the chromosome is to the right.)

As long as one complete set of lactose utilization genes is present within the cell (either on the plasmid or on the chromosome), these diploids always exhibit a *lac*⁺ phenotype (i.e., they make both permease and galactosidase). Other partial diploids were then constructed in which both the F′*lac* plasmid and the chromosome carried *lac*⁻ genes (i.e., neither by itself contained all of the genetic information necessary for the cell to metabolize lactose). When these partial diploids were tested for the Lac⁺ phenotype, it was found that some were now able to utilize lactose, while others were not. All of the Lac⁻ mutants initially isolated fell into two groups: those having a mutation in the *lacZ* gene, and those with a defect in the *lacY* region.

The partial diploids—F′*lacY*⁻ *lacZ*⁺/*lacY*⁺ *lacZ*⁻, and F′*lacY*⁺ *lacZ*⁻/*lacY*⁻ *lacZ*⁺—had a Lac⁺ phenotype (producing both β-galactosidase and permease), but the genotypes—F′*lacY*⁻ *lacZ*⁺/*lacY*⁻ *lacZ*⁺, and F′*lacY*⁺ *lacZ*⁻/*lacY*⁺ *lacZ*⁻—had the Lac⁻ phenotype. That is, they contained two functional copies of one gene, but lacked a functional copy of the other. The existence of two groups was good evidence that the *lac* system consisted of at least two genes (*lacY* and *lacZ*), and that a cell must have a functional copy of each in order to metabolize lactose.

Further experimentation was needed to establish the precise function of each gene. Experiments in which cells were placed in a medium containing ¹⁴C-labeled lactose showed that no lactose would enter a *lacY*⁻ cell, whereas it readily entered a *lacZ*⁻ mutant. Treatment of a *lacY*⁻ cell with lysozyme (an enzyme that destroys part of the cell wall and makes it permeable) enabled labeled lactose to enter a *lacY*⁻ cell, indicating that the product of the *lacY* gene (i.e., lactose permease) is involved with the transport of lactose into the cell through the cell membrane. Enzymatic assays showed that β-galactosidase is present in *lacZ*⁺ cells, but not *lacZ*⁻ cells; this provided the initial evidence that the *lacZ* gene is the structural gene for β-galactosidase. Genetic mapping showed that the *lacY* and *lacZ* genes are adjacent. Another gene called *lacA* is located next to the *lacY* gene (**Figure 11-3**). It codes for a galactoside transacetylase enzyme, whose physiological role is unknown.

These observations show a very sensitive on-off mechanism that allows the production of enzymes involved in lactose metabolism only when the cell needs them; they also provide a classic example of inducible enzyme synthesis:

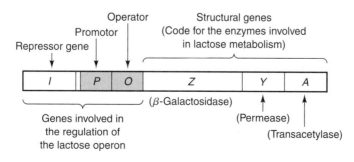

Figure 11-3 Arrangement of genes in the *lac* operon.

1. When a culture of Lac⁺ *E. coli* was grown in a medium that contained no lactose (or any other related β-galactoside compound), the intracellular concentrations of β-galactosidase and permease were extremely low—possibly one or two molecules per bacterium. However, when lactose was present in the growth medium, the number of these molecules was increased about 10^5-fold.

2. When lactose was added to a Lac⁺ culture growing in a lactose-free medium (that also lacks glucose for reasons that will be discussed later), the synthesis of both β-galactosidase and permease began nearly simultaneously, as shown in **Figure 11-4**. Analysis of the mRNA present in the cells before and after the addition of lactose showed that no *lac* mRNA (the mRNA that codes for β-galactosidase and permease) was present prior to the addition of lactose, and that it was, in fact, the addition of lactose that triggered the synthesis of *lac* mRNA.

This kind of analysis is done by growing cells in a radioactive medium in which all newly synthesized mRNA is radioactive, isolating the mRNA, and allowing it to renature with previously obtained *lac* DNA. Since the only RNA that will anneal to the *lac* DNA is *lac* mRNA, the amount of radioactive DNA-RNA hybrid molecules is also a measure of the amount of *lac* mRNA. These two observations led to the view that the lactose system is inducible, regulated at the level of transcription, and that lactose acts as an inducer.

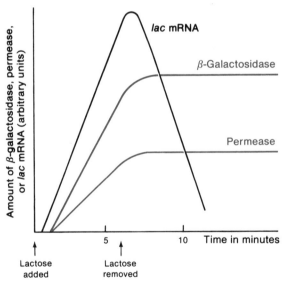

Figure 11-4 The "on-off" nature of the *lac* system. *Lac* mRNA appears very soon after lactose is added; β-galactosidase and permease appear subsequently because of the time required for sequential translation. When lactose is removed, no more *lac* mRNA is made, and the amount of *lac* mRNA decreases owing to the usual degradation of mRNA. Both β-galactosidase and permease are stable. Their concentration remains constant even though no more can be synthesized. A third protein of the *lac* system, β-galactoside transacetylase, is synthesized coordinately with β-galactosidase and permease. This protein, the product of the *A* gene, is not involved in lactose metabolism.

Lactose itself is rarely used in experiments to study the induction phenomenon for a variety of reasons—one is that β-galactosidase catalyzes the cleavage of lactose, resulting in a continual decrease in lactose concentration. Eventually, the cell would use up all of the available lactose and the system would be switched off. This complicates the analysis of many types of experiments (kinetic experiments, for example). Instead of lactose itself, a sulfur-containing analogue called **isopropylthiogalactoside (IPTG)** (**Figure 11-5**) is used; this analogue is an inducer, but not a substrate, of β-galactosidase. That is, IPTG will "switch on" enzyme synthesis, but it will not be metabolized.

Mutants have been isolated in which *lac* mRNA is made in both the presence and absence of an inducer, indicating that regulation of the expression of *lac* genes had been eliminated. Such mutants provided the key to understanding induction. In these mutants, *lac* genes were said to be **constitutively expressed**—that is, they were "on" all the time. Mutations that lead to constitutive expression are called **constitutive mutations.** Tests were performed using partial diploids carrying two constitutive mutations, one on the chromosome and one on a plasmid. These tests showed that the mutants fall into two groups, termed *lacI⁻* and *lacOᶜ*. The characteristics of the mutants are shown in **Table 11-1.** As shown in Table 11-1 (entry 1), in the absence of an inducer (IPTG in this case) the *lacI⁺* cell fails to make *lac* mRNA, whereas this mRNA is made by a *lacI⁻* mutant (entry 2), even in the absence of an inducer. Thus, the *lacI* gene is apparently a regulator gene whose product is an inhibitor that keeps the system turned off unless an inducer is present. Entries 3 and 4 show that the presence of a functional *lacI* gene anywhere in the

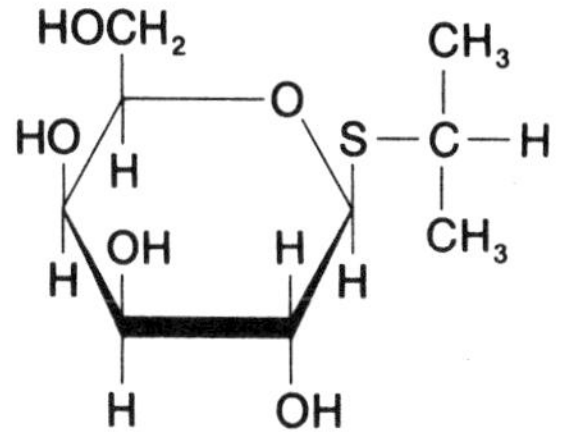

isopropylthiogalactoside (IPTG)

Isopropylthiogalactoside (IPTG)

Figure 11-5 Structure of Iso-propylthiogalactoside (IPTG).

TABLE 11-1	CHARACTERISTICS OF PARTIAL DIPLOIDS HAVING COMBINATIONS OF *lacI* AND *lacO* GENES. PLASMID GENES ARE SHOWN IN BLUE.

Genotype	*lacZ* Expression Inducer Absent (Without IPTG)	Inducer Present (With IPTG)	Phenotype
1. *I⁺ O⁺ lacZ⁺*	−	+	Inducible
2. *I⁻ O⁺ lacZ⁺*	+	+	Constitutive
3. F′ *I⁻ O⁺ lacZ⁺*/*I⁺ O⁺ lacZ⁺ ⁽ᵒʳ ⁻⁾*	−	+	Inducible
4. F′ *I⁺ O⁺ lacZ⁺*/*I⁻ O⁺ lacZ⁺*	−	+	Inducible
5. *I⁺ Oᶜ lacZ⁺*	+	+	Constitutive
6. F′ *I⁺ Oᶜ lacZ⁺*/*I⁺ O⁺ lacZ⁻*	+	+	Constitutive
7. F′ *I⁺ Oᶜ lacZ⁻*/*I⁺ O⁺ lacZ⁺*	−	+	Inducible
8. F′ *I⁺ Oᶜ lacZ⁻*/*I⁻ O⁺ lacZ⁺*	−	+	Inducible
9. F′ *I⁺ Oᶜ lacZ⁺*/*I⁻ O⁺ lacZ⁻*	+	+	Constitutive

Interpretations of the phenotypes observed from the above genotypes:

1 + 2 indicate that the presence (in these cases on the bacterial chromosome) of a functional I gene product (I⁺) is required to "turn off" expression of the *lacZ⁺* gene. In *lacI⁻* cells, the *lacZ* gene is expressed constitutively.

3 + 4 indicate that a functional *lacI* gene anywhere in the cell (on either the bacterial chromosome or a plasmid) is capable of regulating the expression of the *lacZ⁺* gene.

5–9 indicate that mutations in a second gene, the *lacO* gene, can also lead to constitutive expression of the *lacZ⁺* gene. The *lacO⁺* gene can *only* regulate expression of copies of the *lacZ⁺* gene located adjacent to it, on the same chromosome or plasmid.

cell (on the chromosome or on the plasmid) is sufficient to regulate (i.e., make responsive to an inducer) any $lacI^-$ system in the cell, whether it is on the same DNA molecule or not. The $lacI$-gene product must, therefore, be a diffusible entity. Wild-type copies of the $lacI$ gene are present in a $lacI^+/lacI^-$ partial diploid, so the system is inhibited in the presence of IPTG (it would not be if the $lacI^+$ gene were not present). The $lacI$ gene product, a protein molecule, is called the *Lac repressor*. Genetic mapping experiments place the $lacI$ gene adjacent to the $lacZ$ gene, establishing the gene order $lacI, lacZ, lacY$. How the repressor prevents synthesis of lac mRNA will be explained shortly.

The O^c mutations have properties different from the $lacI^-$ mutations. Mapping experiments show them to be located in a very small segment between the $lacI$ and $lacZ$ genes. A significant characteristic of $lacO^c$ mutations that was found by examining the partial diploids shown in entries 6 and 7 is that β-galactosidase is synthesized constitutively in an $O^c lacZ^+/O^+ lacZ^-$ cell, but is made in an $O^c lacZ^-/O^+ lacZ^+$ cell only when an inducer is present. It should be noted that the O^c mutation causes constitutive expression of *lac* genes only when the mutation and the genes are on the same DNA molecule. As only the genes *cis* (adjacent) to the O^c mutation are affected by it (i.e., expressed constitutively), the O^c mutation is said to be *cis*-dominant. Since the presence of an O^+ gene on one DNA molecule cannot alter the constitutive behavior of an O^c lac system on another DNA molecule in a partial diploid, the O^+ region does not code for a diffusible product. Actually, the O region is a noncoding region of DNA, and is involved in regulation. This region is called an **operator;** the repressor protein binds to the O^+ operator and prevents expression of the *lac* structural genes.

operator

Another class of Lac$^-$ mutants (called P$^-$) have been isolated in which no *lac* mRNA is made. These mutations have been mapped to a small region adjacent to the operator, and, in experiments with partial diploids, have been shown to have a *cis*-dominant effect in that they only prevent transcription of structural genes on the same DNA molecule as the mutant. Because of this, the P region, like the O region, must define a noncoding DNA site, and has been determined to be a promoter.

The regulatory mechanisms of the *lac* system were first explained by the operon model, which has the following features (**Figure 11-6**):

1. The lactose-utilization system consists of two kinds of components: structural genes that code for products needed to transport and metabolize lactose, and regulatory elements (the *lacI* gene, the operator (O), and the promoter (P)). Together, the structural genes and regulatory elements comprise the **lac** operon.

2. The products of the *lacZ* and *lacY* genes are encoded in a single polycistronic mRNA molecule. The transacetylase enzyme mentioned earlier is also encoded in this molecule.

3. The *lac* operon promoter is immediately adjacent to the operator region. RNA polymerase binds to the promoter and initiates transcription of the *lacZ, lacY,* and *lacA* genes into the mRNA molecule.

4. The *lacI* gene product (the repressor) is always produced (at about 20 copies per cell) and binds to the operator.

5. When the repressor is bound to the operator, initiation of transcription of *lac* mRNA by RNA polymerase is prevented.

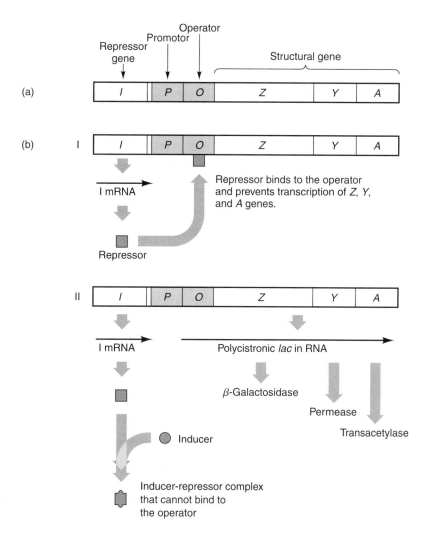

Figure 11-6 (a) Genetic map of the *lac* operon (not drawn to scale). The *P* and *O* sites are actually much smaller than the genes. (b) Diagram of the *lac* operon in (I) repressed and (II) induced states. The inducer alters the shape of the repressor, so the repressor can no longer bind to the operator.

6. Inducers stimulate mRNA synthesis by binding to and deactivating the repressor, a process called **derepression.** Thus, in the presence of an inducer the operator is unoccupied, and the promoter is available for initiation of mRNA synthesis. That is an example of negative regulation. Note that regulation of the operon requires that the operator be adjacent to the structural genes of the operon (*lacZ, lacY,* and *lacA*); however, proximity of the *lacI* gene is not necessary because the repressor is a soluble protein and is, therefore, diffusible throughout the cell.

The operon model is supported by a wealth of experimental data and explains many of the features of the *lac* system, as well as numerous other negatively regulated genetic systems. One aspect of the regulation of the operon—the effect of glucose—has not yet been discussed. Examination of this feature indicates that the *lac* operon is also subject to positive regulation.

β-Galactosidase forms glucose by cleaving lactose. (The other cleavage product, galactose, is metabolized by enzymes of the galactose operon, described later.) Thus, if both glucose and lactose are present in the growth medium, the activity of the *lac* operon is not needed, so no *lac* mRNA is made in the presence of glucose. Another regulatory element is, therefore, needed for initiating *lac* mRNA synthesis, which is regulated by the concentration of glucose. However, the inhibitory effect of glucose on the expression of the *lac* operon is quite indirect, as explained later.

The small molecule **cyclic AMP (cAMP)** is universally distributed in animal tissues, and in multicellular eukaryotes it is important in regulating the action of many hormones (**Figure 11-7**). It is also present in *E. coli* and many other bacteria. Cyclic AMP is synthesized by the enzyme adenylate cyclase, and its concentration is related to glucose concentration. So, when the glucose level in the cell is high, the cAMP concentration is low. Conversely, when the glucose level is low, the cAMP concentration is high. In a medium containing glycerol (or any other carbon source that enters the cell by a channel different from the type used by glucose), or under starvation conditions, the cAMP concentration in bacterial cells is high (**Table 11-2**). It is cAMP that links the *lac* operon activity with the intracellular concentration of glucose.

E. coli (and many other bacterial species) contain a protein called cAMP receptor protein (CRP) that is encoded in a gene called *crp*. Mutants of either *crp* or the adenylate cyclase gene are unable to synthesize *lac* mRNA, indicating that both CRP function and cAMP are required for *lac* mRNA synthesis. CRP and cAMP bind together to form a complex (denoted cAMP-CRP) that is a regulatory element for the *lac* system. The cAMP-CRP complex must be bound to a specific base sequence (called an activation site) in the promoter region in order for transcription of the *lac* operon to occur (**Figure 11-8**). Thus, the cAMP-CRP complex is a positive regulator, in contrast to the repressor, a negative regulator. Therefore, the *lac* operon is regulated both positively and negatively.

Using mixtures containing purified *lac* DNA, repressor, cAMP-CRP, and RNA polymerase, two key points have been established:

1. When there is not enough cAMP-CRP in the cell (due to high levels of glucose), the cAMP-CRP complex is not formed. In the absence of the cAMP-CRP complex (**Figure 11-9**), RNA polymerase binds weakly to the *lac* promoter, so transcription is rarely initiated. However, if cAMP-CRP is bound to the activation site in the *lac* promoter, it enhances the binding of RNA polymerase and its subsequent initiation of transcription. When cAMP-CRP is bound to the activation site, it affects RNA polymerase activity by making direct contact with the enzyme. The cAMP-CRP complex has two domains—one binds to DNA, and the other to RNA polymerase.

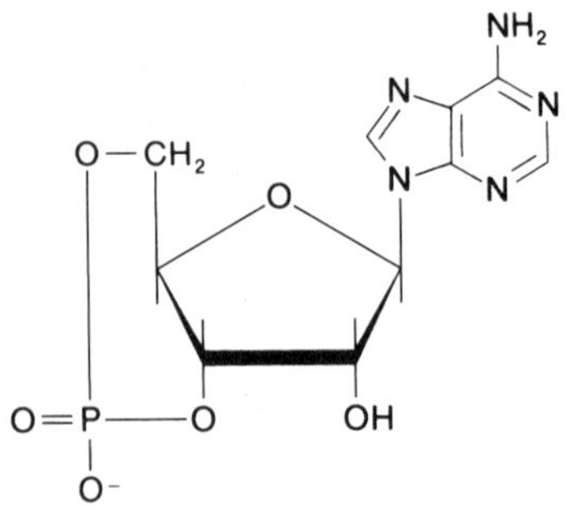

Figure 11-7 Structure of cyclic AMP.

TABLE 11-2 CONCENTRATION OF CYCLIC AMP IN CELLS GROWING IN MEDIA HAVING THE INDICATED CARBON SOURCES

Carbon Source	cAMP Concentration
Glucose	Low
Glycerol	High
Lactose	High
Lactose + glucose	Low
Lactose + glycerol	High

2. Because there is partial overlap in the promoter and operator of the *lac* operon, when the *lac* repressor is bound to DNA at the operator region, it physically blocks the binding of RNA polymerase to the promoter.

The Galactose Operon: Two Promoters and DNA Looping

The galactose operon of *E. coli* is also an inducible system, regulated by cAMP-CRP and a repressor. The *gal* operon has three cistrons—*galE, galT,* and *galK,* which encode enzymes that metabolize galactose—and is transcribed into polycistronic mRNA molecules. The operon contains two promoters, *P1* and *P2.* The cAMP levels in the cell influence the *gal* operon in an interesting way: The cAMP-CRP complex binds to a noncoding DNA site in the operon and regulates transcription initiation from the two promoters in opposite fashions—cAMP-CRP **activates** transcription from promoter *P1,* but **inhibits** transcription from promoter *P2.* Thus, when the intracellular cAMP concentration is high, transcription of the operon is initiated at the *P1* promoter; but at a low cAMP level, it is initiated at the *P2* promoter. This mechanism of dual control insures that the galactose-metabolizing enzymes are synthesized whether the cells are growing in the presence of glucose or in another carbon source. By using a mixture of purified components in an in vitro system and varying the cAMP concentration, it has been shown that the amount

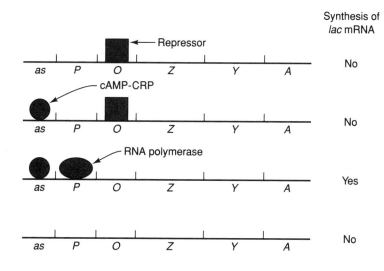

Figure 11-8 Four states of the *lac* operon showing that *lac* mRNA is made only if cAMP-CRP is present and repressor is absent. The symbol *as* represents the activation site.

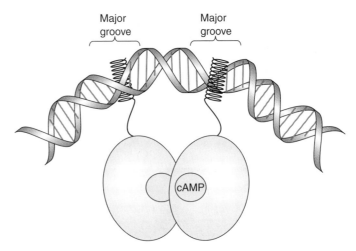

Figure 11-9 Fitting of α-helical segments of a cAMP-CRP dimeric protein into adjacent major DNA grooves. Recall that a similar binding phenomenon was described in Chapter 5 (Figure 5-13). In this example, however, each subunit binds one molecule of cAmp and binding of the protein causes the DNA to kink in minor groove regions and subsequently bend approximately 90 degrees. That bending facilitates subsequent binding of RNA polymerase.

of synthesis of *gal* mRNA from the two promoters is inversely related to the level of cAMP (**Figure 11-10**). Note that the total amount of *gal* mRNA made at any level of cAMP is constant.

The negative regulator, the *gal* repressor, is the product of the *galR* gene. The *galR* gene, unlike the *lacI* gene (its counterpart in the *lac* operon), is located quite far from the operon it regulates. Gal repressor inhibits transcription from both promoters. To act, it must bind to two operator elements, designated O_E and O_I, present in the operon on opposite sides of the promoters (**Figure 11-11**). When cells are exposed to galactose, galactose acts as an inducer and causes derepression of the operon by preventing the binding of the repressor to O_E and O_I. Mutations of *galR* (a *trans*-acting mutation), or to either of the two operators (a *cis*-acting mutation), cause constitutive (i.e., constant) expression of the operon.

Figure 11-10 Regulation of transcription of the *gal* operon from the two promoters, *P1* and *P2*. In the absence of cAMP-CRP, *gal* mRNA is made mostly from the *P2* promoter. At high cAMP-CRP concentrations, *gal* mRNA is made principally from the *P1* promoter.

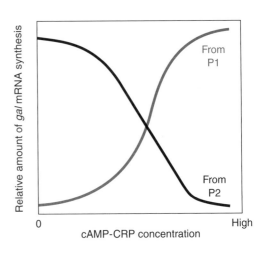

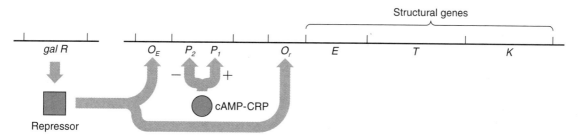

Figure 11-11 The regulation of the *gal* operon of *E. coli*. The two operators, O_E and O_I flank the promoters. The map is not drawn to scale.

The repressor of the *gal* operon functions differently from its counterpart in the *lac* system. Instead of physically blocking the binding of RNA polymerase, the two repressor molecules (one bound to each operator) interact with each other, causing the formation of a loop in the DNA. This loop of DNA, in the region between the two operators, contains the promoters to which RNA polymerase is bound (**Figure 11-12**). This looping prevents the RNA polymerase from initiating transcription. While the exact mechanism of repression is not known, it is believed that the formation of DNA loops permits physical contact between regulatory protein molecules and RNA polymerase bound to areas spatially separated on the DNA molecule; this contact somehow serves to regulate the rate and efficiency of transcription.

The Arabinose Operon

The arabinose operon in *E. coli*, which encodes enzymes involved in the utilization of the sugar arabinose, provides another example of variation in the behavior of regulatory proteins. The *ara* operon, like *lac* and *gal* is inducible, and not normally expressed. The *ara* enzymes are made only when the cells are exposed to arabinose. The expression is regulated by a protein called AraC, which by itself acts as a repressor. When arabinose is present, the sugar binds to AraC and converts it into an activator. The arabinose-AraC complex activates transcription

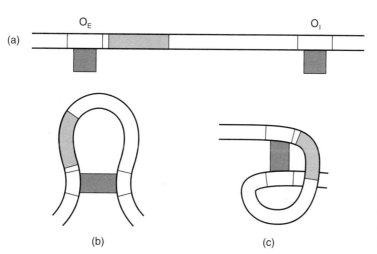

Figure 11-12 DNA looping in the *gal* operon. A repressor dimer binds to each operator (a) and the dimers associate together, generating a loop of the intervening DNA by mechanisms (b) or (c). The *gal* promoter region is shaded.

from the *ara* promoter. If the gene encoding AraC is deleted, the *ara* operon is never expressed—even in the presence of arabinose. Thus, the arabinose operon provides an example of a regulatory protein that acts as a repressor under one condition, and as an activator under another.

The Tryptophan Operon and Attenuation

In some operons, a single molecule does not by itself act as a repressor. The repressor for such an operon requires the presence of a small molecule (called a **corepressor**) in order to bind to the operator. In its absence, the repressor (called an aporepressor) is unable to bind to the operator, and the genes of the operon are expressed. In *E. coli,* the expression of enzymes involved in the synthesis of many amino acids and vitamins are regulated in this fashion. The cell represses the synthesis of the appropriate enzymes when an amino acid or vitamin is present in excess. This system of repression, which is present in the tryptophan operon, is highly economical in its use of energy.

There are five genes that encode the enzymes involved in tryptophan biosynthesis. The arrangement of the five cistrons (named *trpE, trpD, trpC, trpB,* and *trpA*), the promoter, and the operator is shown in **Figure 11-13**. Together, these components form an operon, and, following the definition of an operon, can be transcribed into a polycistronic mRNA molecule. In the *trp* operon, tryptophan acts directly in the repression system.

The regulatory protein of the repression system of the *trp* operon is called the Trp repressor, and is the product of the *trpR* gene (which is located elsewhere on the chromosome). Mutations in either the *trpR* gene ($trpR^-$) or in the operator (O^c) cause constitutive initiation of *trp* mRNA synthesis, just as in the *lac* and *gal* operons. The tryptophan molecule (acting as a corepressor) must bind to Trp repressor (the aporepressor) to achieve repression. The reaction scheme of this process is:

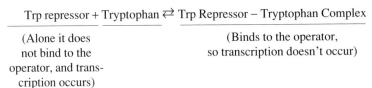

Trp repressor + Tryptophan ⇄ Trp Repressor − Tryptophan Complex

(Alone it does not bind to the operator, and transcription occurs)

(Binds to the operator, so transcription doesn't occur)

corepressor

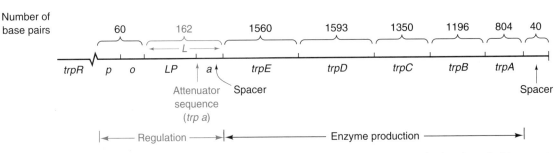

Number of base pairs

60   162   1560   1593   1350   1196   804   40

L

trpR   p   o   LP   ↑a↑   *trpE*   *trpD*   *trpC*   *trpB*   *trpA*

Attenuator sequence (*trp a*)   Spacer   Spacer

|← Regulation →|← Enzyme production →|

Figure 11-13 The *E. coli trp* operon. For clarity, the regulatory region is enlarged with respect to the coding region. The proper size of each region is indicated by the number of base pairs. The regulatory elements are shown in blue. *L* is the leader mRNA segment; *LP,* the leader peptide; and *a,* the attenuator genes. The *trpR* gene codes for the repressor, and is located elsewhere on the chromosome.

Only when tryptophan is present does an active repressor molecule thus inhibit transcription. When the external supply of tryptophan is depleted (or at least reduced substantially), the equilibrium in the preceding equation shifts to the left, the operator is unoccupied, and transcription begins. This is the basic "on-off" regulatory mechanism of the *trp* operon.

Since the *trp* operon codes for a set of biosynthetic (rather than degradative) enzymes, neither glucose nor cAMP-CRP influences the activity of the operon. Furthermore, there is more to regulation of the *trp* operon than the simple on-off system just described. When a small amount of tryptophan is available (but not enough to allow cell growth without the concomitant synthesis of tryptophan), starvation is prevented by a modulating system in which the amount of transcription in the derepressed state is determined by the concentration of tryptophan. In the *on* state, a second level of regulation subjects the expression to a finer control (a control more sensitive to tryptophan concentration). This is provided by two factors:

1. Premature termination of transcription upstream of the first structural gene.
2. Regulation of the frequency of this termination by the intracellular concentration of tryptophan.

Thus, tryptophan synthesis has a *two-part* regulation system. High levels of tryptophan shut off the system at the level of the corepressor. In this case, transcription never begins because the promoter is blocked. If tryptophan levels are not high enough to engage the corepressor, the system becomes subject to "fine-tuning." If tryptophan levels are sufficiently low, the entire operon is transcribed.

Note in Figure 11-13 that the *trp* operon has two additional regions, *LP* and *a*. At the 5'-end, upstream from the first cistron, the *trp* mRNA contains a 162-base segment called leader RNA (designated L in the figure). This leader sequence has several notable features (also see **Figure 11-14**):

1. An AUG codon and a downstream UGA stop codon in the same reading frame define a region encoding a polypeptide consisting of 14 amino acids, called the leader polypeptide (lp) (Figure 11-14(a)).
2. Two adjacent tryptophan codons are located in the *trp* mRNA at positions corresponding to the tenth and eleventh amino acids in the leader polypeptide. We will see the significance of these repeated codons shortly.
3. Four segments of the leader RNA (denoted 1, 2, 3, and 4) are capable of base pairing in two mutually exclusive ways—by forming either the stem loops 1–2 and 3–4 (Figure 11-14(b)), or just the stem loop 2–3 (Figure 11-14(d)).

This arrangement enables premature termination of transcription to occur in the *trp* leader region by the following mechanism.

In the *trp* operon of wild-type bacteria, when tryptophan is present, transcription terminates after making only 140 base-pair mRNA, and before any of the structural genes are transcribed. In the absence of tryptophan, transcription continues, allowing the synthesis of *trp* enzymes. Mutants that have a deletion in part or all of a 28-base-pair-long segment preceding position 160 make *trp* enzymes at a rate that is six to seven times greater than that of a wild-type cell growing in the absence of tryptophan, indicating that this region must have a regulatory role. This region (called an **attenuator**) has all the usual features of an intrinsic transcription termination site (described earlier—see Figure 8-8 in Chapter 8), including a potential stem-loop

attenuator

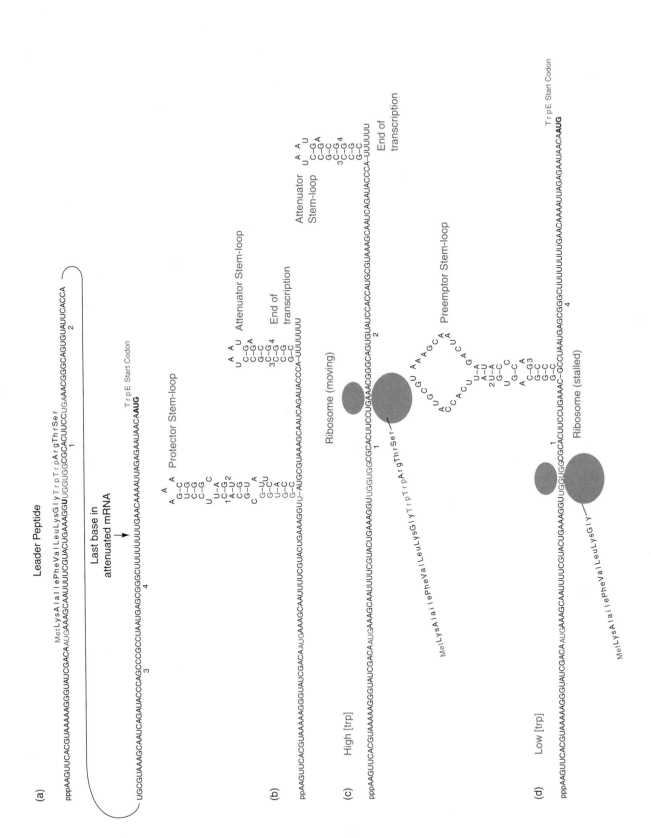

configuration in the mRNA (stem loop 3–4), followed by a sequence of several uridine residues. Transcription terminates after the seventh residue. Features of the leader RNA sense the level of tryptophan in the cell and modulate the amount of transcription accordingly. The mechanism for effecting this premature termination is based upon the potential of the RNA to form the alternative stem loops 1–2 (protector), 2–3 (preemptor), and 3–4 (attenuator). The directive to terminate or read through depends upon whether or not the stem loop 3–4 attenuator structure is formed.

Termination of transcription at the attenuator is mediated through translation of the leader peptide region. Because there are two tryptophan codons in this sequence, the translation of the sequence is sensitive to the concentration of charged tRNA^{trp}. That is, if the supply of tryptophan is inadequate, that amount of charged tRNA^{trp} will be insufficient, and translation by the ribosome will be stalled at the tryptophan codons. Two points should be noted:

1. All base pairing is eliminated in the segment of the mRNA that is in contact with the ribosome.
2. If the stem loops 1–2 and 3–4 are formed simultaneously, then stem loop 2–3 cannot be present.

Figure 11-14(c) shows that the end of the Trp leader peptide is in segment 1. Usually a translating ribosome is in contact with about ten bases in the mRNA past the codons being translated. Thus, when the final codons of the leader are being translated, segments 1 and 2 are partially covered by the ribosome, and so are not base paired. Recall that in a coupled transcription-translation system, the leading ribosome is not far behind the polymerase. Therefore, if the ribosome is in contact with segment 2 when the synthesis of segment 4 by RNA polymerase is being completed, then segments 3 and 4 are able to form the stem loop 3–4 without segment 2 competing for segment 3. The presence of the 3–4 stem-loop configuration allows termination to occur when the terminating sequence of seven uridines is reached. If there is very little tryptophan present, the concentration of charged tRNA^{trp} becomes inadequate, and the translating ribosome stalls at the tryptophan codons (Figure 11-14(d)). These codons are located 14 bases before the beginning of segment 2. Segment 2 is, thus, free before segment 4 has been synthesized, so the stem loop 2–3 can form. In the absence of stem loop 3–4, termination of transcription does not occur, and the complete mRNA molecule is made, including the coding sequences for the *trp* structural genes. Hence, if tryptophan is present in excess,

---

Figure 11-14 The leader region of *trp* mRNA and the model for the mechanism of attenuation control in *E. coli trp* operon. (a) The 5′ terminal region of *trp* mRNA, called the leader RNA. It shows the encoded 14 amino acid leader polypeptide with the 2 Trp codons, the chain terminating UGA codon, and the beginning **AUG** codon of the TrpE protein (bold face). If tryptophan is present, transcription terminates at RNA position 140 as indicated. Four different segments (designated 1, 2, 3, and 4), preceding the termination point potentially participate in alternative stem-loop structures as shown in (b), (c), and (d). (b) Free mRNA showing areas for potential formation of stem-loops 1–2 and 3–4. (c) mRNA structure in presence of relatively high concentration of tryptophan. Ribosome reaches region 2, preventing stem-loop 1–2 but allowing stem-loop 3–4 formation. (d) Relatively low concentration of tryptophan. Ribosome stalls in region 1, allowing formation of stem-loop 2–3 and preempting stem-loop 3–4.

termination of transcription occurs (**Figure 11-15**), and little enzyme is synthesized; if tryptophan is absent, termination does not occur, and the enzymes are made. At intermediate concentrations, the fraction of initiation events that result in completion of *trp* mRNA synthesis will depend upon how often translation is stalled at the tryptophan codons, which in turn depends on the concentration of tryptophan. Keep in mind that this is an example of the coupled transcription-translation processes found in prokaryotes (refer back to Chapter 9, Figure 9-16, for an illustration of these events). This system is exquisitely sensitive to tryptophan levels because transcription depends on the speed of translation of the leader sequence.

Many operons responsible for amino acid biosynthesis (for example, leucine, isoleucine, phenylalanine, and histidine operons) are regulated by attenuators equipped with base-pairing mechanism for competition described in the tryptophan operon. In the histidine operon, which also has an attenuator system (that is, prematurely terminated mRNA), a similar base sequence encodes a leader polypeptide having seven adjacent histidine codons (**Figure 11-16**). In the phenylalanine operon, seven phenylalanine codons are also present in the leader, but they are divided into three groups (Figure 11-16).

Specific Versus Global Regulators

Regulatory proteins, both positive and negative, can be highly specific; for example, the Lac repressor controls only the *lac* operon and no other genes in *E. coli.* Alternatively, regulatory proteins can have a more global effect on gene expression. cAMP-CRP, for example, controls the expression of many genes in *E. coli,* and so does a regulatory protein called LexA. LexA is a negative regulator that controls genes that encode enzymes involved in the repair of DNA damage induced by UV light. These genes are scattered around the *E. coli* chromosome, and LexA protein binds to the operator regions of all these genes. When the cells are exposed to a sublethal dose of UV light, LexA is inactivated, and the DNA-repair genes are expressed to carry out the repair job. A set of genes or operons that is **regulon** regulated by a global regulator—positive or negative—is called a **regulon.** Specific and global regulators are frequent in bacterial cells.

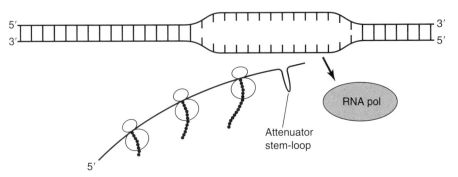

Figure 11-15   Relatively high levels of tryptophan permit normal translation of the leader peptide mRNA, but regions 3 and 4 from an attenuator stem-loop, which results in uncoupling of RNA polymerase (blue) from the DNA open complex. Hence, the downstream mRNA that codes from the *trp* operon is not transcribed.

(a)
  Met Lys His Ile Pro Phe Phe Phe Ala Phe Phe Phe Thr Phe Pro Stop
  5′ AUG AAA CAC AUA CCG UUU UUC UUC GCA UUC UUU UUU ACC UCC CCC UGA 3′

(b)
  Met Thr Arg Val Gln Phe Lys His His His His His His His Pro Asp
  5′ AUG ACA CGC GUU CAA UUU AAA CAC CAC CAU CAU CAC CAU CAU CCU GAC 3′

Figure 11-16 Amino acid sequence of the leader peptide and base sequence of the corresponding portion of mRNA from (a) the phenylalanine operon and (b) the histidine operon. The repeating amino acid is shaded in blue.

Autoregulation

The synthesis of some gene products, whose requirements in a cell vary greatly, is regulated by a mechanism called **autoregulation.** In the simplest autoregulated system, the gene product is also a regulator: it binds to a DNA site at or near the promoter. The autoregulation could be positive or negative. When the concentration of the gene product exceeds what the cell can use, the product molecules occupy the DNA site and transcription is inhibited. At a later time, when the concentration of unbound molecules decreases, the molecule bound to DNA leaves the site, allowing transcription by RNA polymerase from a free promoter. This is negative autoregulation. The synthesis of some regulatory proteins (e.g., CRP) is negatively autoregulated in this way. In a positively autoregulated system, the gene product acts as an activator of its own gene. An elegant example of positive autoregulation is the synthesis of a repressor protein of bacteriophage, λ. This system is described in Chapter 14.

Autoregulation at the level of translation has also been recognized. For example, when protein components of ribosomes become excessive, they negatively regulate their own synthesis by binding to mRNA. This blocks the binding of intact ribosomes, and, thus, translation too.

autoregulation

Constitutive Genes

Many genes in bacteria have no on-off regulatory mechanism. They are expressed all the time, at rates predetermined by the nucleotide sequence of their **promoters** (promoter strength). Such constitutive genes generally code for the living cells' housekeeping proteins, which are needed in a given amount. Examples of such genes are the *lacI* gene that encodes the Lac repressor, or genes encoding glucose-metabolizing enzymes, whose levels remain constant in a changing environment. The strength of promoters of constitutive genes, as well as of inducible or repressible genes, determines the maximum rate at which the cognate genes can be expressed.

Posttranscriptional Control

Frequently, unequal amounts of proteins are made from different cistrons of an operon. For example, the ratios of the numbers of copies of β-galactosidase, permease, and transacetylase made from the *lac* operon are 100:50:20. These differences, which result from posttranscriptional regulation, are achieved in two ways:

1. **Translational control.** The *lacZ* gene is translated first (**Figure 11-17**). The frequency with which it is translated is determined by the efficiency with which a ribosome binds the Shine-Dalgarno sequence (described earlier in Chapter 9, Figure 9-12) around the starting AUG codon in *lacZ* mRNA. Frequently the translating ribosome detaches from the mRNA molecule at the end of *lacZ* translation. The frequency at which the next cistron *lacY* is translated is also determined by the efficiency at which the next cistron *lacY* is translated is also determined by the efficiency at which a ribosome (either dissociated from the end of the *lacZ* cistron or recruited from the cell's surplus supply) binds to the AUG codon of *lacY*. The same rule regulates the translation efficiency at the *lacA* cistron.

   The efficiency at which the ribosome binds to the critical binding sequence on mRNA is also known to be modulated (inhibited) by formation of secondary structures involving this region and either another part of the mRNA or another regulatory RNA molecule with complementary base sequences.

2. **mRNA stability.** mRNA is unstable: In bacteria, the half-life is on the order of a few minutes. We have discussed before that *lac* mRNA quickly disappears if the inducer (Figure 11-5) is removed from the cell. This instability helps cells quickly shift their spectrum of protein synthesis in response to radical changes in the environment. Degradation of *lac* mRNA is initiated from its 3′ terminus. Hence, at any given instant there are more copies of the *lacZ* portion of mRNA than of the *lacY* portion of mRNA, and more copies of the *lacY* mRNA than of *lacA* mRNA. Thus, the amounts of the three *lac* enzymes made also depends upon the availability of the corresponding mRNA segments. In prokaryotes, this mode of regulation occurs frequently.

In some systems, mechanisms also exist by which the degradation of mRNA from its 3′ end is substantially reduced by formation of RNA secondary structures resistant to degradation, which enhances the amount of translation of the mRNA molecule.

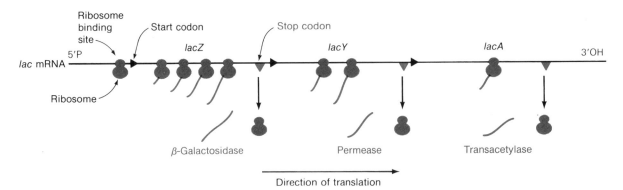

Figure 11-17 Translation of *lac* mRNA into proteins. All ribosomes attach to the mRNA at the ribosome binding site. At each stop codon, some ribosomes detach.

Thus, the overall expression of an operon is regulated by controlling transcription initiation of an mRNA molecule and premature transcription termination of the polycistronic mRNA (in some cases), whereas the relative concentration of the proteins encoded by the mRNA are determined by controlling both the degradation rate of the mRNA and the frequency of initiation of translation of each cistron. These general concepts of gene control have been firmly established by demonstrating that transcriptional or translational reactions with purified components are regulated by regulatory DNA elements and proteins. The chemical and physical basis of most of the gene regulatory macromolecular interactions are not well understood. Currently, they are the focus of intensive research among molecular biologists.

Control by Proteolysis

Unlike mRNA, most proteins are stable during normal growth unless they cease to be useful or become deleterious to a cell and must be removed for cell growth to succeed. An example of the second type of protein in *E. coli* is SulA, encoded by the *sulA* gene. When chromosomal DNA is damaged, the synthesis of SulA is induced along with DNA repair enzymes. The SulA protein inhibits cell division to provide time for DNA repair. When the repair process is completed, and SulA is no longer needed, it is quickly degraded by a protease called Lon. Mutant cells defective in Lon protease do not divide, and ultimately die when subjected to DNA damaging agents.

Feedback Inhibition and Allosteric Control

When a large amount of an amino acid is available in a cell, the synthesis of the enzymes that make the amino acid is repressed because those enzymes are not needed anymore, as previously explained. However, the cell contains a large amount of the enzymes before repression begins, and could continue to catalyze the amino acid biosynthetic reactions. But the cell has a mechanism by which the synthesis of the amino acid can be quickly stopped. Wasteful synthesis almost never occurs. For example, high concentrations of the amino acid isoleucine actually inhibit the activity of the first enzyme of the 5-enzyme pathway that catalyzes the synthesis of isoleucine from threonine (**Figure 11-18**). The bound enzyme is unable to catalyze the reaction. This stops isoleucine biosynthesis immediately. This process of very specific inhibition of catalytic activity of an enzyme in a biochemical pathway by binding to

KEY CONCEPT

The "Protein Levels" Concept

The net amount of a protein in a cell represents a balance between the rate it is produced and also by the rate at which it is degraded or exported from the cell.

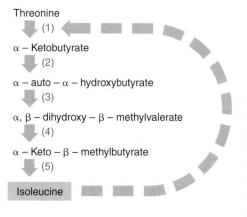

Threonine
↓ (1)
α – Ketobutyrate
↓ (2)
α – auto – α – hydroxybutyrate
↓ (3)
α, β – dihydroxy – β – methylvalerate
↓ (4)
α – Keto – β – methylbutyrate
↓ (5)
Isoleucine

Figure 11-18 Feedback inhibition of the pathway for synthesizing isoleucine. Isoleucine is boxed to indicate that it is the feedback inhibitor of the enzyme (1), as indicated by the dashed arrow.

feedback inhibition

a product of a later enzyme in the same pathway is called **feedback,** or **end product inhibition.** In feedback inhibition, isoleucine binds reversibly to a site different from the active site, and changes the structure of the enzyme protein, which keeps it from recognizing its substrate. Proteins, whose shape or conformation is changed by binding to a small molecule at a site different from the active site, are called **allosteric proteins (Figure 11-19).** Small molecules that bring about the allosteric changes are called **allosteric effectors.** Effectors bind to the protein by weak bonds (hydrogen bonds, ionic bridges, and Van der Waal's attractions), and not by covalent bonds. This makes a reversal of the binding process possible (when the end-product concentration falls to a low level, the enzyme needs to be reactivated).

Some biosynthetic pathways are responsible for the synthesis of two products from a common precursor. A hypothetical example of such a **branched pathway** is the following:

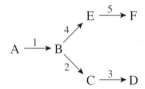

In this type of pathway, it would be undesirable for a single product to inhibit enzyme 1 because both branches would be blocked. In general, the most economical kind of inhibition prevails—namely, D inhibits enzyme 2 and F inhibits enzyme 4. In this way, neither D nor F prevents the synthesis of the other. Conversion of A to B is wasteful if both D and F are present; therefore, in many branched pathways it is found that D and F together inhibit enzyme 1.

The elegance of feedback inhibition in a branched pathway is shown in **Figure 11-20,** which shows a well-studied multibranched pathway in which lysine, methionine, and threonine are synthesized from aspartate. Note how these amino acids inhibit enzymes 4, 6, and 8 immediately after the main branch in the pathway. Furthermore, three **isoenzymes** (different proteins that catalyze the same reaction)—1a, 1b, and 1c—are separately inhibited by lysine, homoserine, and threonine, respectively. Each of the three isoenzymes carries out the reaction at a different rate, each one synthesizing as much aspartyl phosphate as is needed to

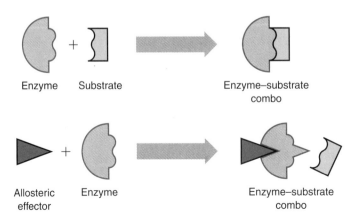

Figure 11-19 Conformational change of an enzyme protein by binding to an allosteric effector results in the inability to bind to its substrate. Compare this conformational change with the models for enzyme action illustrated earlier in Chapter 4 (Figure 4-12).

Enzyme   Substrate   Enzyme–substrate combo

Allosteric effector   Enzyme   Enzyme–substrate combo

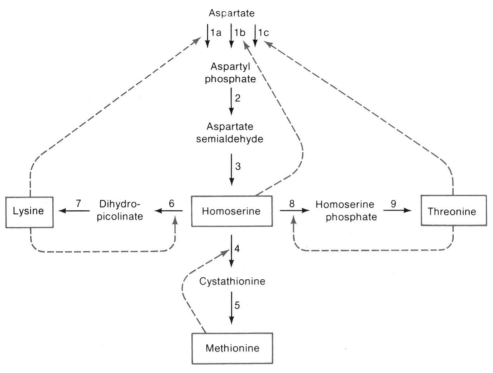

Figure 11-20 Feedback inhibition of the pathway for synthesizing lysine, methionine, and threonine. The boxed molecules are inhibitors. The dashed arrows lead from an inhibitor to the enzyme that is inhibited.

form the appropriate amount of the product that inhibits the isoenzyme. When an isoenzyme is thus inhibited, the amount of aspartyl phosphate required to synthesize the needed amino acids is still made.

Feedback inhibition occurs in both eukaryotes and prokaryotes, but has been more fully described in the latter. It can occur in a variety of ways; for example, binding of the product (which acts as an inhibitor) to the enzyme causes a physical change in the shape of the enzyme and thereby alters its catalytic properties. The changes in the properties of the regulatory proteins we have discussed in this chapter are all examples of **allosteric conformational changes** induced by the binding of specific small molecule effectors to Lac and Gal repressors (from active repressor to inactive repressor), CRP (from inactive activator to activator), AraC (from repressor to activator), and the Trp repressor (from inactive repressor to repressor). Note that allosteric inhibition or activation allows a very rapid response to changes in the cellular environment, whereas induction and repression provide an adjustment mechanism for longer-lasting changes in the environment.

A Future Practical Application?

A variety of commercially important proteins, including hormones, antibodies, and growth-stimulating proteins have been experimentally engineered into bacteria such as *E. coli*. In several instances, the desired protein product is secreted from the cell into the culture medium. That, of course, makes collection of the protein product relatively straightforward.

In many cases, however, the desired "foreign" product is degraded in the bacterial cell before it is secreted. The destruction of those protein products is usually the result of degradation by proteolytic enzymes produced as part of the bacterium's normal metabolism. Perhaps as many as a dozen different proteolytic enzymes may be active in bacterial cells.

Researchers have begun to develop mutant strains of bacteria lacking one or more of those proteases. Many of the bacterial strains lacking those enzymes nevertheless grow well. They also exhibit greatly increased yields of the designed protein product. Thus, increases in the productive capacity of bacterial strains that are genetically engineered for the production of foreign proteins is likely in the future.

SUMMARY

The synthesis of particular gene products is controlled by mechanisms collectively called gene regulation. Gene regulation generally occurs at the level of transcription or translation. Transcriptional regulation can be characterized as positive or negative. In negative regulation, an inhibitor (repressor) bound to the DNA must be removed before transcription can occur. In positive regulation, an activator molecule must bind to the DNA to allow transcription. The *E. coli* lactose system was first explained by the operon model, and is an example of both negative and positive regulation. In this system, lactose is an inducer that alters the shape of a repressor (lacI), so the repressor can no longer bind to its operator. Transcription of the lactose operon, however, also requires the presence of the cAMP-CRP activator complex. Other relatively well understood operons include the galactose operon, the arabinose operon, and the tryptophan operon. Each has its own set of components, and each works in a unique fashion.

Regulatory proteins, both positive and negative, can be highly specific (lacI), or may have a global effect (CRP) and control the expression of several genes. Some gene products can also regulate their own expression (autoregulation). The overall expression of an operon is regulated by controlling transcription initiation and premature transcription termination, whereas the relative concentration of protein is determined by controlling the degradation rate of the mRNA, the frequency of initiation of translation, and the degradation rate of the protein itself. The activity of the protein can also be under the control of allosteric interactions or feedback inhibition.

DRILL QUESTIONS

1. What distinguishes negative from positive regulation? How would you distinguish between negative and positive autoregulation?
2. What are the main features of a repressor, corepressor, and aporepressor?
3. What do the terms *induced* and *derepressed* mean in relation to gene functions?
4. What is the genetic evidence that *lac O^c* mutations are *cis*-dominant?
5. Explain how the regulatory protein AraC can be both a repressor and an activator.
6. Why does attenuation not occur in eukaryotes?
7. List two mechanisms that a bacterial cell uses to control the amount of mRNA present inside the cell.

8. List three mechanisms a bacterial cell uses to control the amount of protein present inside the cell.

PROBLEMS

1. Suppose *E. coli* is growing in a growth medium containing lactose as the sole source of carbon. The genotype is $I^- Z^+ Y^+$. Glucose is then added. Which of the following will happen?
   (a) Nothing.
   (b) Lactose will no longer be utilized by the cell.
   (c) Lac mRNA will no longer be made.
   (d) The lac repressor will bind to the operator.

2. Imagine a mutant lactose repressor (denoted *lac I^x*) that could interact with its operator and prevent transcription initiation, but could not interact with its inducer (lactose or IPTG). What would be the lac Z phenotype (i.e., expressed or not expressed), in the presence and absence of inducer, corresponding to the following genotypes?
   (a) $I^x O^+ Z^+$ $^+$IPTG
                                    $^-$IPTG
   (b) $I^x O^+ Z^-/I^+ O^+ Z^+$ $^+$IPTG
                                    $^-$IPTG
   (c) $I^x O^+ Z^+/I^+ O^+ Z^-$ $^+$IPTG
                                    $^-$IPTG

3. Ribonuclease P is an enzyme needed for the proper maturation of tRNAs. It has been shown that some mutations in ribonuclease P cause increased expression of the *trp* biosynthetic operon. Explain.

4. An *E. coli* mutant is isolated that renders the cell simultaneously unable to utilize a large number of sugars, including lactose, zylose, and sorbitol. However, genetic analysis shows that each of the operons responsible for utilization of these sugars is free of mutation. What are the possible phenotypes of this mutant?

CONCEPTUAL QUESTIONS

1. Imagine an enzyme that must be synthesized in response to an environmental stimulus (e.g., lacZ in response to lactose). What are the advantages and disadvantages for regulating this enzyme at the transcriptional versus post-transcriptional level?

2. Genetic analysis has been fundamental in identifying and characterizing the individual components of a regulatory system. For example, the lactose operon was initially characterized by making hundreds of *lac$^-$* mutants and looking at their genetic interactions. Once a regulator protein (lacI, galR, AraC, trpR, CRP) has been identified, how might the genetic consequences of inactivating that regulatory protein be used to discriminate between negative and positive control?

3. Why is the autogenous type of regulation common for abundant proteins that are incorporated into macromolecular assemblies? (For example, ribosomal proteins in prokaryotes are autogenously regulated at the level of translation.)

in this chapter you will learn

1. That important differences exist between gene regulation in prokaryotes and eukaryotes.

2. About the ways transcription and RNA processing are regulated.

3. About the ways translation is regulated.

4. The special features of the immune system.

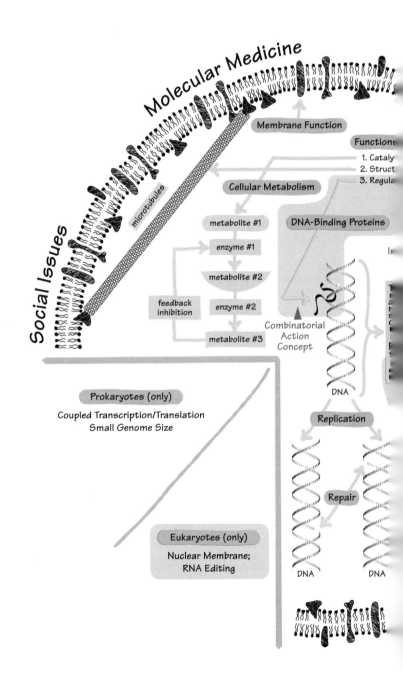

Molecular Medicine

Social Issues

microtubules

Membrane Function

Functions
1. Cataly
2. Struct
3. Regula

Cellular Metabolism

metabolite #1

enzyme #1

metabolite #2

feedback inhibition

enzyme #2

metabolite #3

DNA-Binding Proteins

Combinatorial Action Concept

DNA

Replication

Repair

DNA   DNA

Prokaryotes (only)

Coupled Transcription/Translation
Small Genome Size

Eukaryotes (only)

Nuclear Membrane;
RNA Editing

Regulation of Gene Activity in Eukaryotes

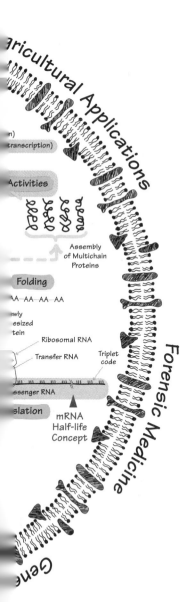

More than 200 cell types have been identified in the human body, including those for muscle, liver, nerve, lung, epithelium, bone marrow (B), and thymus (T) lymphocytes. These cell types differ dramatically in both morphology and function. However, with a few exceptions, the genetic information of the various cells within a given organism is virtually the same. These different cell types arise mainly from the differential expression of an identical genome, which is due to the expression of a characteristic subset of mRNAs and proteins in any given type of cell. The majority of the proteins and mRNAs, however, are produced in all cell types, and their levels do not differ by more than a few fold from one cell to another (i.e., they are responsible for so-called "housekeeping" functions). Eukaryotes have evolved complex regulatory mechanisms to express a specific gene in one cell type, but not in another, without changing the sequence of the gene itself. In principle, gene expression can be regulated at each step in the circular synthesis/modification pathway from DNA to RNA to protein to DNA (**Figure 12-1**). Potential control levels include:

1. Transcription.
2. RNA processing (splicing or other types of processing/modification).
3. mRNA transport.
4. mRNA degradation and storage.
5. Translation.
6. Posttranslational modulation of protein activity.

The expression of a given gene is likely to be regulated at several of these levels. Those levels provide an outline for the first section of this chapter.

Important Differences in the Genetic Organization of Prokaryotes and Eukaryotes

Numerous differences exist between prokaryotic and eukaryotic cellular and genomic organization. The most striking differences, introduced in earlier chapters, are the following:

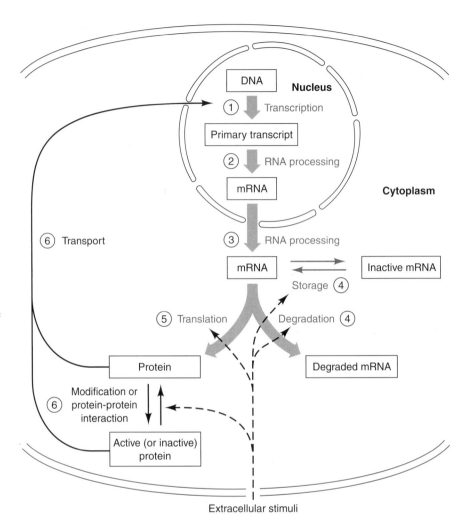

Figure 12-1 Potential control levels for the regulation of gene expression in eukaryotes. Levels 1 and 2 are localized in the nucleus and 4 and 5 in the cytoplasm; 3 and 6 indicate translocation between those two compartments. Note that with 6 only one possible route of protein translocation (from the cytoplasm to the nucleus) is indicated. RNA metabolism is shown in blue.

chromatin

1. The eukaryotic cell contains a membranous nucleus that includes the chromosomes as carriers of genetic information. In addition to the nucleus, the eukaryotic cell contains many other subcellular compartments (e.g., mitochondria) not found in prokaryotic organisms.

2. The chromosomal DNA of eukaryotes is packaged into a highly regular nucleoprotein complex called **chromatin** (see Chapter 5).

3. In eukaryotes (particularly those with a large genome), many DNA segments are repeated hundreds, thousands, or even millions of times. The function of these highly repeated sequences—if there is any—has yet to be elucidated. In contrast, most prokaryotic genomes contain essentially no repetitive sequences, except for a few repeats of ribosomal RNA (rRNA) and transfer RNA (tRNA) genes. Earlier, in Figure 3-9 (Chapter 3), these differences in genomes were described.

4. A typical eukaryotic gene is **monocistronic** (i.e., one transcription unit usually encodes one translation unit). Prokaryotes, however, often have several translation units organized in a single operon comprising one transcription

unit (**polycistronic** gene). The translation units of one operon usually encode different proteins involved in the same regulatory pathway (see Figure 11-13 in Chapter 11). In contrast, eukaryotic genes, even if they belong to the same regulatory pathway, are usually not grouped closely on the same part of the chromosome.

5. A characteristic feature of most eukaryotic genes is that they are split (i.e., the amino acid sequence of the gene product is usually not colinear with the genetic information). The noncoding sequences (so-called **introns**) need to be "spliced out" in a complicated process that generates mature messenger RNA (mRNA) (see Chapter 8).

In the following sections we will see how eukaryotes have incorporated some of these features into particular modes of regulation.

The Regulation of Transcription Initiation

Whether a gene is expressed or not is often determined at the level of transcription (mRNA synthesis). This should not be surprising, since it would appear to be the most economical way for a cell to control the expression of a gene. (However, nature often defies the obvious expectation, e.g., by transcribing a specific gene in several cell types, but converting the transcript (mRNA) into an active, translated form only in one cell type!)

Research on bacterial transcription established early on that genes consist of essentially two parts: the coding region containing the information for the exact composition of the gene product (protein or RNA), and the so-called promoter sequence located closely upstream of the coding region. Promoters serve as "docking sites" for the transcription machinery (see Figure 8-4 in Chapter 8) and determine when and how efficiently a gene is transcribed. The prokaryotic transcription apparatus is relatively simple and comprises only one type of RNA polymerase (with four subunits) and so-called sigma factor proteins that confer specific promoter recognition. Additional promoter-specific DNA binding proteins have been shown to modulate the level of transcription. Some of these proteins activate transcription by specifically interacting with both the promoter and RNA polymerase; most of them, however, are repressors that reduce transcription by sterically interfering with the transcription apparatus (for further details on prokaryotic transcription refer back to Chapter 8).

Initially, researchers assumed that transcriptional regulation would be similar in eukaryotes. However, extensive work over the last 20 years has revealed a surprisingly complex situation, as we shall discuss in this section. The components and steps involved in eukaryotic transcription can be summarized as follows:

1. There are three different *gene class*-specific RNA polymerases, each of which consists of at least 10 subunits.
2. A number of additional proteins, the so-called general, or basal, transcription factors, together with the RNA polymerase, form a minimal complex that recognizes a promoter and can initiate transcription in vitro.
3. The relative level of transcriptional initiation is regulated by various specific **transcription factors.** These bind to upstream regions of promoters,

introns

transcription factors

or to other regulatory sequences called enhancers that are often located very far away from the coding region of a given gene. Furthermore, additional factors, so-called coactivators, which themselves do not bind DNA, may interact with DNA-bound transcription factors.

4. The specific transcription factors bound to distal promoter or enhancer sequences most likely stimulate transcription by directly interacting with the basal transcription apparatus while looping out the intervening DNA. In addition, at least some of these factors seem to also help re-organize chromatin into a more accessible, "loose" form that is experimentally characterized by a higher sensitivity toward DNaseI (see Figure 5-7 in Chapter 5). Conversely, repressing factors binding so-called silencer DNA sequences may instead disturb interactions within the transcription apparatus and/or reorganize the chromatin into a tight, nonaccessible form.

Presumably this increased degree of complexity in eukaryotes is necessary for the fine-tuned transcriptional regulatory circuits that exist in multicellular organisms. During the differentiation of a precursor cell into a specific cell type, for example, specific sets of genes are turned on, while others are turned off. Such genes include regulatory genes that control the establishment and maintenance of a cell type, as well as genes that encode cell type-specific proteins, which may be needed in massive amounts. Examples of the latter type of proteins are the hemoglobins of erythroid cells, insulin of pancreatic beta cells, immunoglobulins of B lymphocytes, and actin and myosin of muscle cells.

Three Different RNA Polymerases Transcribe Eukaryotic Genes

In eukaryotes, three different RNA polymerases (see Chapter 8) fulfill the different transcriptional tasks. Each polymerase consists of at least 10 subunits, of which they share five. RNA polymerase I transcribes only the tandemly repeated genes encoding ribosomal 18S, 5.8S, and 28S RNAs in the nucleolus. RNA polymerase III transcribes some small RNA genes, such as those encoding tRNAs, ribosomal 5S RNA and the U6 small nuclear RNA (U6 snRNA). Transcripts from these genes are produced in all cell types and, therefore, can be regarded as housekeeping genes. In addition to transcribing most small nuclear RNA (snRNA) genes, RNA polymerase II transcribes all the protein-coding genes. These latter genes are perhaps the most interesting ones because their products are often decisively involved in cellular responses to environmental and developmental cues. Our focus in this chapter will be on the transcriptional regulation of protein-coding genes.

The Transcriptional Regulation of RNA Polymerase II-Transcribed Genes

Promoter regions of genes transcribed by RNA polymerase II usually contain an initiator region (Inr) where transcription starts, a TATA box some 30 base pairs upstream of the initiation site, and always additional upstream promoter elements (**Figure 12-2**(a)) as well. The part of the promoter comprising the initiator region

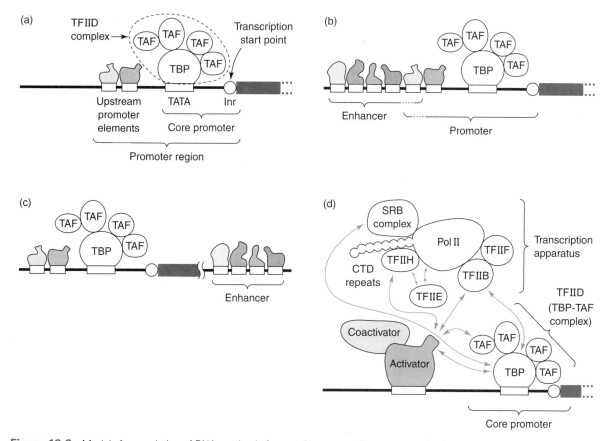

Figure 12-2 Models for regulation of RNA synthesis from various genes that are transcribed by RNA polymerase II. Regulatory sequences (enhancers and promoters), as well as the multitude of proteins comprising a transcription complex, described in the text, are illustrated. Protein coding sequence is shown in blue.

and the TATA box is viewed as the "core promoter," since it can serve as a minimal "landing site" for the transcription apparatus in vitro. While the bacterial RNA polymerase needs just one accessory protein, a sigma factor, to recognize a promoter and specifically initiate transcription, all eukaryotic RNA polymerases need a number of so-called general, or basal, transcription factors (proteins) to initiate transcription. In the case of RNA polymerase II, these general transcription factors have been grouped into six fractions called TFIID, B, F, E, F, H, each containing one or more polypeptides. TFIID, for example, is a complex that consists of a sequence-specific TATA-box binding protein (TBP) and a number of TBP-associated factors (so-called TAFs). (Interestingly, TBP is essential for transcription by all three RNA polymerases; the TAFs, however, appear to be polymerase specific.) As TBP binds to DNA, it causes the DNA to bend (reminiscent of the cAMP-CRP/DNA interaction described in Figure 11-9 of the previous chapter). That interaction distorts the double helix and thereby facilitates the eventual binding of RNA polymerase to the TATA region and the Inr.

The other general transcription factors—TFIIB, F, E, F, H—do not bind directly to DNA. Rather, they form a more or less tight association with RNA polymerase II. This complex is here referred to as the (basal) transcription apparatus. Initiation of transcription at specific DNA sites, albeit at a low level, can actually be experimentally obtained in vitro with the core promoter and this transcription apparatus. Thereby only TBP of the TFIID complex, that is, without the TAFs, needs to be bound to the TATA box, since the TAFs are required to mediate the effect of upstream promoter sequences and their bound factors.

In vivo, however, upstream promoter elements are absolutely required. A promoter contains multiple elements, each about 10 bp in length, that provide binding sites for transcription activators (often also just called "upstream transcription factors," or "specific transcription factors"; Figure 12-2(a)). Transcription activators regulate the frequency with which a gene is transcribed. A larger number of upstream promoter elements in DNA usually induces a higher frequency of transcription (Figure 12-2(b)). In principle these transcription activators can induce transcription in vivo in two ways: (1) by opening up chromatin to expose genes for transcription, and (2) by directly interacting with proteins of the transcription machinery. The compacted structure of chromatin generally has a repressing effect on transcription, and this explains why core promoters are not sufficient to confer any transcription in vivo. However, at least some transcription activators (e.g., glucocorticoid receptor, and Ga14 factor of yeast) have been shown to be able to contact their binding site even when it is wrapped around a nucleosomal core! Such interactions—perhaps in collaboration with a so-called SWI/SNF complex that modifies nucleosomal structures—may alter the structure of chromatin structure and render a promoter more accessible for binding the transcription apparatus.

While all promoter regions transcribed by RNA polymerase II contain one or more upstream promoter elements, they do not always contain a TATA box. Many housekeeping genes, for example, lack TATA boxes and tend to have multiple transcription start sites. The assembly of the transcription machinery on such promoters has been studied less extensively than on TATA-box-containing promoters. Hence, with those promoters the details of assembly are not yet well understood. It is known, however, that the TBP-TAFs complex (TFIID) is also required for accurate transcription with TATA-less promoters. Presumably, the initiator region(s) Inr plays a more prominent role for the assembly of the transcription machinery with such promoters. Accordingly, it has been postulated that Inr-binding proteins might tether TFIID to the promoter.

enhancer

Another type of regulatory sequence, called **"enhancer,"** also increases the rate of transcription initiation. Enhancers seem to be associated with most genes that are "regulated," while they may not be essential for all housekeeping genes, which are expressed in virtually all cells at all times. Interestingly, enhancers can be located either upstream or downstream of the start site—close to the promoter (Figure 12-2(b)) or thousands of base pairs away (Figure 12-2(c)). Enhancers are comprised of DNA sequence motifs similar to upstream promoter regions and, therefore, can bind some of the same transcription factors. In several cases, when a single upstream promoter element is experimentally multimerized, the resulting synthetic segment behaves as a strong enhancer, i.e., it activates transcription from a remote position. This indicates that at least some transcription factors stimulate transcription via a similar mechanism, regardless of whether they are bound to proximal promoter sequences or remote enhancers. Therefore, promoters and enhancers can overlap both physically (Figure 12-2(b)) and functionally.

Although most of the transcription factors so far discovered are transcriptional activators, transcriptional repressors also exist. Again, some of these repressors can bind either to upstream promoter or to distantly located so-called **silencer** sequences. A silencer can be regarded as having the opposite effect of an enhancer.

silencer

The Structure of Specific Transcription Factors and Their Role as Regulators in Cell Type-Specific Transcription and Embryonic Development

In recent years a number of genes that encode DNA-binding transcription factors have been "cloned" (see Chapter 15), and much has been learned about their structure and function. These proteins generally consist of several functional domains that often can be experimentally exchanged between different protein complexes, a fact that has facilitated the elucidation of their functions. Transcription factors contain at least three different domains specifying DNA binding (DNA binding domain), allowing translocation of the protein complex to the nucleus (nuclear translocation signal, NLS), or interacting with parts of the transcription apparatus (activation domain). Some of these activation domains seem to work well only over relatively short distances from the initiation site of transcription, that is, from a promoter position. Others (so-called enhancer-type domains) can, however, also act over large distances, that is, from a remotely located enhancer position. Quite often a transcription factor contains additional domains for interaction with other DNA-binding transcription factors or regulatory proteins that can substantially increase or decrease the activity of the factor (for further details see the section on "Regulation of Protein Activity" later in this chapter). So-called "coactivators" that specifically interact with DNA-bound transcription factors (but not with DNA) consist of functional domains similar to those of bonafide transcription factors. Instead of a DNA binding domain, however, they contain a domain that allows specific "piggyback" binding to a DNA-bound transcription factor (see Figure 12-2(d).

Transcriptional repressors interact with the transcription apparatus in a way that leads to inhibition of transcription. The separation into activating and repressing transcription factors is not an absolute one: Depending on the promoter structure and interactions with other factors, one gene's repressor can be another gene's activator, as is exemplified by the dual role of some regulatory proteins, such as the glucocorticoid receptor.

Interestingly, the overall structure of transcription factors (and other components of the transcription machinery) is very similar in species as phylogenetically distant as yeast and humans. Indeed, it has been shown that several yeast and insect transcription factors work well in human cells! Some transcription factors are found exclusively in one or a few cell types and help to control the establishment and/or maintenance of specific cell types (e.g., the proteins Myf 5 and Myo D found in muscle cells). The experimental expression of the transcription factor Myo D in fibroblasts can induce these relatively ordinary cells to differentiate into muscle cells. Other transcription factors help to ensure the production of abundant cell type-specific proteins. The transcription factor Oct-2, for example, is found mainly in B lymphocytes. Together with the ubiquitous Oct-1 factor and a B cell-specific coactivator (Bob1/OBF1), it contributes to the cell type-specific transcription of immunoglobulin genes, which encode antibodies. The activity of other ubiquitously expressed transcription factor proteins can also be modulated after they are synthesized, depending on extracellular stimuli. As we will see, such modes of regulation provide a major mechanism for regulating transcription (see section on "Regulation of Protein Activity" later in this chapter).

Transcription factors are also responsible for many crucial steps in the embryogenesis of invertebrates and vertebrates. It came as a surprise to find that at least some of the substances (so-called "morphogens") that were postulated decades ago to control tissue or organ formation within the embryo are actually transcription factors. For example, the product of the maternally transcribed *bicoid* gene is a transcription factor with a so-called "homedomain" for specific DNA binding. It forms an anterior-posterior gradient at an early stage of embryogenesis in the fruitfly, *Drosophila melanogaster,* and is referred to as a morphogen.

The Assembly of the Transcription Apparatus and the Mechanism of Transcriptional Activation by Upstream Promoter Elements and Enhancers

The complex of proteins we have referred to as the "transcription apparatus" is very complicated and by no means fully understood (Figure 12-2(d)). It consists of RNA polymerase II and a number of general transcription factors (TFIIB, E, F, H). RNA polymerase II itself consists also of a number of subunits. From yeast the genes of all 12 subunits have been cloned, whereas in mammals not even the exact number of subunits is known. Most recently, yet another class of proteins, called SRBs, has been genetically defined in the baker's yeast, *Saccharomyces cerevisiae*. These SRBs are known to form a large multiprotein complex that can also be associated with the transcription apparatus. So far no mammalian SRBs have been discovered, although their existence seems quite likely due to the high degree of conservation mentioned earlier between the transcription machineries of all eukaryotes.

The process by which the transcription apparatus is assembled is still not well understood. Studies with different experimental systems suggest two possible assembly pathways, whereby in both cases TFIID (or minimally, TBP) binds first to the TATA box: (1) assembly in a DNA-dependent, sequential fashion, as found in mammalian cell-extracts; or (2) assembly in a DNA-independent manner, as observed in yeast extracts. In the first case, the promoter-bound TFIID (TBP/TAF) complex serves as a binding site for TFIIB, which subsequently recruits RNA pol II and TFIIF into the complex. Finally, TFIIE and TFIIH are bound, so transcription can start. In the latter situation (2), only TFIIE and TFIID need to be added to the preassembled transcription apparatus before the actual interaction with the promoter takes place. This mode is schematically indicated in Figure 12-2(d). The still unanswered question of whether both pathways exist in vivo is important, since a sequential assembly (case 1) would allow for the most elaborate regulatory mechanisms. In any case, transcripts obtained with this minimal number of about 50 polypeptides are only detectable in vitro. Even so, the transcription efficiency is quite low in the absence of additional transcription factors that bind to upstream promoter elements.

The boosting of transcription in vitro by the addition of factors that bind to upstream promoter elements requires the association of TAFs with TBP. This observation suggests that TAFs provide at least some of the interaction targets for the transcription activators (see Figure 12-2(d)). As might have been expected from study of the different types of activation domains found in transcription activators (e.g., acidic, proline- and glutamine-rich domains), they appear to interact with different TAFs. Some of the interactions are schematically indicated in Figure 12-2(d), either with direct protein-protein contacts or with arrows.

The further analysis of this network may lead to many more surprises. Keep in mind that each protein-protein and protein-DNA interaction has the potential to serve as the basis for a regulatory mechanism.

Little is known about the transition from transcriptional initiation to elongation, that is, the release of RNA polymerase II and associated factors from the promoter-bound complex. Evidently, this process must involve uncoupling of some protein-protein and protein-DNA interactions. It has been postulated that the carboxy-terminal domain (CTD) of the largest subunit of RNA polymerase II plays a crucial role in the assembly of the initiation complex, as well as the transition to elongation. The CTD domain consists of a repeated heptapeptide with the consensus sequence Tyr-Ser-Pro-Thr-Ser-Pro-Ser. The number of repetitions is constant for each species (e.g., 26 in yeast, 43 in *Drosophila*, 52 in human and mouse). Interestingly, the CTD is underphosphorylated in the assembled complex, but heavily phosphorylated during ongoing transcription. The already-mentioned SRBs and the general transcription factor TFIIH, which exhibits a CTD kinase activity that is modulated by TFIIE, have been shown to interact with the CTD (Figure 12-2(d)). Obviously, the dramatic change in the CTD's charge could affect interactions between assembly partners at the initiation complex. According to one simple model, CTD phosphorylation facilitates elongation by allowing for release of RNA polymerase II from the initiation complex.

Gene Transcription Correlates with Decondensed Chromatin and DNA Hypomethylation

The genome of eukaryotes is packaged into a highly organized nucleoprotein complex called chromatin, whose major protein components are the so-called histones (see Chapter 5). This complex, however, is not identical throughout the entire genome. In chromatin regions that contain actively transcribed genes, the

KEY CONCEPT

The "Combinatorial Action" Concept

The expression of eukaryotic genes is regulated by macromolecular complexes consisting of a variety of proteins that combine with each other and bind to specific nucleotide sequences (e.g., promoters and enhancers).

histones are acetylated and one of them (histone H1) is underrepresented, resulting in a more loosely packed nucleoprotein complex. As a consequence, regions of active chromatin typically exhibit an increased overall sensitivity to experimental treatment with DNaseI. These regions of higher DNaseI sensitivity are often associated with transcriptionally active genes, and also contain a few so-called DNaseI hypersensitive sites. Hypersensitive sites represent binding sites for regulatory proteins in enhancers and also the location of other control regions. These sites may be located in front of, behind, and, sometimes, within the gene, and they often include promoter and enhancer sequences. Transcription factor proteins bound to these regulatory sequences might prevent tight chromatin packaging and, thus, allow more easy access by DNaseI. In addition, some of these transcription factors have been shown to induce DNA to bend, making it more vulnerable to nuclease attack.

Despite the fact that all the cells in an organism contain identical copies of its genome, the locations of hypersensitive sites varies among cell types. Those hypersensitive sites associated with a specific gene are, however, not usually detected in cells that normally do not express the gene. Hemoglobin genes, for example, have hypersensitive sites in both immature and mature erythroid cells, but not in nonproducer cells, such as those found in the liver, brain, or muscles. Most likely, hypersensitive sites are formed as soon as the transcription factors that specifically bind to these sites appear in the cell at a given stage of differentiation.

methylation

Gene activity correlates not only with decondensed chromatin structure, but also with hypomethylation of the gene. The principle of DNA modification by **methylation** of specific cytosines is exploited by both prokaryotes and eukaryotes. The specific methylation pattern of a given gene (or even a large chromosomal region) can be maintained during many rounds of cell division (**Figure 12-3**). In bacteria, methylation of the adenine in G$\underline{A}$TC and cytosine in C$\underline{C}$A/TGG sequences can occur. In higher eukaryotes, only the modified DNA base 5-methyl cytosine has been detected thus far. It is typically present in the $\underline{C}$G dinucleotide sequence (and also as $\underline{C}$NG in plants). Recall, as was mentioned in Chapter 10, methylated cytosines are hotspots of mutation because they are quite vulnerable to transition mutations, which convert a methylated C into a T. This also explains the relative underrepresentation of CG dinucleotides in vertebrate genomes. Nevertheless, the fact that methylated Cs persist in the genome suggests that there must be a strong natural evolutionary selection to maintain cytosine methylation for certain cellular functions.

heterochromatin

In invertebrates (for example, sea urchins), methylation appears to be used mainly as a means to inactivate highly repetitive DNA, which is packaged into dense **heterochromatin.** In vertebrates, cytosine methylation appears to be used in a more sophisticated manner, since the same DNA sequence can be methylated in one cell type, but not in another. The β-globin gene, for example, is methylated in nonproducer cells, but unmethylated in erythroid cells where it is transcribed. Also, the housekeeping genes transcribed by RNA polymerase II are unmethylated, even though they are usually particularly rich in side-by-side CG nucleotides. These unmethylated CG-rich sequences, usually containing enhancer or promoter sequences, are called **CpG islands.**

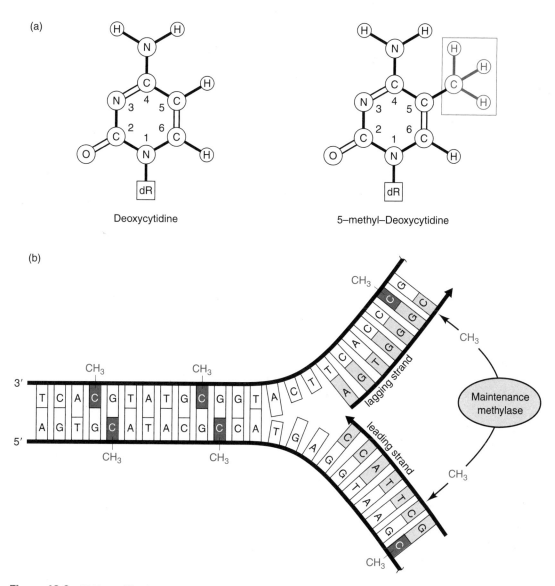

Figure 12-3 DNA modification by methylation. (a) Deoxycytidine and 5-methyl-deoxycytidine are indicated (dR = deoxyribose). (b) The methylation, in which the hydrogen atom at carbon number 5 of deoxycytidine is replaced by a methyl group, is catalyzed by a maintenance methylase enzyme that recognizes hemimethylated CpG dinucleotides and fully methylates them shortly after DNA replication.

Differential methylation of single genes or entire chromosome regions can occur in a cell type-specific (e.g., β-globin) or in a chromosome-specific (e.g., X chromosome) manner. Most interestingly, in the phenomenon called "genomic imprinting," differential methylation patterns of chromosomal regions seem to be established in a germline-specific (i.e., egg/sperm cell) manner. In other words, the methylation pattern and activity of certain genes can differ, depending on whether the genes are inherited from the father's sperm or from the mother's egg (the insulin-like growth factor 2 (IGF2) gene, for example, is paternally unmethylated and maternally methylated).

We are only beginning to understand the meaning of this type of differential methylation. It might explain the lack of parthenogenesis (embryonic development from fatherless eggs that have undergone duplication of the maternal genome) in mammals. Similarly, it is also impossible to create embryos from a duplication of the paternal set of chromosomes. One simple explanation would be that a 1:1 mix of differentially methylated genes inherited from both sperm and egg is required for the proper gene balance and correct development of the embryo. Therefore, differential methylation may prevent parthenogenesis in mammals.

Which regions of a given inactivated gene are methylated? It is obvious that the sites of methylation are more crucial than the overall amount of methylation. For example, methylation of a promoter region can abolish activity, while extensive methylation in the protein-coding part of the gene may have no effect on gene expression. Therefore, methyl groups do not merely act as a "roadblock" to RNA polymerase II-mediated transcription. It is clear, however, that methylation of a promoter can sterically prevent binding of transcription factors. In addition, a more indirect inhibition of gene expression also appears to exist. It results in a tighter chromatin packaging of methylated DNA. While there is no question about a strict correlation between gene inactivity and DNA methylation, the phenomenon has eluded a thorough analysis for many years. Part of the confusion can be attributed to the abnormal methylation patterns often obtained with cells grown long term in culture for laboratory experiments. Also, in rare cases, methylation is positively correlated with gene expression in mammals, whereby methylation of a silencer region may prevent repressing factors from binding to genes.

The question of whether DNA methylation is a cause or an effect of gene inactivation remains a matter of debate. In several cases it was found that gene inactivity preceded the addition of methyl groups to the DNA. Therefore, it was argued that methylation was just perpetuating, rather than initiating, the inactive state of a gene. But, model experiments have shown that methylation of the promoter region, and even of binding sites for a single transcription factor, can result in a complete loss of gene activity. The mechanism that controls gain or loss of a methylation pattern during cell type-specific differentiation is one of the most challenging problems of contemporary research on vertebrate development.

The Regulation of RNA Processing

Alternative Splicing

The discovery of split eukaryotic genes in 1977 was completely unexpected, since previous studies with bacterial genes had revealed strict colinearity between a gene and its product. Recall (from Chapter 8) that in most genes of higher eukaryotes, the coding sequences (**exons**) are interrupted by noncoding intervening

exons

sequences (**introns**). Therefore, the primary transcripts of such a gene need to be processed in order to produce the mature messenger RNAs containing a reading frame that is colinear to the gene product. This processing occurs in the nucleus, followed by the transport of the mature mRNA to the cytoplasm.

While introns can be found in most of the genes in higher eukaryotes, they are rare in lower eukaryotes, for example, yeast. Are introns a relatively recent development in the evolutionary history of the genes of higher eukaryotes? It has been widely believed that the opposite is true: Introns are probably almost as old as life itself, although their structure might have changed through time. Since, in general, organisms with a short generation time tend to have fewer and shorter introns, it would at first glance seem most likely that introns have been widely lost in prokaryotes and lower eukaryotes as an adaptation to their short generation time and pressure to minimize genome size.

Evidence is rapidly accumulating, however, that supports the view that introns have arisen relatively late in eukaryotic evolution. Included in that evidence is information from surveys of intron distributions among diverse taxa. This data favors the notion that introns did not arise early and spread through a growing phylogenetic tree, as the original "introns early" view predicts. Instead, introns appear in mammalian and plant DNA that are lacking in the DNAs of earlier evolving species. It is now, therefore, widely believed that introns were inserted into preexisting genes. In fact, mechanisms involving DNA self-splicing in a "reverse direction" have been postulated to provide an explanation for how introns might become inserted into preformed genes.

Splicing of RNA, that is, the removal of intron RNA from a primary transcript that generates the mature mRNA, is an extremely precise process. Eukaryotes have evolved a splicing machinery (called a **spliceosome**), consisting of several protein/RNA complexes (called **small nuclear ribonucleoprotein complexes** or **snRNPs**), as shown in **Figure 12-4**.

One of the most interesting discoveries made regarding splicing was that certain introns in the organelles (e.g., mitochondria) of lower eukaryotes and plants can be spliced autocatalytically in vitro. That is, splicing can occur in the test tube in the absence of any protein components. Nevertheless, this occurs by a reaction mechanism that is identical to that used for the splicing of nuclear introns. The enzymatic activity actually resides within the intron RNA itself. In the case of self-splicing organelle introns, the splice sites are aligned by intramolecular base-pairing interactions between conserved exon and intron sequences. In the case of nuclear pre-mRNA splicing, this task is fulfilled by the RNA moieties of the snRNPs. They base pair with conserved sequences at the splice sites and also with one another to form a network of interactions that are believed to form the catalytic site(s) for the cleavage and ligation reactions that generate mature mRNA. This phenomenon is reminiscent of the cleavage action of RNase P, an RNA-containing enzyme described earlier (see "ribozymes" in Chapter 3). Several non-snRNP proteins do, however, assist in these interactions in vivo. The base pairing between pre-mRNA and snRNAs also ensures the accuracy of splicing.

A cell cannot, of course, afford to miss any of the splice junctions by even a single nucleotide, because this could result in an interruption of the correct

introns

RNA splicing

snRNPs

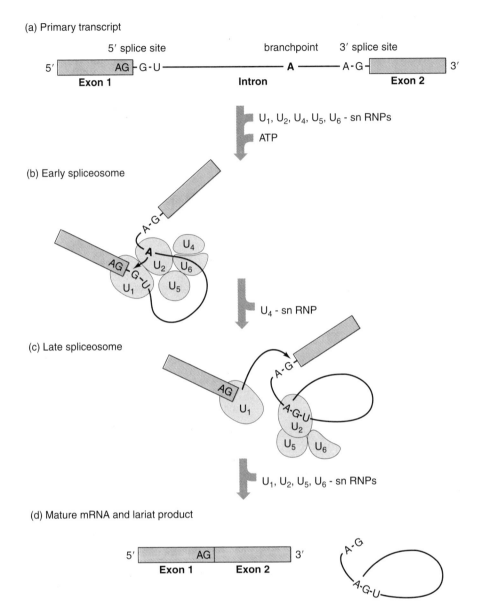

Figure 12-4 The production of mature messenger RNA (mRNA) by splicing of the primary transcript (also called pre-mRNA). Pre-mRNA splicing occurs in large ribonucleoprotein complexes known as spliceosomes, which are assembled from small nuclear ribonucleoprotein particles (snRNP) that recognize specific sequences on the primary transcript (a) and (b). The first step in splicing (b) involves cleavage at the conserved 5′ splice site sequence with the concomitant covalent joining of a conserved guanosine (G) residue at the 5′ end of the intron via a 2′, 5′-phosphodiester bond to an adenosine within the intron known as the "branch point site" yielding exon 1 and intron-exon 2 intermediates. In the second step of splicing (c), cleavage at the conserved 3′ splice site occurs, and the two exons are joined by ligation, creating the mature mRNA and the excised intron (d). Both the intron-exon 2, as well as the intron products, are also called "lariats" because of their branched circular structure. The spliced (mature) mRNA is transported to the cytoplasm, while the lariat product remains in the nucleus and is degraded. Exons are shown in blue.

reading frame, leading to a truncated protein. Skipping of exons, such as direct splicing from the 5′ site of exon 1 to the 3′ site of exon 3, also needs to be avoided. It came as a surprise, therefore, to find that whenever a correct 5′ or 3′ splice site has been eliminated by mutation, splicing nevertheless occurs, albeit to previously undetectable, "cryptic" splice sites within the same transcript. These cryptic splice sites seem to be secondary due to a lower match to the splicing consensus sequence, or to an unfavorable RNA structure. The splicing machinery obviously is able to distinguish between stronger and weaker splice sites and chooses the best one available. In a number of human diseases, however, loss of the function of a necessary protein is a consequence of a mutation in the correct splice site. This is best documented in several thalassemias, where there is an underrepresentation of hemoglobin α or β chains. In some of these genetic diseases, globin production is reduced or eliminated by mutation of a splice site, resulting in an RNA that undergoes erroneous splicing to cryptic sites and can, thus, no longer be translated to the correct protein.

Even though splicing needs, in principle, to be very precise, a given set of splice sites is not used in all cells under all circumstances. In mammalian cells, alternative modes of splicing (frequently cell type-specific) are often used to produce several protein isoforms of a given gene that have overlapping, but distinct, functions. Even though the general splicing machinery is the same in all cells, the eukaryotic cell has evolved a strategy to produce **differentially spliced** versions of a primary transcript. This does not involve just one mechanism, but rather several, depending on the mode of **alternative splicing.** Splicing patterns are actually guided by various specific proteins that are found in the nuclei of cells. Different modes of alternative splicing are summarized in **Figure 12-5**:

alternative splicing

(a) *Differential promoter selection* due to cell type-specific transcription factors can dictate the splicing pattern. In the longer primary transcript, which has been started from promoter P1, the (stronger) 5′ splice site overrides the second one. Thus, the second promoter (P2) is actually included in the primary transcript (Figure 12-5(a)). Genes for a myosin light chain in muscle cells, as well as an amylase (digestive enzyme), are regulated in this manner.

(b) *Differential cleavage/polyadenylation site selection* can also determine the splicing pattern. The selective usage of a particular poly(A) site, in a cell type-specific manner, creates a longer or shorter primary transcript. Accordingly, the splicing apparatus uses the most downstream (stronger) 3′ splice site, as long as it is available. In this way, a differential choice of poly(A) sites results in an alternative splicing of immunoglobulin gene transcripts. The product is an antibody molecule that is either membrane bound or secreted into the serum. The same mode of alternative splicing is found in α and β tropomyosin (cytoskeleton proteins) and calcitonin/ CGRP genes.

(c) An example of the **intron retaining mode** is the cell type-specific splicing of primary transcripts of the P element transposase gene in *Drosophila*. Transposition of this P element significantly contributes to the mutational load of

(a) Alternative selection of promoters (e.g., *myosin* primary transcript)

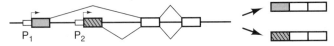

(b) Alternative selection of cleavage/polyadenylation sites (e.g., tropo*myosin* primary transcript)

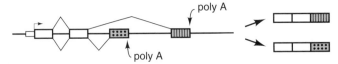

(c) Intron retaining mode (e.g., *transposase* primary transcript)

(d) Exon cassette mode (e.g., *troponin* primary transcript)

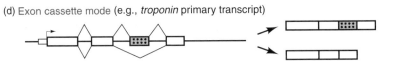

Figure 12-5 Different modes of alternative splicing of primary transcripts.

the organism. Perhaps for this reason, transposition in somatic cells is abrogated. The transposase gene is transcribed in somatic cells, but the third intron remains part of the mature mRNA, which, therefore, codes only for a truncated protein (the third intron contains a stop codon; therefore, the reading frame is shorter). A mature mRNA with a full-length open reading frame is only produced in germ cells by also splicing out the third intron. This leads to the production of competent transposase exclusively in the germline and, therefore, a low level of transposition activity in the fly. The phenomenon of transposition is described in Chapter 14.

(d) In the **exon cassette mode,** some exons can be included or excluded independently of other exons, and usually the same reading frame is maintained whether the exon is spliced out or not. This mode was found in the genes for neural cell adhesion molecules (N-CAMs) and also in troponin-T genes (expressed in muscle cells).

In (a) and (b), alternative splice patterns are obtained not by modification of the main splicing activity, but by differential promoter or polyadenylation site selection. For the "intron retaining" and the "exon cassette" modes (c and d), however, it is believed that cell type-specific factors act to modify the main splicing activity of certain transcripts. For example, a protein could bind at a splice site and, thus, prevent splicing by steric hindrance. The splicing machinery, in an opportunistic manner, would then choose the next available splicing site. The converse situation, namely a factor changing the structure of the transcript such that a previously inaccessible site becomes available for splicing, is another possibility. So far, several factors (proteins) have been found that function in alternative splice site choice in a concentration-dependent manner. These factors appear to be expressed in different amounts in different tissues.

Human genes usually are comprised of large numbers of relatively small exons. Relatively large introns, including some as large as 10 kb are, however, present in some human genes. Included in some of those introns are "exon-like" sequences that likely are—in some types of cells—actually spliced into functional primary transcripts. Thus, although the human genome may contain only 30,000–100,000 genes, many more proteins than would be expected based on those gene counts are likely distributed through the broad spectrum of cell types found in the adult (see Proteomics in Chapter 13).

Trans-Splicing

RNA splicing usually occurs in *cis* configurations. That is, exons of the same transcript are joined. When **trans-splicing,** namely the splicing of an exon of one transcript to another exon that is located on a different transcript, was shown to be possible in model experiments, it was considered a curiosity rather than a natural occurrence. It was originally argued by researchers that if *trans*-splicing were possible under natural circumstances, it would wreak havoc among cellular transcripts, leading to hundreds of thousands of new protein combinations. However, as was seen to be the case with surface antigen proteins of malaria-causing trypanosomes, orderly *trans*-splicing is possible by selective matching of two specific RNAs yielding a mature mRNA. *Trans*-splicing has also been found in nematodes, but not, as yet, in higher eukaryotes. It may well be a remnant of an ancient condition whereby new proteins are generated from different exons by exon shuffling at the *trans*-splicing (RNA) level, as well as the genome (DNA) level.

trans-splicing

RNA Editing

Viruses and lower eukaryotes in particular exhibit several unexpected mechanisms for the control of gene expression. Perhaps the most spectacular one is **RNA editing,** which was first found in the mitochondria of the unicellular trypanosomes, the blood parasite that causes sleeping sickness. For several of their mitochondrial genes, it was shown that the corresponding primary transcripts are extensively modified by insertion (or, less often, by deletion) of uridine nucleotides to yield the final mRNA that is translated into protein. In some cases, these newly inserted uridine nucleotides can make up over half of the total number of nucleotides of the mature mRNA! The initial RNA transcript is specifically altered with the help of so-called "**guide RNAs.**" A simple case is illustrated in **Figure 12-6**.

RNA editing

A cell type-specific form of editing has been found in mammalian cells. The messenger RNA for intestinal apolipoprotein B is modified at a single base-position by specific replacement of a C by a U residue in intestinal (but not liver) cells. This converts a glutamine codon (CAA) to a stop codon (UAA) and, thus, leads to a truncated protein. This base change is caused by a deaminase recognizing some structural feature of the mRNA in a cell type-specific manner. A similar type of editing might occur in many transcripts of plant mitochondria. In this case, bases are also changed from C to U without nucleotide insertion or deletion. Whether specific deaminases or guide RNAs are involved in this process is not yet known.

guide RNA

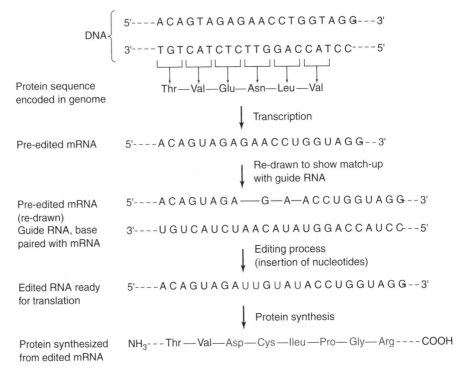

Figure 12-6 Editing of an mRna. The guide RNA base pairs with the pre-mRNA, with some gaps in the base complementarity. Those gaps are filled by insertion of bases (in this example, U), which generates the shift in the reading frame in the mRNA. The protein that is translated from the edited mRNA has a new amino acid sequence, as shown in the bottom illustration.

A different type of RNA editing has been discovered in paramyxoviruses (e.g., measles virus). In this case, "stuttering" of the RNA-dependent RNA polymerase at a particular region results in the addition of G residues that were not coded for in DNA. This shifts the reading frame and yields mRNAs encoding one or more additional proteins.

Combinatorial RNA Processing

In humans, more than 95% of the typical gene is comprised of introns! It has been speculated that more than 25% of human primary transcripts are alternatively spliced, which, of course, greatly increases the information content of the genome.

Considering what we have just learned about alternative splicing, *trans*-splicing, and RNA editing, it should be easy to imagine the extent to which relatively small numbers of so-called "genes" can generate large numbers of protein products. Later in this chapter, we will learn how proteins are themselves alternatively processed to yet further increase the total number of "functional" gene products in a living cell.

As mentioned previously, estimates of the number of genes in the human genome presently range upwards from 30,000. The number of different proteins in the whole human organism is, however, estimated to be between five- and ten-fold greater than the estimated number of genes. A variety of RNA processing mechanisms no doubt account, in part, for the disparity between gene count and protein array. Indeed, several disciplines, including bioinformatics, genomics, and proteomics, have developed the goal of understanding information flow from

KEY CONCEPT

The "Mix and Match" Transcript Concept

Many eukaryotic primary transcripts are "cut and shut" or "cut and pasted" to generate different mRNAs from single long primary transcripts. RNA editing can further change a reading frame.

nucleotide sequence to protein function. Those disciplines are described in the next chapter.

Regulation of Nucleocytoplasmic mRNA Transport

One of the most prominent defining features of eukaryotic cells is the existence of a membranous nuclear envelope that creates distinct nuclear and cytoplasmic compartments. The evolution of local separation of transcription and mRNA processing in the nucleus and translation in the cytoplasm has greatly increased the potential for the regulation of gene expression of eukaryotes. The nucleocytoplasmic translocation of macromolecules, such as RNAs and proteins, provides a further level of gene regulation.

The nuclear envelope consists of two concentric membranes, each based on the bilayer structure described in Chapter 5 (e.g., Figure 5-17). The outer membrane can be regarded as an extension of the **endoplasmatic reticulum (ER).** **Nuclear pore complexes,** which are crucial for the translocation of different types of molecules in both directions, are evenly spread over the nuclear envelope (3,000–4,000 per nucleus). Each nuclear pore complex is highly organized in an eight-fold symmetric structure. It consists of a large number of different proteins with a total molecular weight of $25–100 \times 10^6$ Daltons, and is about 100 nm in diameter. The pore itself is an aqueous channel and has a diameter of about 10 nm and a length of about 15 nm. It allows free diffusion of molecules of up to about 20,000 Daltons. However, larger molecules need to be actively transported in an energy-dependent manner through these pores, probably by receptor-mediated gating of the otherwise too-narrow channels. Specific transport occurs in both directions: some proteins and RNAs are unidirectionally transported, while others seem to be able to shuttle between the two compartments, depending on the type of macromolecule and the physiological state of the cell.

Proteins that are actively translocated from the cytoplasm to the nucleus contain a so-called nuclear translocation signal (NLS), which is typically a short lysine- and arginine-rich peptide sequence. That sequence is presumably recognized by a cytosolic factor (protein) that confers a specific interaction with a component of the nuclear pore complex and, subsequently, the nuclear transport system. As we will see later in this chapter, in the case of the transcription factor NF-κB, specific occlusion of the NLS by an interaction of the NLS-containing protein with another protein (IκB in the case of NF-κB) can cause cytoplasmic retention of an otherwise nuclear-localized protein.

During and immediately after synthesis, the nuclear RNA is covered by RNA binding proteins forming **hnRNPs (heterogenous nuclear ribonucleoprotein particles).** It has been postulated, although not really proven, that hnRNPs bind to receptors thought to be located on the nucleoplasmic side of the nuclear pore complexes, through which they, subsequently, are exported to the cytoplasm. During translocation, at least some of these proteins probably get stripped off the RNA and remain in the nucleus; others seem to shuttle between the nucleus and cytoplasm, consistent with a direct function in the RNA translocation process.

However, little is yet known about these RNA binding proteins and their functional domains. Features of the RNA to be exported, on the other hand, are better understood: Properly modified ends, that is, the cap-structure at the 5′ and the poly-A tail at the 3′ end (described in Chapter 8), are important prerequisites for the export of RNAs to the cytoplasm.

Studies performed primarily with yeast have revealed a relationship between splicing and transport. Primary transcripts without introns leave the nucleus immediately after synthesis (and capping and polyadenylation). Intron containing transcripts, however, need to be spliced first. The stronger the affinity of a pre-mRNA for splicing factors, the less efficient the transport of the unspliced RNA out of the nucleus seems to be. Splicing factors that bind to the RNA appear to prevent export until splicing is completed.

Despite the inverse correlation between spliceosome assembly and nuclear export of intron-containing RNAs, the cell has ways to export mRNAs that still contain one or even several introns. This is understood in terms of the "intron retaining mode" of alternative splicing (Figure 12-5(c)). For example, the AIDS (HIV) virus (see Chapter 16) is known to have evolved the ability to "sneak" unprocessed primary transcripts out of the host cell's nucleus (where they are produced). Short arginine-rich RNA-binding domains in a so-called "Rev protein" recognize RNA hairpins or loops in specific regions of partially spliced primary transcripts. Those RNAs that bind Rev become protected from further action by the spliceosome. This incompletely spliced macromolecular complex is then transported out of the nucleus, where its RNA component is first translated into viral proteins, and eventually built into new viral particles to serve as the AIDS virus's genome.

Interestingly, eukaryotic viruses, due to the relative ease with which they can be cultured in mammalian cells in the laboratory, often provide the researcher with a first glimpse of novel regulatory mechanisms. It remains to be seen, therefore, whether the previously described mechanism has more general relevance to eukaryotic gene regulation.

Regulation of mRNA Stability

Most prokaryotic mRNAs are short-lived. That is, they are degraded within minutes. This allows for rapid responses to changes in the environment by transcription of a different set of genes. In eukaryotes, the longevity of individual mRNA species can be very diverse, but, in general, eukaryotic mRNAs are more stable than prokaryotic mRNAs. The half-life (or time required for 50% of the starting level to disappear) of β-globin mRNA, for example, is more than 10 hours. On the other hand, many mRNAs encoding regulatory proteins, such as growth factors or the products of the cancer-related protooncogenes (e.g., *fos* and *myc*), are rather short-lived with a half-life between 15 and 30 minutes. Such short-lived mRNAs share an evolutionarily conserved AU-rich sequence of about 50 bases in their 3′ nontranslated region. When such a sequence motif is genetically engineered into the 3′ nontranslated region of the β-globin messenger, this mRNA becomes short-lived. Interestingly, eukaryotes have exploited mechanisms to differentially regulate the stability of a given mRNA in response to environmental stimuli or to changes of the physiological state.

Steroid receptors, such as the estrogen or the glucocorticoid receptor, are well known to act as hormone-dependent transcription factors, which stimulate transcription of their target genes upon interaction with hormone. Surprisingly, steroid receptors can also increase the stability of certain mRNAs in a manner that is not understood. For example, the half-life of the mRNA for vitellogenin (the major protein in eggs of frogs) increases about 30 times upon addition of estrogen.

Conversely, the addition of iron to cells decreases the stability of the transferrin receptor mRNA (**Figure 12-7**(a)). As a consequence, less transferrin receptor is made. This is reasonable from a physiological point of view, since the transferrin receptor is an iron-scavenging protein that increases the intracellular concentration of iron. The regulation of the stability of the transferrin receptor mRNA is mediated by an iron-responding protein (IRP). At a low intracellular iron level, this protein binds to stem-and-loop structures at the 3′ end of the mRNA. However, upon addition of iron, the factor is released and unmasks, most probably, a site that then is attacked by an as yet unknown nuclease. Interestingly, IRP also controls translation of ferritin mRNAs (Figure 12-7(b)). Ferritin binds free iron within the cell. With a low level of intracellular iron, IRP binds to a stem-and-loop structure near the 5′ end of the mRNA and blocks translation (for further details see the next section on "Regulation of Translation").

The control of mRNA stability has also been studied extensively for the mRNA that encodes histones (Figure 12-7(c)). During the S-phase (DNA-replication period) of a cell, there is a great demand for histone proteins. Not unexpectedly, therefore, histone mRNA stability (and also transcription) is cell cycle-regulated. In the S-phase, the half-life of histone mRNA is about one hour. As soon as replication is completed or artificially blocked with chemicals, histone mRNA is degraded within minutes. The regulation of histone mRNA stability depends upon a short 3′ stem-loop structure that is unique to these RNAs and replaces the poly(A) tail found in other mRNAs. Experimentally replacing the 3′ region of the β-globin with the 3′ region of a histone mRNA is sufficient to make the β-globin mRNA stability DNA replication-dependent. A "hairpin loop-binding protein" associates with the stem loop and, thereby, blocks RNase action (Figure 12-7(c)). Interestingly, degradation by RNase is dependent on ongoing translation, as is also the case for the previously mentioned instances. If translation stops more than 300 bases upstream of the actual stop codon (due to the experimental insertion of a (mutant) stop codon), the histone mRNA is no longer rapidly degraded in non-S-phases. It has, therefore, been suggested that a ribosome-associated nuclease is somehow involved in the degradation. That hypothesis would explain why most unstable mRNA is selectively stabilized when cells are treated with the protein synthesis inhibitor, cycloheximide.

The regulation of α and β tubulin mRNA stability is also intimately linked to translation (Figure 12-7(d)). The α and β tubulin proteins are the principal subunits of microtubules. The degradation of their mRNA is autoregulated since it is dependent on free α and β tubulin proteins. It has been shown for β tubulin mRNA that the sequence of the mRNA necessary for autoregulated instability is located, unexpectedly, in the first 13 bases to be translated. Autoregulated mRNA destabilization requires β tubulin mRNA to be translated beyond codon 41.

KEY CONCEPT

The "mRNA Half-life" Concept

The contribution an mRNA makes to cellular function is determined, in large part, by how quickly it is degraded.

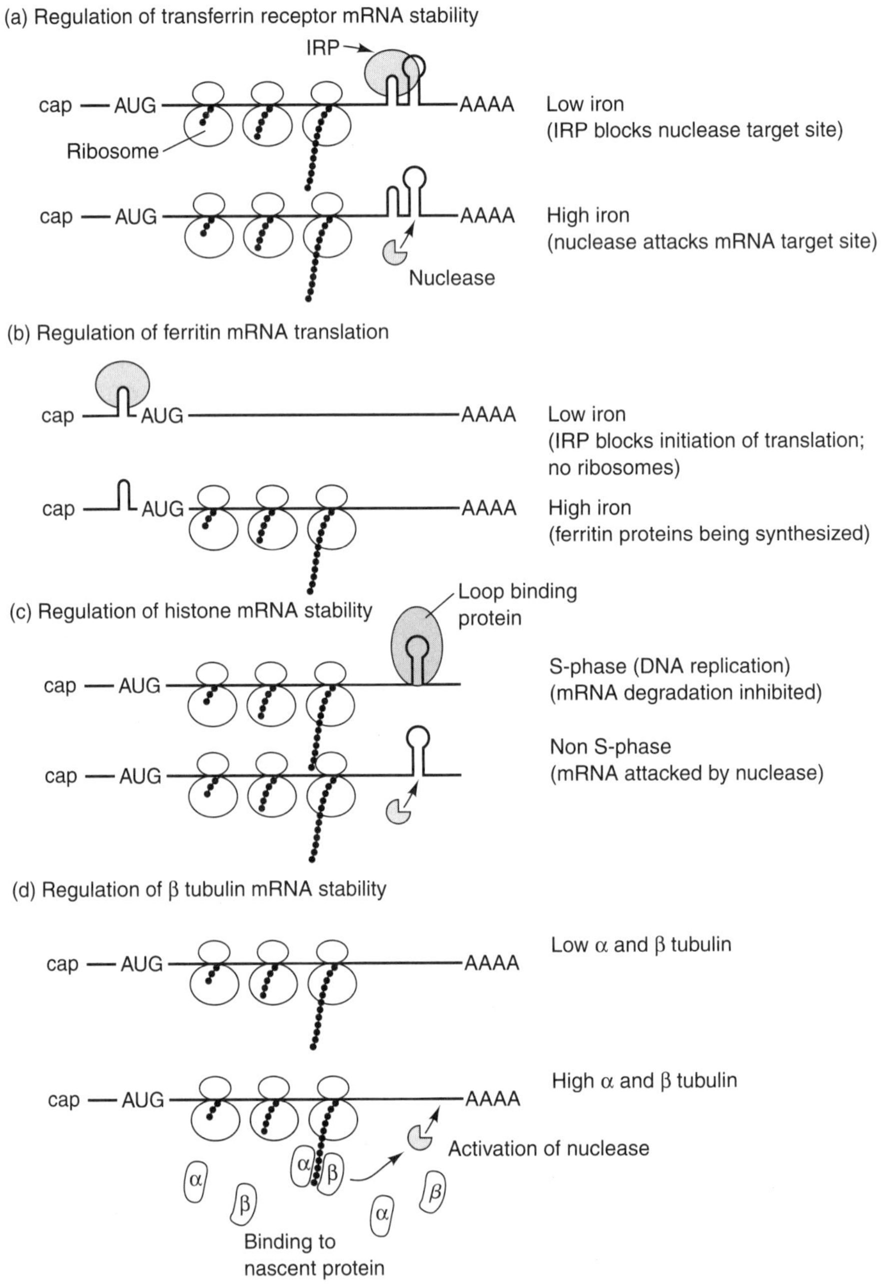

(a) Regulation of transferrin receptor mRNA stability

cap — AUG — AAAA   Low iron
(IRP blocks nuclease target site)

Ribosome

cap — AUG — AAAA   High iron
(nuclease attacks mRNA target site)

Nuclease

(b) Regulation of ferritin mRNA translation

cap — AUG — AAAA   Low iron
(IRP blocks initiation of translation;
no ribosomes)

cap — AUG — AAAA   High iron
(ferritin proteins being synthesized)

(c) Regulation of histone mRNA stability

Loop binding protein

cap — AUG   S-phase (DNA replication)
(mRNA degradation inhibited)

cap — AUG   Non S-phase
(mRNA attacked by nuclease)

Figure 12-7   Different examples for the regulation of mRNA stability are shown in (a), (c), and (d); (b) displays the regulation of the translation of ferritin mRNAs involving the same factor (IRP) used for the regulation of transferrin receptor mRNA stability indicated in (a). Open double circles present ribosomes; filled circles, the IRP protein; notched circles, various nucleases; α and β, the subunits of tubulin.

(d) Regulation of β tubulin mRNA stability

cap — AUG — AAAA   Low α and β tubulin

cap — AUG — AAAA   High α and β tubulin

Activation of nuclease

Binding to nascent protein

In the current model, some cellular component (presumably tubulin itself) binds to the peptide encoded by the first nucleotides of the open reading frame, after which an unknown nuclease is activated, which degrades the β tubulin mRNA.

Regulation of Translation

In contrast to the few well-defined factors required to start translation in *E. coli*, there are a great number of eukaryotic initiation factors (eIFs; see Chapter 9). In prokaryotes, genes of related function, such as enzyme chains for a specific metabolic pathway, are often grouped into operons and transcribed as one contiguous, polycistronic transcript. This transcript is then subject to translational control to produce the appropriate amount of each gene product. Translational control is, therefore, of obvious importance in prokaryotes. Nevertheless, translational control is also widely used in eukaryotes, even though one might have thought that eukaryotic genes, which typically are monocistronic, could be sufficiently regulated at the transcription/processing level.

The typical eukaryotic cell can decrease the overall rate of translation in response to a variety of different conditions, such as viral infection, deprivation of growth factors, or heat shock. These situations have been found to induce the phosphorylation of eIF-2, one of the many eukaryotic translation initiation factors. The protein eIF-2 mediates binding of the methionyl initiator tRNA to the small 40S ribosomal subunit. That activity is abolished when the eIF-2 protein is phosphorylated. Besides this rather *general* mode of negative translational control, there also exist mechanisms that are *specific* for certain mRNA species.

A very well-studied case of this latter class of negative translational control is the ferritin system (see also the previous section on "Regulation of mRNA Stability" and Figure 12-7(b)). As described previously, ferritin is an iron storage protein. Its intracellular concentration is tightly controlled at the level of translation and is dependent on the intracellular iron concentration. If there is a low intracellular level of iron, the iron responsive protein factor (IRP) binds to a stem-and-loop structure at the 5′ end of the ferritin mRNA. In this way, IRP blocks translation, most likely by physically blocking the binding of the translation machinery to the initiation codon AUG. However, upon iron addition, IRP no longer binds to the RNA and translation of the ferritin mRNAs is increased 100-fold.

Negative translational control is very important in embryogenesis of higher eukaryotes. A great number of mRNA species, which are synthesized during development of the oocyte, are deposited as so-called "maternal mRNAs" in the eggs of higher eukaryotes. These mRNAs are stored in an inactive form and, upon fertilization, activated for translation by a mechanism that is not yet understood. Curiously, many of the stored, inactive mRNAs have only 10 to 30 A's at their 3′ ends, a length that apparently is too short for translation. (Poly(A) tails are usually about 200 nucleotides long.) Upon "reactivation," additional A's are added again to their 3′ end, which greatly stimulates initiation of their translation.

The translational controls discussed so far influence the rate of translational initiation. Usually, protein synthesis goes on automatically once translation is initiated. However, nature also has exploited exceptions to this rule (e.g., with the phenomenon of translational frameshifting found in retroviruses and in a coronavirus). In *Rous sarcoma* virus, for example, 95% of the protein translated from the *gag-pol* mRNA is the major glycoprotein antigen (*gag*). However, a few percent are translated into a longer combined protein (*gag-pol*) that encodes the characteristic retrotranscriptase of the virus. How is this 20:1 ratio of *gag* to retrotranscriptase (*gag-pol*) regulated? A ribosome that encounters a specific sequence shortly before the UAG terminator triplet of the gag encoding sequence (and followed by a characteristic structure of the mRNA) is kicked out of its reading frame and moves back by one nucleotide. From there it continues translation of the *pol* (retrotranscriptase)

translational
frameshifting

moiety in a "minus 1" reading frame. A similar process also has been observed with the human immunodeficiency virus HIV. This **translational frameshifting,** as it turns out, is only one way to ensure the creation of two polypeptides from one mRNA. In another retrovirus (Moloney leukemia virus), the ribosome reads across the UAG stop codon in 5% of the cases, inserts glutamine instead of terminating, and continues translation into the polymerase moiety that, in this case, is in the same reading frame.

Regulation of Protein Activity

Two Principal Pathways for Intracellular Protein Translocation

The regulation of protein activity mainly comprises three types of processes: the translocation of proteins into different cellular compartments, protein-protein interactions, and reversible or irreversible enzymatic modifications. These apparently different cellular functions are intimately related to each other.

The subcellular translocation of newly synthesized polypeptides occurs via two major pathways. In the cytosolic pathway, the entire protein is translated in the cytoplasm (membrane-free translation). Subsequently, the product is either taken up by an organelle (nucleus, peroxisome, mitochondrion, or chloroplast), or it remains in the cytoplasm. The uptake into the organelles most probably occurs via organelle-specific receptors that recognize short peptide sequences of the proteins (called **translocation signals**). The proteins remaining resident in the cytoplasm stay there because they do not contain a specific translocation signal.

translocation signals

The second pathway is through the endoplasmic reticulum or ER. The decision for this pathway is made during translation. A protein destined for this pathway usually contains a hydrophobic signal sequence at its N-terminus. This signal sequence is recognized by the **signal recognition particle (SRP),** a protein/RNA complex, which in turn docks the **polysome** (mRNA with the translation machinery) to the SRP receptors of the ER membrane. During ongoing translation (now membrane bound), the newly synthesized polypeptide is translocated into the lumen of the ER. The proteins are either completely secreted into the ER lumen or, if they are transmembrane proteins, remain bound to the ER membrane. Subsequently, most proteins are transported to the Golgi apparatus and from there either to the lysosomes, secretory vesicles, or directly to the plasma membrane. These translocations typically occur via vesicles that fuse with the membrane of their subcellular target and release the protein content into the lumen of the new compartment or into the extracellular space. All plasma membrane proteins (e.g., receptors of peptide hormones), or enzymes and proteins that are secreted (e.g., digestive enzymes of the intestinal tract), are translocated via the ER pathway.

signal recognition
particle (SRP)

polysome

Posttranslational Modifications and the Regulation of Protein Stability

A great number of posttranslational modifications of amino acid side-chains have been described in proteins. For most of these modifications, the function is not known. Some of them may be unimportant chemical side reactions, but others may have important functions, since they are linked to specific enzymes. Any given modification occurs preferentially or exclusively in one or the other of the two translocation pathways and in specific cell compartments previously described. Some modifications are permanently required and are irreversible (e.g., attachment of nonpeptide helper groups to some enzymes). Other covalent modifications are reversible and allow rapid responses to extra- and intracellular stimuli. The most prominent reversible modification appears to be protein phosphorylation. Other frequently found modifications are acetylation, methylation, and glycosylation. The importance of protein phosphorylation was discovered through its ability to regulate enzymes involved in intermediary metabolism. However, it is clear that phosphorylation also is exploited extensively in other cellular events, such as signal transduction or cell-cycle regulation. Phosphorylation can either increase or decrease the activity of the target protein.

Posttranslational modification processes are also evident in the regulation of protein stability. Some cytosolic proteins are stable for several days, and then are randomly degraded and replaced by new ones. Conversely, several rate-limiting enzymes of a metabolic pathway often have half-life times of only a few minutes. Other short-lived proteins include the products of the protooncogenes, such as *fos* or *myc,* which probably play important roles in the control of cell growth. Since these proteins are degraded so rapidly, their intracellular concentration can be changed very quickly by changing the rate of synthesis in response to extracellular stimuli. Usually these types of proteins also have rather unstable mRNA (described in the section on "Regulation of mRNA Stability" earlier in this chapter). Proteins appear to bear not only a translocation signal, but also sequences that determine their half-life. The half-life of a given protein appears to be a consequence of some destabilizing sequences and general features of the secondary structure. Interestingly, in some model experiments, a strict correlation between the instability of a given cytosolic protein and the first amino acid at its N-terminus has been found. The amino acids Met, Ser, Thr, Ala, Gly, Val, and, most probably, also Cys and Pro, have a stabilizing effect on the protein when located at its N-terminus. All other amino acids make the protein vulnerable to a protease that rapidly degrades it.

The proteolytic machinery responsible for selective protein degradation is ubiquitin dependent. **Ubiquitin** is a small protein of 76 amino acids that binds to other proteins. When the ubiquitin-dependent proteolytic machinery recognizes a destabilizing amino acid at the N-terminus of a protein, a ubiquitin molecule is covalently attached either to the N-terminus itself or to a nearby lysine. Subsequently, a series of additional ubiquitins are added, producing a multiubiquinated, branched protein. A specific ATP-consuming protease recognizes such ubiquinated proteins and degrades them while recycling the ubiquitin polypeptides.

As already mentioned, ubiquitin-dependent protein degradation is specified by the first amino acid at the N-terminus of each protein. Although in all eukaryotic protein genes the first codon (AUG) codes for a methionine, very few mature protein species have a methionine at their N-terminus. Usually the N-terminus is cotranslationally modified by trimming and end-addition activities not well understood. Interestingly, proteins entering the ER pathway often bear an unmodified

Ubiquitin

N-terminus, making them highly unstable in the cytosol, but not in the ER. It may be very important in allowing the cell to be able to specifically degrade proteins that have entered the cytosolic pathway by accident.

Posttranslational Regulation of Transcription Factor Activity

Transcription factors (proteins) bind to an enhancer or a promoter in a sequence-specific manner. Subsequently, transcription of the associated gene is stimulated (see the section on "The Regulation of Transcription Initiation" earlier in this chapter). Recently, a number of transcription factors and their corresponding genes have been studied in detail. While some transcription factors, especially those stimulating the transcription of housekeeping genes, always exist in an active state, several transcription factors are known whose activity is regulated at the posttranslational level. This regulation involves the range of differential subcellular translocation, stable interactions with other proteins, and enzymatic modifications.

A well-studied case of posttranslationally regulated transcription factors is the steroid receptor (e.g., glucocorticoid receptor (GR); **Figure 12-8**). In the absence of hormone, GR is complexed to a 90 kD protein first recognized because it appears in cells that have been heat shocked. It is, therefore, called a "heat shock protein" (hsp90), and is located in the cytoplasm. Hsp90 was, therefore, regarded as a cytoplasmic anchorage protein of GR. When the steroid hormone glucocorticoid diffuses into the cell, it binds to the hormone binding domain of GR. This causes the release of GR from hsp90, most probably because of a conformational (shape) change induced by binding of the hormone. The release leads to the exposure of a nuclear translocation signal on the GR. This transcription factor is then translocated into the nucleus via nuclear pore complexes. In the nucleus, GR binds to its target sequence and stimulates transcription of the GR-responsive genes.

KEY CONCEPT

The "Instability → Flexibility" Concept

The inherent instability of many macromolecules (e.g., some proteins and mRNAs) in the living cell facilitates adapting to changing metabolic conditions.

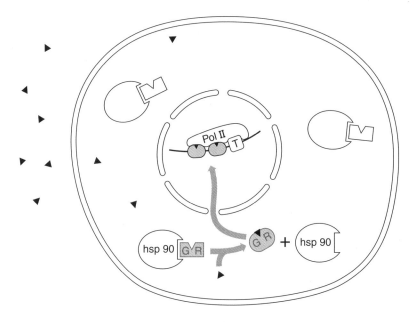

Figure 12-8 The glucocorticoid receptor (blue): a paradigm for the posttranslational regulation of a transcription factor. Solid triangles represent glucocorticoid molecules. T represents the target region in DNA (part of the promoter) to which GR binds.

An equally fascinating mode of regulation has been found with studies of NF-κB, a mammalian transcription factor that induces transcription of the genes involved in immune defense or inflammation. It is found in a number of different mammalian cell types, often in the cytoplasm in an inactive form, interacting with a cytoplasmic anchorage protein, designated as IκB (Inhibitor of NF-κB). NF-κB is posttranslationally activated on addition of agents that impair cellular functions (e.g., agents such as UV, DNA-damaging chemicals, viruses, parasites, or lipopolysaccharides of bacteria). This activation process involves phosphorylation, ubiquitination, and proteolytic degradation of IκB, which unmasks the nuclear translocation signal (NLS) of NK-κB and, hence, allows translocation of NF-κB to the nucleus. There it activates the transcription of a set of genes. Recent experiments have indicated a way by which a great number of different stimuli could initiate this cascade of events. H_2O_2 is a potent inducer of NF-κB and also appears to be commonly produced on addition of many (if not all) NF-κB inducers. Therefore, H_2O_2 (or a derivative) has been postulated to serve as a second messenger that leads to enhanced IκB phosphorylation.

Other transcription factors can be located in the nucleus but nevertheless remain inactive. The regulation of the activity of the yeast transcription factor Ga14 is an example. Ga14 has the potential to stimulate genes that encode proteins involved in galactose metabolism. Ga14 always binds to its DNA target, but, in the absence of galactose, it cannot activate transcription. This is due to the Ga180 protein, which stably interacts with Ga14 and covers its transcription-activating domain. With the addition of galactose, Ga180 remains bound to Ga114, but in a different manner that enables Ga14 to stimulate transcription. In contrast, the activity of the mammalian transcription factor Oct-1 can be specifically augmented by the protein VP16, which is encoded by herpes simplex virus and contains a very strong activation domain. VP16 can form a complex with Oct-1 and, thus, stimulate transcription by virtue of its own very strong activation domain. Neither Ga180 nor VP16 bind to DNA by themselves.

The transcriptional activity of a gene can also be influenced by direct enzymatic modification such as phosphorylation. The yeast heat-shock transcription factor HSF can stimulate transcription of genes essential for heat-shocked cells. HSF always binds to its target HSE gene (heat shock responsive element). HSF, however, is only active when the cells have been incubated at a high temperature. The activation of HSF exhibits a strong correlation with its phosphorylation. So far it is not clear if this modification helps to expose an activating domain of HSF, or if the phosphorylated region itself acts as an activating domain.

KEY CONCEPT

The "Activation of a Transcription Factor" Concept

The activity of proteins that facilitates initiation of transcription (so-called "transcription factors") is often up or down regulated by the action of steroid hormones, small molecules, or even other proteins.

Gene Rearrangement: Joining of Coding Sequences in the Immune System

The life of a multicellular organism is under constant threat by all kinds of microorganisms and viruses. Most of these are warded off by the skin, or by secretions of the digestive tract, such as hydrochloric acid in the stomach, and digestive enzymes in the intestine. Quite often, however, a virus enters the body via the lining of the gut or small lesions in the skin. Once inside the body, the invader might initiate a deadly race with the immune system of the infected organism. The immune system of vertebrates offers a selective defense against all possible infectious agents, and is ready even against those that might appear on earth in the future. How is this possible? Several million different protein molecules, **immunoglobulins** or **antibodies** (see Chapter 4), are produced, whereby each can bind to a distinct infectious target molecule or **antigen** in a specific manner. If each antibody molecule were encoded by a separate gene, a major fraction of the entire genome would have to be devoted to antibody synthesis. The number of antibody genes (about 10^{12}) would even exceed the estimated total number of human genes (about 30,000). However, the immune system has evolved a special way to produce such a huge variety of antibodies. In general, a mature antibody-producing cell (B-lymphocyte) makes only one type of antibody. Hence, there must exist some mechanism that is responsible for programming each individual B-cell. That mechanism involves specific gene rearrangements, and is, therefore, very different from all other modes of gene regulation discussed so far, none of which change the genome. However, the mechanism for generation of antibody diversity is not the same for all vertebrates. For example, birds have evolved a system rather different from the one in mammals. In this section, we will focus on the system found in most mammals.

immunoglobulins
antibodies
antigen

Different Immunoglobulin Gene Segments

Immunoglobulin G (IgG)

Immunoglobulin G (IgG) is one of several classes of immunoglobulins. It is a tetrameric protein containing two L chains and two H chains (for the structure of IgG, see Chapter 4). Both the L and H chains consist of two regions, the constant and the variable region. The variable region is responsible for recognizing and binding a specific antigen. Recall that the antigen-binding site consists of amino acid residues from both heavy and light chains. There are two types of L chains, called κ and λ. We will only describe how the κ type is produced; the basic mechanism for synthesis of the λ type is similar, differing only in detail.

Three different classes of gene segments are used to form a κ-type L chain, namely V, J, and C segments, which are all located on the same chromosome (**Figure 12-9**(a)). There are about 100 different V gene segments (which are responsible for the synthesis of the first 95 amino acids of the variable region), 4 different J gene segments (which encode the final 12 amino acids of the variable region and join the V and C regions), and 1 copy of the C gene segment (which encodes the constant region). In the genome of embryonic cell, the V segments form a tight cluster, the J segments form a second tight cluster far downstream from the V segment cluster, and the C segment follows not far after the J segment cluster. Note that each V gene segment is preceded by a promoter from which transcription can potentially be initiated, and that an enhancer is located between the J_4 and the C segment (Figure 12-9(a)).

Regions encoding particular IgG molecules have been cloned by the recombinant DNA techniques to be described in Chapter 15. From various mouse cell lines, each producing a particular IgG molecule, several have been characterized. For each clone obtained from an antibody-producing cell line, it has been found that a large segment of the embryonic DNA sequence is absent, and that the miss-

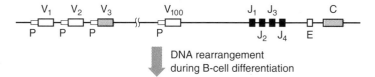

(a) Germline DNA

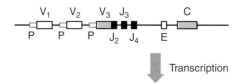

DNA rearrangement
during B-cell differentiation

(b) Rearranged DNA of specific B-lymphocyte

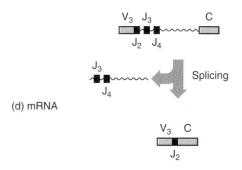

Transcription

(c) pre mRNA

Splicing

(d) mRNA

Figure 12-9 The production of the mature κ-type L chain mRNA of a particular IgG molecule (blue) involves processing at the DNA and the RNA levels. During B-cell differentiation, the DNA between a V and J segment (in our example, between V_3 and J_2) is excised, i.e., the genomic sequences are specifically changed (compare (a) and (b)). With this step, the enhancer (E) is moved closer to the promoter (P) of V_3 and the given B lymphocyte is able to produce a specific κ-type L chain pre-mRNA (c). This pre-mRNA undergoes splicing, whereby the C segment is joined to V_3J_2. This results in the mRNA (d), which is ready for translation to produce the specific κ-type L chain protein. Note that (a) and (b) occur only once in any cell, whereas (c) and (d) occur every time the gene in (b) is transcribed.

ing DNA is always a sequence between the particular V gene segment and the J gene segment. This can be explained by a **gene rearrangement** mechanism in which DNA is deleted between the V and J segments chosen from recombination. Many different genomic sequences, each encoding a particular κ chain, have been cloned, and each clone lacks a different DNA segment. For example, in Figure 12-9(b), the entire DNA between V_3 and J_2 is absent, while in another clone there might be a V_{86}-J_1 junction instead. Please note that the step from 12-9(a) to 12-9(b) occurs only *once* for any cell, while the steps from 12-9(b) to 12-9(d) occur every time the gene (in 12-9(b)) is transcribed.

Any of the approximately 100 V segments can be joined to any of the 4 J segments, so that at least 400 (100×4) different variable regions (or V-J combinations of the κ chain) can be potentially encoded. The H (heavy) chain genes are

gene rearrangement

organized in a similar, but not identical, way. In the mouse, for example, there are three types of gene segments: 200 V, 12 D, and 4 J segments. Therefore, about 9,600 $(200 \times 12 \times 4)$ heavy chain variable regions can be potentially encoded upon appropriate DNA rearrangements. These rough calculations (the exact number of V segments in these clusters of segments is not known) give about 4×10^6 different antigen-binding sites, assuming both the heavy and light chain variable regions equally contribute to the antigen-binding site. The number of potential antigen-binding sites is even higher, since the junctions produced by two particular gene segments (e.g., by linking a given V with a given J segment) are variable. A variable number of nucleotides often is lost (or sometimes added) from the ends of the two recombining segments (junctional diversification). This mechanism increases the possible number of different antigen-binding sites once more by a factor of about 1,000. Following such addition or deletion of nucleotides, only in one of three cases will the correct reading frame be restored to yield a functional gene.

How Are Immunoglobulin Gene Rearrangements Controlled?

What is the enzymatic machinery that creates all of these combinatorial variants of immunoglobulin genes? As shown in **Figure 12-10**(a), the signal sequences for

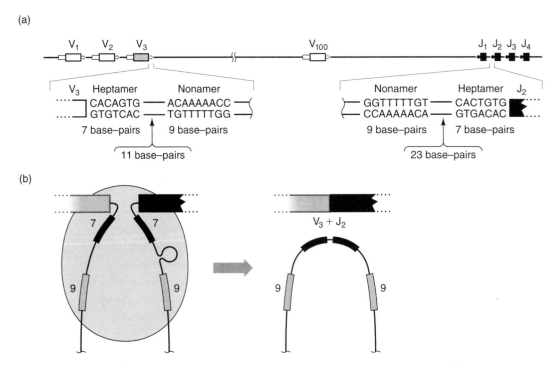

Figure 12-10 (a) The joining of V and J DNA segments during B-cell differentiation is a highly precise process involving conserved DNA sequences called heptamer (7) and nonamer (9) sequences. (b) A site-specific recombinase is thought to recognize these sequences, after which the randomly selected V segment and J segment are joined.

the V-J joining are a pair of specific base sequences 3′ of each V segment (heptamer/spacer/nonamer) and 5′ of each J segment (nonamer/spacer/heptamer). It is presumed that a site-specific **recombinase** enzyme recognizes these sequence motifs, cuts the DNA, and joins randomly selected V and J segments (Figure 12-10(b)). The same sequence motifs are used in the heavy chain pool to join V, D, and J segments. The products of two lymphoid-specific genes, RAG 1 and 2, are components of the recombinase. The evidence for their importance is several fold: (1) mice, which are artificially made deficient for the RAG 1 and RAG 2 genes, do not develop mature B and T cells; (2) ectopic expression of RAG 1 and RAG 2 in nonlymphoid cell lines renders these cells competent for immuno-globulin gene rearrangements; and (3) most recently, immunoglobulin recombination assays in vitro have indicated that recombinant RAG 1 protein greatly increases rearrangement activity and also complements an inactive extract from a RAG 1 (-/-) cell line.

Perhaps the most striking aspect of these gene rearrangements is their precision. In order to avoid a scrambling of the entire genome, the recombinase has to be targeted specifically to the immunoglobulin locus. Furthermore, the time window of enzyme activity is quite narrow, being restricted to a specific phase during the embryonic development of B lymphocytes.

Figure 12-9(b) indicates that the joining of segments cannot fully generate an L chain reading frame because the spacer between the J and C gene segment remains, and the actual L chain derived from only one V gene segment and one J segment and recombined DNA still usually contains many V and J segments. The recombination process, however, brings the enhancer (E) much closer to the promoter of V_3 (P), allowing transcription to start. This leads to the production of primary transcripts (pre-mRNA, see Figure 12-9(c)). The correct reading frame is obtained by a final RNA splicing event, as shown in Figure 12-9(d). Note how the particular V-J joint determines the splicing pattern. The RNA removed by splicing is always between the C segment and the right end of the J segment, which joined to the V segment.

Allelic Exclusion, Clonal Selection, and Affinity Maturation

Whether the site-specific rearrangement processes happen in advance of, or in response to, the occurrence of a specific antigen has been a major question. Today it is generally accepted that the events occur at random in the course of development, and when embryogenesis is complete there are several million different antibody-producing B lymphocytes. Each of these cells produces only **a single type of antibody,** due to the strict feedback regulation of the site-specific DNA rearrangements. The feedback mechanism designated as **allelic exclusion** allows the production of only one type of rearranged H and L chain gene per cell. However, very little is known about this phenomenon at the molecular level.

How does an organism know when to produce a huge amount of a particular antibody in response to a particular antigen? Each antibody-producing cell makes a small amount of a specific antibody. Some of this antibody becomes bound to

recombinase

the cell surface. When such a cell is exposed to the specific antigen that can react with its antibody, a complex antigen-antibody network forms. This event stimulates cell division, and extensive production of specific antibodies. This process is called **clonal selection.**

Another interesting mechanism that allows the immune system to increase the affinity of antibodies to a given antigen, again by modifying genomic sequences, deserves mentioning. Repeated immunization of an animal with the same antigen results in the production of antibodies having, on average, a higher affinity to the antigen. This phenomenon is called **affinity maturation.** By somatic hypermutation (the molecular details of which are not understood), specific sequences in the V region accumulate point mutations with repeated exposure to a given antigen. Since B lymphocytes are stimulated to proliferate in response to the binding of antigens, those cells that bear hypermutations yielding an antibody with higher affinity to the antigen will proliferate in response to the binding of antigens and outgrow the cells producing antibodies with lower affinity. The mutation frequency over an immunoglobulin variable region can be as high as 10^{-3}/cell cycle/base pair. This frequency would completely inactivate a genome if it were not restricted to the variable regions of active immunoglobulin light and heavy chain genes.

A Future Practical Application?

Tissue engineering is developing as a new force in biotechnology. The goal of this endeavor is to produce, in the laboratory, genuine replacement parts for the human body. So far, attempts to construct, in the research laboratory, sheets of skin and bone pieces have provided much of the initial focus of research efforts.

Attention is being directed toward discovering (1) the signal molecules or transcription factors (e.g., "bone maturation protein"—BMP), which activate the expression of the genes that give specific tissues their unique identities; (2) the gene structure, including the promoters and enhancers, associated with those genes; (3) ways to artificially regulate the genes responsible for the three-dimensional patterning (i.e., shape) of specific tissues and organs; and (4) the source of the mechanical forces holding a tissue or organ together.

Once those components of a tissue system are characterized, it is expected that the next phase of research will involve growing appropriate cells, in the presence of those growth- and differentiation-guiding components, in rubber molds for imparting the appropriate shape (in the case of bone) to the artificial tissue.

Such artificial tissues and organs are expected to play important roles in medicine: first, as replacement parts (e.g., artificial skin for healing burns); and second, as valuable laboratory systems for the discovery of new drugs. Drug therapy is often constrained by the inability of a potentially active drug to reach its target tissue. Artificial tissues are expected to provide a promising avenue for experimenting with novel ways to modify drugs so that they will pass across the gut, or the blood-brain barrier, or other difficult tissue boundaries.

SUMMARY

Regulation of genetic systems in eukaryotes is accomplished in a variety of ways, most of which are quite different from the mechanisms observed in prokaryotes. In eukaryotes, three different RNA polymerases fulfill transcriptional tasks that, in prokaryotes, are carried out by a single RNA polymerase. It is at this level of transcription that decisions are often made with regard to gene expression. Sequences upstream of eukaryotic genes known as promoters are required for initiation of transcription. Activation of transcription is carried out by binding of transcription factors to these sites. Additional regulatory sequences, known as enhancers, are often associated with genes. They also stimulate transcription, but, in contrast to promoters, are often located several kilobases away from the coding region, upstream or downstream. Once mRNA is transcribed, further regulation can occur during splicing events that remove eukaryotic introns and create mRNA sequences that can be translated into functional proteins. For example, alternate splice sites can be chosen, resulting in the appearance of several different mRNA species encoded in one gene. The nucleocytoplasmic translocation of these RNAs can then provide a further level of gene regulation. Once the mRNA has arrived in the cytoplasm, the variety of their half-lives, as compared with the short-lived nature of prokaryotic mRNA, allows for further regulation. Eukaryotes have also developed mechanisms whereby the stability of mRNA can be changed in response to environmental stimuli or physiological states. Gene expression can also be controlled at the protein level through translational regulation, posttranslational modification, and differential protein stability.

Antibody diversity is the consequence of a complex set of gene rearrangements, in which various gene segments are joined together. The mRNA transcribed from those "joined" genes is further modified by splicing, so that the final mRNA that is translated into the antibody protein is only distantly related to the gene sequences in the B cells where differentiation occurred. Those rearrangements and splicings generate the large numbers of different antibody proteins typically found in eukaryotic organisms.

DRILL QUESTIONS

1. What RNA polymerase(s) transcribe eukaryotic genes? Name the polymerase(s) and the type of gene(s) it transcribes.

2. In prokaryotes, regulatory elements are fixed positions with respect to the gene(s) regulated. How does the situation differ in eukaryotes?

3. Do the V and J segments of an antibody gene join to form the variable or the constant region of an IgG molecule?

4. List several mechanisms a cell uses to increase the concentration of a particular mRNA molecule to a very high value.

5. How does the organization of chromatin differ with regard to gene activity? What experimental technique can be used to find such regions of activity?

6. Give the two types of splicing that can occur in the processing of eukaryotic mRNA, and state the differences between them.

7. How might a cell be signaled to synthesize one particular protein, but no other?

8. Outline the mechanism for generating antibody diversity.

PROBLEMS

1. (a) If you wanted to redesign the mouse genome to increase antibody diversity, but add the least amount of DNA, which component would you increase?

   (b) If an organism has 150 different V gene segments, 12 J gene segments, and 3 possible V-J joints for L chains, and can make 5,000 different H chains, how many different antibodies can be made?

2. If an enhancer were moved from a position near gene *A*, where it has strong enhancing activity, to a position 50 nucleotides upstream from gene *B*, which is constitutively transcribed, would transcription of gene *B* be increased?

3. In prokaryotes, positive regulatory elements often provide a binding site for RNA polymerase, which strengthens the binding of the polymerase to the transcription start site. What is the evidence that suggests that most eukaryotic regulatory elements function differently?

4. Suppose that a tripeptide (Q) is made in a mammalian cell culture in response to an external effector (E). In the absence of E, an 800-nucleotide-long primary transcript can be identified that hybridizes to a cloned sample of the *Q* gene. That RNA in the cell is eventually capped and poly(A) tailed. In an in vitro protein-synthesizing system, it directs the synthesis of the Q tripeptide. If E is added to the growth medium, very little of that primary transcript can be detected. Instead, however, a 700 nucleotide-long mRNA molecule can be detected in the cytoplasm. This mRNA directs the synthesis of a peptide, which is different from the Q tripeptide. How might E regulate the synthesis of Q?

5. Several cases are known in which a single effector molecule regulates the synthesis of different proteins encoded in distinct mRNA molecules 1 and 2. Give a brief possible molecular explanation for each of the following observations made when the effector is absent.

   (a) Neither nuclear nor cytoplasmic RNA can be found that hybridizes to either of the genes encoding molecules 1 and 2.

   (b) Nuclear (but not cytoplasmic) RNA can be found that hybridizes to the genes encoding molecules 1 and 2.

   (c) Both nuclear and cytoplasmic RNA (but not polysome-associated RNA) can be found that hybridize to the genes encoding molecules 1 and 2.

6. Suppose that two proteins are synthesized initially at the same time because the two RNAs encoding each protein are made in response to the same signal. At a later time, when the transcription signal is no longer present, one of the proteins is still made at nearly the same rate, and the other is not detected. Suggest a means for this temporal programming.

CONCEPTUAL QUESTIONS

1. Since there are a number of diseases resulting from loss of function of a gene product due to a mutation in a correct splice site, how might such diseases

be treated using molecular biology approaches? What, if any, negative consequences could these approaches have?

2. Almost nowhere in the biological kingdom, with the exception of the gene rearrangement associated with antibody synthesis, are there mechanisms for regulating gene expression that involve permanent changes in the nucleotide sequence of DNA observed. Why do you suppose such mechanisms do not exist?

3. Phosphorylation of proteins is another method used by eukaryotic cells to regulate gene expression. Proteins called kinases phosphorylate amino acids such as tyrosine and serine on other proteins to cause either increased or decreased activity. Similarly, many kinases control their own activity level through autophosphorylation. Two such kinases are c-src and its oncogenic counterpart, v-src. The src kinase phosphorylates tyrosine residues. A c-src kinase contains two tyrosine residues, while v-src contains only one. It is known that the presence of c-src in cells does not lead to a high level of cancer transformation, while the presence of v-src is highly transforming. Propose an explanation for this observation.

in this chapter you will learn

1. About studies which focus on DNA sequences as a starting point for discovery.

2. About using large-scale gene expression data to gain insight into cell and tissue function.

3. About the importance of characterizing the total collection of proteins in normal and diseased tissues.

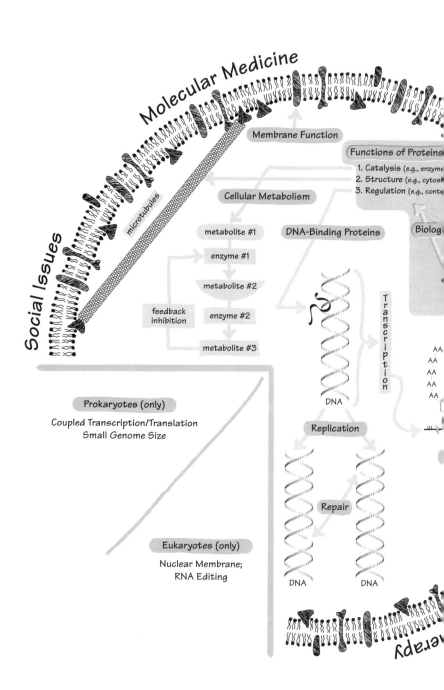

Molecular Medicine

Social Issues

Membrane Function

Functions of Proteins

1. Catalysis (e.g., enzyme
2. Structure (e.g., cytos
3. Regulation (e.g., cont

microtubules

Cellular Metabolism

DNA-Binding Proteins

Biolog

metabolite #1

enzyme #1

metabolite #2

feedback inhibition

enzyme #2

metabolite #3

Transcription

DNA

AA
AA
AA
AA
AA

Prokaryotes (only)

Coupled Transcription/Translation
Small Genome Size

Replication

Repair

Eukaryotes (only)

Nuclear Membrane;
RNA Editing

DNA

DNA

DNA

erapy

Genomics and Proteomics Drive Information-Age Biology

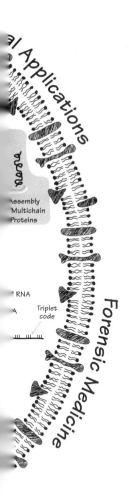

al Applications

Assembly
Multichain
Proteins

RNA

Triplet
code

Forensic Medicine

genomics
bioinformatics
proteomics

Completion of the nucleotide sequencing of the genomes of a variety of organisms, including—most importantly—the human genome, stands as an amazing achievement for molecular biology. That accomplishment has led to a rapid-pace buildup of the subdisciplines of genomics, bioinformatics, and proteomics. **Genomics** is the study of size, physical structure, and sequence information contained in an organism's DNA; **bioinformatics** represents interpretations of the meaning of information contained in the sequence data; and **proteomics** provides a comprehensive view of how the structures and functions of all the endpoint gene products (proteins) in a specific type of cell are integrated. These new subdisciplines build on the structural and functional information described in the previous several chapters. Progress in a variety of technologies have converged to drive these endeavors: automated nucleotide sequencing of DNA, x-ray crystallography, mass spectrometry and nuclear magnetic resonance spectroscopy of proteins, enhanced separation methods, and inexpensive super computing power.

The stage for making dramatic progress was actually set several decades ago. In the mid-1960s, a complete genetic map of a bacteriophage was compiled. That feat was a landmark for molecular genetics, for it represented an inventory of each and every gene that comprised a life form. With that accomplishment, the imagination and drive of molecular biologists was enhanced. The next logical step—deciphering the complete nucleotide sequence of an organism—came into focus. Indeed, many of the scientists who witnessed that first triumph are still active, and had the opportunity to share in the joy of the completion of the project of the century—the nucleotide sequence of the human genome. What those "eyewitnesses" have observed is the revelation that, even at the nucleotide sequence level, a "unity of nature" exists. Genome sequences and protein functions are, in many instances, remarkably conserved over the phylogenetic spectrum. That fact has served as a catalyst for blurring the traditional boundaries between many biological research areas.

Thus, a single laboratory engaged in structural genomic studies or proteomics likely analyzes data from a variety of model organisms, including insects, plants, and mammals. The scope of individual research projects has, therefore, enlarged considerably. Collaborations between scientists in distant laboratories, and between academic and corporate laboratories, has become the norm. The landscape of molecular biology displays frontiers that are being pushed further into the distance at an ever accelerating rate.

Our goal in this chapter is to provide an overview of these newest areas of molecular biology research. The types of projects presently being pursued in each of those subdisciplines will be listed, some of the technologies explained, and examples of successful research endeavors provided.

Genomics—The Use of DNA As a Starting Point for Discovery

Traditionally, molecular biology made progress by first identifying specific macromolecules or physiological functions and then characterizing them with one or another biochemical or physical method. Next, the features of the molecule were integrated into a model for a specific cell function, and hypotheses devised and experiments run to test the validity of that model. Finally, the role the molecule plays in the broad spectrum of cell behavior patterns was traced back to specific genes.

Nowadays, faster progress can often be made by inverting that scenario. The genome itself can be used as the starting point for many analyses. Once a certain gene or region of chromosomal DNA is recognized as a potential candidate for storing information for carrying out a specific process or building a particular structure, ways in which that information content is converted into cellular behavior patterns can be explored. Although all organisms store information in DNA, the size of a genome and arrangements of the nucleotide sequences in the genome do, however, vary from one organism to another. That fact necessitates a discussion of genome sizes as a prelude to developing an understanding of genome function.

Genome Sizes in the Biological Kingdom Vary Considerably

It has been known for several decades that the DNA content of a eukaryotic cell varies considerably—in a seemingly random fashion in some instances—through the biological kingdom. This phenomenon is referred to as the **C-value paradox.** That is, the apparent lack of correlation between DNA content in an organism's genome and the size and level of complexity of the organism, as illustrated for laboratory model organisms in **Figure 13-1**, originally baffled molecular biologists. However, it has recently been recognized that larger genomes do not necessarily contain more genes than smaller genomes. Instead, they contain more repetitive DNA (DNA in multiple copies—see Figure 3-10 in Chapter 3). Retrotransposons—DNA sequences that relocate within a genome (see Chapter 14)—comprise the major portion of the repetitive DNA sequences. Thus, what is often referred to as "junk DNA" (since it has no apparent protein-coding function) may represent retrotransposon sequences that are not ordinarily transcribed. For example, approximately 40% of a typical mammalian genome is comprised of retrotransposon DNA. Virtually all other eukaryotic genomes also contain retrotransposons in varying amounts.

Those retrotransposons represent short stretches of nucleotide sequences that are relocated within a genome through a process of RNA-mediated duplication. The

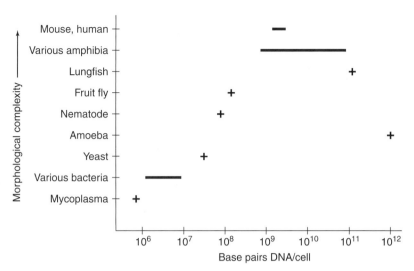

Figure 13-1 Total DNA per cell varies widely among eukaryotes, and does not relate to the level of morphological complexity (so-called C-value paradox). Among plants even more spread exists. For example, corn cells contain 20× more DNA than mustard plant (*Arabidopsis*) cells.

retrotransposon sequence is first transcribed into RNA, then reverse-transcribed into DNA. That complementary DNA is then inserted back into the genome at a new site. Although no function has yet been formally ascribed to retrotransposons, we will learn in the next chapter that movement of nucleotide sequences from one site to another in a chromosome can have profound effects on the regulation of the expression of neighboring genes. Indeed, a few cases of hemophilia and muscular dystrophy have, for example, been traced to genetic changes in protein-coding genes caused by retrotransposon movement within the human genome. Retrotransposons, like mutations, may, therefore, provide a mechanism for genetic variation within a population, and thereby contribute evolutionary adaptation advantages.

The DNA content variations shown in Figure 13-1 include dramatic variations among closely related organisms (e.g., various amphibia). It has been proposed that (1) some species have accumulated duplicate copies of retrotransposon DNA relatively rapidly, and, thus, possess large genomes; and conversely (2) other (often closely related) species have evolved mechanisms for the rapid elimination of retrotransposon DNA and, therefore, possess smaller genomes.

Genome Sequencing Offers Novel Insights Into Cellular Function

Genome sequencing projects have revitalized exploration in many areas of molecular biology research. Traditionally, genes were discovered through mutations and enzyme characterizations. The numbers and variety of mutants were used to estimate how many genes contributed to this or that cellular behavior pattern. With the advent of genome sequencing, a more direct approach became available. The nucleotide sequences of many genomes can be read in an all-inclusive fashion, much the same way this text can be read (from its very beginning to its very end). The automated version of the sequencing method described in Chapter 3 (Figure 3-15) permits rapid and accurate collection of sequence information (**Figure 13-2**). Once archived and stored in a public database, it is available for general use. Internet search engines can be used—even by students reading this textbook—to access some of those computer programs.

The availability of an ever-increasing quantity of DNA sequence data presents challenges for its interpretation. As will be described in the following section

KEY CONCEPT

The "Genome Size vs. Gene Number" Concept

Organisms with very different genome sizes may nevertheless contain similar numbers of protein coding genes.

Figure 13-2 Sequence information is collected by preparing a collection of DNA fragments, sequencing them individually, and using a computer program to align the fragments based on nucleotide sequence overlaps. Viral, bacterial, nematode, insect, plant, and mammalian (including human) genomes have been completely sequenced in this manner. Databases containing those sequences can be accessed with Internet search engines (key words: genome sequences).

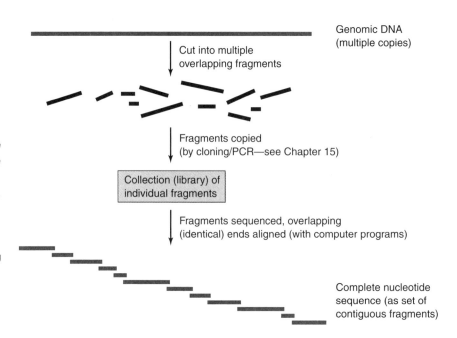

Genomic DNA (multiple copies)

Cut into multiple overlapping fragments

Fragments copied (by cloning/PCR—see Chapter 15)

Collection (library) of individual fragments

Fragments sequenced, overlapping (identical) ends aligned (with computer programs)

Complete nucleotide sequence (as set of contiguous fragments)

(**Table 13-1**), studies that compare genome sequences between organisms offer great promise for unraveling mysteries surrounding the C-paradox mentioned earlier, mechanisms that regulate gene expression described in Chapters 11 and 12, gene number (see the following section), as well as various other issues. Indeed, since direct comparisons of genome sequences between organisms are expected to be very informative, scientists are naturally eager to sequence as many genomes as possible.

Which Genomes Should Be Sequenced?

Approximately 1.7 million known species represent—in principle—candidates for complete genome sequencing. Yet large-scale sequencing projects are expensive! Institutional collaborations that engage in cost-sharing are, therefore, becoming the norm. International committees are attempting to establish criteria for developing priorities for moving forward with several sequencing efforts. Included are the following considerations: ease of obtaining the sequence, which includes factors such as genome size and breadth of interest in the scientific community; usefulness of sequence data as a starting point for experimentation; unique features an organism exhibits for testing specific hypotheses; and general relevance to understanding phylogenetic relationships of organisms.

Research projects that employ genome sequence information as a starting point most often are initiated by sorting through the sequence data (so-called "data mining") for similarities between species (e.g., nematode and fruit fly), patterns of sequence organization (e.g., arrangement of repeated sequences), or intron/exon configurations (among similar genes in widely diverged species). This type of research is not necessarily guided by a hypothesis. That is, data mining research collects facts that can be used to construct a model or scheme for testing, or which later can be used to initiate so-called "hypothesis-driven" research projects. In the same conceptual fashion in which genomics reverses the traditional linear se-

TABLE 13-1	SAMPLING OF GENOMIC BIOINFORMATICS PROGRAMS
Program	**Example of Type of Research Task**
Gene counting	Identify nucleotide sequences that serve protein-coding function
	Estimate gene numbers to understand levels of complexity of regulatory mechanisms
Functional genomics	Identify function of isolated genes by placing them in a living cell
	Analyze gene expression patterns for large numbers of genes at the same time, with microarrays (see text below)
Structural genomics	Establish three-dimensional folding parameters of proteins for which gene's nucleotide sequence is known
	Combine three-dimensional analysis of structure of proteins with genome sequences to deduce function of proteins
Medical genomics	Compare disease-causing genes in model organisms (e.g., mice) with their human counterparts to identify structural homologies at both DNA and protein level
	Identify candidate human genes for drug discovery
Plant genomics	Determine reason plant genomes encode so many "transcription factors" (proteins that facilitate transcription)
	Compare plant gene expression mechanisms with animal counterparts
Evolutionary genomics	Identify genome changes among organisms in terms of addition or loss of DNA sequences
	Compare common ancestors to identify timing of genetic changes
Genome modeling	Design novel bacteriophage genomes for fighting antibiotic resistant bacteria
	Design minimalist genomes for understanding basic principles of cellular metabolism

quence of gene discovery steps (protein > mRNA > DNA) by putting DNA first in the sequence, data mining reverses the discovery flow. Rather than devising a hypothesis or model and then testing it, data is interpreted with the aim of building a model or developing a hypothesis for subsequent experimentation. Computational methods are a key component of data mining, and have been employed so successfully with genome data that bioinformatics has emerged as a field of its own.

Bioinformatics—Using DNA Sequence Information to Build Knowledge

Bioinformatics actually represents an extension of **computational biology,** which has existed as a field of investigation for decades. The development of appropriate algorithms (formulae for solving complex, highly specific problems) and modern computing power have, however, accelerated the rate at which computational biology has progressed. In addition, special themes in molecular biology have emerged, including evolutionary genomics, functional genomics, and structural genomics. Thus, the label "bioinformatics" is now routinely applied to those molecular biology endeavors that rely heavily on highly sophisticated computations. Table 13-1 briefly reviews several genome research projects that are largely dependent on computational methods for their development.

Simple two-dimensional diagrams of the type used throughout this textbook—although necessary to develop an overview of the essential features of a macromolecular structure or metabolic pathway—are greatly simplified. A review of Table 13-1 and **Table 13-2** (below) will reveal the inherent complexities of gene expression and protein function. Three-dimensional illustrations on which a fourth dimension (timing) is superimposed (e.g., by animation on a computer screen) will be required to accurately display that complexity. For example, proteins change shape, very deliberately, during the course of functioning. Indeed, those changes facilitate the high degree of sophistication associated with living processes! Thus, computational assessment necessarily plays a major role in interpreting genomic and proteomic data.

TABLE 13-2 SAMPLING OF PROTEOMIC RESEARCH PROJECTS	
Project	**Experimental Task**
Identification of sample protein	Use amino acid composition of newly discovered protein to compare with known proteins
	Use the nucleotide sequence of its gene to compare with known genes
	Translate a nucleotide sequence into an amino acid sequence
Quantification of protein levels	Determine the relative amounts of specific proteins in various cells/tissues
Protein interactions	Elucidate the macromolecular complexes in which specific proteins participate
Predict posttranslational modifications	From amino acid sequence, determine candidate residues for phosphorylation, potential sites for linked oligosaccharide, etc.
Recognize processed proteins	Compare length of protein with known, larger MW precursors
Elucidate phylogenetic relationships	Compare protein sequences among diverse organisms
Understand three-dimensional folding	Combine three-dimensional structural information of proteins with whole genome sequences to deduce the function of proteins and the control of their expression
	Search for novel protein folding motifs in proteins from unusual microorganisms (e.g., thermophilic bacteria)
Protein trafficking	Identify location and movement patterns of proteins, especially those associated with cell membranes
Disease therapy	Understand the specific protein interactions that cause disease states

Review of the entries in Table 13-1 should reveal the large gap that exists between reading a genome's nucleotide sequence and understanding the basic physiology and behavioral features of an organism.

Gene Counting Exercises Challenge Our Definition of the Term "Gene"

It is the protein-coding genes rather than the genes that code for ribosomal or transfer RNA that are most relevant in gene counting endeavors. Those genes are expected to make the most significant contribution to the overall features of each and every organism. In prokaryotes, gene counting is a relatively straightforward task, since the prokaryotic genome is almost entirely composed of single-copy genes (see Figure 3-10 in Chapter 3), each of which is easily recognized by a computer software scan of the nucleotide sequence of a prokaryotic genome. A gene's promoter region, start/stop translation codons, or transcription-termination sequences reveal the boundaries of a typical prokaryotic gene. Thus, the gene count for bacteria such as *E. coli* (3,600 protein-coding genes) is relatively reliable.

Understanding the eukaryotic genome is, however, a much more complex endeavor. As mentioned previously, in addition to protein-coding sequences, many eukaryotic genomes contain repetitive DNA (so-called "junk DNA"). Thus, there appears to be no "direct" correlation between gene number and genome size. Even among closely related species, vast differences in genome size exist. The genome of the Van Dyke's salamander, for example, contains approximately four times more DNA than does its relative, the Shenandoah salamander. Gene counting is further complicated by the alternative use of promoters, exons, and transcription termination sequences during processing of primary mRNA transcripts (Figure 12-5 in Chapter 12). Thus, a single stretch of nucleotide sequence may participate in coding for the amino acid sequence of more than one protein. In addition, DNA rearrangements (e.g., of the immunoglobulin genes (see Figure 12-9 in Chapter 12)) further complicate attempts at precisely defining what constitutes a gene, and exacerbates the difficulties associated with developing accurate gene counts.

Nevertheless, when complete nucleotide sequence data are available, gene numbers can be estimated. A hierarchy of approaches can be taken: first, use of computer programs to search the nucleotide sequence for protein-coding regions (i.e., reading frames); second, counting the number of different RNA transcripts observed in comprehensive RNA microarray analyses (see the following); third, comparing genomes among different species for overlaps with DNA microarrays. Using those strategies, the following estimates have been made for the common eukaryotic model organisms: yeast = 6,000; nematode = 19,000; and fruit fly = 15,000.

For the human genome, gene counting is especially difficult because of the much larger size of introns and the presence of so-called pseudogenes (DNA regions that, although resembling regular genes, are not expressed due to subtle differences in nucleotide sequence). Estimates for the human genome range from 30,000 to 65,000, with some estimates as high as 120,000. That wide spread largely reflects the use of the different computation strategies mentioned previously.

Review of those gene counts has led to a derivative issue: How are those genes organized along the length of a genome? In a clustered arrangement? Widely

separated by surplus DNA? In a random pattern? With series of repeats closely linked? Specialists in DNA structure and computational biology are presently reviewing genome sequence data in an attempt to answer such questions. Those types of analyses can actually be traced back to earlier attempts at understanding genome organization that used DNA-DNA renaturation as an experimental method (e.g., Figure 3-10 in Chapter 3). It is expected that interpretations of these kinds of sequence data will be valuable for gaining insight into the ways in which the expression of various genes is regulated.

Functional Genomics Employs Methods Which Monitor Expression of Thousands of Genes at Once

Direct comparisons of raw nucleotide sequence data among species can be very informative. By determining which genes are shared by various organisms, and deleting from the comparisons genes shared with prokaryotes or archaebacteria, the subset of genes that are unique to eukaryotes provides insight into why the typical eukaryotic cell is capable of carrying out more complex functions. Eukaryotic-only genes include, for example, those that are involved in membrane transport, chromosome organization, and the regulation of cell division.

Incomplete explanations should, however, be anticipated from simply comparing nucleotide sequence information among organisms. Estimates of the number of genes in some eukaryotes (e.g., fission yeast = approx. 4,900) overlap with preliminary estimates for some prokaryotes (e.g., *Pseudomonas* bacterium = approximately 5,000). Furthermore, many functional genomic studies, including several of those included in Table 13-1, have so far relied heavily on "strong inference" as a "logic" (see Chapter 1) for gaining insight into the regulation of gene expression. *Direct* proof regarding the action of a newly identified gene is often lacking, so instead correlations are employed to "infer" a function. **Figure 13-3** summarizes the conceptual features of this approach. However, studying the expression of one gene at a time, as has traditionally been done with mutant versions of specific genes, often yields only partial insight into the complex regulatory schemes that guide eukaryotic cell behavior patterns.

Understanding the complexity of eukaryotic cell activity can be approached in a more comprehensive fashion using strategies that survey the expression of large numbers of genes simultaneously. The expression of a single, specific gene can therefore be analyzed in the context of the large number of other genes which are typically expressed in the same cell, at the same time. This approach is described in the next section.

Microarray RNA Expression: The Study of Gene Expression on a Genomic Scale

DNA microarray

Southern blot

The **DNA microarray** procedure included in **Figure 13-4** provides insight into the complex gene expression patterns of eukaryotic cells by using a systematic, global strategy. The **Southern blot** (see Figure 8-13 in Chapter 8), originally described in 1975, is actually the forerunner of the microarray. The power of DNA microarrays as experimental tools derives from the high affinity and specificity of complementary base pairing described in Chapter 3.

Natural selection has generated regulatory mechanisms that restrict the expression of most genes to specific cells, at certain times, under defined conditions.

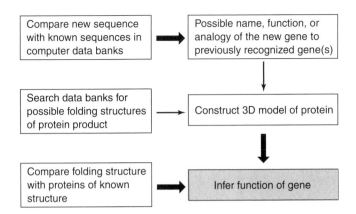

Figure 13-3 Beginning with nucleotide sequence information, it is often possible to eventually infer the function of a new gene. It is probable that a relatively small number of ancestral protein domains were linked in various combinations and thereby evolved into present-day proteins. Recognizing folded structures in the gene's protein product that are also present in known proteins (e.g., enzymes) can, therefore, assist in gene identification.

RNA microarray analyses, therefore, provide important insights into the genetic control of phenotypic characteristics. For example, the role a gene plays in metabolism can often be inferred by observing when and in which cells it is expressed. Extensions of that strategy (**Figure 13-4**) include comparing gene expression profiles between different types of cells (e.g., muscle and brain), the same tissue at

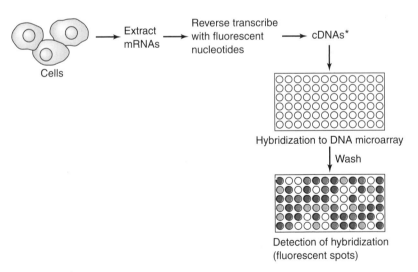

Figure 13-4 Microarray analyses begin with glass slides onto which DNAs or oligonucleotides representing portions of selected genes are spotted. By using technologies developed for the manufacture of computer chips, thousands of individual genes can be spotted on a single slide (or silicon chip). Cellular RNAs (e.g., mRNAs from a specific type of eukaryotic cell) are fluorescently labeled by converting them to cDNAs with the enzyme reverse transcriptase in the presence of fluorescent nucleotides.* They are then hybridized directly to the DNAs on the glass slide. Each square on the grid represents a single gene. With automated scanning methods, hybridizations to specific DNAs can be recognized by their fluorescence. Thus, in one step, gene expression profiles for thousands of genes can be obtained.

various times in the life cycle of an organism (e.g., embryo vs. adult), or even normal and diseased (e.g., cancerous) tissues. In addition, by totaling up the number of genes expressed on microarrays prepared from mRNAs extracted from a large variety of cells/tissues of a specific organism, gene counts can be made.

Extension of Microarray Technology Reveals DNA Variation

Hybridization of DNAs (rather than mRNAs) to genes that are arrayed on glass slides can be employed to identify mutations and recognize genetic variations among populations. It is expected that soon it will be possible to use arrays to routinely test individual patients for a variety of genetic diseases. Human genetic disorders that result from the action of multiple genes may one day also be recognized by hybridizing a patient's DNA to an array of the genes that contribute to complex genetic disorders—such as schizophrenia, alcoholism, and heart disease—that are caused by the action of several genes.

Genomics and Bioinformatics Are Necessary, But Not Sufficient for Understanding Cell Behavior Patterns

As was mentioned in the opening sentence of this textbook (Chapter 1), the goal of molecular biology is to fully understand the range of cell behavior patterns. Genome sequence information provides a valuable place to start, as the list of projects included in Table 13-1 reveals. It should be kept in mind, however, that, although cell behavior patterns certainly reflect information stored in the genome, it is proteins that are actually responsible for the specific characteristics of different types of cells. Furthermore, just because a gene is present in a genome does not mean that it is actually transcribed and translated into a functioning protein. Study of proteins, therefore, brings the researcher closer to being able to understand and/or manipulate the function of individual cells. Fortunately, efforts at recognizing specific proteins and studying the way they interact with one another, as well as appreciation for their role in diseases, are contributing to the detailed knowledge of "why cells behave the way they do"!

Unraveling the inherent complexity of cells and the sheer numbers of proteins in the typical (e.g., human) cell demands both sophisticated instrumentation and immense computational power. A new subdiscipline of biology—proteomics— has emerged to undertake this task. In many instances, universities and private enterprises (e.g., pharmaceutical and biotechnology companies) have combined forces to undertake this immense task.

Proteomics Focuses on the Totality of Proteins: Their Numbers, Structures, Interactions, Locations, and Functions

proteome

The term **proteome** was invented to describe the **prote**in complement expressed by a gen**ome.** Proteomics attempts to study the sum total of all the proteins in a cell from the point of view of their individual functions, and how the interaction of specific proteins with other cellular components (e.g., nucleic acids, membranes, organelles, etc.) affects the function of those proteins.

The conceptual basis of proteomics is fundamentally different from genomics. While the genome of a cell is essentially static and can be well defined for all the cells in an organism, the proteome continually changes in response to external and internal conditions. This holds true even for prokaryotes. For example, *E. coli* expresses different proteins and, thus, has a different proteome when cultivated in a minimal medium (containing just carbohydrate and nitrogen sources and various minerals), rather than a rich, complete growth medium (such as beef broth). Similarly, during mammalian development, specialized cells express different proteins, develop dissimilar but characteristic proteomes, and ultimately differentiate into diverse tissues. Table 13-2 lists various proteomic research projects.

For the proteomic projects listed in Table 13-2, the tremendous quantity of data necessary for comprehensive analysis requires utilization of modern automated sample handling instrumentation and data handling software. Robotics are, therefore, often employed to guide protein samples through a procedure or to carry out processing reactions. For example, literally thousands of peptides can be prepared for mass spectroscopy in a matter of hours with advanced "high-throughput" methods.

Those endeavors are driven by multiple forces: genomics and its revelation of the large numbers of proteins encoded in most genomes; powerful protein characterization methods such as two-dimensional electrophoresis (e.g., Figure 2-11 in Chapter 2); advanced sample preparation methods coupled to high resolution mass spectrometry (for identifying amino acid sequences); in vivo functional tests; and innovative computational tools. The relationships of those various technologies are illustrated in **Figure 13-5**.

Proteomics Uses Protein Structure As a Starting Point for Identifying Functions and Interactions

Historically, the function of a specific protein was found using extensive genetic and biochemical analyses, or by comparing the amino acid sequence of a newly discovered protein with one of known function. With complete genome sequences and microarray mRNA expression patterns now available, general biochemical functions of proteins can be inferred by associating them with other proteins that participate in a common macromolecular complex, metabolic pathway, biological process, or closely related physiological function.

The size and complexity of the task facing proteomics can be appreciated by assuming between five and 50 modifications (e.g., phosphorylation, glycosylation, etc.) or interactions (e.g., association with cell membranes) per protein. The result? Between 30,000 and 300,000 variations can be expected to exist for the 6,000 proteins coded for in a single yeast cell! Although experiments have characterized (i.e., identified by name) about 30% of yeast proteins, methods are frequently not rapid, inexpensive, or complete enough to reveal the identity of many of the remaining proteins. There is a need, therefore, to assign function for most proteins with the aid of sophisticated computational methods. **Structural proteomics** has emerged as a subdiscipline for accomplishing those tasks.

structural proteomics

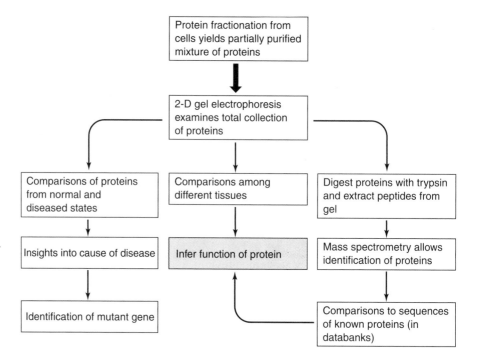

Figure 13-5 Convergence of diverse methods (e.g., protein purification, gel separation, mass spectroscopy) and concepts (e.g., comparisons between normal and diseased cells) characterizes proteomic research.

Structural Proteomics Uses Folding Patterns to Deduce a Protein's Function

Protein folding patterns, which can be recognized from the analysis of x-ray crystallographic data (**Figure 13-6**), often can be used to understand the function of a specific protein. Although this method generates elegant data, the production of protein crystals requires relatively large amounts of pure protein. Nevertheless, for many well-characterized proteins, atomic models for an ever-increasing list of proteins—archived in the Protein Data Bank (mentioned later in this chapter)—are being developed.

An alternative strategy for developing three-dimensional models of newly discovered proteins can, however, be employed that circumvents the need for crystal samples of the test protein. The method involves, first, cutting a protein into small pieces and, second, employing nuclear magnetic resonance spectroscopy for recognizing the folded structure of each of the pieces. With those folds revealed, sophisticated computational methods are employed to combine the folding patterns

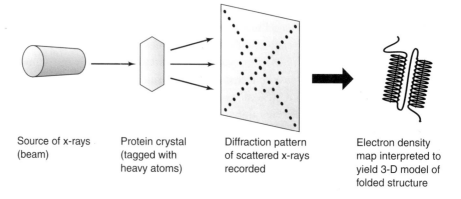

Figure 13-6 Determination of the folded (tertiary) structure of a purified protein by interpreting x-ray diffraction patterns of its crystals.

Source of x-rays (beam)   Protein crystal (tagged with heavy atoms)   Diffraction pattern of scattered x-rays recorded   Electron density map interpreted to yield 3-D model of folded structure

of the individual fragments into a single three-dimensional model. In the case of an enzyme, function can often be deduced (**Figure 13-7**) from NMR data describing the active site region.

Expression Proteomics and Chemical Proteomics Offer Insights Into Disease Therapy

Expressional proteomics involves the study of proteins in comparative tissue samples. In order to comprehend how best to treat a particular disease, it is necessary to identify the proteins associated with that disease and to understand how they function. In some instances, this task involves recognizing an altered or missing protein. For example, sickle cell anemia is known to be caused by a point mutation in the globin gene (see Figure 16-3 in Chapter 16). In other cases, especially when a particular protein functions in many different tissues, a more broad-based characterization is called for. Such features that may require elucidation include the following: the cellular localization of the specific protein; its quantification, tissue distribution, and posttranslational modification state; its interaction history; and its binding properties (e.g., for enzymes—binding to substrates). Genetic diseases that are caused by the action of multiple genes (e.g., cancer) will certainly require this more comprehensive strategy.

A typical question posed by a researcher might be: What types of protein changes just reviewed correlate with the transformation of a normal cell into a cancerous cell? With that information, subsequent analyses can be directed toward establishing cause/effect relationships between the affected protein(s), their genes, and cancer. In addition, drug design strategies that target a known protein are likely to be more successful.

Chemical proteomics involves at least three approaches. For one, a protein revealed as contributing to a diseased state and known to have a particular size, shape, and catalytic activity is matched up with a small metabolite (e.g., drug) that binds to it. Identification of drugs that inactivate the protein, without causing toxic side effects, is the goal of this task. Successes include antifungal drugs that bind to CYP51, a fungal enzyme that—when inhibited—diminishes the effects of common athlete's foot infections; and cyclosporin, which binds to a phosphatase enzyme and suppresses the immune response (important for preventing rejection in organ transplant patients).

chemical proteomics

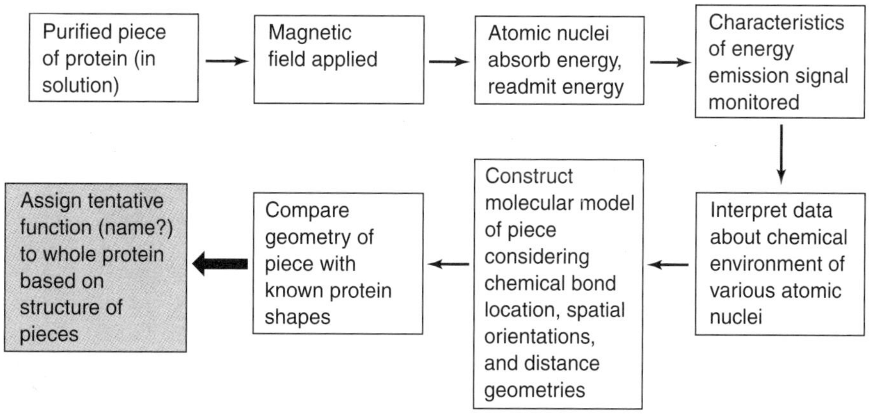

Figure 13-7 Nuclear magnetic resonance spectroscopy provides an opportunity to reveal the details of specific regions (e.g., active site) of proteins without needing to understand the entire three-dimensional structure of the protein.

For another, searches for protein involvement in specific diseases can be carried out by testing proteins expressed in diseased tissue against a variety of small molecules. Discovery of the binding capabilities of the protein might lead to the identification of the role of the protein in disease.

Finally, if a three-dimensional model of the structure of a disease-causing protein is available, clefts on its surface can be identified. Then drugs can be designed for binding and inactivating the protein. An example of this strategy is illustrated in **Figure 13-8**. As straightforward as those three procedures may appear, complications do, however, abound. For example, special efforts are often required to identify proteins that are present in cells in low abundance, as well as those that are embedded in cell membranes (and, thus, difficult to extract in high yields). In addition, serum proteins bind and sequester many drugs that are administered to patients, diminishing their effectiveness, and, thus, requiring increased doses.

The Emergence of a New "Logic of Molecular Biology"?

In Chapter 1, the thought processes and reasonings often employed to solve problems in molecular biology were reviewed. Those "logics" are frequently used to develop hypotheses for testing, and have proved to be very powerful for generating discoveries in molecular biology. For some projects, especially very large undertakings such at those discussed in this chapter, a new strategy is, however, emerging. A move is occurring away from "hypothesis-driven research" and toward a system of solving structures or collecting data first and then asking questions later, or even retrofitting a hypothesis to the data that was collected. That is, projects are designed without a formal hypothesis, and with no model to be validated. The goal is to first make discoveries about phenomena that we neither knew about beforehand nor expected to exist, and then, second, to understand relationships and connections among the data points. This is the approach taken by the Human Genome Project: Sequence data was first collected, and now various types of genomics studies (Table 13-1) are being designed to interpret that data.

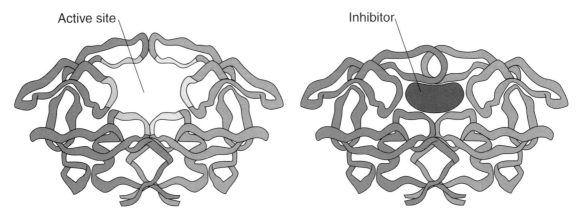

Figure 13-8 Once the three-dimensional structure of the protease that is required for the production of HIV (AIDS virus) copies was determined (left), inhibitors that resemble the natural target (aspartyl residues) were designed to fit into the enzyme's active site and block its action (right).

Important strategic considerations are beginning to replace "hypothesis design" as a guiding force for making progress in genomics and proteomics. Which genomes to sequence? Which proteins to target for structural studies? Which cells to choose for gene expression analyses? Those are all valid questions that often serve as a starting point for research projects. Answers usually are derived by balancing such features as cost, level of technical difficulty, potential commercial payoff, and possible impact on furthering our understanding of basic biological phenomena. For example, human membrane proteins represent an important target for structural genomics studies. Many of those proteins are potential drug targets, for they play key roles in signaling both within cells and between cells. Yet, membrane-associated proteins are difficult to purify. Hence, their analyses are expensive, and are proceeding slowly.

This new "logic" might be termed the **strategic and technological resource** logic, for it represents using rationales for choosing a project or research direction based largely upon considerations such as cost, technological advantages, data quantity, and potential for commercial exploitation. This logic has been used to develop large-scale projects designed, for example, to characterize the three-dimensional structure of the 1,000 different folding motifs believed to account for the entire array of possible tertiary structures of all known proteins.

Perhaps this paradigm shift should be viewed in a historical perspective, for the recent emphasis on the logic of "strategic and technological resource" may be temporary. At present, serious gaps exist in our predictive power at the atomic level. Thus, **strong inference** is not an especially useful logic for designing experiments that can predict, for example, the *interactions with other proteins that a DNA-binding protein undergoes as it binds to double-stranded DNA*. However, once sufficient data on those complex issues is collected, the design of appropriate models or hypotheses may return to fashion, and the logics that were explained in Chapter 1 may be restored to preeminence.

Data Exchange Is an International Endeavor

Rapid progress in genomics, bioinformatics, and proteomics is facilitated by the free and rapid exchange of data. A variety of databases—each of which can be located by Internet search engines—have been established to store information. For example, the Protein Data Bank, originally established in 1971, is thriving as a result of the revival of interest in protein structure and function and the advent of the Internet. Efforts are under way to interpret nucleotide sequence data for whole genomes in terms of the structure and function of specific proteins. Some of those databanks are updated daily.

Those types of databases employ formatting standards that permit quick access and easy interpretation of information. Consequently, laboratories in different parts of the world can share information quickly and inexpensively. Therefore, the opportunities that are available for collaboration provide an occasion for a comprehensive and fresh view of important issues in virtually all fields of biology.

Future Practical Application

Attempts to bridge the gap between gene discovery and therapeutic treatment for human diseases provide a focal point for many pharmaceutical companies. For example, the *tub* gene was discovered in laboratory mice to be responsible for an obesity phenotype. The protein product of that gene has been identified as playing a role in the regulation of transcription. The three-dimensional structure of the core of the *tub* protein revealed the presence of a positively charged groove. That groove likely plays a role in binding to DNA (which is negatively charged). Opportunities are now available for designing small molecules (drugs) that restore the function of the mutated protein by binding to it and correcting defects in its conformation.

Similar attempts to produce "designer drugs" will be undertaken as genomic studies reveal additional potential drug targets.

SUMMARY

The advent of automated genome sequencing machines has generated an unprecedented quantity of data concerning genome size, organization, and potential function. Subdisciplines of molecular biology have emerged—based on earlier studies in computational biology—that examine that data in an attempt to understand how information flows from the genome to functional proteins. Genomics, which focuses on an organism's genome; bioinformatics, which searches for meaning in the genomic sequences; and proteomics, which analyzes the function of proteins represent major driving forces in information-age molecular biology.

Representative research projects from each of those areas are described, and include attempts to understand gene regulation by comparing nucleotide sequences among organisms, identification of disease-causing genes (as potential candidates for drug therapy), and characterization of protein function on the basis of structural features. Consideration of those aspects of genome structure and function has generated a reconsideration of the meaning of the term "gene," for several different proteins can often be traced back to a single (classical) "gene." Thus, one gene is capable of generating sufficient information for several proteins with distinctly different functions.

With microarrays, which consist of DNA hybridization platforms, it is possible to comprehend some of the complexity associated with the expression of thousands of individual genes in a single type of cell. Likewise, with two-dimensional protein gel electrophoresis methods, it is possible to study the synthesis of large numbers of proteins simultaneously, and to gain insight into the function of individual proteins. For analyzing function, rather than employing the traditional approach, which begins with mutant genes or enzyme assays, pieces of individual proteins are isolated (usually extracted directly off two-dimensional gels) and the structure of the fragments determined with NMR spectroscopy. Using sophisticated computer programs, those structures are then compared to known protein structures contained in protein data banks. This approach, from structure to function, shows promise for accomplishing the formidable task of determining the function of up to 100,000 (or more) proteins that may be present in a higher eukaryotic organism.

Many of those research projects depend heavily upon sophisticated technologies, and are initiated with the goal of first collecting data, then explaining it. This "data mining" approach contrasts sharply with traditional "hypothesis-driven" research of molecular biology.

DRILL QUESTIONS

1. Define the following terms: *genomics, bioinformatics,* and *proteomics.*
2. What is the so-called "C-paradox?" What explanations account for it?
3. What is meant by the term "gene counting?" Of what value is a gene count?

4. What is the approximate numerical relationship between gene count and protein count?

5. Explain the advantage of using DNA microarrays for studying gene expression.

6. How has the discipline of proteomics changed the old idea that *one gene is responsible for one protein?*

7. What role does "strong inference" (Chapter 1) play in these new sub-disciplines of molecular biology?

8. How does "data mining" differ from "hypothesis-driven research?"

PROBLEMS

1. List several considerations for prioritizing the genomes that should be sequenced in the near future.

2. Design a conceptual strategy for determining the human gene count.

3. Develop a conceptual model for the human genome that assumes that less than 5% of the nucleotide sequences actually code for proteins. Focus on the remaining 95% of the sequence.

4. Genome "complexity" (gene number and arrangement) as the predominant manner in which genomes are compared is slowly being surpassed by genome "connectivity" (regulatory circuits or networks) as a more meaningful way in which to compare the genetic makeup of two different organisms (e.g., fruit fly and human). Write a brief essay that validates that paradigm shift.

5. Rather than identify individual proteins by traditional genetic (mutant analyses) and biochemical (activity assays) methods, as has already been done for a few thousand different proteins (mostly enzymes), "identification by structure" has been proposed as a more efficient manner in which to identify the function of the remaining tens of thousands of different proteins. List several of the limitations of the latter (structure-based) approach.

6. What experimental strategy do you suppose was employed to devise HIV protease inhibitors? What are several important considerations for preparing clinically useful doses of those inhibitors?

CONCEPTUAL QUESTIONS

1. What role will (synthetic) organic chemists play in the development of the aspect of proteomics that deals with disease therapy?

2. How could the medical genomics project included in Table 13-1 be exploited for use in a clinical laboratory, so that a patient might be tested—in a physician's office—for genetic diseases, or susceptibility to drug side effects?

3. Since proteomics predicts that there potentially exist >100,000 different proteins in a higher eukaryotic organism, how might the methods that are employed to determine the structure of protein fragments be scaled up so that, rather than analyzing them at the rate of, say, one fragment per day, hundreds could be analyzed in that time?

Experimental Manipulation of Macromolecules

in this chapter you will learn

1. About transposable elements: how they relocate, and their effects on gene expression.

2. About the basic properties of plasmids and how they are transferred from cell to cell.

3. About the life cycles of diverse bacteriophages.

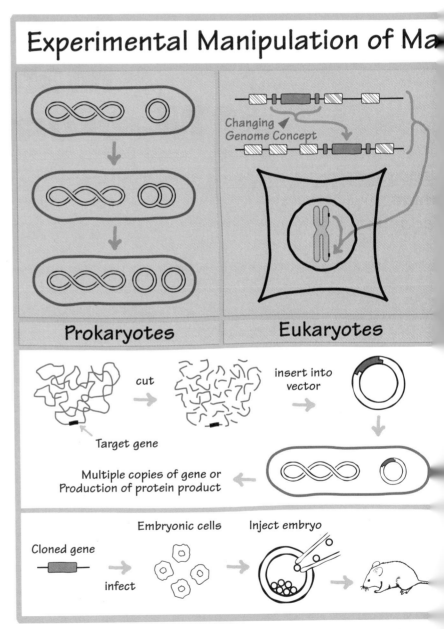

Experimental Manipulation of Ma

Changing
Genome Concept

Prokaryotes

Eukaryotes

cut

insert into
vector

Target gene

Multiple copies of gene or
Production of protein product

Cloned gene

Embryonic cells

Inject embryo

infect

olecules

sposons,
mids, and
eriophage

enetic
ineering

ctical
ications

Transposons, Plasmids, and Bacteriophage

These seemingly diverse genetic forms will be considered in order of their intrinsic complexity. Transposons, which consist of small numbers of contiguous genes that are integrated into the host's chromosome, represent the simplest of these forms. Plasmids are more complex, for they are larger, exist in the cell independent of the host's chromosome, and can be transferred (directly) from cell to cell. Bacteriophages are much more complex, for they are comprised of not only nucleic acid, but also a protein coat (which endows them with the ability to be released from their host cell and survive in hostile environments).

Their evolutionary histories remain unclear, and whether they were actually functionally related at an earlier time is difficult to determine. Constructing a phylogenetic tree for these particular genetic elements presents a large puzzle to molecular biologists for many reasons. First, it is unclear whether all manner of those natural genetic elements (besides the conventional "chromosome") have already been recognized. The Postscript Section of this textbook emphasizes the notion that revelations await diligent research. Therefore, it remains to be seen whether other natural genetic elements will be uncovered, that will fill what appear to be gaps in our understanding of the relationships of these genetic forms. Second, the rapid rate of change these forms are capable of means that earlier phylogenetic relationships between them may be obscure. Finally, it is possible that some forms may have emerged more than once, and others may have become extinct, during the course of the two billion years of evolution on planet earth.

For the beginning student of molecular biology, it might be frustrating to attempt to develop a comprehensive view of these diverse genetic elements, realizing that a seamless connection between them does not necessarily exist. Yet it should be recognized that, in their own right, transposons, plasmids, and bacteriophage warrant study. Some (transposons and plasmids) alter gene expression patterns of their hosts and thereby play a major role in enhancing genetic fitness. Others (certain plasmids and select phages) are useful as tools in genetic engineering endeavors. We will begin this chapter by learning about transposable elements.

Transposable Elements—Their Discovery Surprised Molecular Biologists

Geneticists are able to construct genetic maps because of the stability of the linear array of genes on the chromosomes of individuals of a given species. These maps tend to give the impression that the genomes are rather static, suffering only the occasional base-pair changes due to replication errors or other mutagenic insults. However, this is not the case. A special class of opportunistic DNA sequences has evolved in all organisms. These DNA sequences have the unique ability to move as a discrete unit from one position in the genome to another. This ability to relocate is called **transposition,** and the elements themselves are called **transposable elements** (abbreviated **Tn elements**). Transposable elements found in bacteria are simply called **transposons.**

    Tn elements are found in plasmids and viruses, as well as in the genomes of prokaryotic and eukaryotic organisms. Tn elements were first detected in bacteria in the *gal* operon as mutations that eliminated the function of all the genes in the operon. These mutations were caused by the insertion of specific DNA sequences into the *gal* operon. Certain insertions were about 800 base pairs in length; others were about 1,300 base pairs long. Moreover, in wild-type revertants derived from these mutations, the entire block of inserted DNA was missing. These inserted DNA segments are called **insertion sequences.**

    A variety of different IS sequence elements have been described in *E. coli* (see **Table 14-1**). These elements range in size from several hundred to several thousand base pairs long, and are characterized by their ability to move from one point in the genome to another. They do not, however, carry a detectable phenotype; genetically, they can be detected only by the effects of their insertion on the function of a gene in the chromosome or on the plasmids discussed later in this chapter. Physically, their presence can be detected at a particular position, either by hybridization experiments with a suitable probe or by direct nucleotide sequencing.

    IS sequences contain promoters, and transcription and translation-termination signals. Depending on the position and orientation of the IS insertion, IS control signals can interfere with proper expression of the target genes at either the transcription or translation level. Most often, the protein gene product is inactivated because the added sequences interfere with proper protein folding. If the gene into which the insertion has occurred is part of an operon, the expression of downstream genes may also be blocked. However, if the insertion point of the IS sequence is upstream

transposition

transposable elements

Tn elements

transposons

insertion sequences

TABLE 14-1	PROPERTIES OF SEVERAL *E. coli* INSERTION ELEMENTS	
Element	Number of Copies and Location	Size in Base Pairs
IS1	5–8, chromosome	768
IS2	5, chromosome; 1 in F plasmid	1,327
IS3	5, chromosome; 2 in F plasmid	Approximately 1,400
IS4	1 or 2, chromosome	Approximately 1,400
IS5	Unknown	1,250
γλ	1 or more, chromosome; 1 in F	5,700

of a gene or operon, the control of expression of these genes can be changed—they may be expressed at a considerably higher (or lower) rate. Alternatively (as in *gal* operon mutations), expression of all genes of the operon may be blocked by an appropriately placed IS element. Such changes are consequences of transcriptional control signals in the IS sequence itself.

What properties of the IS sequences are responsible for their remarkable ability to transpose? Analysis of the DNA sequences of many different IS elements has revealed a striking feature: All elements contain a **terminally inverted sequence** (**Figure 14-1**). The size of the inverted terminal sequence may vary from one type of IS element to another; however, all members of the same type carry the same sequence. For example, IS1 has 23 base-pair terminal repeat sequences; IS2 has a 41 base-pair repeat. The two copies of the terminal sequences are perfect, or nearly perfect, repeats. Moreover, the integrity of the inverted terminal repeats is essential for IS elements to transpose (i.e., relocate).

All IS elements are capable of encoding a special DNA binding protein called a **transposase.** Transposase is the key protein component responsible for carrying out the transposition process. Transposases are DNA-binding proteins that have the ability to specifically recognize the terminally inverted sequences of the IS sequence. This ability to recognize the borders of the IS element is an important feature of the mechanism, permitting the IS element to move as a discrete unit. In general, transcription of transposases are under the control of promoters within the IS element. Given the importance of the reaction they carry out, it is not surprising that the expression of transposases is carefully controlled, or that the level of transposases in cells is usually very low. We will discuss later how transposase catalyzes transposition.

transposase

A second important sequence feature involving IS elements concerns the site of insertion of the element. Analysis of large numbers of different insertion points for a given IS element revealed that insertion generates a short **direct repeat** flanking the element. The length of the direct-flanking repeat is characteristic of the type of IS element. The sizes range from about 3 to 13 base pairs. The repeat sequence is not part of the IS sequence itself. Rather, it is a repeat of sequences present in the target that was attacked by the transposase enzyme. The fact that a **target sequence** is found to be repeated after the insertion process is completed provides an important clue to the mechanism by which the IS element transposes. It is most likely that the direct repeat of the target sequence is a consequence of a staggered cut in the target DNA made by the transposase (see **Figure 14-2**).

direct repeat

The DNA-recognition properties of the transposase enzyme coded for by the transposon are most clearly responsible for determining where the IS element will insert. Most transposases have very broad sequence recognition properties, and

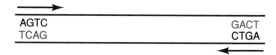

Figure 14-1 An example of a terminal inverted repeat in a DNA molecule. The arrows indicate the inverted base sequences. Note that the sequences AGTC and CTGA are *not* in the same strand. In a direct repeat, the sequence in the upper strand would be AGTC . . . AGTC.

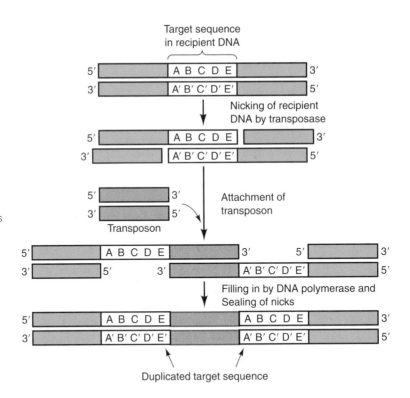

Figure 14-2 A schematic diagram indicating how target sequences might be duplicated by formation of a staggered cut at the target sequence. Capital letters represent the nucleotides of the target sequence (A′ represents the nucleotide that is the complement of A). Following attachment, the free 3′ ends on the transposon can serve as primers for the DNA synthesis, which duplicates the target sequence. Note: a terminal repeat sequence (Fig. 14-1) is located at both ends of the transposon.

mediate transposition into a wide variety of target sequences. For many IS elements, the sequence of the site chosen for insertion is close to random; however, transposases usually have some preference for insertion into sites with certain sequence features. Such sites are known as **hot spots** for insertion.

Complex Transposons

IS1, IS2, IS3, and so on are examples of relatively simple transposable elements. They contain terminal inverted repeats and each encodes a "specific location" transposase. They are, therefore, capable of movement from one location to another, but they carry *no detectable* phenotype. Two types of more **complex transposons** have been observed as natural constituents of plasmids. The first class of complex Tn elements consist of a gene or multiple genes, usually encoding resistance to an antibiotic, sandwiched between two IS elements or IS-like elements. For example, the transposon Tn9 consists of a gene conferring resistance to chloramphenicol, flanked by IS1 elements in direct orientation. Tn10 contains a gene conferring tetracycline resistance, flanked by IS10 elements in an inverted orientation. Regardless of the orientation of the flanking IS elements, the termini of the composite element as a whole are inverted because each of the component IS termini are inverted (**Figure 14-3**). Both IS elements flanking the antibiotic resistance gene can be capable of independent transposition. In some Tn elements, however, one of them has become mutated and can be considered vestigial, providing only a terminal repeat to the transposition process. Like their simpler relatives, the IS sequences, transposition of the complex Tn elements generates a short direct repeat in the target genome. The length of the repeat is characteristic of the Tn element.

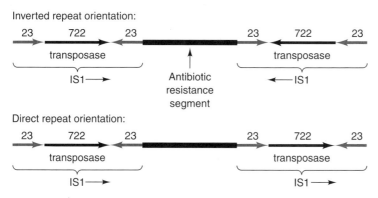

Figure 14-3 Two types of complex transposons showing an antibiotic resistance gene flanked by two IS1 insertion sequences in inverted and direct repeat orientations. (The numbers indicate the sizes of various portions of the transposon in base pairs.) Blue arrows indicate the inverted repeats of the IS1 elements. Note that the figure is not drawn to scale.

Thinking about how such elements arose, one could imagine that complex elements were selected under conditions of exposure to antibiotics. Prior to selection, two IS elements happened to flank the antibiotic resistance gene. Selection pressure resulted in the capture of the gene by the two IS elements. The transposon provided a means for horizontal transmission of resistance to other members of the population. Genes other than those conferring antibiotic resistance can be part of such elements. Tn1681 contains a gene encoding an enterotoxin that plays a role in causing human diarrhea. It is likely that in response to appropriate selective pressures, a complex transposon containing essentially any bacterial gene can be detected. The ability of this type of mechanism to respond to environmental changes provides genetic plasticity, and has profound implications for bacterial evolution.

More Complex Transposons Exist

The second type of complex transposon has a more complicated genetic structure. The best known of these elements is a 5 Kb transposon called Tn3 (see **Figure 14-4** for a schematic outline). Elements of this class are not flanked by IS sequences, but instead have shorter inverted repeats; the inverted repeats of Tn3 are 38 base pairs long. Between the inverted repeats are three genes, two of which are similar in function to those already discussed. The *bla* gene encodes an enzyme, called β lactamase, that confers resistance to ampicillin. The second gene, called *tnp*A, encodes a transposase that specifically recognizes the inverted repeats of Tn3 during the transposition process. The third gene, called *tnp*R, encodes a bifunctional protein, called **resolvase,** that functions in the transposition process. Resolvase is a specific DNA-binding protein that recognizes a short DNA sequence called the **internal resolution site (IRS).** The resolvase (TnpR) also represses synthesis of its own transcript and the *tnp*A transcript by virtue of its binding affinity for sites I, II, and III. The consequences of resolvase binding to the IRS will be discussed later.

resolvase
internal resolution site
(IRS)

The Process of Transposition

Transposition involves the movement of a defined segment of DNA from one spot in a genome to another site either in the same genome or in another, such as a viral

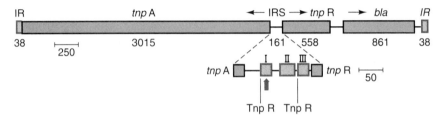

Figure 14-4 Organization of the Tn3 transposon. The solid blue boxes (IR) indicate the terminal inverted repeats, the grey boxes indicate open reading frames; *tnp*A, transposase; *tnp*R, resolvase; bla, β-lactamase (the gene for ampicillin resistance). The black arrows show the direction of transcription. Numbers indicate the length in base pairs. The total length of the element is 4,957 base pairs. IRS stands for internal resolution sequence (this region is also shown magnified five-fold). This region contains promoter elements for the *tnp*A and *tnp*R genes, as well as three sites for resolvase binding (I, II, III) indicated by blue boxes. The blue arrow shows the site at which recombination between two TN3 elements of a cointegrate occurs, a process mediated by the enzyme resolvase.

or plasmid, genome. A convenient way to measure the frequency of transposition is to take advantage of the fact that movement can occur between a chromosome and a conjugative plasmid, such as the F plasmid (described later in this chapter). If this occurs, the F plasmid acquires the ability to confer an easily detectable phenotype: antibiotic resistance. The frequency at which the F plasmid is used as a transposition target can be determined by measuring the frequency at which females become antibiotic-resistant males after conjugation. Only if the transposon has moved from the chromosome to the F plasmid will the F plasmid confer antibiotic resistance on the female after conjugation. Experiments have indicated a frequency of movement of about 10^{-5}–10^{-7} per element per generation for transposons such as Tn10. This frequency is in the same range as the spontaneous mutation rate seen for a typical bacterial gene.

Conservative Transposition

Transposition occurs via two types of mechanisms. Which mechanism is employed depends on the type of transposable element. In the simpler of the two mechanisms, transposition involves excision of the Tn element through cleavages at its inverted termini by transposase. This excision is followed by insertion of the element at a new site. Such a cut-and-paste mechanism is a **conservative** process, that is, it involves no replication of the original element, except for the small amount needed to create the direct repeats flanking the element at its new location. The donor molecule (having suffered a double-strand DNA break at both ends of the element) can be lost in the process. To enter the recipient molecule, the transposase enzyme makes a staggered double-stranded cut. The length of the stagger is characteristic for each type of element. For example, the 9-base-pair direct repeats that flank Tn10 are generated by the Tn10 transposase making a 9-base-pair staggered cut in target DNA. The two ends of the element are then ligated to the cut ends of the target DNA by transposase. The ligation creates 9-base-pair, single-strand gaps flanking the element—these are filled in by DNA polymerase of the host (Figure 14-2). This conservative mechanism of transposition is used by Tn10 and Tn9.

Replicative Transposition

The existence of a second transposition mechanism was shown by observing that the transposition of certain transposons, such as Tn3, involved a special intermediate DNA structure called a **cointegrate.** The cointegrate form is a result of fusion of the donor and target replicons mediated by transposase. An important feature of the cointegrate is that it contains two copies of the transposon in a direct orientation at the junctions of the donor and target sequences. Thus, in the cointegrate, the transposon has been replicated. The cointegrate is an intermediate structure in transposition. It can be resolved by a special recombination process into two separate DNA molecules, each containing a copy of the transposon (**Figure 14-5**). This step is carried out, not by the transposase, but by the product of the Tn3 *tnp*R gene, resolvase. The recombination step, called **resolution,** occurs at the IRS sites in the paired copies of Tn3 in the cointegrate called *res* sites (Figure 14-5). Transposition of Tn3 is, thus, a **replicative** process in which a second copy of the element is produced. Tn3 produces 5-base-pair direct repeats in its target sequence, implying that its transposase makes a staggered cut in the target DNA, like Tn10 transposase.

cointegrate

Control of Transposition Frequency

It is clearly important to cell viability that the frequency of movement of transposable elements be tightly controlled to prevent the accumulation of additional copies of the elements in the genome. One way transposition is controlled is by limiting production of transposase. The Tn10 transposase is produced at an extremely low level—less than one molecule per cell per generation. This low level of production is accomplished by control both at the level of transcription of the

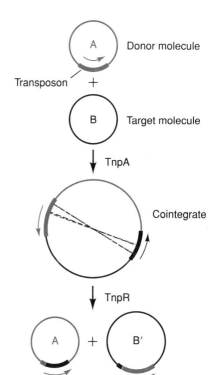

Figure 14-5 A simple two-step model for replicative transposition. In the first step, transposase (TnpA) encoded by the donor molecule (A) mediates fusion with the target molecule (B) to generate a cointegrate molecule. Note that the transposon has been duplicated in the cointegrate. In the second step, called resolution, recombination between specific sites in the two transposon copies (blue lines) is mediated by the TnpR protein (resolvase) to generate two separate molecules. One molecule is identical to the donor; the other (B′) is the target molecule now containing a copy of the transposon.

transposase gene and translation of the transposase mRNA. Thus, in most cells carrying Tn10, transposition is prevented simply by the absence of transposase.

A second way in which transposition frequency is controlled is that the process of transposition is largely confined to a small time window in the bacterial cell cycle. *Transposition occurs preferentially immediately after DNA replication of the element itself.* E. coli DNA normally is **methylated** at specific sites by the action of an enzyme called DNA adenine methylase. However, when a replication fork passes over a DNA region, the newly made DNA strand is transiently unmethylated, so the recently replicated part is only partially methylated. Shortly after replication, the *dam* methylase methylates the newly made DNA strand. This converts the daughter DNA molecules to the fully methylated state. The transposase, however, interacts poorly with fully methylated DNA, but interacts strongly with partially methylated DNA. During that short time-interval before the DNA is fully methylated, transposition occurs. Thus, the short time available for transposase action serves to control transposition frequency. (Recall that mismatch repair (Chapter 10) also uses this transient undermethylated state to determine which strand to repair.)

Transposable Elements in Eukaryotes

The existence of transposable elements was first detected in Barbara McClintock's studies of maize in the 1940s.* These elements are fully comparable to those in bacteria and were detected by their effects on the expression of certain maize genes. Maize contains several types of transposable elements. These elements have a sequence organization similar to their bacterial counterparts. They have inverted terminal repeats, encode a transposase, and cause short direct repeats in their targets when they transpose. They are called **autonomous** because they have the ability to transpose. Other members of the same transposon family are often present in maize, but are unable to transpose on their own because internal deletions have destroyed their transposase gene. These copies are called **nonautonomous,** and are incapable of movement unless transposase is supplied by another transposable element. Nonautonomous elements retain the inverted terminal repeats at which transposase acts. The best characterized maize Tn element is a 4.5 Kb element called **Ac** (for **activator**); defective versions of Ac are called **Ds** (for **dissociation**).

As in bacteria, transposons in maize are associated with a variety of structural aberrations of chromosomes. Deletions, duplications, and inversions may occur at sites where autonomous elements are present. The transposition of maize elements is regulated in part by the development of the plant; unknown signals generated at certain periods of plant development affect the frequency of movement of these elements. It is possible for an autonomous element to become reversibly inactivated, that is, incapable of transposition. Presumably, such a change results from secondary modifications to the DNA sequence of the element. Cycles of methylation and demethylation of certain DNA sequences may be involved in this process.

---

*She was awarded the Nobel Prize in 1993.

Transposons have been detected in the genomes of other eukaryotes, including those of yeast, *Drosophila melanogaster,* and humans. They are probably present in all genomes. Researchers are exploiting them in model organisms for basic studies on the regulation of gene expression. The so-called *P* **element** in *Drosophila* represents a long (~3 Kb) sequence with inverted repeats at its ends. Modified versions that lack the transposase enzyme activity have been constructed in the laboratory. Thus, once inserted in the fly's genome, it remains in a fixed position and can be used to generate frameshift mutations, or to serve as a genetic marker. As well, a *P* element can be engineered in the laboratory to contain a foreign gene. When inserted into a bacterial plasmid and subsequently injected into a fruit fly egg, the *P* element is integrated into the embryo's chromosomes, creating a transgenic animal.

Like the transposons described for bacteria, eukaryotic transposons cause unstable mutations in genes that revert to wild type when the element is deleted. In addition, these transposons cause short direct repeats at the site of their insertion. However, some of these elements have a fundamentally different genetic organization and mechanism of transposition than the bacterial elements (such as Tn10 or Tn3) and are described next.

KEY CONCEPT

The "Changing Genome" Concept

The sequence or arrangement of nucleotides in genes is constantly evolving, either by random mutation, by occasional recombination events, or from the action of transposable elements.

Retroviruses Infect Eukaryotic Cells and Insert in Host Genome

These eukaryotic transposons are similar in organization to the proviral form of **retroviruses,** which are eukaryotic viruses that contain an RNA genome. The most famous example is the human immunodeficiency virus (HIV), which causes AIDS (acquired immune deficiency syndrome) (see Chapter 16). Retroviruses introduce an RNA molecule into cells that can be converted into double-stranded DNA by a unique viral-encoded enzyme called **reverse transcriptase.** The double-stranded form of the virus can integrate randomly into the genome of the infected cell, where it resides as a **provirus.** After integration, the provirus is replicated as part of the host genome. The proviral form can express viral genes as mRNAs, which are translated by the host into viral proteins that make up the capsid. Capsid proteins and the mRNA itself, which constitute the viral genome, assemble into infectious viruses that are then shed from the cell. The integration of the retroviral provirus into the host genome generates short direct repeats flanking the viral sequences (like those seen in bacterial transposition). The retroviral infectious cycle, therefore, has features in common with transposons.

retroviruses

reverse transcriptase

Imagine a situation in which a genome has acquired a number of retroviral proviruses integrated at various points. The proviral genes continue to be transcribed by the host RNA polymerase. Over time, mutations may occur that inactivate or alter the proviral DNA sequence so that functional viral proteins can no longer be produced. The proviral element could then be regarded as functionally defective. However, as long as a functional reverse-transcriptase-like activity is produced by the element or its relatives, the mRNA can be converted into double-stranded DNA. It then will be able to integrate into new genomic locations by a process resembling bacterial transposition. The important features of this pathway are that the appearance of the proviral DNA form, at new genomic locations, amounts to a transposition event, and that the transposition process involves an RNA intermediate. This "relocated" gene is called a **retrotransposon.**

Retrotransposons Make Up a Large Proportion of Human Genome

Retrotransposons in the human genome are believed to represent vestiges of retroviral infections that occurred during an earlier era in human genome evolution. Presumably, as the result of mutation, most of the original retroviral DNA sequences eventually lost their ability to transpose and, hence, now remain in a relatively fixed position. A relatively small percentage of them have, however, retained the ability to relocate. As mentioned earlier in Chapter 13, such retrotransposons occupy a huge proportion (approximately 40%) of the total human genome's DNA nucleotide sequence. Such retrotransposons comprise the moderately repetitive DNA described in Chapter 3 (Figure 3-10). There is no evidence that they code for proteins, and, hence, they merely occupy—in a physical sense—a major portion of the human genome. Some examples do, however, bear a close resemblance to legitimate genes, but since they do not generate protein products, they are referred to as **pseudogenes.**

Why they have persisted during the evolution of the human genome sequence is unclear. At a first glance it might appear that retrotransposons merely represent parasitic DNA that has not yet been eliminated from the genome. Conversely, it may be the case that, during the phylogenetic history of the human species, transposons played a key role in creating new genes and/or regulatory sequences (e.g., enhancers, silencers, etc.) and have not yet been deleted. Indeed, several functional genes that have sequences resembling ancient transposons have been recognized in the human genome. These include telomerase, an RNA-containing enzyme that is responsible for the synthesis of the so-called "telomeres" (chromosome ends), and splicing genes involved in gene rearrangements of the human immune system. Yet, one additional consideration is often put forth: Perhaps retrotransposons now function to facilitate the structuring/restructuring of eukaryotic chromatin (including folding and looping), which characterizes transitions in form that occur during the DNA replication and gene expression phases of chromosomes.

Similar considerations concerning "persistence of seemingly nonfunctional DNA" also apply to introns. As Human Genome Project data is interpreted, a more complete understanding will be achieved of what, if any, functional significance retrotransposons (and introns) possess.

Plasmids

Plasmids represent the extrachromosomal genetic material (DNA) segments capable of autonomous replication in cells. They are found in most bacterial species. Most plasmids are circular, double-stranded DNA molecules that are isolated as supercoiled molecules; however, other forms, such as linear double-stranded molecules, have also been described. Plasmids are not confined to bacteria; they have been isolated from yeast, protozoa, and plants. In humans, certain viruses (including human papilloma virus and Epstein-Barr virus) can also exist as plasmids within cell nuclei. In this form, the viruses replicate along with the nuclear chromosomes.

The molecular biology of plasmids has been most intensively studied in bacteria. **Table 14-2** summarizes some of the key features of several well-studied plasmids.

Certain bacterial plasmids have the ability to transfer themselves physically from one cell to another, thereby spreading through a population. Special properties of certain plasmids even permit interaction between two bacterial cells of different species. These transfer abilities are central to the role that plasmids play in bacterial evolution. The genes carried by plasmids represent a mobile reservoir of genetic information that can endow the host bacterial cell with the ability to grow in otherwise hostile environments. An example of immediate relevance to humans is that bacterial antibiotic resistance is often conferred on bacterial cells by plasmid genes. Certain plasmids are capable of efficient transfer to other antibiotic-sensitive cells, resulting in rapid horizontal conversion of an entire bacterial population to drug resistance. Moreover, it is common for a single plasmid to confer resistance to multiple antibiotics.

Bacterial plasmids span a wide range of sizes, from those containing one or two genes (several Kb in size) to those carrying several hundred genes (more than 500 Kb). Because certain plasmids are so large (and capable of carrying so many genes), it is possible for them to greatly alter the metabolic capabilities of their host cell. In the absence of selective pressure, however, having the plasmid present is irrelevant to the host, and the plasmid is genetically dispensable. Under these circumstances, the plasmid could conceivably represent a genetic burden to the cell and would, therefore, likely be lost from populations. The evolutionary mission of the plasmid, however, is to propagate itself and insure its own stable transmission in the population. Despite being nothing more than naked DNA molecules incapable of survival outside a cell, bacterial plasmids have evolved quite sophisticated mechanisms to ensure their own survival.

An important property of bacterial plasmids is that they exist in characteristic copy numbers per cell. A central problem faced by all plasmids is the need to coordinate their replication with that of the host chromosome, so that they are not diluted by cell reproduction. Bacteria are capable of reproducing at widely different rates. In the lower intestine of a mammal, bacteria might divide once per 24 hours; in a rich medium in a culture flask, they can divide every 20 minutes. Bacterial plasmids must be able to sense the division rates of their hosts and alter their own replication rates accordingly. Plasmid replication must also be regulated so that it does not compromise host chromosome replication and the life of the cell.

TABLE 14-2	FEATURES OF SELECTED PLASMIDS OF *E. coli*			
Plasmid	**Size (Kb)**	**Copy Number**	**Conjugative**	**Other Phenotype**
ColE1	6.6	10–20	No	Colicin production and immunity
F	95	1–2	Yes	*E. coli* sex factor
R100	89	1–2	Yes	Antibiotic-resistance genes
P1	90	1–2	No	Plasmid form is prophage; produces viral particles
R6K	40	10–20	Yes	Antibiotic-resistance genes

Since plasmids are stably inherited DNA molecules, they have served as valuable research models to investigate the molecular mechanisms that control DNA replication. Because of their small size and genetic dispensability, they are easy to manipulate with the techniques of recombinant DNA technology. As will be described later, plasmids have played a key role in the development and use of recombinant DNA technology (see Chapter 15).

In this chapter, we will briefly describe some of the central features of plasmids, concentrating on bacterial plasmids because these are most widely understood at present. We will focus on the replication properties of plasmids and the important property of plasmid transfer.

Plasmid-Borne Genes

Bacterial plasmids play an important role in human disease. This was illustrated dramatically by the discovery during the 1960s of bacterial strains that had acquired simultaneous resistance to four antibiotics in heavy use in Japanese hospitals. The multiple drug-resistance traits were transferable to sensitive strains and were responsible for an epidemic of bacterial dysentery. The antibiotic-resistant bacteria harbored a large plasmid, called an **R factor** (for resistance), carrying genes that encoded proteins that destroyed or inactivated the antibiotics. Other R plasmids have been discovered that render bacteria resistant to heavy metals found in the environment, such as mercury and lead. The presence of these plasmids in bacteria allows for bacterial growth in heavily contaminated environments.

R factor

A bacterial cell's ability to resist the action of an antibiotic or to neutralize toxic environmental substances usually depends upon the production of large amounts of a few key enzymes. These enzymes act by attacking the harmful substance. By virtue of encoding these key enzymes on plasmids, which function essentially as mini-chromosomes, the bacterium is well suited to produce large amounts of the required enzyme quickly because the plasmid is often present in multiple copies (Table 14-2). With multiple copies of the relevant genes, transcription quickly generates sufficient enzyme to combat the antibiotic or chemical challenge the bacterium faces.

E. coli is a normal resident of the human large intestine. Certain plasmids endow *E. coli* with the ability to grow in new environments, and thereby convert an otherwise harmless bacterial resident into a pathogen. Such plasmids encode proteins that can alter the bacterial cell surface, permitting adhesion of the bacteria to the lining of the small intestine. A second gene sometimes present on a plasmid encodes a secreted protein toxin (called an **enterotoxin**), which can damage intestinal cells and is directly responsible for the symptoms of dysentery (**Table 14-3**).

Other plasmids, called **Col plasmids,** provide a selective growth advantage to cells carrying them because the plasmid encodes a secreted antibacterial protein (called a **colicin**) that kills cells lacking the plasmid. Immunity to the effects of the colicin is conferred on the host cell by a special protein also encoded by the plasmid. The mechanisms by which colicin proteins act to kill cells are varied. Certain types (such as colicin E1) produce holes in the membranes of sensitive cells, permitting ions to flow out. Others, such as colicin E3, enter cells and specifically cleave ribosomal RNA, blocking protein synthesis.

TABLE 14-3	PLASMID-ENCODED PROTEINS
Protein	**Function**
Colicin	Secreted protein, kills bacteria lacking plasmid that encodes colicin-immunity protein.
Enterotoxin	Secreted protein, alters ion balance of eukaryotic cells. Responsible for water loss from cells.

Plasmids can provide advantages to their hosts by other means, such as carrying **DNA restriction/modification genes.** These genes encode sequence-specific DNA binding proteins that recognize and degrade DNA, which is foreign to the cell (restriction), while physically altering the host DNA so that it is not degraded (modification). Several hundred distinct restriction/modification systems have been described in bacteria. Their natural function is to protect the host bacterium from infection by an unwanted virus. In certain instances, the viral DNA is chemically modified so that its ability to replicate is impaired. In other cases, the DNA of the infecting virus is destroyed, thus short-circuiting the phage replication cycle (see Figure 6-6 in Chapter 6). Destruction of the virus DNA is achieved by specific nuclease enzymes. Those enzymes recognize specific nucleotide sequences and cleave DNA at those sites. These systems are the source of restriction endonucleases, which are essential molecular tools of recombinant DNA technology, described in the next chapter. Many of these restriction/modification systems are plasmid borne, while others are carried in the host genome.

Plasmid Transfer

There has been speculation that bacterial viruses evolved from plasmids. Small fragments of the bacterial cell's chromosome may have existed in the form of primitive plasmids, which eventually evolved into complete bacteriophage. The ability to transfer genes from one bacterial cell to another, a property possessed both by plasmids and viruses, no doubt contributes to the evolution of bacterial species.

From this evolutionary perspective, perhaps no plasmid-borne functions are more important than those permitting plasmid transfer from cell to cell, thereby mediating the flow of genes through a bacterial population. This process of plasmid transfer is called **conjugation. Figure 14-6** illustrates the conjugation process. Conjugative plasmids carry a large block of transfer genes responsible for the specialized cell structures and enzymes required to physically move the plasmid genome from a donor cell to a recipient. Such plasmids are said to be **self-transmissible.** The prototype conjugative plasmid of *E. coli* is the **F** plasmid (or **sex factor**), a large circular plasmid of approximately 95 Kb present in 1–2 copies per cell (**Figure 14-7**). Cells carrying the F plasmid are called **males,** and are capable of mating with cells lacking the F plasmid **(females).** Cells lacking the F plasmid are often represented by the notation of F^-, while F^+ is used to represent cells having an F plasmid.

conjugation

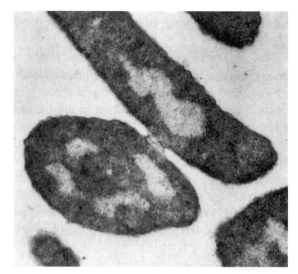

Figure 14-6 Electron micrograph of two *E. coli* cells during conjugation. The small cell is an F⁻ cell; the larger cell contains F'*lac*. (With permission, from L. Carol. *J. Mol. Biol.*, 16: 269. Copyright 1966: Academic Press.)

Conjugation is a complex process involving the gene products of a large gene complex within the plasmid, called the *tra* (for transfer) operon, consisting of more than 20 genes (Figure 14-7). (As was explained in more detail in Chapter 11, an operon is a set of contiguous genes coordinately expressed through a large poly-cistronic mRNA.) When male cells are mixed with female cells, mating pairs form, joined together by a special structure called a conjugation bridge. Certain F genes encode proteins for a specialized male surface structure called a **pilus.** The F pilus (essentially a tubular extension of the bacterial cell surface) probably interacts with specific receptors on the female outer membrane surface to establish a conjugation bridge. Following cell-cell contact, one strand of the supercoiled F DNA is nicked at a specific sequence called ***ori*T.** This is accomplished by a special DNA endo-nuclease encoded by the F plasmid. Other proteins (encoded by *tra*) then bind to the exposed 5'-terminus and guide that DNA strand into the female cell. As the DNA strand enters the female cell, it is replicated coordinately with the other plas-

pilus

Figure 14-7 A map of the sex plasmid F. Points are given in kilobases. The single capital letters refer to the midpoints of the locations of the corresponding *tra* genes. The insertion sequences, γδ (Tn1000), IS2, and IS3, are shown in blue. The blue arrow (*ori*T) shows the location of the origin for transfer replication, as well as the direction of transfer. *ori*V shows the location of the origin for vegetative (nontransfer) replication.

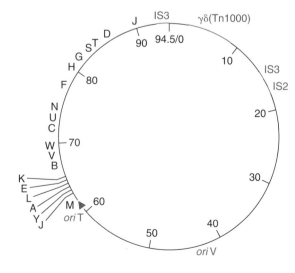

mid parent strand remaining in the donor. Following DNA replication in both donor and recipient cells to convert the single strand into double-stranded DNA, the new F plasmids are circularized. The closed circular DNA is then converted to a supercoiled form by the enzyme DNA gyrase. The process is shown diagrammatically in **Figure 14-8**. The result of conjugation is the semiconservative replication of F in both cells. Initially a female, the daughter cell is converted into a male cell by this process. It then is capable of transferring the F plasmid. Thus, the F plasmid can be thought of as an infectious agent, capable of rapid spread through a population of female cells.

The ability to mediate conjugation is not confined to the F plasmid. Other plasmids, particularly the R plasmids, also carry the genetic information required for conjugation. When these plasmids are transferred, the female recipient becomes resistant to antibiotics, which can be inactivated by the products of certain R plasmid genes. As with F plasmid transfer, the recipient female becomes a donor male capable of subsequent transfer. Plasmids such as ColE1 (mentioned later in this chapter) are not self-transmissible by themselves. They have, however, evolved the ability to take advantage of the genetic free-ride offered by a co-residing conjugative plasmid. When the conjugative plasmid is transferred, smaller plasmids such as ColE1 can also move into the recipient, and are said to be **mobilizable.**

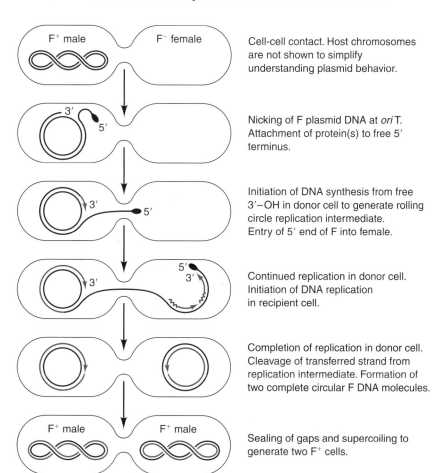

F⁺ male   F⁻ female

Cell-cell contact. Host chromosomes are not shown to simplify understanding plasmid behavior.

3′
5′

Nicking of F plasmid DNA at *ori* T. Attachment of protein(s) to free 5′ terminus.

3′
5′

Initiation of DNA synthesis from free 3′−OH in donor cell to generate rolling circle replication intermediate. Entry of 5′ end of F into female.

5′
3′

Continued replication in donor cell. Initiation of DNA replication in recipient cell.

Completion of replication in donor cell. Cleavage of transferred strand from replication intermediate. Formation of two complete circular F DNA molecules.

F⁺ male   F⁺ male

Sealing of gaps and supercoiling to generate two F⁺ cells.

Figure 14-8 A model for transfer of F plasmid DNA from an F⁺ male (donor) cell to an F⁻ female (recipient) cell using a rolling circle replication mechanism of the type described in Chapter 7. Black lines indicate parental DNA strands. Smooth blue lines indicate newly synthesized DNA. Sawtooth blue lines indicate RNA primers that are subsequently removed. The bacterial chromosomal DNA is omitted for clarity.

F Plasmids Can Transfer Chromosomal Genes by Recombining with the Host Genome

Conjugation as previously described is genetically manifested only through the conversion of the recipient cell into a donor. However, conjugation was initially discovered in a successful search for genetic processes in which chromosomal genes were transferred from one cell to another. We must, thus, account for the ability of the F plasmid to transfer genes other than its own. This ability was understood when it became clear that the F plasmid could exist in another form in certain cells *in addition to its autonomous existence as a separate plasmid. The F plasmid was found to be able to undergo recombination with the host chromosome,* and since both molecules are circular, a single recombination event results in integration of the plasmid into the chromosome. The conversion of free F into integrated F is a rare event, occurring in about one in 10^5 cells. These rare cells are called **Hfr** cells (for *H*igh *f*requency of *r*ecombination). The F plasmid can integrate into the chromosome at many different locations, so a variety of bacterial strains can be generated, each containing a single integrated F plasmid at different spots in the host genome. The molecular basis for the entry of the F plasmid into the chromosome at specific locations (IS sequences) was described earlier in this chapter.

The F plasmid genes responsible for making the cell male are expressed in Hfr cells. The Hfr cell, thus, can mate and transfer DNA (**Figure 14-9**). When the integrated F DNA in an Hfr strain is nicked by the endonuclease and transfer of the DNA strand into the recipient cell begins, DNA from the host chromosome can also be transferred by virtue of its physical continuity with the F DNA. This transfer of host DNA into the recipient potentially can have profound genetic consequences. DNA from the donor is transferred at a rate of about 50 Kb/min. About 100 minutes are required for transfer of the entire bacterial chromosome. In most cases, however, the mating pairs are not stable long enough for the entire chromosome (see Figure 5-2 in Chapter 5) to be transferred, and the cells break apart, leaving the recipient with a partial copy of the F plasmid and a segment of the donor genome. Most often, therefore, only a subset of chromosomal genes is transferred. Because the origin of transfer is within the F genome, one part of the F plasmid is the first to enter the recipient, but the rest of the F plasmid will be last to enter. Therefore, a complete F copy is not present in the recipient unless the entire chromosome is transferred. As a result, the recipients of most Hfr transfers are not capable of being donors. As with autonomous F plasmid transfer, replication (in both cells) accompanies single-strand transfer.

Left to its own devices, the piece of chromosomal DNA transferred from an Hfr donor has little chance of survival and propagation in the recipient. However, because of sequence homology with genes of the recipient chromosome, it can undergo recombination mediated by the homologous recombination system of *E. coli*. The recombination event can be detected by appropriate selection or screening. For example, in a mating between certain Hfr *leu*$^+$ cells and a F^-*leu*$^-$ recipient, F^-*leu*$^+$ cells can be detected. In these cells, the recombination event results in the replacement of the *leu*$^-$ gene with the incoming *leu*$^+$ gene.

Hfr

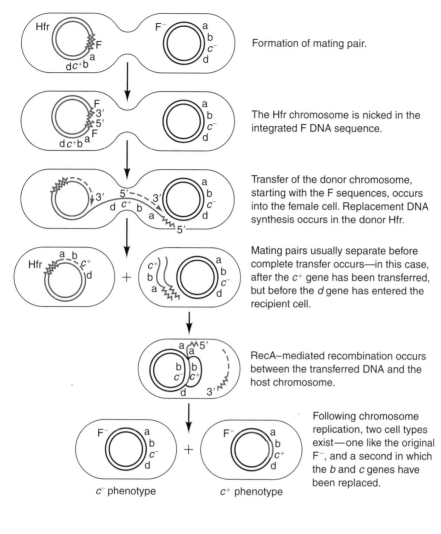

Figure 14-9 A diagram showing conjugation between an Hfr male (donor) bacterium and an F⁻ female bacterium. The solid blue and black circles represent the chromosomal DNA of the Hfr and F⁻ cells, respectively. Dashed blue lines indicate DNA synthesis. Sawtooth lines represent the integrated F plasmid. Lower case italic letters represent genes on the chromosome. The F⁻ cell carries a mutation in the *c* gene (c^-), while the Hfr strain has a wild-type copy of this gene. Following Hfr transfer and replication, a population of F⁻ cells has been converted to a c⁺ phenotype. These genes and the F plasmid DNA are not drawn to scale relative to the chromosomal circles. Supercoiling is omitted for simplicity.

Formation of mating pair.

The Hfr chromosome is nicked in the integrated F DNA sequence.

Transfer of the donor chromosome, starting with the F sequences, occurs into the female cell. Replacement DNA synthesis occurs in the donor Hfr.

Mating pairs usually separate before complete transfer occurs—in this case, after the c^+ gene has been transferred, but before the *d* gene has entered the recipient cell.

RecA–mediated recombination occurs between the transferred DNA and the host chromosome.

Following chromosome replication, two cell types exist—one like the original F⁻, and a second in which the *b* and *c* genes have been replaced.

c^- phenotype c^+ phenotype

Hfr Transfer Can Be Used to Investigate Gene Location

The analysis of many Hfr strains revealed an important feature concerning transfer. Different Hfr strains reproducibly transfer different bacterial genes at high frequency. For example, the hypothetical strain Hfr1 might transfer genes *A B C* at a high frequency and genes *X Y Z* at a low frequency, while another strain, Hfr2, may transfer genes *X Y Z* at a high frequency and genes *A B C* at a low frequency. This behavior is due to different insertion points for the F plasmid in the Hfr donor's chromosome. Those bacterial genes oriented to one side of the ***ori*** T transfer point are the first bacterial genes to enter the recipient, and will be transferred at a higher frequency than genes further away from the *ori* T.

Since genes are transferred in a linear order to the recipient in an Hfr transfer, the time of transfer of a particular gene reflects its relative position on the bacterial chromosome to the origin of transfer. Hfr strains, thus, are important tools for determining a map of the bacterial chromosome. That is, the order of the linear array of genes on the chromosome can be established. By determining the time of transfer of a particular gene relative to the transfer of other genes, the relative physical locations of the genes can be determined.

F Plasmids Insert Into Specific Locations

What determines the F plasmid insertion point? The F plasmid carries several special DNA sequences also present in several copies in the bacterial chromosome. These short **insertion sequences** (abbreviated as **IS** sequences) were described earlier in this chapter. The transfer properties of most Hfr strains can be understood as a consequence of a recombination event between an IS sequence on the F plasmid and the homologous sequence on the chromosome.

F′

We have seen that the F plasmid can display two distinct lifestyles in a cell: either as a free plasmid or integrated into the chromosome to form an Hfr. A third type of F plasmid has also been observed. This form called **F′** (F prime), results from an imprecise excision of an F from the chromosome and reestablishment in a plasmid form. The excision can also remove bacterial genes adjacent to F, which are then carried in a plasmid form. The excision itself is the result of a recombination event between sequences within F and sequences outside. Many different F′ plasmids have been described; some carry very large blocks of the bacterial chromosome. Like F plasmids, F′ plasmids retain the ability to transfer themselves to female recipients, as long as they have not lost any of the genes necessary for transfer.

F′ plasmid transfer results in the efficient introduction of the associated chromosomal genes into the recipient. However, the genetic consequences are different from Hfr transfer. After transfer of the F′, the F′-associated bacterial genes can be stably maintained by virtue of the ability of the F plasmid replication origin to function. Thus, recipient F′ cells carry two copies of certain genes: one of the F′ plasmid and one on the bacterial chromosome, and are genetically described as **merodiploid** (i.e., partially diploid), in contrast to the normal haploid state of the bacterial genome. Since F′ plasmids are much smaller than the bacterial chromosome, their complete transfer usually occurs before mating pairs spontaneously break apart. Consequently, the female recipients are converted into F′ males. F′ plasmids are usually designated by the chromosomal genes they carry—for example, F′*lac pro* carries the genes for lactose utilization and proline biosynthesis.

Plasmid DNA Replication

As mentioned earlier, a fundamental property of bacterial plasmids is that they exist at characteristic copy numbers in cells. Plasmid copy numbers are usually expressed as a ratio of the number of plasmid molecules to the number of chromosomal copies per cell. Naturally occurring plasmids exist at a wide range of copy numbers. Certain plasmids, such as the sex factor F or the prophage of the *E. coli* cirus P1, exist at copy numbers of 1–2, while those of the ColE1 family exist at copy numbers of 20–40.

Plasmids are generally dependent on host enzymatic machinery for their own replication. The ColE1 plasmid, for example, carries no genetic information specifying replication-related enzymes—it relies entirely on host components. Other plasmids may encode certain enzymes involved in particular steps of replication, but in no known case is a plasmid completely independent of host replication enzymes. *E. coli* contains three different enzymes that catalyze DNA synthesis—DNA polymerases I, II, and III. Most *E. coli* plasmids, as well as the *E. coli* chromosome itself, use DNA polymerase III as the central replication enzyme. (An important exception is the ColE1 plasmid family, in which DNA polymerase I plays the key replication role.)

Plasmid Copy Number

Plasmid copy number is regulated by controlling the frequency of initiation of DNA replication. Initiation of plasmid replication occurs at an *ori*. Replication forks starting at *ori* move either unidirectionally or bidirectionally around the circular plasmid.

Most plasmid replication forks are bidirectional. The ColE1 plasmid, however, replicates using a unidirectional fork. Two questions of importance are:

1. What special sequence features of *ori* enable it to serve as a replication starting point?
2. How is the frequency of initiation events controlled so as to maintain a specific copy number?

Many plasmid *ori* sequences are organized similarly to *ori*C, the bacterial replication origin. A key functional feature of origins in general is that the DNA double helix must open in that region, permitting access to the bases by DNA polymerases and other enzymatic replication machinery. The site-specific opening of the helix is accomplished in several steps.

The first step is the recognition of the *ori* region by a specific DNA-binding protein. For *ori*C of the bacterial host, the key protein involved in replication initiation is encoded by the *dna*A gene (see Figure 7-7 in Chapter 7). Many plasmids encode an analogous protein, which is required specifically for their replication. Called a Rep protein, it plays a key role in the initiation of plasmid replication. If the *rep* gene is missing or mutated, plasmid replication cannot occur from the plasmid origin. Moreover, plasmid copy number is related to the intracellular level of the Rep protein. The function of the Rep protein is to recognize specific short DNA sequences at the plasmid *ori*.

At the *ori,* other host-encoded proteins interact with the bound Rep protein to assemble a DNA replication enzymatic machine called a **replisome.** Once bound to the *ori,* the replisome causes local unwinding of the helix, followed by RNA primer formation and the replication of leading and lagging DNA strands. The specificity for assembling the replisome at *ori* comes not from sequence-specific binding of the replisome itself, but rather from an initial DNA sequence recognition by the Rep protein, which directs subsequent replisome assembly at the *ori.*

Based upon that knowledge, molecular biologists used intuitive reasoning (see Chapter 1) to speculate that intracellular levels of the Rep protein would be carefully regulated. In fact, plasmids have evolved highly sophisticated molecular mechanisms to control the production of this critical protein. These control mechanisms sense the concentration of plasmid DNA in the cell, and regulate the production of the Rep protein accordingly. When a plasmid is introduced into a cell, initiation of plasmid replication occurs quickly to establish the plasmid at its steady-state copy number. This requires relatively high-level expression of the plasmid *rep* gene.

After the plasmid is established, Rep protein levels must be controlled to prevent runaway plasmid replication, which can be lethal to the cell. The amount of active Rep protein is controlled at the level of translation, rather than by posttranslational alteration, for example. The expression of *rep* genes of some plasmids, such as the *rep*A gene of the R factor, R1, is controlled both at the transcriptional and posttranscriptional levels by separate plasmid-encoded gene products that repress *rep* gene transcription.

Bacteriophage

Bacteriophage, or phage, have played an important role in the development of molecular biology. Because of their relative simplicity, the availability of enormous numbers of mutants, and ease of propagation, phage have been extraordinarily useful in the study of basic processes such as replication, transcription, translation, and regulation. In this portion of the chapter, we shall examine some basic features of phage biology and provide a few examples of reproductive strategies by looking at particular stages of the life cycles of certain phages.

A bacteriophage is a bacterial parasite. By itself, it can persist, but a phage can reproduce only within a bacterial cell. Although phage possess genes encoding a variety of proteins, all known phage use the ribosomes, protein-synthesizing factors, amino acids, and energy-generating systems of the host cell.

Phage must perform four minimal functions for continued survival. These are:

1. Protect their nucleic acid from environmental chemicals that could alter the molecule (for example, break the molecule or cause a mutation).
2. Deliver their nucleic acid to the inside of a bacterium.
3. Convert an infected bacterium to a phage-producing system that yields a large number of progeny phage.
4. Release phage progeny from an infected bacterium.

These problems are solved in a variety of ways by different phage species.

Phage particles also differ in their physical structures from species to species, and often certain features of their life cycles are correlated with their structure. The three basic structures are shown in **Figure 14-10**.

The most common type of nucleic acid in phage is double-stranded linear DNA; however, double-stranded circular DNA, single-stranded linear and circular DNA, and single- and double-stranded linear RNA are also found (**Table 14-4**). The molecular weight of phage nucleic acid has a one hundredfold range of variation among species, in sharp contrast with bacteria, for which the weight rarely varies more than about 10%. Phage with larger nucleic acid molecules possess more genes, and, consequently, can have more complex life cycles and are less dependent on bacterial enzymes for their reproduction. An unusual feature of the DNA of some phage is the presence of bases other than the standard A, T, G, C, as shown in Table 14-4. For

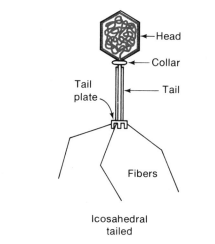

Icosahedral tailless

Icosahedral tailed

Filamentous

Figure 14-10 Two-dimensional view of three basic phage structures. The nucleic acid is shown in blue. See Figure 1-2 (Chapter 1) for an electron micrograph of a tailed phage.

example, T4 contains glucosylated 5-hydroxy-methylcytosine instead of cytosine, and SPO1 has hydroxymethyluracil instead of thymine. Phage nucleic acid is always isolated from the extracellular environment by an enclosing protein shell called either the **coat** or the **capsid,** and is thereby protected from harmful substances.

capsid

TABLE 14-4	PROPERTIES OF THE NUCLEIC ACID OF SEVERAL PHAGE TYPES				
Phage	**Host**	**DNA or RNA**	**Form**	**Weight, $\times 10^6$**	**Unusual Bases**
φX174	E	DNA	ss, circ	1.8	None
M13, fd, f1	E	DNA	ss, circ	2.1	None
PM2	PB	DNA	ds, circ	9	None
186	E	DNA	ds, lin	18	None
B3	PA	DNA	ds, lin	20	None
Mu	E	DNA	ds, lin	25	None
T7	E	DNA	ds, lin	26	None
λ	E	DNA	ds, lin	31	None
N4	E	DNA	ds, lin	40	None
P1	E	DNA	ds, lin	59	None
T5	E	DNA	ds, lin	75	None
SPO1	B	DNA	ds, lin	100	HMU for thymine
T2, T4, T6	E	DNA	ds, lin	108	Glucosylated HMC for cytosine
PBS1	B	DNA	ds, lin	200	Uracil for thymine
MS2, Qβ, f2	E	RNA	ss, lin	1.0	None
φ6	PP	RNA	ds, lin	2.3, 3.1, 5.0*	None

Note: Abbreviations used in this table are: E, *E. coli;* B, *Bacillus subtilis;* PA, *Pseudomonas aeruginosa;* PB, *Ps. aeruginosa BAL-31;* PP, *Ps. phaseolica;* ss, single-stranded; ds, double-stranded; circ, circular; lin, linear; HMU, hydroxymethyluracil; HMC, hydroxymethylcytosine.

*φ6 contains three molecules.

Stages in the Lytic Life Cycle of a Typical Phage

lytic
lysogenic

virulent

temperate

Phage life cycles fit into two distinct categories: the **lytic** and the **lysogenic** cycles. A phage in the lytic cycle converts an infected cell to a phage factory, and many phage progeny are produced. Figure 6-6 (Chapter 6) illustrated the basic features of the lytic cycle. A phage capable of only lytic growth is called **virulent.** The lysogenic cycle, which has been observed only with phage containing double-stranded DNA, is one in which no progeny particles are produced; the phage DNA is usually inserted into the chromosome of the host bacterium. A phage capable of such a life cycle is called **temperate.*** In this section, the lytic cycle is outlined.

There are many variations in the details of the life cycles of different phage. There is, however, what may be called a basic lytic cycle, which is the following (**Figure 14-11**):

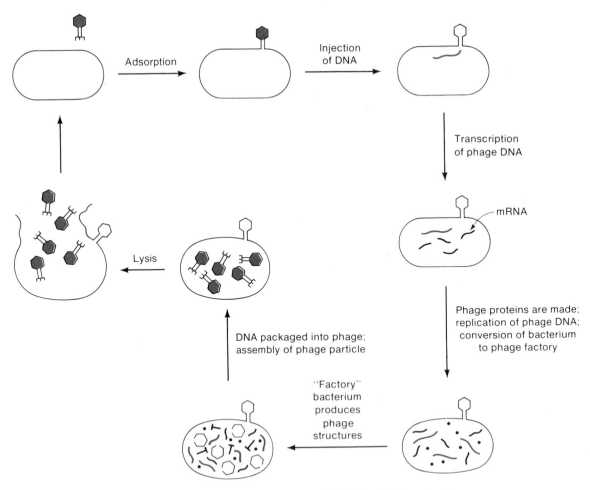

Figure 14-11 Schematic diagram of the life cycle of a typical phage.

*Most temperate phage also undergo lytic growth under certain circumstances.

1. *Adsorption of the phage to specific receptors on the bacterial surface.* These receptors serve the bacteria for purposes other than phage adsorption. (For example, the receptor for phage T6 is involved in transporting nucleosides into the bacterial cell.)

2. *Passage of the DNA from the phage through the bacterial cell wall.* Some types of tailed phage use an injection sequence, shown schematically in **Figure 14-12**.

3. *Conversion of the infected bacterium to a phage-producing cell.* Following infection by most (but not all) phages, a bacterium loses the ability either to replicate or to transcribe its DNA; sometimes it loses both. This shutdown of host DNA or RNA synthesis is accomplished in many different ways, depending on the phage species.

4. *Production of phage nucleic acid and proteins.* The phage redirects the biosynthetic pathways of the infected cell to make copies of phage nucleic acid and synthesize phage proteins. This reprogramming is accomplished by transcribing and then translating phage genes. Transcription of phage genes is almost always initiated by the bacterial RNA polymerase, but after the first set of transcriptions, the bacterial polymerase is usually modified to recognize phage-specific promotors, or a phage-specific RNA polymerase is synthesized. Transcription is regulated, and phage proteins are synthesized sequentially in time as they are needed. Distinct classes of mRNA are made at various times after infection; the major division is between early mRNA and late mRNA. Early mRNA usually encodes the enzymes required for takeover of the bacterium for phage DNA replication, and for the synthesis of late mRNA. Late mRNA encodes the components of the phage particle, proteins needed for packaging of nucleic acid in the particles, and enzymes required to break open the bacterium. The more complex phage usually synthesize several classes of early mRNA, which are made in a particular time sequence. We will study an example of this in phage T7.

5. *Assembly of phage particles.* This process is often called **morphogenesis.** Two types of proteins are needed for the assembly process: **structural proteins,** which are present (possibly in modified form) in the phage particle, and **catalytic proteins,** which participate in the assembly process, but do not become part of the phage particle. A subset of the latter class consists of the maturation proteins, which convert intracellular phage DNA to a form appropriate

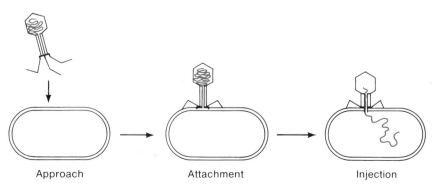

Approach   Attachment   Injection

Figure 14-12 Injection sequence of a tailed phage. In the injection stage, the tail sheath contracts and drives a core protein tube through the cell wall like a hypodermic syringe.

lysozyme

lysis

for packaging in the phage particle. Usually 50–1,000 particles are produced; the number depends on the phage species.

6. *Release of newly synthesized phage.* With most phages, a phage protein called a **lysozyme** or an **endolysin** is synthesized late in the cycle of infection. This protein causes disruption of the cell wall so that phage are released to the surrounding medium. This disruptive process is called **lysis.**

The events just described occur in an orderly sequence, as exemplified by the life cycle of *E. coli* phage T4 listed in the following (times in minutes at 37°C):

$t = 0$	Phage adsorbs to bacterial cell wall. Injection of phage DNA probably occurs within seconds of adsorption.
$t < 1$	Synthesis of first phage mRNA begins.
$t = 2$–3	Synthesis of host DNA, RNA, and protein is turned off.
$t = 5$	Degradation of bacterial DNA begins. Phage DNA synthesis is initiated.
$t = 9$	Synthesis of late mRNA begins.
$t = 12$	Completed heads and tails appear.
$t = 13$	First complete phage particle appears.
$t = 25$	Lysis of bacteria; release of about 300 progeny phage.

The total duration of the cycle is typical, most phage having a life cycle of 20–60 minutes, which is comparable to the generation times of most bacteria. The life cycles of animal viruses are considerably longer—24–48 hours—but this is again comparable to the life cycle of an animal cell.

Specific Phage

On the scale of biological complexity, phage are relatively simple forms of life. However, phage are sufficiently complex that there is not a single phage type for which all of the molecular details of its life cycle are completely understood. The best known are a small number of phage species that grow in *E. coli, Bacillus subtilis,* and *Salmonella typhimurium.* Special features of each phage have resulted in particular phage being more suited to the study of certain processes. For example, regulation of transcription is best understood in phage λ and T7, the study of morphogenesis has been more successful with phage T4 and λ than with other phage, and phage P1 and P22 gave the first clues to understanding transduction. T5 has yielded the best information about DNA injection. T4 has been most profitable in the study of DNA synthesis, and the study of T7 has shown clearly how a phage takes over a bacterium.

In the following sections, several well-understood features of a few phage will be described.

E. coli Phage T4

E. coli phage T4 is large by bacteriophage standards (Table 14-4). Its genome contains 166,000 base pairs of DNA, sufficient coding capacity for about 200 average-sized genes. The large genome of T4 allows it to be less dependent on host

KEY CONCEPT

The "Simple Genome" Concept

Viral genomes are extraordinarily simple, in part because viruses borrow functions from their host cells.

functions and have a more complex life cycle than is possible for smaller phages. In this section, we will consider examples of increased independence and complexity in T4 DNA metabolism.

Unlike other phage, T4 DNA replication proceeds normally in thymine-requiring hosts in the absence of added thymine because T4 encodes its own thymidylate synthetase. In fact, T4 specifies a number of enzymes involved in nucleotide metabolism (dihydrofolate reductase, nucleotide reductase, thioredoxin reductase). These enzymes do more than duplicate *E. coli* functions; they are responsible for the high rate of replication in infected cells.

A second complexity of T4 DNA metabolism is that T4 DNA contains a modified form of cytosine, called **5-hydroxymethylcytosine (HMC),** which base pairs with guanosine (**Figure 14-13**). This base is further modified by glucosylation—that is, a glucose molecule is coupled to the −OH group of HMC. As we shall see, HMC plays an important role in the T4 life cycle.

E. coli does not encode enzymes for HMC synthesis, but rather, its biosynthesis is directed by two T4 enzymes: dCMP hydroxymethylase and dNMP kinase

$$\text{dCMP} \xrightarrow{\text{dCMP hydroxymethylase}} \text{dHMP}$$

$$\text{dHMP} \xrightarrow{\text{dHMP kinase}} \text{dHDP}$$

The *E. coli* enzyme, nucleoside diphosphate kinase, forms all nucleoside triphosphates in *E. coli,* then converts dHDP to dHTP, the immediate precursor of HMC in DNA.

T4 also encodes nucleases that attack cytosine-containing DNA and, thus, degrade the host chromosome to mononucleotides. If T4 DNA did not contain HMC, it too would be degraded by these nucleases.

Most dCMP (whether produced *de novo* or from degradation of the host DNA) is converted to dHMP. However, some will be converted to dCTP, which can be incorporated into replicating T4 DNA. Of course, this DNA would be susceptible to attack by the phage-encoded nucleases. To prevent dCTP incorporation, another phage-specified enzyme, dCTPase, hydrolyzes both dCTP and dCDP to dCMP, the substrate for dCMP hydroxymethylase.

Although HMC-containing DNA is resistant to many DNases (including most restriction enzymes (see Chapter 15)), *E. coli* does possess an endonuclease that attacks certain nucleotide sequences containing HMC. To avoid this damage, HMC is glucosylated. This is accomplished by two phage enzymes, α-glycosyl transferase (α*gt*) and β-glycosyl transferase (β*gt*), each of which transfers a glucose from uridine diphosphoglucose (UDPG) to HMC that is already in DNA. Glucosylation is an example of postreplicative modification.

The *E. coli* endonuclease is inactive against glucosylated DNA. A simple genetic experiment shows that the protection of HMC DNA from endonucleolytic attack is the only essential function of glucosylation. A T4 α*gt*⁻ β*gt*⁻ double mutant cannot carry out glucosylation, so its newly synthesized DNA is attacked by the *E. coli* HMC endonuclease. However, T4 α*gt*⁻ β*gt*⁻ double mutants do grow on an *E. coli* mutant (*rglB*⁻), which lacks this nuclease, even though the phage DNA is nonglucosylated.

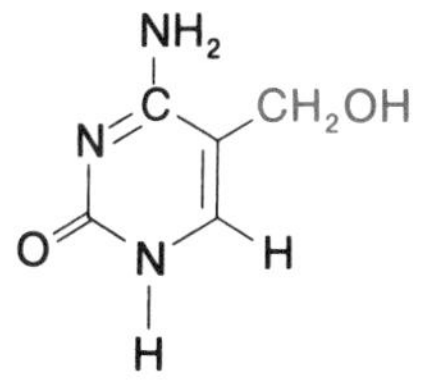

5-hydroxymethyl-cytosine (HMC)

5-Hydroxymethylcytosine (HMC)

Figure 14-13 Hydroxymethylcytosine. If the CH₂OH (blue) in HMC were replaced by hydrogen, the molecule would be cytosine.

An important property of T4 DNA molecules packaged in phage particles is that they are terminally redundant (**Figure 14-14**)—that is, a sequence of bases (about 1% of the total) is repeated at both ends of the molecule. Terminal redundancy is a feature of the DNA of many phage species and can be generated in several ways, as will be described below.

Another feature of T4 DNA is that, although each phage particle contains one DNA molecule, the molecules differ from phage to phage, even if the population was produced by replication of a single phage particle and its progeny. A sample of T4 DNA molecules is **circularly permuted,** the meaning of which is shown in **Figure 14-15**. The figure indicates schematically that the termini of the DNA molecules in the population can be found at many different bases within an overall sequence—in reality, probably at any point in the base sequence. Note that circular permutation is a property of a phage population, whereas terminal redundancy is a property of an individual phage DNA molecule. Both terminal redundancy and circular permutation are consequences of the mechanism by which DNA is packaged into the phage head.

The mechanism of packaging T4 DNA in a phage head is not yet fully understood. The basic problem is how the long strand of DNA is tightly folded so that it will fit in the capsid. It is thought that packaging begins by the attachment of one end of a DNA molecule to a protein contained in a precursor of the phage head. Then, either a condensation protein or a small, very basic molecule (such as a polyamine) induces folding (**Figure 14-16**). The best-understood feature of the process is that the molecule is **cut from** long concatemers. The cuts are not made in unique base sequences in the DNA, because if they were, T4 DNA could not be circularly permuted. Instead, the cuts are made at positions that are determined by the amount of DNA that can fit in a head. Presumably, a free end of the DNA molecule enters the head, and this continues until there is no more room, then the concatemer is cut. This is known as the **headful mechanism,** and it explains how both terminal redundancy and circular permutation arise (**Figure 14-17**). The essential point is that the DNA content of a T4 particle is greater than the length of DNA required to encode the T4 proteins. Thus, when cutting a headful from a concatemeric molecule, the sequence at the end of the packaged DNA is a duplicate of the sequence packaged first—that is, the packaged DNA is terminally redundant. The first segment of the second DNA molecule that is packaged is not the same as the first segment of the first phage. Furthermore, since the second phage must also be terminally redundant, a third page-DNA molecule must begin with still another segment. Thus, the collection of DNA molecules in the phage produced by a single infected bacterium is a circularly permuted set.

E. Coli Phage T7

Phage T7 is a midsized phage (Table 14-4). Its DNA has a molecular weight of about 26×10^6 and has no unusual bases. It has a terminal redundancy of 160 base

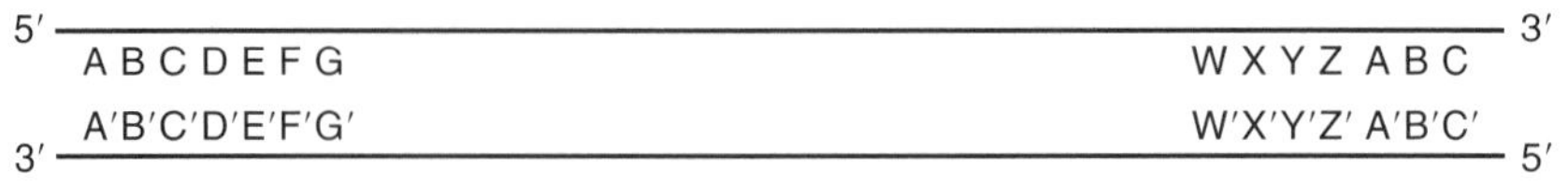

Figure 14-14 A terminally redundant molecule.

```
5' ─────────────────────────────────────────────  3'
   A B C D E F G                    W X Y Z  A B C
   A'B'C'D'E'F'G'                   W'X'Y'Z' A'B'C'
3' ─────────────────────────────────────────────  5'
```

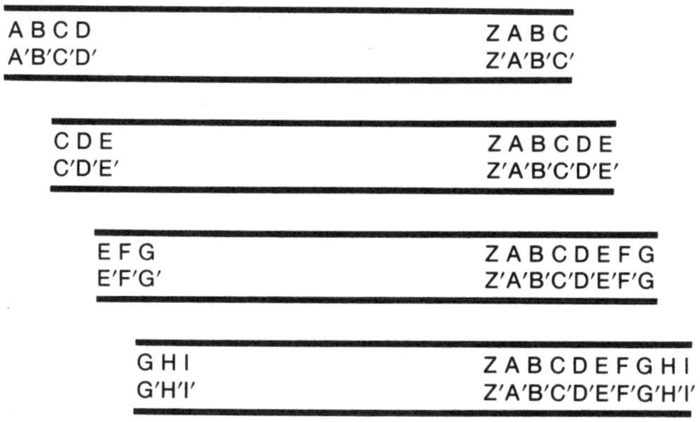

```
A B C D                    Z A B C
A'B'C'D'                   Z'A'B'C'

  C D E                    Z A B C D E
  C'D'E'                   Z'A'B'C'D'E'

    E F G                  Z A B C D E F G
    E'F'G'                 Z'A'B'C'D'E'F'G

      G H I                Z A B C D E F G H I
      G'H'I'               Z'A'B'C'D'E'F'G'H'I'
```

Figure 14-15 A circularly permuted collection of terminally redundant DNA molecules.

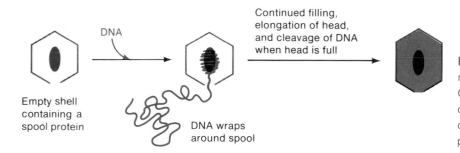

Figure 14-16 Proposed model for filling a T4 head. Cleavage and rearrangement of head proteins is known to occur at several stages of the process.

Empty shell containing a spool protein

DNA wraps around spool

Continued filling, elongation of head, and cleavage of DNA when head is full

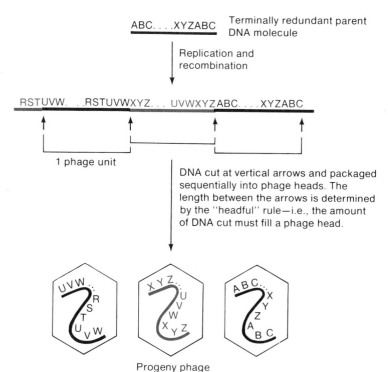

ABC. . . .XYZABC   Terminally redundant parent DNA molecule

Replication and recombination

RSTUVW. . .RSTUVWXYZ. . . UVWXYZABC. . . .XYZABC

1 phage unit

DNA cut at vertical arrows and packaged sequentially into phage heads. The length between the arrows is determined by the "headful" rule—i.e., the amount of DNA cut must fill a phage head.

Progeny phage

Figure 14-17 Origin of circularly permuted T4 DNA molecules. Alternate units are shown in different colors for clarity only.

pairs, but it is not circularly permuted. The DNA molecule is encased in a head that is attached to a very short tail.

T7 has fewer genes than T4 (about 50), which simplifies analysis of its life cycle. Furthermore, most of the gene products have been identified by gel electrophoresis, and many have been sequenced by chemical means. The complete sequence of 39,930 base pairs of T7 DNA has been determined; this information allows direct inspection to locate all of the promoters, termination sites, regulatory sites, spacers, leaders, initiation codons, termination codons, and genes. A great deal is known about the molecular biology of T7; however, we shall only outline how it regulates transcription.

T7 transcription occurs in three temporal stages (**Figure 14-18**), with Class I transcripts occurring first. Class I transcripts originate at three promoters (*pI*) located very near the left end of the T7 DNA molecule. These transcripts end at the termination site, T_E, and are made by *E. coli* RNA polymerase. As expected, the *pI* sequences strongly resemble those of *E. coli* promoters.

Two class I gene products are essential for the transition to class II and III transcription. One of those is a protein kinase enzyme that phosphorylates and thereby inactivates *E. coli* RNA polymerase (though inactivation is not complete). The second enzyme is T7 RNA polymerase. Only T7 RNA polymerase recognizes the promoters (*pII* and *pIII*) for class II and III transcripts. The sequences of these T7 specific promoters are very different from those recognized by *E. coli* RNA polymerase.

T7 RNA polymerase does not recognize the terminator, T_E. Hence, class II transcripts proceed past the site to the terminator T_ϕ (Figure 14-18). Class II transcripts encode replication proteins, the major structural protein of the phage head, and a second inhibitor of *E. coli* RNA polymerase. This latter protein completely inhibits the remaining activity of *E. coli* RNA polymerase.

Termination at T_ϕ is only 90% efficient, meaning that 10% of the transcripts proceed to the right end of the DNA molecule. The extended region encodes the tail, maturation, and lysis proteins, which are required in small quantities compared to the head proteins.

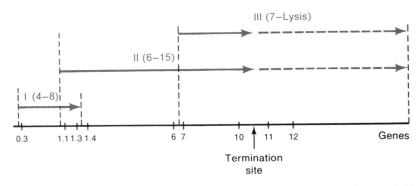

Figure 14-18 A simplified transcription map of T7. The DNA and gene numbers are in black; the three classes of transcripts are in blue. The blue numbers indicate the time interval (in minutes at 37°C) during which the transcript is made. The termination site indicated by the black arrow is about 90% efficient; thus, 10% of the class II and III molecules are extended further, as indicated by the dashed lines.

Since DNA replication proteins are enzymes, they are needed in only very small amounts compared to the head and tail proteins. Late in infection, when there is no need to continue making these enzymes, class II transcription shuts off and class III transcription begins.

Class III transcripts are initiated at seven *pIII* promoters to the left of T_ϕ, and encode both structural and lysis proteins. These transcripts, like those of class II, terminate at T_ϕ with 90% efficiency. However, there are three additional *pIII* promoters to the right of gene 12 (Figure 14-18), so that only gene 11 and 12 expression is entirely dependent on read through.

The mechanism of temporal regulation of class II and III transcription is unusual and rather unsophisticated in comparison to phage λ. Shutoff of class II transcription depends on promoter strength and the synthesis of a transcription inhibitor. Class II promoters are weak compared to those of class III. One of the proteins encoded in class II transcripts reduces all transcription. When this protein is made, transcription is initiated only at strong (*pIII*) promoters.

The mechanism that delays class III expression is quite uncommon. Although most phage species inject their DNA within a few seconds of adsorption, injection of T7 DNA takes about 10 minutes. Thus, there is a period when *pII* promoters are the only promoters available to T7 RNA polymerase. Entry of *pIII* promoters into the cell occurs around 7 minutes after adsorption.

Let us now summarize how the three classes of transcripts are regulated. At first, *E. coli* RNA polymerase makes class I transcripts. An inhibitor, translated from class I transcripts, prevents further synthesis of these transcripts. T7 RNA polymerase, translated from class II transcripts, prevents their further synthesis. Slow injection delays entry of *pIII* promoters into the cell; however, once *pIII* promoters are available, class III transcripts are made. The net result is early synthesis of the DNA replication proteins and late synthesis of the structural proteins of the phage particle.

E. Coli Phage M13

E. coli phage M13 (**Figure 14-19**) differs in many ways from other bacteriophage discussed in this chapter. It is one of the smallest known phage (Table 14-4). Its genome is composed of only 10 genes (**Figure 14-20**), all of which are essential. The phage particles contain a circular DNA molecule (the viral or (+) strand) that is only 6,407 nucleotides long. M13 is a rod-shaped, filamentous phage that can package circular DNA molecules that are several times longer than its viral strand. The phage adsorbs only to male strains of *E. coli*, but does not lyse infected cells. The infected cells continue to grow and divide, producing 100–200 phage per cell, per generation.

In this section, we will consider M13 DNA replication and its role in recombinant DNA technology.

The first step of (−) strand replication is the synthesis of an RNA primer by *E. coli* RNA polymerase at the (−) origin on a single-stranded (+ strand) viral DNA molecule. The primer is 30 nucleotides long, and is extended from its 3′-OH end by *E. coli* DNA polymerase III around the circle until it reaches the 5′-P end of the primer. *E. coli* DNA polymerase I then degrades the primer and replaces it with deoxynucleotides. The two ends of the (−) strand are immediately

Figure 14-19 Electron micrograph of phage M13.

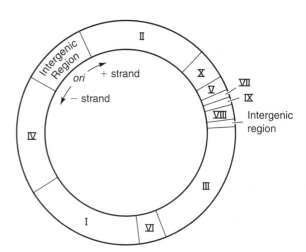

Figure 14-20 Genetic map of M13. Genes are designated by roman numerals: II-replication protein, V-single-stranded DNA binding protein, VIII-major coat protein, IX-minor coat protein. The origins of replication are in the intergenic space between genes II and IV.

sealed by *E. coli* DNA ligase. The resulting supercoiled, double-stranded molecule is known as **RFI (replication form I).**

M13 gene II protein is required for viral "daughter" strand synthesis. These strands are made from RFI templates by a variation of the rolling-circle model (**Figure 14-21**). Synthesis is initiated when M13 gene II protein nicks the (+) strand at the (+) origin. Using the *E. coli* replication proteins (Rep. ssb, and DNA polymerase III), the (+) strand is extended from its 3'-OH end at the nick. As the old (+) strand is displaced by the elongating new (+) strand, it is coated with ssb. When the new (+) strand reaches the origin, it is nicked again by gene II protein, freeing the old (+) strand whose 3'-OH and 5'-P ends are joined by gene product II.

Note that replicating RFs do not directly produce double-stranded daughter molecules. The products of one round of replication are a single-stranded (+) circle, and a double-stranded RF molecule.

At early times during infection, the daughter (+) strands serve as templates for the synthesis of (−) strands, accounting for RF → RF replication. During this period, RFs are also transcribed, producing a coat protein (gene VII product) that associates with the cytoplasmic membrane, and gene V product, a single-stranded DNA binding protein, which is necessary for the final stage of replication.

The switch from RF → RF to RF → ss occurs when sufficient gene V protein has accumulated to bind the new viral strands and prevent them from serving as templates for (−) strand synthesis. The gene V protein and viral DNA complex then move to the cell membrane, where the gene V protein is displaced by coat protein. The completed virus particles are excreted through the cell membrane without lysing the cell.

If the M13 DNA molecule is enlarged by inserting foreign DNA at a site near the minus origin, no phage functions are altered. The enlarged molecule replicates and is packaged into a larger coat (unlike T4, the size of the DNA molecule determines the size of the phage particle). M13 is, therefore, a very useful tool in DNA cloning. Using recombinant DNA techniques, foreign DNA is replicated as a part of the M13 molecule. The DNA packaged in phage particles is an abundant source of single-stranded DNA for sequencing, site-directed mutagenesis (see Figure 15-9 in Chapter 15), and other procedures.

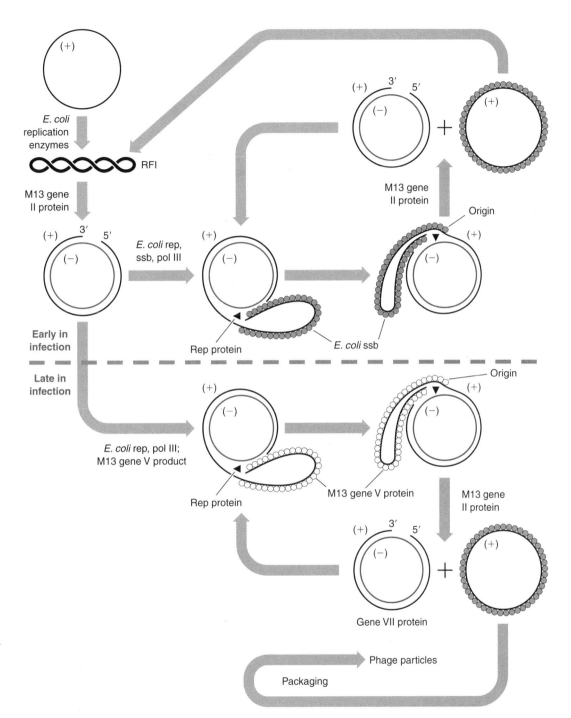

Figure 14-21 A diagram of the replication of phage M13. *E. coli* replication enzymes convert the single-stranded phage DNA into the RFI form, which is then nicked by the M13 gene II protein. Rolling circle replication (see Figure 7-9 in Chapter 7) ensues to generate a daughter strand (blue) and a displaced (+) single strand that is coated with *E. coli* ssb (filled circles) early in infection, or M13 gene V protein (open circles) late in infection. When the entire (+) strand is displaced, it is cleaved from the daughter (+) strand and circularized by the joining activity of the M13 gene II protein. The cycle is ready to begin anew. Note that the (−) strand is never cleaved. Viral (+) strands coated with M13 gene V protein are packaged into phage particles after the gene V protein is displaced by gene VII protein.

E. Coli Phage λ: The Lytic Cycle

λ is a temperate *E. coli* phage. It can engage in both a lytic cycle, in which progeny phage are produced, and a lysogenic cycle, in which a phage-DNA molecule is inserted into the bacterial chromosome. Special features of the elegant regulatory systems of λ determine which pathway is used in any given infection.

The DNA of λ has two components: a double-stranded segment containing 48,489 nucleotide pairs of known sequence, and a 12-nucleotide single-strand extension at the 5′-P terminus of each strand. The two terminal single-strand extensions have complementary base sequences, and are called **cohesive ends.** Immediately following injection of the DNA, the cohesive ends base pair, yielding a circle (**Figure 14-22**). The adjacent 5′-P and 3′-OH groups are quickly sealed by *E. coli* DNA ligase, and the circle is then supercoiled by host topoisomerase enzymes (see Chapter 7). No phage gene products are involved in this process.

The arrangement of the genes of phage λ is shown in **Figure 14-23**. As is typical for phage, the genes are clustered according to function. For example, the head, tail, replication, and recombination genes form four distinct groups. Clustering and transcription of the members of a cluster from the same DNA strand allow λ to regulate transcription with a small number of regulatory sites (Figure 14-23).

In this section, we will consider the regulatory events of the lytic cycle, and replication and maturation of λ DNA. In the next section, we will examine the regulatory events of the lysogenic pathway and the factors that determine which pathway will be followed.

Transcription of λ occurs in three temporal stages: very early, early, and late. Very early and early transcription precede the decision to follow the lytic or lysogenic pathway. Hence, they are the same in both types of infections. Very early mRNA encodes regulatory proteins, early transcripts encode replication and re-

cohesive ends

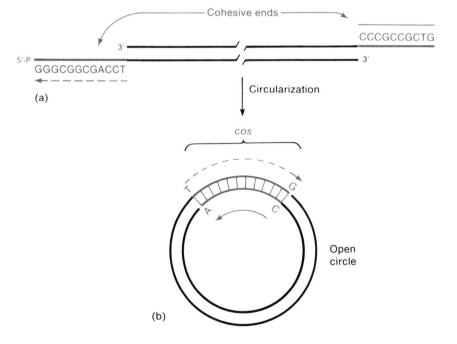

Figure 14-22 (a) A diagram of a λ DNA molecule showing the complementary single-stranded ends (cohesive ends). Note that 10 of the 12 bases are G or C. (b) Circularization by means of base pairing between the cohesive ends. The double-stranded region that is formed is designated *cos.*

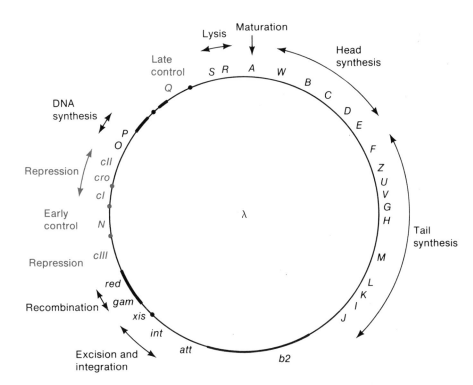

Figure 14-23 Genetic map of phage λ. Regulatory genes and functions are indicated in blue. All genes are not shown. Major regulatory sites are indicated by black solid circles. Regions nonessential for both the lytic and lysogenic cycles are denoted by a heavy line.

combination enzymes, and late lytic mRNA encodes the structural proteins of the phage particle and lysis proteins.

In marked contrast to T7, λ uses *E. coli* RNA polymerase to transcribe all of its mRNA. Qualitative control of λ transcription is accomplished by regulating both initiation and termination. Initiation is regulated by λ-encoded activators (*cI, cII*) and repressors (*cI, cro*—described in Figure 5-13 in Chapter 5) that control promoter accessibility. Termination is controlled by phage-specific **antiterminators** (*N* and *Q*), proteins that allow RNA polymerase to transcribe beyond specific termination sites.

antiterminators

Temporal (i.e., over time) regulation is achieved by sequential expression of regulatory proteins. Consider the following example. A DNA segment contains the sequence

$$p \quad A \quad B \quad t1 \quad C \quad t2$$

in which *p* is a promoter, *A B C* are genes, and *t1* and *t2* are terminators. Initially, RNA polymerase will attach to *p* and transcribe *A* and *B*. Assume that *B* encodes an antiterminator so that RNA polymerase ignores *t1*. In this case, before the *C* product can be made, the *B* product must be made; thus, the *C* product will be synthesized at a later time than the *A* product. For further details please refer to a microbial genetics textbook.

As was mentioned previously, λ DNA circularizes shortly after infection and the cohesive ends base pair to form the double-stranded *cos* region. Clearly, at some stage of the life cycle, the single-stranded termini must be regenerated, since the DNA is linear within a phage head. This is accomplished by a cutting system called **terminase** (or **Ter**), which cuts within the *cos* region of progeny DNA. However, the cuts are not made in progeny circles; instead, DNA replication and cutting are coordinated in a way that is found with many phage species. The earliest stage of replication of λ DNA is the bidirectional θ mode described earlier in Chapter 7 (see Figure 7-4). Progeny circles do not continue to replicate in this way, but switch to the rolling circle mode, yielding a long linear branch (**Figure 14-24**). Progeny λ DNA molecules are cut from the branch of the rolling circles by terminase. The circular portion of the rolling circle is not cut (this would prevent further replication) because terminase requires two *cos* sites, or one *cos* site plus one free, single-stranded terminus. Thus, excision of λ units occurs by sequential cleavage from the free end of the branch, as shown in **Figure 14-25**. This mechanism for determining the amount of DNA to be packaged is fundamentally different from the T4 headful mechanism. For λ, like M13, the length of the phage chromosome determines how much DNA is packaged. Thus, like M13 phage, λ can be used as a vector for genetic engineering. That is, because it can accommodate extra, inserted DNA (see Table 15-2 in the next chapter).

E. Coli Phage λ: The Lysogenic Life Cycle

There are two types of lysogenic cycles. Phage λ is the prototype for the more common one, the essential features of which are shown in **Figure 14-26**.

1. A DNA molecule is injected into a bacterium.
2. After a brief period of transcription, needed to synthesize an integration enzyme and repressor of phage transcription, transcription is turned off.
3. A phage DNA molecule is integrated into the DNA of the bacterium forming a **prophage.**
4. The bacterium continues to grow and divide and the prophage is replicated as part of the bacterial chromosome.

prophage

The less common type, for which *E. coli* phage P1 is the prototype, differs from the preceding one in that the phage DNA molecule becomes a plasmid (an independently replicating circular DNA molecule) rather than a segment of the host chromosome. In this section, we will consider only the λ lysogenic cycle.

Two important properties of lysogens are the following:

1. A lysogen is *immune* to infection by a phage of the type that lysogenized the cell. Although the DNA of the superinfecting phage is injected and circularized, it is neither replicated nor transcribed.
2. Even after many cell generations, a lysogen can initiate a lytic cycle; in this process (called **induction**), the prophage is excised as a circular DNA molecule (**Figure 14-27**) and initiates lytic growth.

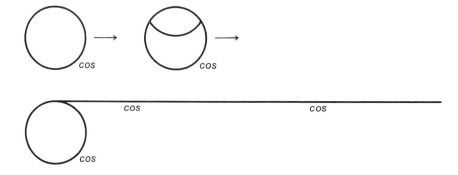

Figure 14-24 The replication sequence of phage λ.

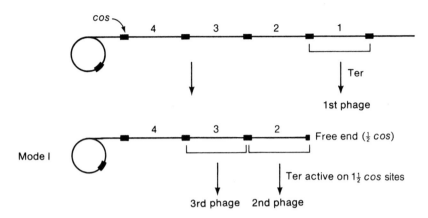

Figure 14-25 The mode of cutting λ units from a branch of a rolling circle. The units are cut sequentially from the free end by terminase.

What determines whether a λ infection will be lytic or lysogenic? The principle determinant seems to be the level of the so-called *cII* protein. It is a transcription activator that binds to two promoter regions of genes that are required for lysogeny. High levels of *cII* favor lysogeny, while low levels favor the lytic pathway. The instability of the *cII* protein is due to degradation by proteases. Under favorable growth conditions, proteases are plentiful, the *cII* protein is degraded, and, hence, the lysogenic cycle is not activated. Under poor growth conditions, proteases are less abundant and there is less degradation of *cII,* so those promoters are activated and lysogeny is favored.

Transducing Phage

Some phage species, known as **transducing phage,** are able to package bacterial DNA as well as phage DNA. Particles containing bacterial DNA, called **transducing particles,** are not always formed and usually constitute only a small fraction of a phage population. Such particles can arise by two mechanisms, aberrant excision of a prophage and packaging of bacterial DNA fragments. We begin with a discussion of the first mechanism.

One mechanism by which transducing particles form is shown in **Figure 14-28**, which depicts the formation of the galactose- and biotin-transducing forms of phage λ—namely, λ*gal* and λ*bio.*

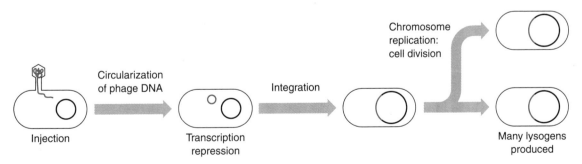

Figure 14-26 The general mode of lysogenization by insertion of phage DNA into a bacterial chromosome.

Figure 14-27 An outline of the events in prophage induction. The prophage DNA is in blue. The bacterial DNA is omitted from the third panel for clarity.

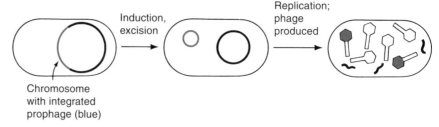

Chromosome with integrated prophage (blue)

When λ prophage is induced, an orderly sequence of events ensues in which the prophage DNA is precisely excised from the host DNA (Figure 14-28). In the case of phage λ, this is accomplished by the combined efforts of the *int* and *xis* genes acting on the left and right prophage attachment sites. At a very low frequency—namely, in about one cell per 10^6–10^7 cells—an excision error of two incorrect cuts is made—one within the prophage, and the other cut in the bacterial DNA. The pair of abnormal cuts will not always yield a length of DNA that can fit in a λ phage head—it may be too large or too small. However, if the spacing between the cuts produces a molecule between 79% and 106% of the length of a normal λ phage-DNA molecule, packaging can occur. Since the prophage is located between the *E. coli gal* and *bio* genes, and because the cut in the host DNA can be either to the right or the left of the prophage, transducing particles can arise carrying the *bio* genes (cut to the right) or the *gal* genes (cut to the left). Formation of the λ*gal-* and λ*bio*-transducing particles entails loss of λ genes. The λ*gal* particle lacks the tail genes, which are located at the right end of the prophage; the λ*bio* particle lacks the *int, xis, . . .* genes from the left end of the prophage. The number of missing phage genes, of course, depends on the position of the cuts that generated the particle and, thus, correlates with the amount of bacterial DNA in the particle. The missing phage genes come from the prophage ends, but because of the permutation of the gene order in the prophage and the phage particle, the deleted phage genes are always from the central region of the phage DNA, as shown in Figure 14-28.

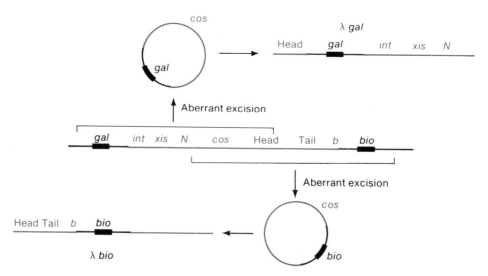

Figure 14-28 Aberrant excision leading to production of λ*gal* and λ*bio* phages.

Since these transducing particles lack phage genes, they might be expected to be nonviable. This is, indeed, true of the λ*gal* type, which lacks genes essential for synthesis of the phage tail and, in the case of the larger deletion-substitutions, also the genes for the phage head. These particles are, thus, incapable of producing progeny; they are defective and this is denoted by the symbol d—a *gal*-transducing particle is written λd*gal* (and sometimes λd*g*). The λd*gal* particle contains the cohesive ends, and all of the information for DNA replication and transcription; it, therefore, goes through a normal life cycle, including bacterial lysis. In fact, if the head genes are not deleted, the concatemeric branch produced by rolling circle replication is cleaved by the Ter system. However, tails are not added to the filled head, no viable particles are produced in the lysate, and plaques are not formed on a host lawn. Note that there is no discrepancy between the formation of a λd*gal*-transducing particle and its ability to reproduce itself. A λd*gal* particle fails to reproduce only because it lacks the tail genes, but such a particle arises from a normal prophage having a full set of genes.

The situation is quite different with λ*bio* particles, for these usually lack only nonessential genes—*int, xis,* and so on. These genes are needed for the lysogenic cycle, but not for the lytic cycle, so these particles are able to replicate and to form plaques. To denote this, the letter p, for plaque forming, is added; that is, the particle is called λp*bio.*

Transducing particles are quite useful in genetic experiments. For example, they can be used to transfer a bacterial gene from one cell to another. They have also been used as a means of isolating particular segments of a bacterial chromosome.

KEY CONCEPT

The "Levels of Sophistication" Concept

Prokaryotes such as viruses and bacteria, under heavy natural selection pressure, have evolved more stream-lined mechanisms for regulating gene expression than higher organisms (e.g., humans).

Transposons, Plasmids, and Bacteriophage in Genetic Engineering

Each of those forms of DNA has found widespread application in genetic engineering manipulations. Plasmids are used most routinely as cloning vectors for small fragments of genes. Bacteriophages can accommodate larger DNA segments, and, thus, are commonly used to make "libraries" of genomic DNA fragments. In addition, M13 phage can be employed to generate "mutations" in isolated segments of genes. Those applications are described in more detail in the next chapter. Transposons, as mentioned earlier in this chapter, provide opportunities in some model organisms (especially *Drosophila*) for introducing foreign genes into laboratory strains. **Figure 14-29** summarizes those applications. Thus, in addition to being significant components of the biological kingdom in their own right, and, therefore, worthy of study, these genetic elements have been exploited by researchers for laboratory manipulations.

A Future Practical Application?

The use of plasmids in genetic engineering presents a wide variety of opportunities for converting basic molecular biology knowledge into practical applications. Chapter 16 will review many of those applications.

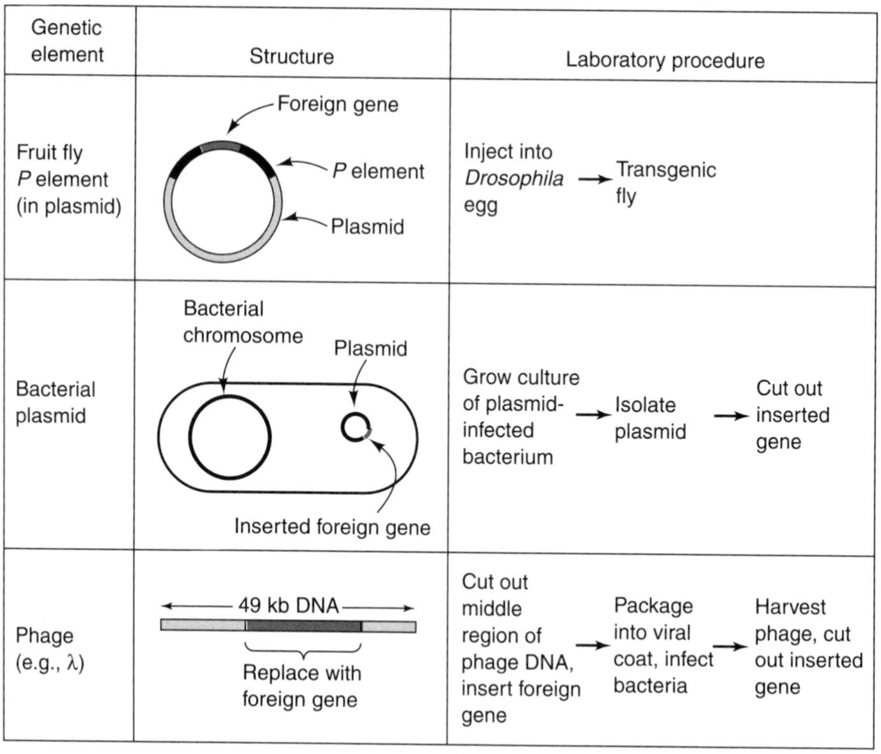

Genetic element	Structure	Laboratory procedure
Fruit fly *P* element (in plasmid)	Foreign gene / *P* element / Plasmid	Inject into *Drosophila* egg → Transgenic fly
Bacterial plasmid	Bacterial chromosome, Plasmid, Inserted foreign gene	Grow culture of plasmid-infected bacterium → Isolate plasmid → Cut out inserted gene
Phage (e.g., λ)	← 49 kb DNA → / Replace with foreign gene	Cut out middle region of phage DNA, insert foreign gene → Package into viral coat, infect bacteria → Harvest phage, cut out inserted gene

Figure 14-29 Use of various genetic elements for recombinant DNA procedures.

Bacteriophage, especially the filamentous types (e.g., M13), can be employed in the laboratory to express a vast collection of antibody molecules on their surfaces. This method is based upon the fact that filamentous phage can incorporate foreign proteins into their coats and still replicate normally. Therefore, a potentially powerful method for producing novel artificial antibodies for treatment of human diseases is available.

As a first step, a collection of genes that code for a vast array of antibody molecules is inserted into the viral genomes. The individual viruses, depending on the particular antibody gene they contain, replicate and incorporate that specific antibody molecule into their protein coat. Initially, a population of viruses are employed, and large numbers of antibody genes (up to 10^8 different ones) are inserted. The phage system is, therefore, referred to as a "display library." Such a library consists of a population of phage containing antibody molecules as part of the structure of their protein coat.

In a second step, the display library is treated with the disease-causing antigen (e.g., a cancerous tumor protein), and those phage containing an antibody that reacts with the antigen are recognized. The phage are then replicated and the antibody extracted from the viruses' protein coat. With this method, it is expected that in the future it will be possible to harvest antibodies that react with very rare antigen targets, and side-step conventional antibody-preparation tasks usually requiring relatively large quantities of antigen.

Hence, this artificial antibody production system that is "piggybacked" on the M13 phage replication system offers promise as a commercial method for producing highly specific medicinal agents.

SUMMARY

Transposons are mobile genetic elements. Insertion sequences or IS elements consist of a transposase gene flanked by inverted repeats, representing simple transposons. More complex transposons consist of an antibiotic resistance gene flanked by IS elements, or genes for antibiotic resistance, transposase, and resolvase flanked by inverted repeats. Transposition may be conservative via a "cut-and-paste" mechanism or replicative (with a duplication of the transposon). Both types may share an early common intermediate. Transposase recognizes and cleaves the inverted repeats and makes a staggered cut in the target DNA, which leads to short flanking direct repeats of this DNA. Resolvase carries out recombination between the two transposons of a cointegrate molecule.

Eukaryotes also contain transposons. They are similar to retroviruses. Retroviral infections during the early phylogenetic history of eukaryotes probably generated the massive amounts of retrotransposon sequences present in nuclear DNA (~40% of the human genome).

Plasmids are extrachromosomal DNA molecules containing an origin of replication and often one or more genes. R factors are plasmids that carry genes conferring resistance against antibiotics or other toxic substances. Col plasmids carry genes for the secreted antibacterial protein colicin, and an immunity protein. F plasmids carry genes for plasmid transfer or conjugation. In Hfr cells, the F plasmid can integrate into the host chromosome and cause transfer of chromosomal DNA. An F′ plasmid is an excised F plasmid containing host DNA. Plasmid replication is carried out by a replisome consisting of host replication enzymes and, often, a plasmid-encoded replication origin binding protein (Rep).

A bacteriophage, or phage, is an extracellular nucleoprotein particle that can infect and replicate within a bacterium. A typical phage consists of a protein shell in which is enclosed a molecule of nucleic acid. Phages contain either single- or double-stranded DNA or single- or double-stranded RNA. The most common phage structure is an icosahedral head, to which a tail is attached. Less common are tailles heads and filamentous structures.

Phages have two distinct life cycles—the lytic and lysogenic cycles. In the lytic cycle, progeny phage are made during the infection and released from the cell shortly after their assembly. A complete lytic cycle generally occurs in 20–60 minutes at 30°–37°C. In the lysogenic cycle, a replica of a phage DNA molecule becomes incorporated into the bacterial chromosome. The bacterium survives the infection and the phage DNA replicates as part of the bacterial genome. At a later time, if the bacterium is damaged, a lytic cycle ensues.

Once phage DNA is in a cell, a variety of strategies are used to produce progeny particles, depending on the phage species and the life cycle selected. There are three types of phage proteins made: catalytic, regulatory, and structural. The catalytic proteins (enzymes) are responsible for DNA replication, further transcription, preparation of DNA for packaging into a coat, cell lysis, and, occasionally, production of DNA precursors, and destruction of bacterial DNA. Regulatory proteins may inactivate bacterial enzymes, but are primarily responsible for causing the various stages of phage production to occur in an orderly and efficient way. Structural proteins form heads, tail, and other components of the finished phage particle. If the lytic cycle is selected, the cycle ends with the bacterium being broken up by lysing enzymes and release of the progeny phage to the surrounding medium.

DRILL QUESTIONS

1. What is meant by the terms direct repeat and inverted repeat? Use the sequence A B C D as an example.
2. What is the difference between conservative and replicative transposition? What base sequence is duplicated in both types?

3. In plasmid transfer, what are the roles of DNA synthesis in the donor and the recipient?
4. The following questions concern F factor transfer.
   (a) What genes endow F-plasmid-containing bacteria with the ability for conjugation?
   (b) What bacterial structure is induced to allow transfer?
   (c) What mode of DNA replication is used?
   (d) What early enzymatic step is needed in transfer replication, but not in normal replication?
5. The following questions are concerned with the recombination functions of the F plasmid.
   (a) What is a bacterial strain called that contains an F plasmid within the chromosome?
   (b) What feature of the F plasmid allows insertion?
   (c) When an integrated F initiates transfer, why is the entire *E. coli* chromosome not transferred?
   (d) What is an F′ plasmid?
6. What are the three morphological forms of phage particles?
7. What kind of nucleic acids are found in the following phages: T4, T7, λ, M13, MS-2? Also discuss circularity.

PROBLEMS

1. A hypothetical transposon has inserted at the target sequence 5′-TTAGCA-3′. Its left inverted repeat sequence is 5′-GCAATGGCA-3′. This now serves as the donor molecule for a transposition event to the new target sequence 5′-GATCCA-3 in a recipient molecule. Show the structure of the DNA (both strands) of the following molecules after all steps of transposition. (Assume that the inverted repeats on the transposon are perfect.)
   (a) The recipient molecule.
   (b) The donor molecule, if transposition occurred by the conservative pathway.
   (c) The donor molecule, if transposition occurred by the replicative pathway.
2. The transposition of a wild-type Tn3 and several mutant Tn3 elements is being followed by observing their transfer to a plasmid and screening for ampicillin resistance (don't worry about the details). Briefly describe the effect of transposition on the frequency, and the types of products produced by transposition for the following mutations relative to wild type. Assume that each mutation completely abolishes at least one particular function, but keep in mind that some proteins are bifunctional. A RecA⁻ bacterial strain is used to eliminate homologous recombination.
   (a) A resolvase point mutation.
   (b) A resolvase promoter mutation.
   (c) A transposase point mutation.
   (d) A transposase promoter mutation.

(e) A point mutation in the *bla* gene.

(f) A mutation in the *res* site.

3. An *F′* (ts)*lac*⁺ plasmid has a temperature-sensitive 40° mutation in its replication system. The host genome also carries a *lac*⁺ gene.

(a) What is the phenotype of an *F′* (ts)*lac*⁺/*lac*⁻ cell at 42°C?

(b) An *F′* (ts)*lac*⁺/*lac*⁻ *gal*⁺ strain is grown for many generations at 37°C and then plated at 42°C. Some *Lac*⁺ colonies form at 42°C. How have these formed?

(c) Some of the *Lac*⁺ colonies in (b) are *Gal*⁻. How have these formed?

4. Suppose that a particular bacterial mutant cannot utilize lactose as a carbon source. If a phage adsorbs to such a bacterium and the infected cell is put in a growth medium in which lactose is the sole carbon source, can progeny phage be produced?

5. What is the source of the RNA polymerase that makes the first phage mRNA in the life cycle of a phage containing double-stranded DNA?

6. What DNA intermediate seems to be obligatory in the replication of single-stranded DNA phages?

CONCEPTUAL QUESTIONS

1. How do you suppose transposons and plasmids evolved? What advantages do they confer on their host to allow their continued existence?

2. How have plasmid-borne genes for antibiotic resistance and toxins affected medical treatment and pharmaceutical development?

3. Since infection of a bacterium by a phage is usually lethal to the bacterium, why have bacteria not evolved to lose their phage receptors?

4. Suppose that a new protein X appears in infected cells. Describe various experiments that you might perform to prove that the gene coding for X is phage encoded and is not encoded in host DNA.

in this chapter you will learn

1. How restriction enzymes work and why they are essential to DNA technology.

2. About various procedures such as cloning and isolating DNA, and detecting recombinant DNA molecules.

Experimental Manipulation of Macror

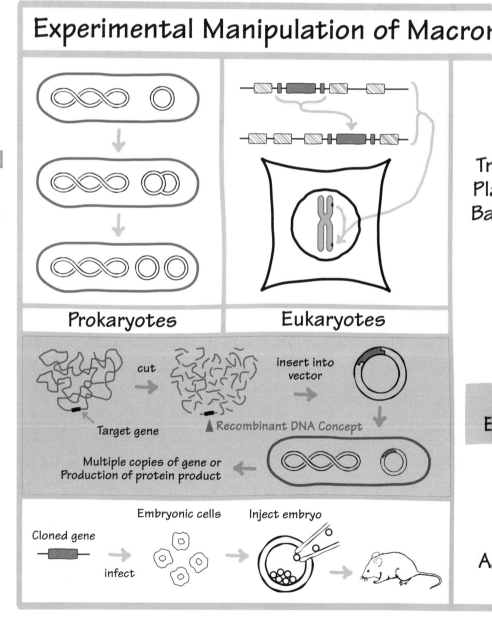

Prokaryotes

Eukaryotes

Tr
Pla
Ba

cut

insert into vector

Target gene   ▲ Recombinant DNA Concept

Multiple copies of gene or
Production of protein product

E

Cloned gene   Embryonic cells   Inject embryo

infect

A

Recombinant DNA and Genetic Engineering: Molecular Tailoring of Genes

es

recombinant DNA
genetic engineering

restriction enzymes

g

conjugation

Technical developments often lead to quantum jumps forward in science. In the past two decades, a recently developed technology, **recombinant DNA** technology, or **genetic engineering,** has revolutionized genetics in both basic research and its practical aspects. The discovery of two naturally occurring and quite remarkable classes of biological molecules made possible the development of methods to isolate and manipulate specific DNA fragments. These two classes of molecules are **plasmid DNA** and **restriction enzymes.** With recombinant DNA methodology, a DNA fragment—even an entire gene and its controlling elements— can be isolated, the fragment coupled with a plasmid or phage, and the hybrid inserted into a bacterium. In addition, since bacteria can be replicated in vast quantities, this process can be used for large-scale production of the foreign DNA insert itself; if the bacterial host is able to express or synthesize the protein product of the foreign DNA, the hybrid plasmid can direct the production of large quantities of otherwise scarce or expensive proteins, or even of proteins containing specific mutations created in the laboratory.

In this chapter, we will learn how segments of a cell's DNA can be cut, modified, joined together, and replicated. The relative ease with which those manipulations can be carried out is remarkable. Recombinant DNA technology has provided spectacular discoveries in virtually all areas of modern biology. We will review some of those recent advances in medicine, agriculture, and commercial processes in the next chapter.

Plasmids Act As Nature's Interlopers

In the early 1950s, it was discovered that bacteria transfer genes to other bacteria by a process called **conjugation** (see Figure 14-6 in Chapter 14). One bacterium attaches small projections (**pili**) on its surface to those on the surface of an adjacent bacterium. DNA from the donor bacterium is passed to the recipient through the pili. The ability to form pili and to donate genes to neighbors is a genetically controlled trait. The genes controlling this trait are not located on the bacterial chromosome, but rather on separate genetic elements called **plasmids,** described in Chapter 14. Plasmids are ideal **vectors** to accept and carry pieces of foreign DNA (including entire genes) into a bacterial host, where the foreign DNA can be replicated along with the plasmid. As vectors, plasmids are, therefore, one of the major tools of genetic engineering.

vectors

Restriction Enzymes Function As Nature's Pinking Shears

Restriction enzymes, many of which have been found in bacteria, recognize and degrade DNA from foreign organisms. Bacteria have confronted the invasion of foreign DNAs for million of years, and have evolved these enzymes as protective mechanisms that preserve their own DNA while destroying the invading DNA. Each restriction enzyme recognizes only one short sequence, usually (but not always) 4 to 6 base-pairs long. For example, EcoR I, one of the first restriction enzymes to be isolated (from *Escherichia coli*) cuts DNA only at the sequence:

$$\downarrow$$
G-A-A-T-T-C
C-T-T-A-A-G
$$\uparrow$$

In this case, the target sequence is called a palindrome, since it reads the same backwards as forwards. As will be mentioned shortly, many (but by no means all) restriction enzymes recognize palindromic target sequences. If this sequence occurs once in a circular invading plasmid, the enzyme will open the circle by cutting it once. If the sequence occurs at several sites, the DNA will be cut into several pieces. In general, the names of restriction enzymes are derived from the first letter of the genus followed by the first two letters of the species name of their bacterial source. (For example, EcoR I isolated from *Escherichia coli,* Hind III from *Hemophilus influenzae,* Taq I from *Thermus aquaticus*—several hundred of these enzymes have been isolated from hundreds of species of microorganisms.) For the molecular biologist, these enzymes allow plasmids to be opened up in vitro so that foreign DNA can be inserted. Restriction enzymes also offer a way of isolating predictable fragments of any DNA molecule. With the discovery of plasmids and restriction enzymes, it became technically simple to obtain a large quantity of exact copies of any chosen DNA fragment by **cloning.** This was done by using restriction enzymes to isolate "the fragment," inserting it into a plasmid, and inserting the hybrid plasmid into a host bacterium, which then reproduced many copies as the bacteria proliferated. This rearrangement of DNA in a living organism is genetic recombination, hence the name **recombinant DNA technology.**

cloning

The process of generating recombinant DNA can be divided into four steps: first, the production of the desired DNA fragments; second, the insertion of these fragments into a suitable vector such as a plasmid or phage; third, the introduction of that vector into an appropriate host (usually a strain of *E. coli*); fourth, the identification, selection, and characterization of recombinant clones. Some of the methods used to accomplish these steps are described in this chapter. Though not exhaustive, the description provides a basic outline of some of the major methods used in recombinant DNA technology.

Isolation and Characterization of DNA Fragments

Owing to their specificity, restriction endonucleases were the first nucleases found to cleave DNA at specific sequences, and have proven useful in genetic analyses from the oligonucleotide to the chromosomal levels. At the oligonucleotide level, they are used to provide specific fragments for cloning, sequencing, and custom-made mutations. Segments of DNA thought to contain regions that control gene expression can, if desired, be cut away selectively, and the effect of their loss on gene expression can be monitored. At the chromosomal level, they permit analyses of entire chromosomes by "walking" or "jumping" (described in Figure 15-4).

Most restriction enzymes make two single-strand breaks, one in each strand (close to, but not opposite each other), thus generating two 3'-OH and two 5'-P groups at each restriction enzyme site. A technically useful property of restriction enzymes was detected by electron microscopy: because the complementary 3'-OH to 5'-P base-pair overhanging termini (called cohesive or "sticky" ends because they can overlap one another) produced by many restriction enzymes, fragments circularized spontaneously. These circles could be relinearized by heating, but, if after circularization they were also treated with *E. coli* DNA ligase (which covalently joins 3'-OH and 5'-P groups), circularization became permanent. This observation was the first evidence for two important characteristics of restriction enzymes:

1. Restriction enzymes make breaks in symmetric sequences.
2. The breaks are usually not directly opposite one another; therefore, sticky (or cohesive, or complementary) ends are produced.

These properties are illustrated in **Figure 15-1**. The sequences recognized by many restriction enzymes are **palindromes,** as shown here:

A B C	C' B' A'		A B	B' A'		A B X	B' A'
A' B' C'	C B A	or	A' B'	B A	or	A' B' X'	B A

Capital letters represent bases (A, B, and C may be the same or different), the ' indicates the complementary base, and the vertical line is the axis of symmetry.

Examination of a very large number of restriction enzymes showed that the breaks are usually in one of two distinct arrangements: staggered, but symmetric around the line of symmetry, forming two different cohesive ends—a single-stranded extension with a 5'-P terminus and a 3'-OH extension; or both cuts at the center of symmetry, forming blunt ("flush") ends. Fragments with blunt ends cannot circularize spontaneously. These arrangements and their consequences are shown in **Table 15-1**, which lists the recognition sequences and cleavage sites for several restriction enzymes. Given its sequence specificity, each restriction enzyme generates a unique set of fragments for a particular DNA molecule.

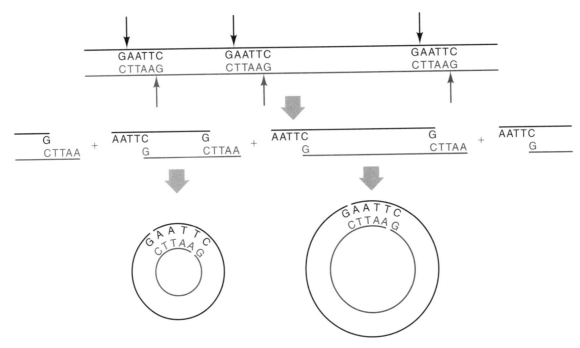

Figure 15-1 Restriction fragments that have overhanging cohesive ends following digestion with some restriction enzymes (here, at the 5′ ends) can circularize. Small arrows indicate cleavage sites of EcoR I; large arrows indicate the next step in the process.

In the years since the usefulness of restriction enzymes was recognized (by Werner Arber, Hamilton Smith, and Daniel Nathans, for which they were awarded the 1978 Nobel Prize), literally hundreds of such enzymes have been characterized. Although most attack short palindromic target sequences, as illustrated in Table 15-1, many others recognize other types of specific sequences. Included are enzymes that recognize target sequences 8 to 10 bases in length (or even longer). In addition, in contrast to the classical examples described herein, some restriction endonucleases attack nonpalindromic sequences. The availability of such a wide range of enzymes provides the research scientist with important tools for the creative manipulations described in the next chapter!

Fragments obtained from a DNA molecule from one organism will have the same cohesive ends as the fragments produced by the same enzyme acting on DNA molecules from another organism. This unsurprising fact is one of the cornerstones of recombinant DNA technology. Furthermore, since most restriction enzymes recognize a unique sequence, the number of cuts made in the DNA from an organism by a particular enzyme is limited. The DNA of a typical bacterial chromosome, which contains roughly 2×10^6 base pairs, is cut into several-hundred to several-thousand fragments, and nuclear DNA of mammals ($\sim 3 \times 10^9$ base pairs) is cut into more than a million. These numbers are large, but still relatively small compared to the total number of bases in the DNA of an organism. Of great usefulness in cloning are the less complex DNA molecules, such as bacteriophage λ or plasmids, which have only 1 to 10 sites of cutting (or even none) for partic-

TABLE 15-1	SOME RESTRICTION ENDONUCLEASES AND THEIR CLEAVAGE SITES

Name of Enzyme	Microorganism	Target Sequence and Cleavage Sites
Generate flush ends		
Bal I	*Brevibacterium albidum*	↓ TGG|CCA ACC|GGT ↑
Generate cohesive ends		
EcoR I	*Escherichia coli*	↓ GAA|TTC CCT|AAG ↑
BamH I	*Bacillus amyloliquefaciens H.*	↓ GGA|TCC CCT|AGG ↑
Hind III	*Haemophilus influenzae*	↓ AAG|CTT TTC|GAA ↑
Pac I	*Pseudomonas alcalignes*	↓ TTAA|TTAA AATT|AATT ↑

Note: The vertical dashed line indicates the axis of dyad symmetry in each sequence. Arrows indicate the sites of cutting.

ular restriction enzymes. Plasmids having a single site for each of a number of restriction enzymes are especially valuable in cloning, as we will see shortly.

If restriction sites common to two cloning partners (a plasmid and a foreign DNA) are not available, an alternative strategy can be employed. Each of the two DNAs can be cleaved with an enzyme that generates protruding termini identical to those generated by the other restriction enzyme. These **cohesive, compatible ends** can then be ligated (see Figures 15-6 and 15-7).

Map of Enzyme Cutting Sites Is a Useful Tool

Restriction maps, which indicate the cutting sites for specific restriction enzymes, provide a convenient method for comparing DNA fragments. Whether two fragments are identical, or whether they share some of the same nucleotide sequence, can be quickly established by comparing their restriction maps. Recall that the cutting sites represent specific nucleotide sequences, as indicated in Table 15-1. Thus, DNA fragments that exhibit different restriction maps can be said to contain at least some differences in their nucleotide sequences. Without investing a great amount of time and effort to perform a complete sequencing reaction of the type described in Chapter 3 (e.g., Figures 3-18 and 3-19), a quick determination of the extent of relatedness of different DNA fragments can often be made by simply comparing restriction maps.

cohesive, compatible ends

restriction maps

The procedure is illustrated in **Figure 15-2**. An actual example of gel electrophoresis of digested DNA, and the resulting restriction map, are shown in **Figure 15-3**. Restriction maps provide a key tool for the isolation of large segments of DNA (e.g., whole eukaryotic genes that contain both introns and exons), to be described next.

Large Maps Require Gene Libraries and Chromosome Walks

chromosome walks

In order to map and/or purify a large segment or an entire genome, it is essential to isolate and put in linear order experimentally manageable (i.e., smaller) mapped segments. To achieve this potentially overwhelming task, **chromosome walks** are taken on large (10–20 Kb) DNA fragments (**Figure 15-4**). That is, repeated cycles of hybridization (described later in Figure 15-8) yield—in each successive cycle—an additional overlapping fragment, akin to "walking," with each identified overlapping fragment representing one step in our "walk."

Many restriction enzymes would produce fragments too small to be useful. Large fragments are obtained by digesting genomic DNA with restriction enzymes whose cleavage sites are rarely found in DNA, or by doing partial digests so that no single DNA molecule is cut at all the potential cutting sites. The restriction fragments are then cloned into suitable hybrid plasmids (as described later in Figure 15-6), grown up (amplified) in bacteria, and recovered. The collection of these fragments, containing pieces of DNA from the entire genome of the organism, is called a **genomic library.**

genomic library

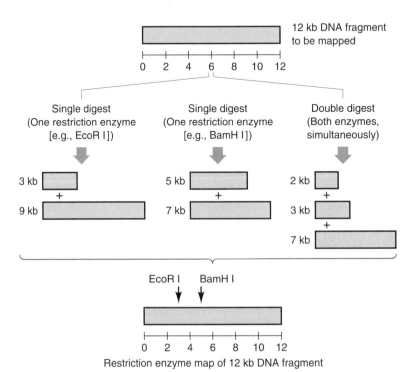

Figure 15-2 Procedure for preparing a restriction enzyme map for a DNA fragment. The cutting sites for the two enzymes (EcoR I and BamH I) used for this simplified example of a 12-Kb DNA fragment are established by comparing results of single and double digests. Sizes of the fragments produced by digestion are determined with gel electrophoresis (see Figure 14-3(a)).
*Kb = kilobase = 1,000 bases

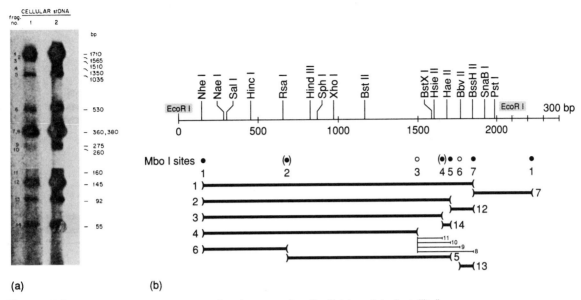

(a)   **(b)**

Figure 15-3 Restriction mapping: an example using a complex, G + C-rich, cellular "satellite" DNA of the Bermuda land crab. This type of DNA is present in very similar, but not identical, multiple copies in the genome. (a) Autoradiogram of [^{32}P]end-labeled Mbo I digest. Lane 1, normal exposure; lane 2, overexposed to visualize minor fragments. Sizes of the 14 fragments are indicated. (b) Restriction map of one repeat unit beginning at an EcoR I site and extending 2,100 base pairs to the next EcoR I site, which signals the beginning of the next repeat unit. The map indicates positions of sites of restriction enzymes that cut this repeat unit only once. However, Mbo I cuts the repeat unit at seven sites indicated by the circles numbered from 1 to 7. Numbered solid bars (1–14) represent Mbo I fragments in (a); thick bars, major fragments; thin bars, minor fragments. ● = sites present in almost all repeat units of the satellite; cuts at these sites yield the major (dark) bands seen on the gel in part (a). (●) = sites in some repeats; thus, for example, cuts between sites 1 and 2, and 1 and 4 yield bands of intermediate density of 530 and 1,510 base pairs, respectively. ○ = sites in very few repeats, yielding correspondingly very few fragments (8–11) seen clearly only in overexposed lane 2 in (a).

The fragment of interest at the beginning of a chromosome walk (containing the known sequence or gene) is cut with a number of other restriction enzymes to map it, and its ends are sequenced. A radioactive probe (commonly about 50 base-pairs long) that will hybridize to the end sequence is used to identify other fragments with overlapping sequences that are adjacent to the first probe, and these are similarly mapped (Figure 15-2) and sequenced at their ends. The orientation of the adjacent probes can be established by comparing overlapping restriction maps. The probe adjacent to the first can then be analyzed in a similar way to find the next fragment in the walk. The process is experimentally slow, but has been effectively used to align as many as 10^6 adjacent base pairs. Obviously, repeated DNAs scattered throughout the genome cannot be used for identifying overlapping sequences.

Chromosome jumping is just that: Instead of the shorter steps of ~10 or 20 Kb taken on walks along chromosomes, jumps of 100 Kb or more are taken. The very

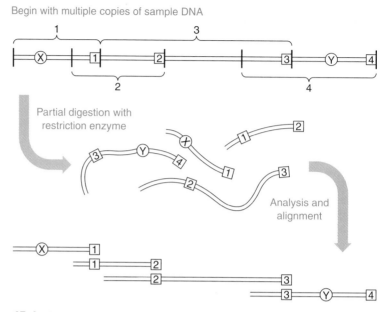

Figure 15-4 Chromosome walking and jumping. Chromosome walking using partial DNA digests with a single restriction enzyme. *Top:* Original (unknown) DNA sequence showing sequences X and Y, with dark vertical bars representing restriction sites (shown only in the top stick figure). The goal is to localize Y with respect to X. Keep in mind that many copies of the "unknown" sequence would be cut in this initial step (this insures that overlapping fragments will be generated). *Middle:* Restriction fragments. The DNA has been cut with the restriction enzyme so that each piece of the original DNA is cut at only some of the sites (i.e., a partial digest). In the example diagrammed, four overlapping fragments are generated, each ending in a sequence indicated by a box with the corresponding fragment number. *Bottom:* Analysis and alignment. The fragment containing X is found to end in sequence 1. A probe hybridizing to sequence 1 also hybridizes to fragment 2, which ends in sequence 2. Similarly, a probe hybridizing to sequence 2 also hybridizes to fragment 3, which ends in sequence 3, and a probe hybridizing to sequence 3 hybridizes to fragment 4, which contains Y. Knowing the orientation from restriction maps of each fragment (Figure 15-2), the sequences can be aligned and the relation of Y to X determined.

It will be apparent that fragments other than the four diagrammed will be generated in the restriction digestion. For clarity, these have been omitted. They will either be silent, that is, will not hybridize with any of the probes, or they will be sub- or superfragments that can be identified by analyses and alignment similar to that depicted.

Instead of cutting with a single restriction enzyme, two different restriction enzymes can be used, provided that overlapping fragments are generated.

large fragments needed for the jumps can be recovered by electrophoresis of partial digests of restriction enzymes. Other than the size of the fragments analyzed (for which cosmids, or large plasmids, are utilized—see **Table 15-2**), the technique is basically the same as chromosome walking. The ends of the fragments are restriction mapped or sequenced, and matching overlaps on other DNA fragments are identified by computer analyses of the assembled sequences or restriction maps. If the chromosomal location of one of the fragments or a gene contained within the original fragment is known, then the chromosomal location of the adjacent DNA

TABLE 15-2	A VARIETY OF CLONING VECTORS ARE AVAILABLE
Type of Vector	**Advantage**
Plasmids	Easy to use and store. Recombinant plasmids readily selected with antibiotics.
Lambda phage	Useful for cloning large (15–20 Kb fragments).
Cosmids	Combined plasmid/phage vector permits cloning of even larger (e.g., 45 Kb) DNA fragments.
Yeast plasmids	Permit direct studies on eukaryotic gene regulation.
Plant plasmids	Bacterial (*Agrobacterium*) infection of plant transfers Ti plasmid into host plant cells.

sequence or gene can also be deduced. By using such methods, a number of genes have been localized to specific chromosomes.

PCR (Polymerase Chain Reaction) Yields Large Quantities of a Specific Sequence

In 1983, a new technique, the **polymerase chain reaction (PCR),** was devised. It has made the detection and cloning of rare DNA sequences possible. The method is based on the amplification of target DNA sequences in genomic DNA. Basically, it is a method for producing large numbers of copies of small or medium-length DNA segments belonging to a larger piece of DNA. One distinct advantage of the PCR procedure: a minute amount of starting material is sufficient for carrying out the reaction (see applications in next chapter). Two specific oligodeoxynucleotides are synthesized, one complementary to the 3′ end on one DNA strand, and one to the 3′ end on the opposite strand of the target DNA to be amplified. The oligodeoxynucleotides are hybridized to the denatured, single-stranded target DNA containing the segment of interest (**Figure 15-5**). The oligonucleotides act as primers. Their 5′ ends serve as initiation points for replication of the original strands by *Taq* polymerase (a heat stable DNA-dependent DNA polymerase isolated from a thermophilic bacterium, *Thermus aquaticus*). After replication of the original strands, the products are denatured by melting, the reactants cooled to annealing temperatures to allow the primers to reanneal to the synthesized products, as well as to the original strands, and another round of replication is initiated. This process, which takes less than 10 minutes, is repeated as many times as desired. After the first few rounds, the vast majority of the products will be the amplified fragment between the 3′ sites of the original DNA, as shown in Figure 15-5. The extensions of the original DNA beyond these sites will be diluted to a negligible proportion. The *Taq* polymerase is used because it is not inactivated by the heating steps used for strand separation after each round. PCR is currently carried out with commercially available machines programmed to perform the temperature cycling automatically.

The products of the amplification can then be used to detect or clone the gene of interest. This technique, and many elegant variations on it, have been used to identify rare genes, such as viral DNA sequences, and to recover genes in very

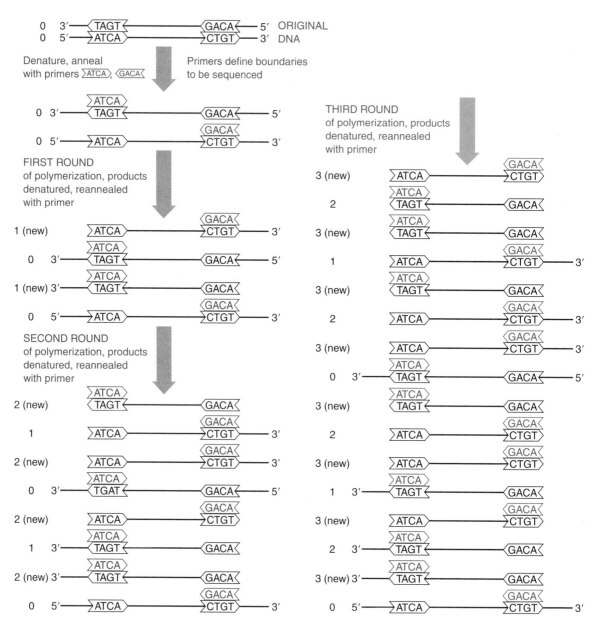

Figure 15-5 Polymerase chain reaction. Top diagram represents the original double-stranded DNA. The sequences at the ends of the segment to be amplified are shown in arrow-shaped boxes pointing in the 5′ → 3′ direction, that is, the direction of polymerization. In practice, these sequences are 15–30 base pairs in length, although only four bases are shown in the diagram. The DNA is denatured by heating, the primers are annealed to the end sequences of the target DNA, and the first round of polymerization is carried out. The numbers in the left-hand column indicate the polymerization round in which a given strand is synthesized, with 0 representing the original strands, and NEW representing newly synthesized strands. At the end of the first round, there are 4 strands, the 2 original strands *with extensions beyond both ends of the target sequence,* and the 2 copied strands *with extensions beyond only the 3′ ends of the target sequence.* (*continued on next page*)

small amounts of DNA, even from a single cell. If an altered base (mutant) is deliberately introduced into the primer, the mutated primer will be replicated and amplified after the first round. This modification of PCR, called "site directed mutagenesis" has been successfully used as one way to generate large numbers of copies of engineered genes. Another way, which employs a phage cloning vector, is described later in this chapter.

Genetic Interlopers: Vectors Function As Vehicles for Transferring Genes

Several types of vectors and some of the procedures for cloning DNA molecules are described in this section.

To be useful, a vector must have three properties:

1. It must be able to enter a bacterial host.
2. It must have an origin of DNA replication that allows it to replicate in the host.
3. Cells thus transformed must be selectable, preferably by growth of host cells on a solid medium containing the selecting agent (typically an antibiotic).

Two types of vectors are commonly used in cloning. One type is derived from bacterial viruses. Phages act like hypodermic needles; they inject DNA into the bacterial host where it replicates. The phages most commonly used for cloning are λ and M13 (see earlier descriptions in Chapter 14). Lambda phage have the unique feature of being able to accept large fragments (up to 20 Kb) of inserted foreign DNA. Vectors developed from M13 (which can accept fragments up to 10 Kb) have unique promoters that facilitate the sequencing of the inserted foreign DNA.

The second type of vector is derived from bacterial plasmids (often pBR322), which are relatively small (**Figure 15-6**). Antibiotic resistance is one of the most important features for plasmids used as vectors. Many plasmids have been genetically engineered to serve as specialized vectors. Polylinkers (also called multiple cloning sites) are synthetic DNA oligomers that provide one or more specific restriction sites, yield cohesive ends, and/or contain a bacterial origin of replication. Polylinkers may have been inserted into these engineered plasmids. Vectors that have received a eukaryotic transcriptional promoter can be expressed in both bacteria and eukaryotes. Since they can shuttle between bacterial and eukaryotic cells,

---

Figure 15-5 (*Continued*) Annealing and polymerization are repeated, yielding 8 strands at the end of the second round, of which 6 have extensions at one or both ends, and 2 encompass the target sequence only. At the end of the third round there are 16 strands, of which 8 have extensions and 8 have the target sequence only. Because the original strands are always present, they are replicated with 3' extensions on every round, and after n rounds there are $2n + 2$ strands with extensions ($2n$ is the number replicated with 3' extensions, 2 are the original with both 5' and 3' extensions). The total number of strands increases exponentially, after n rounds being 2^{n+1}. Thus, after 10 rounds there are 2,048 strands, of which only 22 have extensions, and after 20 rounds there are more than 2 million strands, of which 42 have extensions. In other words, *the contribution of strands with extensions beyond the target sequence rapidly becomes negligible as the target sequence is amplified.* Up to 60 rounds can be carried out successfully. That provides enough DNA to identify a target sequence from a single cell by gel electrophoresis.

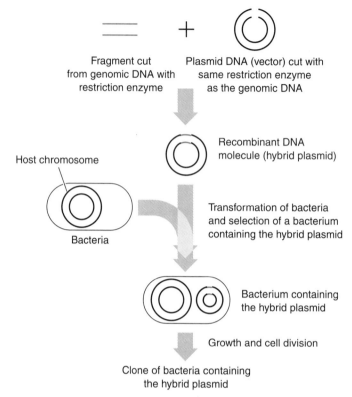

Figure 15-6 Cloning foreign DNA. A restriction fragment of DNA and a plasmid vector linearized by digestion with the *same* restriction enzyme used to cut the DNA sample are ligated with the aid of a DNA ligase enzyme (see Chapter 7). The hybrid (or chimeric) plasmid transforms a bacterium. Henceforth, when the plasmid replicates, so does the inserted DNA.

they are called **shuttle vectors.** Still other vectors have unique promoters useful for sequencing DNA or synthesizing RNA in vitro. Because of their small size, and the ease of handling them, plasmids are the most commonly used cloning vectors. See Table 15-2 for a summary of cloning vectors.

Insertion of a DNA Fragment Into a Vector

The strategy used to ligate a fragment of foreign DNA to a vector depends on the nature of the termini of the DNA. As previously described, restriction enzymes produce three basic types of termini (3′-OH, 5′-P overhangs, and blunt). A strategy for cloning DNA fragments with these termini and some of the problems or advantages associated with their cloning are explained in **Table 15-3**.

Given the large variety of vectors available, the optimal cloning strategy is to select a vector with two restriction sites that are compatible with the restriction sites on the 5′ and 3′ termini of the fragment of foreign DNA. The fragment of foreign

directional cloning

DNA can then be inserted into the vector by **directional cloning (Figure 15-7)**. For example, the vector pUC19 (UC for University of California, where these plasmids modified from pBR322 were constructed) can, for example, be cleaved with EcoR I and Hind III in the multiple cloning site. The linearized plasmid vector can then be ligated to a segment of foreign DNA that has been digested by the same enzymes and, therefore, contains cohesive termini that are compatible with those of the vector. Since two restriction enzymes are used in the construct, the two termini of the fragments are different, and the foreign DNA can, thus, be inserted in only

TABLE 15-3 PARAMETERS OF THE LIGATION REACTION

Ends of DNA Fragments	Cloning Requirements	Features
Blunt ends (ex: Bal I) ↓ ... TGG|CCA ... ... ACC|GGT ... ↑ | Separation ↓ of fragments ... TGG$^{3'}$ $^{5'}$CCA ... ... ACC$_{5'}$ $_{3'}$GGC ...	DNA and ligase present in high concentrations	Restriction sites at junctions of DNAs may be lost; number of wild-type plasmids without inserts may be high; multiple tandem copies of the foreign DNA may be inserted
Identical overhanging cohesive sticky ends (ex: EcoR I) ↓ ... GAA|TTC ... ... CTT|AAG ... ↑ | Separation ↓ of fragments ... G$^{3'}$ $^{5'}$AATTC ... ... CTTAA$_{5'}$ + $_{3'}$G ...	Phosphatase treatment of linearized plasmid DNA reduces religation of wild-type plasmids	Restriction sites at junctions of ligated ends are constituted; foreign DNA inserted bidirectionally; multiple tandem copies of the foreign DNA may be inserted
Different cohesive ends (directional cloning; see Figure 15-7)	Purification of linearized plasmid increases cloning efficiency	Restriction sites at junctions of ligated DNA are usually reconstituted; recircularization of wild-type plasmid is low; insertion of foreign DNA is unidirectional

one orientation—hence, the name **directional cloning.** The resulting circular recombinant plasmid can then transform *E. coli* to antibiotic resistance by expression of the gene that is on the vector. Thus, most of the bacterial cells resistant to ampicillin will contain plasmids that carry a foreign DNA insert between the EcoR I and Hind III sites.

After a fragment of DNA is cloned into a plasmid, inserted into a bacterium, and grown to large amounts in a selecting (antibiotic-containing) medium, the fragment must be recovered from its hybrid plasmid. Since the restriction sites in the plasmid vector are the same as those on the fragment of foreign DNA, the foreign DNA can be recovered by lysing the bacteria to release the plasmid and digesting the plasmid with the restriction enzymes used in cloning.

Insertion into a Vector of a DNA Molecule That Is Complementary to an mRNA

Often the genes that are candidates for cloning are organized in such a way that cloning is difficult. Some genes do not produce selectable products that can serve as an indicator of their presence. Many genes are rare, or are difficult to isolate. In those cases, it is sometimes preferable to enrich or partially purify the fragment containing the gene to be cloned prior to ligating it to the vector.

Directional Cloning

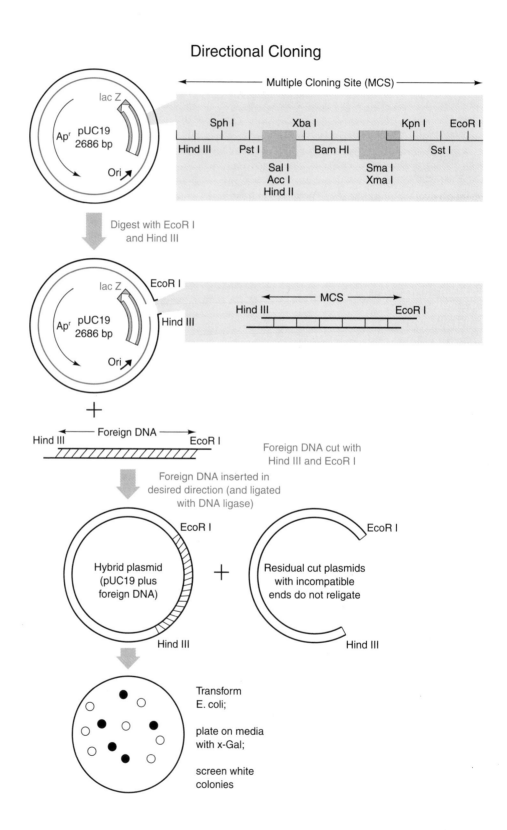

If the restriction map of the DNA (Figure 15-2) of interest is known, then it is possible to predict the size of the DNA fragment that will be generated when that DNA is digested with a particular restriction enzyme. Fragments of the predicted size can be isolated from an agarose gel after electrophoresis and joined to an appropriate vector. However, eukaryotic cell DNA contains about 10^6 cleavage sites for a typical restriction enzyme. Direct isolation of a eukaryotic gene from a mixture of fragments separated by electrophoresis can enrich the frequency of occurrence of the gene, but is possible only if the size of the restriction fragment containing the gene is known in advance.

In this section, another procedure for cloning a particular segment of DNA that encodes an mRNA is described. Using this technique, *any coding sequence whose mRNA can be isolated in almost pure form can be cloned.*

Let us assume that the chicken gene for ovalbumin (the protein in egg white) is to be cloned into a bacterial plasmid. A recombinant plasmid could be formed by treating both chicken cellular DNA and the *E. coli* plasmid DNA with the same restriction enzyme, mixing the fragments, annealing, ligating, and transforming *E. coli* with the DNA mixture containing the recombinant plasmid. However, identification of a bacterial colony containing this plasmid could not be carried out by a test for the presence of ovalbumin because a bacterium containing the intact ovalbumin gene will not synthesize ovalbumin. Introns, sequences that do not code for proteins, will be interspersed in the gene, and bacteria lack the enzymes needed to remove these noncoding sequences (see Chapter 8).

Some specialized cells of higher organisms, such as those producing ovalbumin in chickens, make large amounts of a single protein (in some plants, for example, the CO_2-fixing enzyme ribulosebisphosphate carboxylase/oxygenase can comprise ~50% of the total protein). In such cells the mRNA molecules coding for the abundant protein (whose introns will have been removed by processing enzymes when the mRNA is isolated from the cytoplasm) constitute a large fraction of the total mRNA synthesized in the cell. Consequently, mRNA samples can usually be obtained consisting predominantly of a single mRNA species—in the example of the chicken cells, ovalbumin mRNA. If genes of this class (those whose gene products are the major cellular proteins) are to be cloned, the purified mRNA

Figure 15-7 Directional cloning into a plasmid vector (pUC19). Ap^r indicates the position in the vector of the ampicillin resistance gene; *lacZ,* that of the *lacZ* gene; Ori indicates the origin of replication. pUC19 can be cut with any of the 13 restriction enzymes with sites in the multiple cloning site (MCS, shown expanded on the upper right of the diagram). To achieve directional cloning, both the vector and the foreign DNA to be inserted into the vector are cut by the same two enzymes, here Hind III and EcoR I. The excised segment of the MCS is shown. The foreign DNA can then be inserted in only one direction determined by the compatible cohesive ends. When bacteria transformed with the hybrid plasmid are grown in X-Gal medium, if there are untransformed bacteria present (viz., those lacking an insert), their *lacZ* gene is expressed, and these colonies are blue. An insertion of foreign DNA into the *lacZ* gene interrupts the gene that is then not expressed; these colonies are white. Only white colonies need be screened for the insertion of foreign DNA. Another advantage to directional cloning is that the linearized plasmids that have not ligated to a fragment of foreign DNA cannot circularize. The number of false positives is thereby markedly decreased.

reverse transcriptase

can serve as a starting point for creating a collection of recombinant plasmids, many of which should contain only the protein-coding sequences of the gene of interest. In cloning such sequences, the enzyme **reverse transcriptase** is used.

Many RNA-containing animal tumor viruses (**retroviruses**) contain reverse transcriptase. This enzyme can use a single-stranded RNA molecule (such as mRNA) as a template to synthesize a double-stranded DNA copy (called **complementary DNA** or **cDNA**). If the template RNA molecule is an mRNA molecule with the introns removed from the primary transcript, the corresponding full-length cDNA will contain an uninterrupted coding sequence. This sequence, thus, will not be identical to that of the complete original eukaryotic gene, since intron sequences will be lacking. If, however, the purpose of constructing the recombinant DNA molecule is to synthesize a eukaryotic gene product in a bacterial cell, and the processed mRNA can be isolated, then cDNA copied from it is the material of choice for insertion into the vector. The joining of cDNA to a vector can be accomplished by using special procedures for joining blunt-ended molecules.

complementary DNA (cDNA)

Cloning Strategies for Identifying Specific Genes

The isolation of a specific gene usually begins with the preparation of a gene library from DNA of the appropriate organism. For example, the isolation of human disease-causing genes begins with the preparation of a gene library from a sample of human DNA. In some cases, it is more advantageous to begin with a **cDNA library,** which is prepared from an RNA sample.

cDNA library

When a genomic DNA library is prepared, many copies of an organism's total DNA are cut and inserted into a vector. Thus, the genomic library of a eukaryote will include not only coding regions, but also introns and regulatory sequences.

In the preparation of a cDNA library, mRNA is extracted, purified, and treated with the enzyme reverse transcriptase. Complementary DNA (cDNA) analogs of the isolated mRNAs are thereby obtained. The collection of those cDNAs is treated like the genomic DNA used to make a genomic library: It is cut with restriction enzymes and inserted into vectors. The vectors will collectively contain cDNA that corresponds to all of the mRNA found in the original cell. Since mature mRNA contains no introns or regulatory regions, a cDNA library is composed of coding regions.

Using those libraries, virtually any DNA fragment or gene, if it can be recognized within the collection of perhaps thousands of other genes, can, subsequently, be isolated. Usually, however, in the case of large eukaryotic genes that contain substantial numbers of introns, the entire gene is not inserted into a single vector. Instead of isolating the whole gene in a single vector all at once, small, more manageable individual segments that constitute a series of overlapping fragments, and that comprise the entire gene, are collected with the chromosome walking procedure (Figure 15-4).

Detection of Recombinant DNA Molecules

When a vector is cleaved by a restriction enzyme and allowed to ligate to fragments of the DNA from a particular organism cut with the same restriction enzyme, several types of molecules result. These include:

1. A re-ligated vector that has not acquired any fragments of foreign DNA (methods for avoiding this have already been described).
2. A vector with one or more foreign DNA fragments.
3. A molecule without a vector, consisting only of joined fragments or the organism's DNA. Molecules in this class do not contain an origin of replication, and, therefore, cannot be replicated in bacteria.

To facilitate the isolation of a bacterial colony containing a vector with a particular gene, a method is needed to insure, first, that the bacteria in the colony contain a vector, second, that the vector possesses an inserted foreign DNA fragment, and third, that it is the DNA fragment of interest. In this section, several useful procedures for detecting recombinant vectors will be described.

In the cloning procedure as described so far, a foreign DNA fragment obtained by digestion with a restriction enzyme is ligated to a cleaved vector molecule, yielding a large number of hybrid vectors containing different fragments of foreign DNA. If a particular DNA segment or gene is to be cloned, the vector possessing that segment must be isolated from the set of all vectors possessing foreign DNA. For many genes, simple selection techniques are adequate for recovery of a vector containing that gene. For example, if the gene to be cloned were a bacterial leucine biosynthesis gene, a Leu⁻ bacterial host (incapable of synthesizing its own leucine) would be used, and a Leu⁺ colony selected and grown in a medium lacking leucine. This procedure is useful in cloning bacterial genes that can be selected, or genes that encode products of intermediary metabolism. Often, however, the gene to be cloned is a eukaryotic gene whose product does not change the phenotype of the bacterial host.

Recombinant clones with no easily recognized phenotype are most commonly identified by determining if the DNA in the colonies **hybridizes** with a radioactive probe to the foreign DNA. Bacterial colonies are replica-plated onto a solid support, usually nitrocellulose or nylon filters. The bacteria are lysed with alkali that also denatures their DNA, neutralized, and allowed to hybridize to the ^{32}P-labeled probe. The location of the ^{32}P-hybridized probe is then determined by exposing x-ray film to the filters. Bacterial colonies releasing DNA hybridizing to the DNA of the probe can then be identified by aligning the x-ray film with the original bacterial plates. In this way it is possible to screen for many hundreds of colonies simultaneously, and recognize colonies that carry recombinant plasmids (see **Figure 15-8**). This method is called **colony hybridization**.

colony hybridization

Site-Specific Mutagenesis Using Bacteriophage M-13 Vector

Until recently, it was very difficult to study the effects of specific base-pair changes on the function of a gene. Mutagenic agents such as chemicals or radiation act randomly, and in order to obtain a mutant with a desired phenotype, a researcher might have to screen several thousand mutants. Even then, much additional work is required to determine the precise nature of the mutation that has occurred.

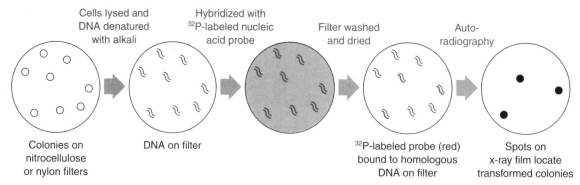

Cells lysed and DNA denatured with alkali

Hybridized with ^{32}P-labeled nucleic acid probe

Filter washed and dried

Auto-radiography

Colonies on nitrocellulose or nylon filters

DNA on filter

^{32}P-labeled probe (red) bound to homologous DNA on filter

Spots on x-ray film locate transformed colonies

Figure 15-8 Detection of transformed cells by colony hybridization with a radioactive probe. Only eight of thousands of colonies on a reference plate (not shown) are included.

site-specific mutagenesis
site-directed mutagenesis

Today, a technique known as **site-specific mutagenesis** (or **site-directed mutagenesis**) is revolutionizing the study of mutations. Once we have cloned a gene and determined its base sequence, we can selectively alter one (or a few) bases within the gene or its regulatory region. This genetically engineered DNA is inserted into an appropriate cloning vector and used to transform host cells in which the effects of the mutation can be studied. How do we actually alter the base sequence of the gene of interest? One commonly used procedure is outlined in **Figure 15-9**.

Let's suppose that it is believed that a certain amino acid (e.g., glutamic acid) is suspected of playing a crucial role in the function of a protein (the one encoded by the gene in which we are interested). Further suppose that we would like to determine how the activity of this protein would be affected if this positively charged glutamic acid (codon GAG) was replaced with a nonpolar amino acid such as valine (codon GUG). (We will assume that we have already cloned and sequenced our gene.) First, for the purpose of mutagenesis, it will be useful to have the DNA of our gene present in single-stranded form. To facilitate this, we will use the M13 phage (discussed in Chapter 14) as our cloning vector. As was explained in Chapter 14, during the early stages of an M13 infection, the single-stranded (+) phage DNA is converted to a double-stranded replicative form (RF) comprised of a (+) and (−) strand. The (−) strand subsequently serves as the template for production of additional (+) strands, which will be packaged into the new phage particles. (See Figure 14-21 in Chapter 14.)

The procedure begins by cloning the wild-type version of our gene into a double-stranded M13 phage cloning vector (Figure 15-9(a)). This recombinant DNA can be used to transform *E. coli* cells. The recombinant phage is allowed to replicate, then single-stranded (+) DNA from the phage progeny (Figure 15-9(b)) is collected. Now we are ready to alter our gene. Using an automated DNA synthesizer (see Figure 3-18 in Chapter 3) a 15–25 base-long oligonucleotide is synthesized that is complementary to our cloned gene, except for the substitution of a single base. (In this instance, we are interested in studying the effects of replacing a glutamic acid with a valine, so our synthetic oligonucleotide should contain a C*A*C instead of its normal C*T*C.) This oligonucleotide will anneal with the (+) strand phage DNA molecule (Figure 15-9(c)), and can be used as a primer

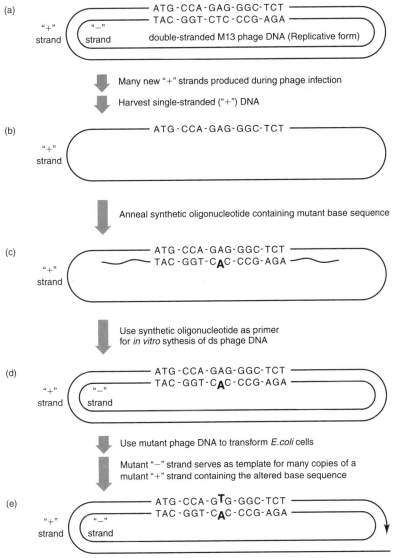

Figure 15-9 (a–e) Site-specific mutagenesis of a gene cloned into an M13 phage cloning vector.

for the synthesis of double-stranded phage DNA molecules in a test tube, using DNA polymerase and other essential enzymes (Figure 15-9(d)). As before, we can use this mutant phage DNA to transform *E. coli* cells. However, this time, the phage progeny will contain our *altered* gene sequence (Figure 15-9(e)). This newly synthesized mutant DNA can be collected from purified phage particles. The mutant gene itself can be excised from the phage DNA by means of restriction enzymes, recloned into an appropriate expression vector, and used to transform host cells derived from the type of organism in which the gene normally occurs. The end result? The transformed cells will contain a copy of the mutant gene, complete with the desired base substitution, thus, enabling us to study the effects of this specific mutation in vivo.

KEY CONCEPT

The "Recombinant DNA" Concept

DNA is DNA, regardless of the organism from which it is isolated. Hence, genes from diverse species can be recombined to construct novel genes, or, in certain instances, even new organisms.

Applications of Genetic Engineering

Unquestionably, recombinant DNA technology has revolutionized biology. At present, its main uses are:

1. Facilitating the production of useful proteins.
2. Creating bacteria capable of synthesizing economically important molecules.
3. Supplying DNA and RNA sequences as research tools.
4. Altering the genotype of organisms (both animals (**Figure 15-6**) and plants).
5. Potentially correcting genetic defects in animals (gene therapy).

It is used in research of all types: basic, medical, agricultural, commercial, and industrial; and can be expected to contribute to filling some of the most fundamental needs of humankind. At the same time, the technology arouses concerns about potential effects on the environment, and risks to health and society. In the next chapter, the current and future potential applications (and a few of the issues these applications have raised) will be discussed.

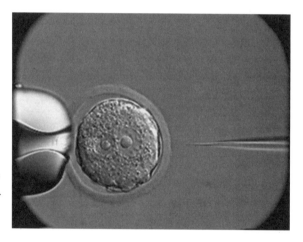

Figure 15-10   Direct injection of foreign genes into animal eggs and embryos (in this example, a newly fertilized mouse egg, displaying both egg and sperm nuclei) with a micropipette is one way in which researchers are attempting to make transgenic animals. Photo courtesy of Niles Fox.

SUMMARY

Recombinant DNA technology (or genetic engineering), which has revolutionized biological research, has been made possible by the use of restriction enzymes and cloning vectors. Restriction enzymes allow the cutting of DNA molecules at specific sites. The locations of various restriction sites along a DNA molecule constitutes a restriction map. Cloning vectors, such as plasmids, phage, and cosmids, allow DNA to be amplified, modified, or expressed. The polymerase chain reaction may first be used in vitro to amplify rare DNA sequences before insertion into the cloning vector. DNA fragments inserted into vectors can transform a suitable host. Transformants are often selected by antibiotic resistance conferred by the vector. The correct insert is usually screened by colony hybridization. Genomic libraries are made from restriction digests of total genomic DNA, while cDNA libraries are made from mRNA that has been transcribed to DNA by the enzyme reverse transcriptase. The order of sequences in a genomic library may be determined by chromosome walking or jumping. Once cloned, genes may be altered by site-specific mutagenesis. The mutant gene may then be expressed and studied.

1. These questions concern restriction endonucleases.
   (a) What is a restriction enzyme?
   (b) What is the biological function of a restriction enzyme?
   (c) For what main purpose do scientists use restriction enzymes?
   (d) What common feature is present in most base sequences recognized by a restriction enzyme?

2. Two types of cuts are made by restriction enzymes. What are they? These two types of cuts produce three possible types of DNA termini. Describe them.

3. What property must two different restriction enzymes possess if they yield identical patterns of breakage?

4. List some properties of an ideal cloning vector.

5. Explain insertional inactivation and (how it is) used in cloning?

6. Once a series of recombinant plasmids or phages has been produced, describe several ways in which plasmids or phage containing a particular insert may be identified.

7. Suppose that a Kan-r/Amp-r plasmid is treated with the Bgl II enzyme that cleaves the *amp* gene. The DNA is annealed with a Bgl II digest of *Drosophila* DNA and then used to transform *E. coli*.
   (a) What antibiotic would you put in the agar to insure that a colony has a plasmid?
   (b) What antibiotic-resistant phenotypes will be found on the plate?
   (c) Which phenotype will have the *Drosophila* DNA? How would you screen for this phenotype?

8. When cloning into plasmid vectors, re-ligation of vectors without inserts is always a problem. Describe two ways in which this problem may be lessened.

9. What is the difference between a genomic library and a cDNA library? What are the major differences in the structure of a gene cloned into either type of library? Give an advantage of each type of clone.

10. Explain the concept of "site-specific mutagenesis." Include mention of its advantages and disadvantages.

1. The restriction enzyme, EcoR I, was used to cut (1) a linear DNA molecule containing ten EcoR I sites and (2) a circular DNA molecule also containing 10 EcoR I sites.
   (a) How many fragments are generated from the linear molecule and from the circular molecule?
   (b) Can all the fragments of the linear molecule and the circular molecule circularize? If not, how many can? Which ones cannot and why?

2. A plasmid contains a single BamH I site within an antibiotic gene. The recognition sequence for BamH is G↓GATCC, where the sequence is written 5′ to 3′ and the arrow indicates the point of cleavage.

(a) A fragment of DNA containing a gene of interest is to be cloned into the BamH I site of the plasmid previously described. Unfortunately, the gene sequence contains a BamH I site. The gene is flanked by Bgl II sites $(A^{\downarrow}GATCT)$, however, as follows:

$$5'\text{----AGATCT----}\overrightarrow{\text{gene}}\text{----AGATCT----}3'$$

Diagram the cuts made by BamH I on the plasmid DNA and the cuts made on the gene-containing fragment by Bgl II. Show both strands and the cohesive ends produced. Are the BamH I and Bgl II cohesive ends compatible (i.e., can the Bgl II fragment be cloned into the BamH I site)? If so, show the resulting construct. Are the BamH I and/or Bgl II sites regenerated?

(b) Repeat the above analysis for the BamH I digested plasmid and the following insert sequence:

$$5'\text{----CGATCC----AGATCG----}3', \text{ cut with Sau3A } (^{\downarrow}GATC).$$

(c) Repeat the above analysis for the BamH I digested plasmid and the DNA sequence in (b) cut with DpnI $(GA^{\downarrow}TC)$.

3. Restriction enzymes A and B are used to cut the DNA of phages T5 and T7, respectively. A particular T5 fragment of size 2.0 Kb is mixed with a particular T7 fragment of size 1.5 Kb and treated with low concentrations of DNA ligase. Three major circular forms are produced, and all three can be cut with both restriction enzymes A and B, producing one, one, and two fragments, respectively. Diagram the most likely structures of these circles. What can you conclude about restriction enzymes A and B?

4. Plasmid pBR607 DNA is circular, double-stranded, and has a molecular weight of 2.6×10^6. This plasmid carries two genes whose protein products confer resistance to tetracycline (Tc^r) and ampicillin (Ap^r) in host bacteria. The DNA has a single site for each of the following restriction enzymes: EcoR I, BamH I, Hind III, Pst I, and Sal I. Cloning DNA into the EcoR I site does not affect resistance to either drug. Cloning DNA into the BamH I, Hind III, and Sal I sites abolishes tetracycline resistance. Cloning into the Pst I site abolishes ampicillin resistance: Digestion with the following mixtures of restriction enzymes yields fragments with the sizes listed in the table below. Position the Pst I, BamH I, Hind III, and Sal I cleavage sites on a restriction map, relative to the EcoR I cleavage site. Show also the approximate locations of the *amp* and *tet* genes, and the nucleotide distances between restriction sites.

Enzymes in Mixture	Molecular Weights of Fragments (millions)
EcoR I, Pst I	0.46, 2.14
EcoR I, BamH I	0.2, 2.4
EcoR I, Hind III	0.05, 2.55
EcoR I, Sal I	0.55, 2.05
EcoR I, BamH I, Pst I	0.2, 0.46, 1.94

5. You have the cDNA clones for two human genes, A and B, which you believe may be closely linked on a human chromosome, since these two genes are known to be close together in rats. Briefly describe the tools and procedures you would use to answer this question.

CONCEPTUAL QUESTIONS

1. How far can genetic engineering go? Will it be possible to do Jurassic Park-type experiments by isolating DNA of extinct animals from fossils and incorporating it into present-day organisms? What, if any, benefits would likely accrue from it if such a procedure was successful?

2. How might the understanding of DNA sequences and gene arrangements, which is developing from recombinant DNA technology studies, permit design of advanced computer programs based on principles of genetic information stored in genomes?

3. Understanding human behavior is the goal of many research-oriented psychologists and sociologists. Can recombinant DNA technology help those research efforts?

CHAPTER

16

in this chapter you will learn

1. How recombinant DNA technology can be used to alter the genome of many organisms

2. Applications of molecular biology to forensic medicine

3. Production of commercially useful plant and animal products using genetically engineered strains

Experimental Manipulation of N

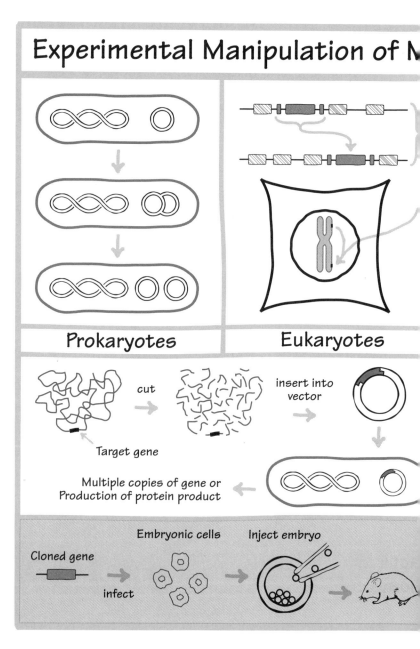

Prokaryotes   Eukaryotes

cut   insert into vector

Target gene

Multiple copies of gene or
Production of protein product

Cloned gene   Embryonic cells   Inject embryo

infect

lecules

sposons,
nids, and
eriophage

enetic
jineering

actical
lications

Molecular Biology Is Expanding Its Reach

As we enter the twenty-first century, molecular biology is extending its influ-ence into areas distant from the laboratory. Ten years ago, it was unlikely that the average citizen would have had a lively dinner-table discussion about the admissibility of DNA fingerprint evidence in the courtroom, or pondered the advisability of purchasing a genetically engineered tomato at the supermarket, or read about the latest gene therapy experiment in their daily newspaper. Yet today all of these scenarios are occurring on a regular basis. Molecular biology and recombinant DNA technology have clearly revolutionized our world! At the same time, this technology arouses concerns about potential effects on the envi-ronment, and risks to health and society. In this chapter, we will examine the impact that molecular biology is having on research, medicine, agriculture, industry, and society as a whole. We will also consider some of the ethical issues that have been raised in response to this "intrusion" of genetic engineering into our lives.

Uses of Recombinant DNA Technology in Research

Unquestionably, recombinant DNA technology has revolutionized biological research. At present, some of its main uses include:

1. Creating bacteria and other organisms capable of synthesizing both useful and economically important molecules.
2. Mapping the genomes of humans and of organisms utilized in research.
3. Supplying DNA and RNA sequences as research tools.
4. Altering the genotype of organisms (both plants and animals).
5. Potentially correcting genetic defects in animals (gene therapy).

It is used in research of all types: basic, medical, agricultural, commercial, and industrial; and can be expected to contribute to filling some of the most fundamen-tal needs of humankind. Let's look at several of these research applications.

Uses in Research: Mammals Made to Order

The analysis of bacterial mutants has been an extraordinarily powerful means of untangling cellular functions, including metabolic pathways. A major stumbling block to extending this approach to higher organisms has been the diploid nature of the eukaryotic genome; it is difficult to isolate an experimentally induced change in both copies of a given gene. One current approach to overcoming this difficulty is the use of targeted mutagenesis via homologous recombination. The technique was employed first in yeast, and, more recently, in mammals. Using this technique, it is possible to replace an endogenous allele (copy) of a gene with one that has been genetically engineered. Moreover, this new gene is introduced into an appropriate genetic environment under normal genetic conditions, including those operative during development and differentiation. This technology has been utilized in several different ways. In some instances, new, and sometimes novel, genes have been introduced into the organism. In other cases, a specific gene has been inactivated, then introduced into the organism in order to learn more about the normal role of the gene in embryonic development, tissue differentiation, the development of cancer, or the functioning of the immune system. This latter approach has been used to produce a unique group of transgenic mice known as **knockout mice,** which are now widely used in mammalian research.

knockout mice

In order to create a transgenic mouse, fertilized mouse ova are obtained from a female mouse early in embryonic development, prior to implantation of the embryo in the uterus (**Figure 16-1**(a)). At this stage of embryonic development, the embryonic stem cells (**ES cells**) making up the embryo are undifferentiated and are capable of giving rise to all types of cells ultimately found in the organism (e.g., they are **totipotent**). ES cells can be removed from the blastocyst (early-

ES cells

totipotent

Figure 16-1 Mammals made to order: production of transgenic mice. In this example, totipotent embryonic stem cells (ES cells) are (a) removed from the blastocyst of an agouti (brown) donor, and (b) proliferated in culture by growth on a layer of metabolizing cells that have been treated (radiation, mitomycin, or other irreversible DNA synthesis inhibitor) to prevent the growth of the feeder cells. Genetically engineered gene(s), into which an antibiotic resistance gene (*neo*) has been incorporated, are introduced into the ES cells by electroporation (c, d). Transformed ES cells in which the engineered gene has been incorporated into the mouse chromosome via homologous recombination (e, f) are selected by growing the cells in a culture medium (G418 medium) that contains a neomycin-like substance (g). (Nontransformed cells are killed.) Neomycin-resistant clones are selected and assessed for euploidy and the presence of the mutated allele of the desired gene (h). Ten to 15 cells meeting the criteria are injected into a blastocyst from a white mouse (i), where they become incorporated into the inner-cell mass, and the blastocyst is reimplanted into a pseudopregnant female mouse. Offspring in successful experiments are chimeric for coat color (j) (white from the maternal embryo and agouti from the transformed ES cells). Chimeric animals are mated to white tester mice (k); if the genetically engineered ES cells have entered the germline, the traits from the ES cells (e.g., coat color) will also be present in the offspring of the chimeric animals. DNA from such offspring is analyzed by Southern blotting (see Figure 8-13) for the presence of the targeted allele. (l) Heterozygous animals are mated, with one-quarter of their offspring expected to be homozygous for the targeted mutation. Homozygosity can be identified by further analysis.

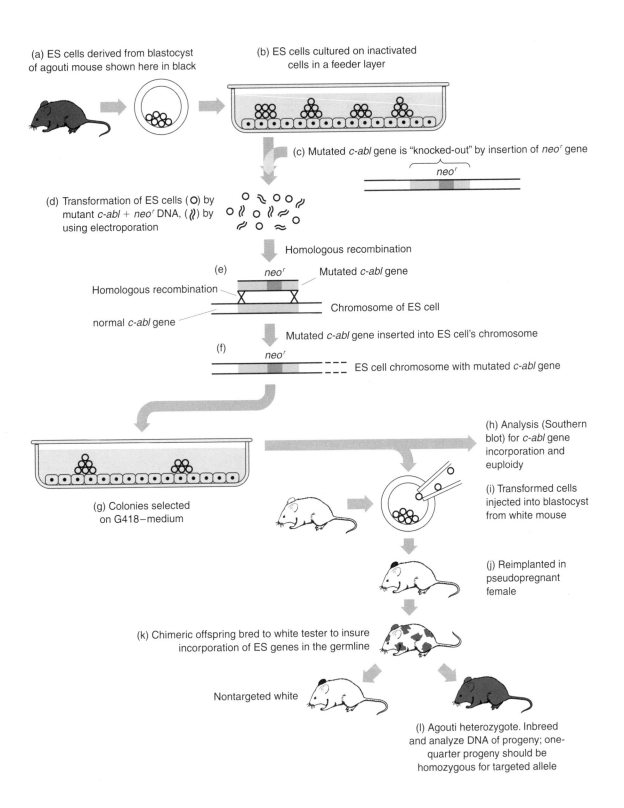

(a) ES cells derived from blastocyst of agouti mouse shown here in black

(b) ES cells cultured on inactivated cells in a feeder layer

(c) Mutated *c-abl* gene is "knocked-out" by insertion of *neo*[r] gene

neo[r]

(d) Transformation of ES cells (O) by mutant *c-abl* + *neo*[r] DNA, (⌇) by using electroporation

Homologous recombination

(e) *neo*[r]   Mutated *c-abl* gene

Homologous recombination

normal *c-abl* gene

Chromosome of ES cell

Mutated *c-abl* gene inserted into ES cell's chromosome

(f) *neo*[r]

ES cell chromosome with mutated *c-abl* gene

(g) Colonies selected on G418−medium

(h) Analysis (Southern blot) for *c-abl* gene incorporation and euploidy

(i) Transformed cells injected into blastocyst from white mouse

(j) Reimplanted in pseudopregnant female

(k) Chimeric offspring bred to white tester to insure incorporation of ES genes in the germline

Nontargeted white

(l) Agouti heterozygote. Inbreed and analyze DNA of progeny; one-quarter progeny should be homozygous for targeted allele

stage mammalian embryo) and grown in culture on a "feeder layer" of living cells (Figure 16-1(b)). This feeder layer nourishes the ES cells and allows them to divide, but doesn't permit them to undergo differentiation. Each cultured ES cell will, thus, give rise to a clone of identical ES cells. These ES cells can be transformed using a genetically engineered gene (Figure 16-1(c)–(f)), and selection for the transformed cells can be carried out in culture (Figure 16-1(g)). The ES cells that have successfully incorporated the modified "foreign" gene into their genomes (Figure 16-1(h)) can then be injected into a blastocyst obtained from a different female mouse, where the mixture of cells will develop into the complete embryo (Figure 16-1(i)–(j)). Let's take a closer look at how this all works in practice.

Suppose that we wanted to learn what effect the disruption of a gene known as **c-abl** has on the mouse. The c-abl gene is the cellular homolog of an oncogene from the Abelson murine leukemia virus. (Oncogenes are genes that, in their nonmutant "cellular" form, are involved in the normal regulation of cell division. When disrupted or altered, however, they are associated with development of certain forms of cancer.)

First, we would need to construct a plasmid vector containing the inactivated gene. This is engineered in vitro to make the desired alterations in the gene, either by base substitution, deletion, or insertion. In our example, we could disrupt the c-abl gene by inserting a promoterless neomycin resistance gene (neo^r) near the 3′ end of its coding region ("knocking it out") (Figure 16-1(c)). Drug resistance markers with controllable promoters have also been used effectively in this context. The modified gene would be inserted into an appropriate plasmid (or other vector, such as a genetically modified virus).

The ES cells used in this experiment would be homozygous for distinctive phenotypic markers, such as coat color or rare electrophoretic variants of tissue enzymes, which can be readily identified in the eventual progeny. The recombinant plasmid could be introduced into ES cells by **electroporation** (reversible permeabilization of cells by electric shock in the presence of the plasmid). Then the cells are plated onto the feeder layer (Figure 16-1(d)–(f)). After 1–2 days (to allow the integration of the recombinant gene into the host DNA), the selection for transformed cells would begin (Figure 16-1(g)). The culture medium would be replaced with fresh medium containing G418, a neomycin-like substance that kills non-neomycin-resistant mammalian cells. Since the neo^r gene lacks its own promoter, neomycin resistance would be expressed only in those cells in which the gene was situated next to a cellular promoter, including, of course, cells in which the engineered construct had replaced one of the endogenous c-abl alleles by homologous recombination.

Next, we would isolate drug-resistant ES colonies and analyze their DNA by Southern blotting (see Figure 8-13 in Chapter 8). A probe derived from the c-abl oncogene, outside the area of homologous recombination, would be used to detect the presence of the normal (cellular) and inactivated c-abl alleles in restriction digests of DNA from the ES cells. Since the parent cell line contained its own c-abl gene, one criterion for homologous integration would be the presence of two bands on the Southern blots: one band in the same position as in the parent, and the other at a position 0.6 Kb larger. The difference is due to the 0.6 Kb neo^r insert in the integrated gene. Only the parental band is observed in nonhomologous

electroporation

recombinants; these are recovered from the drug-selection step in far greater numbers than homologous recombinants.

There are other variations on the technology up to this point, each appropriate to the corresponding details of constructing the original plasmid or other vector. When clones with the desired characteristics have been identified, karyotype analysis of the selected cells is performed to ensure that they are euploid (i.e., have a normal set of chromosomes) (Figure 16-1(h)). Cells (10–15) from selected clones are next injected into blastocysts of embryos that are homozygous for dissimilar markers; for example, if the ES cells are derived from black or agouti (grizzled brown) mice, they are injected into embryos from white or nonagouti stock (Figure 16-1(i)). The injected blastocysts are then reintroduced into pseudo-pregnant females (prepared by mating 2–3 days previously with vasectomized males) (Figure 16-1(j)). In successful experiments, the totipotent ES cells become incorporated into the inner cell mass of the recipient blastocyst and colonize various tissues, including the germ cells of the developing embryo. The resulting transgenic animals can be preliminarily identified by chimeric (mixed) coat color patterns (Figure 16-1(k)).

Male chimeras are then allowed to develop to sexual maturity and are crossed with tester females to determine whether the engineered gene has been incorporated into the germ line. The progeny are again examined for properties, such as hair pigmentation or electrophoretic variants of enzymes derived from the ES cells. Finally, these mice are tested by Southern blot analysis for the presence of the property that was of interest in the first place—in this example, the *c-abl* gene with the *neo^r* insert (Figure 16-1(l)).

Methods closely similar to this are currently being used to develop mouse models for human disorders such as Alzheimer's disease, cystic fibrosis, alcoholism, epilepsy, a number of different forms of cancer, and a variety of other conditions. As of the end of the year 2001, scientists had successfully knocked out more than 4,000 of approximately 35,000 mouse genes, and were examining more than 500 knockout mouse-based models of human diseases.

There are still several major concerns and problems associated with the use of this technology. First, and most obvious, is that if the engineered gene is essential to embryonic development, homozygous offspring may die in utero. Second, it is apparent that the recipient (or host) organism plays an important role, not yet fully understood, in the successful incorporation of the donor cells into the developing embryo and subsequent germ-line transmission. Third, in the current methodology, nonhomologous recombination is far more frequent than homologous recombination, and rigorous selection and careful analysis of cloned, transformed ES cells are necessary before they can be used as donors. The latter point is particularly significant if homologous recombination is eventually to be used in human gene therapy. Having an otherwise normal insulin gene in the wrong location could lead to its inappropriate expression, or even to no expression at all, which could be more harmful than helpful. It has already been observed that in at least two human cancers (chronic myelogenous leukemia and acute lymphocytic leukemia) the human *c-abl* gene has been inactivated as a result of a chromosomal translocation. The danger of recombination of an introduced gene in the wrong location is evident.

There is an obvious way in which such problems can be avoided in instances in which the intended host is an adult organism, for example, rather than a pre-implantation embryo. The cells to be genetically altered can first be removed from the body and the genetic alteration can be carried out in vitro. Altered cells having the desired genotype can be selected in the laboratory, then reintroduced into the original donor or other suitable recipient. This procedure is often called **gene therapy.** One can envision that defective bone marrow stem cells could be removed from a patient and their genetic problems corrected in vitro. Then those corrected cells could be reimplanted into the patient. This procedure is often called "cell therapy." Similarly, sheets of living skin could be maintained in hospitals for ready use in severe burn cases.

An alternative approach that has generated much controversy is the therapeutic use of human embryonic stem cells to correct both hereditary and degenerative disorders. Human embryonic stem cells obtained from the inner cell mass of the blastocyst are **pluripotent.** This means that they can give rise to nearly every type of cell and tissue found in an organism, but cannot develop into a viable embryo. (Cells derived from the inner cell mass of the blastocyst lack the ability to form the placenta and supporting tissue essential for embryonic development within the uterus.) If human ES cells could be induced to undergo differentiation into selected adult tissues, such as nerve cells, heart muscle, pancreas or blood cells, they might prove to be useful in treatment of spinal cord injuries, heart attacks, Parkinson's disease, or diabetes. It has even been proposed that it might be possible to treat AIDS by genetically altering T cells to make them resistant to the AIDS virus, and using them to replace the HIV-infected T cells.

The controversy surrounding stem cell research arises because of the sources from which certain types of human stem cells are commonly obtained. (1) Many ES cells are the product of in vitro fertilization of a donated human egg by a human sperm. (2) Embryonic germ cells (EG), which represent a slightly later stage of development, are derived from fetal tissue from terminated pregnancies. Embryonic tissues are not the only source of stem cells. (3) More specialized, multipotent stem cells, such as blood or skin stem cells, are present in both adults and children. These adult stem cells are less versatile than their embryonic counterparts and differentiate into cells with closely related functions. (For example, blood stem cells, found in human bone marrow, give rise to blue blood cells, white blood cells, and platelets.)

Are there other potential sources of ES cells? It has been suggested that it might be possible to produce ES cells by somatic cell nuclear transfer. One could take a normal, unfertilized, human egg, remove its nucleus, then fuse the enucleated egg with a human somatic cell. If the totipotent fused cell were capable of normal development to the blastocyst stage, pluripotent ES cells could be obtained from its inner cell mass.

Whatever the source, stem cells are not a cure-all for every debilitating disease. Stem cells derived from a genetically unrelated individual would differentiate into tissues that were immunologically different from those present in the intended recipient, and which might be rejected in much the same way as a transplanted organ could undergo rejection. Autologous stem cells would be genetically identical to the cells of the recipient, but they would also possess the same genetic defects originally present in the cells, and might need to be genetically altered before they would be useful. Furthermore, we have not been able to identify and isolate stem cells for all types of tissues, nor do we have the ability at this time to stimulate adult stem cells to differentiate into the myriad of different tissues present in an adult human being. So, while stem cells show promise as part of our arsenal of therapies of genetic and degenerative disorders, much research remains to be done before this potential is likely to be realized.

The long-range goal, however, remains the development of a method that would result in only homologous recombination with high efficiency in virtually any cell type. Such an achievement would offer the utopian potential of using normal genes to correct genetic disorders that may not be recognized until after the birth of a fully differentiated individual.

Uses of Recombinant DNA Technology in Medicine

The goal of medicine is to alleviate and ultimately eliminate human disease. More than one hundred heritable human diseases have been identified and characterized, and well over 9,800 single-gene disorders are currently recognized. This does not even include a large number of disorders having multiple genetic and nongenetic components. When we examine some of these so-called single gene defects, a complex picture emerges. In some cases, a mutation has occurred in a gene that codes for an enzyme or a structural protein; in others, the regulation of gene expression has been altered. Once again, because of the development of recombinant DNA methodologies, the ability to detect genetic defects linked to specific diseases has increased.

One of the most useful applications of recombinant DNA technology is to provide novel and precise methods for early detection of a genetic disease. A variety of techniques are currently being used for this purpose, including simple enzymatic assays, gel electrophoresis of gene products, immunological methods, chromosomal analysis, and restriction analysis. If a deletion, insertion, or point mutation alters a restriction site within or near a gene, it may be possible to detect this change directly on a Southern blot with an appropriate probe, or indirectly, through linkage to another genetic marker associated with the genetic defect. A simple example of this is shown in **Figure 16-2**. In sickle cell anemia, a single base substitution in the β-globin gene ($GAG \rightarrow GTG$) results in a glutamic acid normally found in the β-globin chain being replaced by a valine. This base change eliminates a restriction site recognized by the restriction enzyme *Mst* II, which recognizes the base sequence CCNAGG (N = A, T, G, or C). In individuals with normal hemoglobin (and a normal β-globin allele), digestion with *Mst* II generates two DNA fragments (1.1 and 0.2 kb) that hybridize with a β-globin cDNA probe. In individuals with the sickle cell allele, the A $\rightarrow$ T base substitution eliminates one restriction site, and only a 1.3 kb restriction fragment is evident. (This is an example of a **Restriction Fragment Length Polymorphism, or RFLP**.) It is possible to distinguish whether an individual has one or two copies of the mutant allele by examining the types of bands present: individuals afflicted with sickle cell anemia will have only the 1.3 kb band, while carriers of the sickle cell allele (heterozygotes) will have two normal bands (1.1 and 0.2 kb) plus the abnormal band (1.3 kb).

Restriction Fragment Length Polymorphism (RFLP)

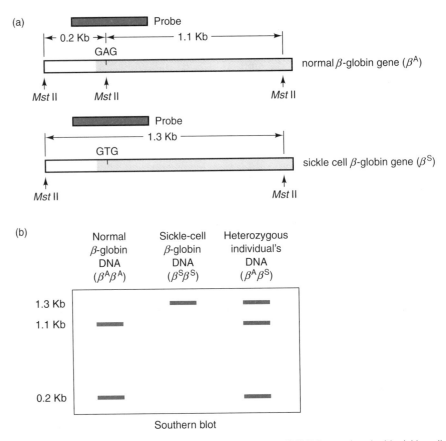

Figure 16-2 Restriction fragment length polymorphisms (RFLPs) associated with sickle cell anemia. (a) In individuals suffering from sickle cell anemia, a base substitution mutation in the β-globin gene (**GAG → GTG**) eliminates a target site recognized by the restriction enzyme *Mst* II. (b) DNA from normal individuals (left lane), individuals suffering from sickle cell anemia (center lane), and heterozygous individuals (right lane) is digested with *Mst* II, separated via gel electrophoresis, and Southern blotted using a probe that hybridizes with the region of the β-globin gene containing the base substitution. DNA from normal individuals will contain two bands (0.2 and 1.1 Kb), DNA from individuals suffering from sickle cell anemia will contain only a single band (1.3 Kb), and that from heterozygous individuals will contain all three bands (0.2, 1.1, and 1.3 Kb).

allele-specific oligonucleotides (ASO)

More specific probes, called **allele-specific oligonucleotides (ASO),** hybridize only with completely complementary base sequences, and can be used to distinguish between sequences differing by as little as a single nucleotide. When used in combination with DNA microarrays, such as a GeneChip® probe array (see Figure 13-4 in Chapter 13), the researcher tests for the presence of many different mutant alleles of a particular allele simultaneously, and can determine whether the individual being tested is homozygous or heterozygous for these alleles.

In some cases, it has not been possible to identify an altered restriction site residing within the mutant allele, but it has been possible to demonstrate that the mutant allele tends to be coinherited with a nearby (linked) gene. Linkages between genes can be demonstrated initially by using the separate genetic markers as probes and determining whether they hybridize to the same restriction fragment on a Southern blot. For linkage to be used as a diagnostic tool, it must first be demonstrated that DNA from patients with a particular genetic disorder shows a characteristic restriction pattern that is not observed in healthy persons. Obviously, this indirect approach is not foolproof, since it is possible for recombination to occur between linked genes, and some individuals who inherit the adjacent marker may fail to inherit the gene associated with the disorder. RFLP analysis has been successfully used to develop indirect tests for genetic disorders such as Huntington's disease, cystic fibrosis, Duchenne muscular dystrophy, and hemophilia B, to name a few.

RFLPs, VNTRs, and DNA Fingerprinting

Genomes, including those of humans, are polymorphic (i.e., they differ in base sequence). Between any two individuals, there is a difference of about one base pair per thousand. Some of these differences may represent point mutation differences (although many are phenotypically "silent" because they occur in the variable third coding base, and do not alter the amino acid being coded). Probably a more common case involves differences in base sequences found in the region situated *between* genes. This polymorphism has well-recognized implications for sequencing the human genome. When these differences occur in restriction sites, they generate restriction fragments of different sizes. This **Restriction Fragment Length Polymorphism (RFLP)** is readily observed by gel electrophoresis of DNA that has been cut by restriction enzymes and probed with an appropriate marker, as we have already seen (Figure 16-2).

Scattered throughout the genome of any individual eukaryote, there are also different numbers of copies of tandemly repeated DNA sequences called **Variable Number Tandem Repeats** or **VNTRs.** VNTRs are short base sequences occurring in variable numbers within tandemly repeated clusters (i.e., they are found adjacent to each other). The number of "repeats" found in a group may vary from one location to the next within the genome of a particular individual, and will also vary between individuals. These repeats are of two basic types: **microsatellite repeats** (also known as **short tandem repeats** or **STRs**), which consist of 2–5 nucleotide-long tandem repeats occurring at many locations throughout the entire genome, and **minisatellite repeats,** which consist of longer repeating units (30–35 bp) containing a shorter, common core sequence. The longer, minisatellite repeats tend to occur at fewer locations within the genome of an individual.

In order to detect VNTR patterns (**Figure 16-3**), the genomic DNA of an individual is cut with a restriction enzyme that does not cut the VNTR array itself, but which cuts to include the entire repeat section (providing a fragment whose length depends upon the number of repeats). The restriction fragments thus produced are separated via gel electrophoresis. Southern blotting is then carried out using a radioactive probe that hybridizes with the basic repeated unit. Autoradiography will reveal the presence of a large number of DNA fragments of different sizes that hybridize to the probe. Since the number of tandem repeats will differ from one individual to the next, what emerges is a highly individual pattern of DNA fragments known as a **"DNA fingerprint."** In addition, the restriction

Variable Number Tandem Repeats (VNTRs)

DNA fingerprint

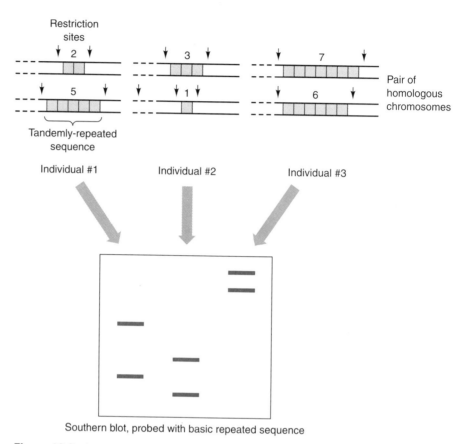

Figure 16-3 Simplified example of a DNA fingerprint resulting from the presence of variable numbers of tandem repeats (VNTRs). Three different individuals (#1, #2, and #3) possess different numbers of copies of a specific tandemly repeated base sequence. DNA from each individual is cut with a restriction enzyme that does not cut the VNTR array itself, the resulting DNA fragments are separated via gel electrophoresis, and Southern blotting is carried out using a probe that hybridizes with the basic repeated units. Different patterns of bands are observed in each individual. Actual DNA fingerprint patterns are, however, much more complex than those illustrated here (see Figure 16-4).

pattern of an individual, like a gene, is inherited in a strictly Mendelian pattern. (This makes sense because half of our DNA, as well as half of our chromosome set, is contributed by each parent.) Consequently, a DNA fingerprint can also provide strong evidence for the paternity of a child, half of whose DNA bands should be attributable to each biological parent.

DNA fingerprinting is currently being utilized in a number of different ways. Recent high-profile legal cases have shown its applicability in demonstrating a link between a crime suspect and a crime scene (**Figure 16-4**). Minute DNA samples present in a drop of blood, a skin scraping, or a hair follicle can be amplified via PCR (see Figure 15-5 in Chapter 15) and used to exonerate or incriminate a suspect. As important as its ability to suggest that a particular suspect was present at the scene of a crime is its ability to prove the innocence of a wrongly accused or convicted individual. It has been used in kidnapping cases to demon-

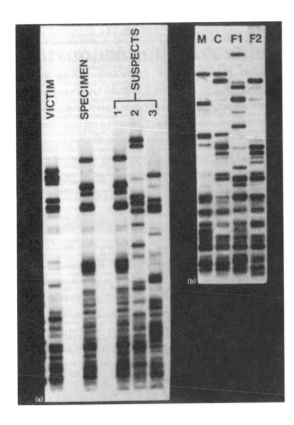

Figure 16-4 Use of DNA fingerprints in forensics and in paternity determination. (a) DNA fingerprints obtained from a rape victim, a semen specimen collected from the victim, and three potential suspects. (b) A mother **(M)** gave birth to a child **(C),** whose paternity was disputed. DNA fingerprints from the two possible fathers are shown in the lanes labeled **F1** and **F2**. (From J. M. Connor and M. A. Ferguson-Smith, *Essential Medical Genetics*, 3rd ed., Blackwell Scientific Publications (1991), p. 129.) Used by permission.

strate that a particular child was, in fact, the offspring of a set of parents whose child had been kidnapped years earlier, and not the biological child of the kidnappers. In cases of wildlife poaching, DNA fingerprints obtained from the head of a lion mounted on a wall of a big game hunter have been used to link the hunter to the remains of a lion found illegally slaughtered on a wildlife preserve. DNA fingerprinting is also being used in other ways to help protect endangered species. The DNA from members of wild populations of animals can be compared to determine their degree of genetic relatedness. This can be important for the purposes of determining whether a species is endangered, whether different populations of a particular species can be interbred successfully, or whether a proposed captive-breeding program is likely to result in excessive inbreeding of the animals involved. When one takes into consideration the diverse applications of DNA fingerprinting and the amount of publicity they have received in the media, it is not surprising that this once relatively unknown technology has emerged from the laboratories to become one of the most discussed technological issues in our society today.

Baa, Baa Black Sheep, Have You Any . . . Alpha-1-Antitrypsin? (Production of Pharmaceutical Products)

Recombinant DNA technology is rapidly moving from the laboratory to the pharmacy. Presently, several different therapeutic approaches are available for the treatment of genetic diseases. One of these is **drug** or **dietary treatment** to circumvent a metabolic block. In this case, the missing hormone or metabolite is administered to the patient. Therapeutically important products are currently being synthesized using recombinant DNA technology. In brief, a gene that codes for a particular peptide, along with an appropriate promoter and a sequence that instructs the cell to secrete the gene product, are inserted into an expression vector (e.g., phage λ gt11); the gene is oriented in such a way that its coding strand is linked to the strand containing a promoter in the vector. Bacterial cells can be prepared that contain several hundred copies of the gene, and, in such cells, synthesis of the genetically engineered gene product may account for 1%–5% of the total cellular protein. The gene product, which is secreted by the bacteria containing the recombinant DNA, can be harvested from the medium in which the bacteria have grown. Somatostatin, insulin, human growth hormone, interferon, opioid peptides (enkephalins and endorphins), interferons, thymosin, tissue plasminogen activator, urokinase, and hemophilia factor VIII have all been synthesized in this way. The list continues rapidly to lengthen. Methods to deliver these protein products to the patient slowly or in sequential doses have also improved. In certain cases, it is now possible to re-place the defective gene product with a functional protein at concentrations that mimic the normal physiological dose of the protein.

If the gene producing the therapeutic product is of eukaryotic origin, special problems must be taken into consideration:

1. Eukaryotic promoters are not usually recognized by bacterial RNA poly-merases; hence, the eukaryotic gene must be linked to a bacterial promoter.
2. The mRNA transcribed from the eukaryotic gene may not be translatable on bacterial ribosomes.
3. Introns may be present, and bacteria are unable to excise eukaryotic introns. In such cases, cDNA, or a synthetic DNA with a sequence deduced from the amino acid sequence of the desired protein product, is used rather than the desired eukaryotic gene.
4. The protein itself often must be processed (for example, insulin), and bacteria cannot recognize processing signals in eukaryotic gene products.
5. Eukaryotic proteins may be recognized as foreign material by bacterial pro-teases and degraded.

yeast artificial chromosome (YAC)

There are several approaches to solving these problems, in addition to those already mentioned. One procedure uses a **yeast artificial chromosome (YAC)** as a cloning vector for eukaryotic genes. Recent studies suggest that when a recom-binant YAC containing a eukaryotic gene is put into a yeast cell (which is also a eukaryote), some of the problems previously described no longer exist.

An even more exciting technology involves the use of transgenic animals (primarily sheep and cows) to produce pharmaceutically important human gene products, which are subsequently secreted into the animal's milk. This technique has already been used successfully to produce both hemophilia factor IX and an-other protein known as alpha-1-antitrypsin. Individuals who have an often-fatal hereditary form of emphysema are deficient in this second protein. The human alpha-1-antitrypsin gene was cloned into vector at a site adjacent to a sheep DNA sequence that regulates the expression of sheep milk proteins. Specifically, this promoter sequence limits the expression of the genes it controls to the mammary gland of the sheep. The engineered gene was then injected in vitro into fertilized sheep oocytes, and the eggs were implanted in ewes. Ultimately, several transgenic lambs were born. When the transgenic females were mated as adults, they produced milk containing high concentrations of alpha-1-antitrypsin, nearly 0.3 lb of human protein per gallon! Results with factor IX production have been less spectacular, but are still quite promising. Someday soon we may see large herds of transgenic sheep and cows filling many of our pharmaceutical needs. In 1997, a quantum leap in animal cloning technology was made with the first successful cloning of a lamb from a cell (a fibroblast) taken from the udder of an *adult* sheep. Scientists were able to achieve this remarkable feat by first growing the sheep fibroblast cells in culture, then serum-starving them to help synchronize the cell cycles of the donor (fibroblast) cell and the recipient (egg) cell. Nuclei from these fibroblast cells were then transplanted into enucleated sheep eggs, and each egg was implanted into the uterus of a female sheep. The result: a lamb (named "Dolly") whose development was directed by the genetic information provided by the transplanted fibroblast nucleus! Similar experiments utilizing donor nuclei from sheep embryos and fetuses met with even greater success. Building upon these latter experiments, Scottish scientists went one step further and genetically engineered *human genes* into the fetal sheep cells that served as the source of donor nuclei. The resulting lambs carried a *human* transgene in their cells.

Since that time, scientists have successfully cloned sheep, goats, cows, mice, and pigs, employing techniques similar to those used to clone Dolly. One group of recently cloned pigs were genetically engineered as potential organ donors to humans. One of the two copies of an important pig gene that causes the human immune system to reject pig organs, the alpha 1,3 galactosyl transferase gene (GATA1), was "knocked out" in these animals. Scientists anticipate producing genetically engineered piglets that lack both copies of this gene, then transplanting organs from these pigs into nonhuman primates to determine whether they will prove to be suitable donor organs. If these experiments are successful, the supply of potential donor organs (including hearts, lungs, and kidneys) could increase dramatically.

Are there any potential problems associated with cloning animals, or with using organs from these cloned animals for transplantation into humans? Unfortunately, there are several major difficulties. First, many of the cloned animals produced to date have been grossly abnormal, failed to survive, exhibited significant growth abnormalities, aged prematurely, or developed degenerative diseases such as arthritis at an unusually early age. Second, knocking out one or both alleles of the GATA1 gene is not likely to eliminate the problem of tissue and organ rejection. The GATA1 gene is only one of approximately 600 genes involved in tissue matching, and it might be necessary to inactivate many porcine genes before pig organs will function as suitable donor organs for human recipients. Finally, potentially fatal viruses may be transmitted from pig to human during transplantation. One such virus, Porcine Endogenous Retrovirus (PERV), frequently infects pigs cells without causing harm to the host. It has been shown that the virus can be transmitted from pigs cells to human cells in the test tube, and such transmission might sicken or even kill the recipient. Moreover, once this pig virus jumped into another species, it might have the potential to cause an epidemic in humans, with devastating consequences.

Genetic engineering is also playing an important role in producing other types of medically important substances. Many vaccines are now produced via recombinant DNA technology. These recombinant vaccines have the advantage of inducing the production of protective antibodies in the immunized individual without exposing them to a potentially pathogenic organism. In addition, since individuals involved in the production of the recombinant vaccine are working with cloned genes that encode surface antigens of the pathogen, and not with the pathogenic organism itself, the entire process of vaccine production becomes immeasurably safer. A recombinant vaccine against the deadly hepatitis B virus is already in use, and someday this technology will undoubtedly play an important role in the development of vaccines against other deadly diseases such as AIDS. In hopes of attaining this goal, genes encoding several different HIV surface antigens have been cloned, including the gene for the gp120 protein found in the protein spikes on the surface of the virus and the p24 protein, found in HIV's inner core. This particular protein is involved in the attachment of the HIV virus to CD4 receptors on the surface of the helper T cells, a step that precedes entry of the HIV virus into its host cell. It is hoped that a vaccine directed against the gp120 protein might not only stimulate the production of antibodies against the HIV virus, but also help block attachment of the virus to the CD4 receptors. AIDSVAX, the gp120-based vaccine, has entered Phase III clinical trials, and is currently being tested for efficacy in humans.

In other areas of medicine, recombinant tissue plasminogen activator (TPA) is being used to save the lives of heart attack victims; interleukin-2 and gamma-interferon are being employed in cancer treatment; diabetics are receiving recombinant insulin; and children suffering from hypopituitary dwarfism are receiving human growth hormone (hGH). Prior to the development of a recombinant hGH, the only source of this hormone was the pituitary glands of human cadavers. Treatment of a single child for 1 year required hormone from 70 cadavers, making treatment difficult to obtain and prohibitively expensive. With the cloning of the hGH gene, and production of recombinant hGH, treatment is now accessible to all who need it.

Supportive Therapy and Gene Therapy

A second approach to treating genetic disorders is **supportive** or **replacement therapy.** In this case, the malfunctioning cells are replaced with normal cells. This is precisely what occurs during the transplantation of bone marrow cells. A patient with bone marrow cancer can be treated with radiation or chemotherapy to destroy all endogenous bone marrow cells. The bone marrow can then be reconstituted from cells preselected from the patient before radiation therapy, or with cells from a matched donor such as a close relative (e.g., sister or brother).

Although these specific therapies may benefit the patient, they often serve to alleviate the symptoms, rather than cure the disease. For many other diseases, neither drug nor replacement therapy is currently available. In addition, some genetic diseases (i.e., cancer) are thought to involve mechanisms controlling the expression of a number of genes (e.g., so-called "oncogenes"), rather than the loss or malfunction of a single protein. The conventional treatments (chemotherapy, radiation, immunological intervention) are ineffective in these cases (i.e., malignant melanoma). One promising new approach to curing such diseases is **gene therapy,** or restitution of the normal gene in vivo.

gene therapy

Several recent advances have enhanced the potential to eliminate human diseases through gene therapy. One is the ability to identify mutant genes that cause disease, and to clone them without having to isolate the protein they encode. This powerful technique has been called **reverse genetics.** Considering the rapid rate at which genes are now being assigned to chromosomes, it is likely that there will be a dramatic increase in the number of genetic diseases identified as being caused by specific defects in single known and cloned genes. (At the first Human Gene Mapping Workshop in 1973, about 75 genes had been localized to 19 autosomes and the X chromosomes by 1998, over 30,000 genes had been mapped, dispersed among 22 autosomal and two sex chromosomes.

Another important advance is the development of techniques for the efficient transfer of DNA into human cells. This achievement is the result of an increased understanding of the biology of cells, as well as the development of a number of viral vectors. Several exciting human gene therapy trials are currently in progress. The very first, which began in 1990, involved a young girl who was suffering from a fatal genetic disorder, **severe combined immunodeficiency (SCID),** one form of which is caused by a defect in the gene that produces the enzyme **adenosine deaminase (ADA).** In individuals who lack this enzyme, the nucleotide 2′-deoxyadenosine accumulates in the bloodstream, killing both the B and T lymphocytes of the immune system. Unless these youngsters are kept in a germ-free environment, they generally succumb to relatively minor infections at an early age. In the world's first human gene therapy experiment, T lymphocytes were removed from the young SCID victim and mixed with a genetically engineered retrovirus containing a normal copy of the ADA gene. When the experiment goes according to plan, the retrovirus introduces a functional ADA gene into the genome of the T cells. Researchers tested these modified T cells to verify that the ADA gene was indeed functioning, then transfused them into the patient. The young girl's condition showed a dramatic improvement: her serum ADA levels rose, her white-blood-cell count became relatively normal, and, for the first time in her life, she was able to attend school and interact with other children. Since T cells are relatively short-lived, this gene therapy procedure must be repeated periodically.

Gene therapy has also been used in the treatment of **familial hypercholesterolemia.** Individuals suffering from this disorder have a defect in the gene that codes for the **low density lipoprotein (LDL) receptor.** This results in the inability to transport cholesterol-rich LDL into the cells. As a result, serum cholesterol levels skyrocket, atherosclerotic plaque builds up in the arteries, and victims may suffer fatal heart attacks before the age of 20. In a somewhat successful gene therapy experiment, part of the liver was removed from a 30-year-old hypercholesterolemia victim, and a functional copy of the LDL receptor gene was introduced into the separated liver cells by means of a genetically modified mouse virus. The cells were reintroduced to the liver, leading to a 25% decrease in serum LDL levels. Despite early promise, these experiments have not led to a permanent "cure" for familial hypercholesterolemia. The duration and level of gene expression of the transferred gene tends to be low. More recent studies are focusing on the feasibility of in vivo transfer of the functional LDL receptor gene to the patient's cells.

Another exciting gene therapy experiment that began in 1995 has been billed as "bypass-without-a-surgeon." A gene has been discovered that stimulates growth of collateral blood vessels around the area of an arterial blockage. A polymer impregnated with this *vegf* (**vascular endothelial growth factor**) gene has been used to coat an angioplasty balloon. The balloon was threaded into the area adjacent to the arterial blockage and inflated, where the *vegf* genes were taken up by the cells of the arterial wall. The transformed cells began to secrete the vegf protein, stimulating new collateral blood vessels to form around the area of the blockage, and thereby relieved the arterial blockage. Direct administration of growth factor proteins has proven successful in treating arterial blockage in the leg, but has been disappointing when used to treat blocked blood coronary arteries. A more recent approach to treatment of coronary artery blockage involves injecting a genetically modified adenovirus, containing the fibroblast growth factor-4 gene, directly into the heart. Preliminary results of these studies suggest that this approach may provide a safe and effective alternative to cardiac bypass surgery.

A number of problems still remain to be resolved before gene therapy can be used to treat a wide variety of genetic disorders. First, it is not possible to remove certain types of cells from the body in order to genetically modify them, nor can all types of cells be grown successfully in culture. (Cells of the lung are a prime example of this problem.) Second, there are major problems inherent in attempting to introduce foreign DNA into specific types of human cells in vivo. Certain commonly used viral vectors infect only certain types of cells, or lack selectivity of infection, or insert their DNA at random locations within the human genome. In order for gene therapy to be successful, the normal gene must reach the tissue in which the gene is normally expressed, enter the cell, undergo homologous recombination with the defective gene, and be expressed at an appropriate time, and at an appropriate level. All of this must be accomplished without disrupting the function of any other genes. Obviously, much basic research still remains to be done. Nevertheless, the advances that have been made increase the probability that genetic treatment can eventually lead to cures for a number of devastating inherited human diseases. The development of efficient gene therapy is a compelling goal because although some genetic disorders occur in only a small percentage of the population (SCID for example), others, such as sickle cell anemia and the thalassemias, affect millions of people around the world.

In an exciting turn of events, in December of 2001, researchers reported the successful use of a new gene therapy procedure to correct sickle cell anemia in mice. This novel approach involved using a lentiviral gene vector (a genetically altered relative of the HIV-1 virus), to introduce a normal version of the hemoglobin gene into mouse bone marrow cells in vitro. The mice were irradiated to kill their remaining, abnormal bone marrow cells, and the genetically modified bone marrow was reinjected. The new gene was rapidly expressed in 99% of the blue blood cells in circulation, and all clinical symptoms of sickle cell anemia disappeared. Before this procedure can be employed in humans, scientists must address two potential safety issues: (1) the possibility that the lentivirus could undergo recombination within the host genome, and give rise to an AIDS-causing virus; and (2) the possibility that it might recombine with host DNA and activate tumor-causing genes. If these problems can be averted, this new therapy might offer the first real hope for a cure for sickle cell anemia to its millions of victims worldwide.

Uses of Recombinant DNA Technology in Agriculture

For decades, genetics has played an important role in plant breeding—that is, in the development of new strains of plants with desirable characteristics. For the most part, plant breeders have had to rely upon the selection of mutants, and the hybridization between different varieties of the same species. Although these approaches have been highly successful, they are tedious. Organisms specifically constructed by recombinant DNA technologies can have much more precisely tailored genes, and, therefore, this technology is likely to become the preferred approach in the future. Creating new varieties of plants by direct alteration of the genotype is an increasingly important application of genetic engineering. Genes for fixing nitrogen, improving photosynthesis, and providing resistance to pests, pathogens, and herbicides have been transferred and successfully expressed in plant hosts. Introduction of herbicide resistance genes into economically important plants such as corn, cotton, and tobacco will permit farmers to apply herbicides to their fields throughout the growing season. *(Before we become elated about this development, we need to consider whether we really want to take steps that will undoubtedly increase the levels of toxic herbicides in the environment!)* Another gene, from the bacterium *Bacillus thuringiensis* (Bt), encodes a protein that kills insect pests such as the cutworm, but which is nontoxic to mammals. When introduced into plants, it will become possible to grow certain crops without heavy application of nonspecific insecticides. This will not prove to be a universal panacea, however. Such insecticidal proteins tend to be relatively selective with respect to the insects they kill. Nevertheless, larvae of beneficial species of insects, such as the Monarch butterfly, can also be killed by the Bt toxin. In a laboratory study conducted at Cornell University, it was shown that 44% of the Monarch butterfly larvae died after being fed milkweed leaves dusted with Bt corn pollen, but were unaffected by pollen from nonaltered corn plants. This finding is significant because milkweed, the food of the Monarch butterfly larva, frequently grows next to cornfields. Moreover, just as resistance to chemical insecticides has arisen within insect populations, so might there be similar resistance to genetically engineered insecticides. With the introduction of Bt crops, insects are exposed to the Bt toxin throughout the growing season, rather that sporadically, as was the case when Bt sprays were used for pest control. Bt resistance can develop through a variety of different mechanisms. The toxin acts by binding to receptors located in the insect gut. In order to bind, the toxin must recognize certain sugars linked to the intestinal wall. Some mutants are Bt resistant due to the lack of a galactosyltransferase, an enzyme responsible for modification of carbohydrates and sugars. Other resistant mutants may possess mutant receptors. Researchers have already noted that certain insect pests have developed heritable resistance to multiple Bt toxins that target different receptors. Widespread development of Bt resistance could result in the loss of Bt from the arsenal of nontoxic pesticides in the near future.

In other related developments, a bacterium has been developed that increases the tolerance of crops to frost. Normal bacteria of the genus *Pseudomonas,* which are found on the leaves of strawberry and potato plants, produce a protein that initiates the formation of ice crystals on the leaves when temperatures drop below freezing. Bacteria have been genetically engineered to inactivate this gene, thus, reducing frost damage. Other plants are being engineered for increased resistance to drought and for tolerance to increased salinity.

One of the novel developments in plant genetic engineering involves crown gall, a plant tumor associated with infection by the bacterium *Agrobacterium tumefaciens.* When a segment of a plasmid carried by the bacterium is inserted into the DNA of its plant host, tumor formation is induced. The plasmid, which is, therefore, called a **tumor inducing (Ti) plasmid,** has been exploited as a vector to transfer foreign genes into host plants. Unfortunately, *A. tumefaciens* only infects broad-leaved flowering plants (dicots). Alternative strategies have been developed for introducing foreign genes into other types of plants. One of the more imaginative involves the use of a microprojectile gun that actually "shoots" small tungsten particles coated with DNA into plant embryos. (The prototype was modified from a standard, bullet-firing gun!) Along more conventional lines, electroporation has been used to transform plant **protoplasts** (cells whose cell walls have been removed enzymatically).

Genetic engineering is being put to a number of creative uses in the areas of agriculture and aquaculture. Plants may soon be synthesizing plastic packing materials, thanks to the efforts of scientists who have cloned two bacterial genes necessary for the synthesis of polyhydroxybutyrate (PHB) into a plant called *Arabidopsis.* PHB is a biodegradable plastic material that is naturally synthesized by certain strains of bacteria. Transgenic *Arabidopsis* plants containing these bacterial genes are able to synthesize and accumulate small granules of PHB in their cytoplasm and vacuoles. Before this system can be used commercially, additional work will need to be done to improve the growth characteristics of these "plastic plants." Their growth tends to be stunted, possibly because PHB biosynthesis utilizes a metabolic intermediate required for another metabolic pathway.

In another interesting development, tobacco plants have been genetically engineered to produce antibody molecules, so-called "plantibodies." Although these "plantibody" molecules are not completely identical to antibodies produced in the human body, they may be useful as supplements to infant formulas. One of the more whimsical experiments produced tobacco plants that glow in the dark. Scientists achieved this by cloning the firefly luciferase gene into a tobacco plant (**Figure 16-5**). (If this technology were expanded to other plants, one might imagine seeing a field full of glowing pumpkins at Halloween!)

Much of the effort in plant genetic engineering has been directed toward the production of fruits and vegetables with higher nutritional value or improved postharvest qualities. Seeds such as beans and corn are naturally low in certain essential amino acids. This can create serious nutritional problems for citizens of developing countries who rely upon a single major crop for nourishment. Scientists are attempting to develop methionine-rich beans and lysine-rich corn to compensate for the low levels of these essential amino acids in "normal" beans and corn, respectively. If such seeds were to become widely available, they could help to provide a complete, economical source of protein for the citizens of less-wealthy countries. Other plants such as tomatoes, which are normally picked when underripe in order to prevent spoilage and bruising during the trip to market, have been modified to produce fruit that ripens

tumor inducing (Ti) plasmid

protoplasts

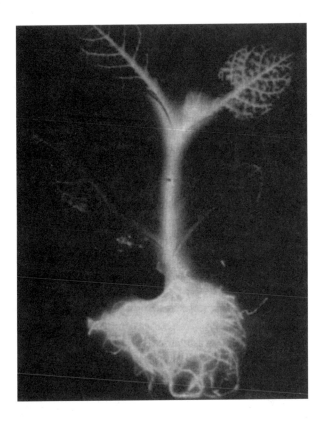

Figure 16-5 A transgenic tobacco plant that has been transformed with a firefly luciferase gene. The transgenic plant emits green light, indicating that the luciferase gene is active. (From D. W. Ow, Keith Wood, Marlene DeLuca, Jeffrey R. DeWet, Donald R. Helinski, and Steven Howell. "Transient and Stable Expression of the Firefly Luciferase Gene in Plant Cells and Transgenic Plants." *Science*, November 1986; 234:856–859, Figure 5.) Used by permission.

more slowly. In one such tomato, the MacGregor's FlavrSavr®, scientists took the gene that codes for an enzyme responsible for softening of the tomato (polygalacturonase) and reversed its orientation in the strand of the DNA normally transcribed. The gene is still transcribed, but now codes for a messenger RNA that is complementary to that produced by the normal copy of the gene (i.e., an **antisense mRNA**). The antisense mRNA will hybridize with the normal mRNA and prevent it from being translated. As a result, these tomatoes ripen more slowly and can be picked when redder and more flavorful. Higher-starch potatoes have been developed for the production of crisper potato chips and frozen french fries, and high-solids tomatoes have been developed for making thicker ketchup and tomato paste. It is becoming increasingly possible to tailor the properties of a crop to fit its intended use.

Recombinant DNA technology is being utilized to produce farm animals that reach maturity more rapidly, produce more milk or leaner meat, or have other economically desirable characteristics, such as the ability to produce pharmaceuticals. These endeavors have met with mixed success, and with a very mixed response from the public. When cattle were treated with a genetically engineered hormone to increase milk production, bovine somatotropin (BST), much concern was expressed publicly about the safety of such milk for human consumption. Transgenic animals are not always healthier animals. Researchers attempted to create a transgenic "superpig" by injecting human and bovine growth hormone genes into fertilized pig ova. Although these pigs were leaner and utilized feed more efficiently, they were also prone to a myriad of health problems, ranging from arthritis and stomach ulcers to heart enlargement and kidney disease.

antisense mRNA

Similar experiments in salmon have met with greater success. Growth hormone genes and promoters from sockeye salmon have been inserted into coho salmon eggs. The transgenic coho grow to market size (10 lb.) in about half the time required by their nonrecombinant siblings. Scientists still don't know how large they might become if not harvested; unlike mammals, salmon continue to grow throughout their entire lifetime. This could prove to have a detrimental effect upon the health of the salmon in the long run. In addition, if such "supersalmon" were to escape into the natural environment and mate with wild salmon, one might imagine major ecological consequences.

Just such a scenario took place when as many as 100,000 genetically modified salmon escaped near the Bay of Fundy. These transgenic fish are voracious feeders, due to their accelerated growth rate, and are likely to out-compete native salmon species for food. Second, even though many of the genetically modified salmon are triploid and, therefore, sterile, they are still capable of mating with native salmon, and thereby decreasing the number of offspring produced by the native species. Finally, the supersalmon are prone to deformities that limit their mobility, as a consequence of their excessive size. If fertile transgenic salmon mate with native salmon and produce offspring, these deformities could spread throughout the population of wild salmon, with serious consequences.

Other Commercial and Industrial Applications

For many years, variance of plants, animals, and microorganisms has been exploited for a variety of commercial applications—such as developing tetraploid corn, breeding faster race horses, and selecting suitable yeasts for brewing and baking—usually by identifying and isolating naturally occurring or artificially produced mutants.

Apart from the obvious applications in the pharmaceutical industry, there are a number of other uses of recombinant DNA technology. For example, attempts are being made to broaden the range of substrates of biological fermentation to include cellulose or even plastic waste. It may be possible one day to engineer organisms that can ferment at high temperatures, so that fermentation and part of the distillation process may take place simultaneously. A DNA polymerase isolated from a thermophilic bacterium, *Thermus aquaticus,* is already being used successfully in PCR at high temperatures. Furthermore, several genes from oil-metabolizing bacteria have been inserted into a plasmid, which was then inserted into a marine bacterium to yield an organism capable of metabolizing petroleum at sea. These organisms, naturally occurring oil-eaters, were used to help clean up the Alaskan shoreline after an 11-million-gallon oil spill in 1989, as well as a 1990 open-ocean spill off the coast of Texas. These bacteria metabolically convert the oil to an emulsion of fatty acids that can then be consumed by marine organisms. Although there is certainly a need for similar organisms to degrade other toxic products in the environment, the problems of the approach are numerous. For example, the concentration of the pollutants is often too low to maintain the growth of the engineered microorganism. The growing concern for cleaning up our planet may provide strong incentives to persuade industry to develop the technology required to overcome these obstacles.

In the food industry, there is the likelihood of using engineered organisms to convert waste into usable food for animals or humans. There is also the possibility of using recombinant DNA technology to devise tests designed to detect harmful or unwanted substances in food and other products. In the future, genetic engineering will affect many industrial and commercial ventures, and offer many applications as yet unimagined.

Molecular Biology: On the Front Line of the Battle Against AIDS

In the 20-year period since the first cases of **AIDS** were recognized in the U.S., this deadly disease has become the second leading cause of death among 25–44-year-olds in the United States. According to recent statistics, 40 million people worldwide are living with HIV or AIDS, one out of every 100 people in the 15–49-year-old age range worldwide is HIV-positive, and 48% of the infected adults are women.

AIDS

The virus responsible for AIDS, **HIV,** belongs to a class of viruses known as retroviruses. The replication of HIV in a host cell is mediated by the reverse transcriptase encoded in the RNA genome of the virus. Upon infection, the virus makes a DNA copy of the RNA genome, forming an RNA/DNA hybrid. The enzyme next makes a DNA/DNA double-helix copy, using the RNA/DNA hybrid as a template. That DNA/DNA copy of the viral genome then integrates into the host's chromosome using recombination enzymes encoded in the genome of the virus. From its location in the host's chromosome, the viral genome is repeatedly transcribed. Some of the transcripts are translated to form capsid proteins for the virus. Eventually, full-length transcripts are packaged, along with several viral enzymes, into mature virus particles.

HIV

HIV, or the "human immunodeficiency virus," derives its name from the fact that it primarily kills one specific type of cell, the so-called "helper T cell," which plays a central role in activating and stimulating the replication of the other cells essential for the immune response. The virus also attacks certain other white blood cells, including the macrophages, which play a critical role in presenting antigens to the helper T cells and alerting them to the presence of infecting microorganisms. By destroying the precise cells required for the mobilization of the immune response, the AIDS virus prevents the activation of B cells (which produce antibodies) and cytotoxic T8 cells (which kill virus-infected cells). Lacking a functional immune system, AIDS patients usually die of infections such as pneumonia, which in normal healthy individuals are routinely eliminated by natural antibodies.

Several factors combine to frustrate attempts of molecular biologists to develop drugs, vaccines, or other treatments for combating the AIDS virus. These include the following: the virus inserts itself into the host genome, and, therefore, is not easily removed from host cells; a long latency period permits the spread of the virus before infected individuals are identified; frequent mutations alter the virus, so that its coat protein, the usual target for vaccines, changes often; the target of the virus—the immune system—represents the body's main line of defense against viral infections. Nevertheless, recent developments in AIDS research suggest that promising new methods of treating and preventing HIV infection may be on the horizon.

A number of new antiviral agents have been or may soon be added to the arsenal of antiviral agents being used to treat HIV-infected individuals. Many of the "traditional" agents, such as AZT (azidothymidine), and its close relatives ddI (dideoxyinosine) and ddC (dideoxycytidine), are nucleoside analogues acting both as an inhibitor of reverse transcriptase and as a substrate for the same enzyme. When incorporated into a growing DNA molecule, they result in chain termination. One significant problem associated with nucleoside analogue therapy is the emergence of resistant HIV strains. This problem is especially common because the reverse transcriptase is highly error prone. As a consequence, the AIDS virus exhibits an unusually high mutation rate. This problem can be circumvented, to a large extent, by using combinations of two or three different nucleoside analogues. Some of the newer nucleoside and nucleotide analogues currently undergoing clinical testing include Tenofovir (an adenosine nucleotide analogue), DAPD, and Emitricitabine (a relative of 3TC). For the moment, at least, all appear to be effective against HIV that is resistant to other, more conventional analogues. A second and equally troublesome problem associated with the use of nucleoside analogues is the occurrence of serious side effects in certain individuals, including bone marrow suppression.

A second group of compounds also show promise as inhibitors of the reverse transcriptase. These so-called non-nucleoside inhibitors (which include tetrahydrobenzodiazepine or TIBO and some related compounds) act by binding to a hydrophobic pocket close to the active site of the reverse transcriptase. Several of the newer, non-nucleoside reverse transcriptase inhibitors (NNRTIs), including (+)-Calanolide A, Capravirine, DPC083. Emivirine, and PNU142721, are more effective in treating resistant HIV infections, and have the added advantages of being able to cross the blood-brain barrier, and persisting in the bloodstream over a relatively long period of time.

Once the viral genome has been converted to double-stranded DNA, it enters the nucleus of the cell, where it is integrated into the host chromosome by a viral enzyme known as **integrase.** Another new class of antiviral agents currently under investigation are the integrase inhibitors. These agents are small oligonucleotide molecules that block the essential integration reaction. One integrase inhibitor, AR-177, recently entered Phase I/II human clinical trials.

protease inhibitors

Two other types of antiviral agents block later steps in the HIV replication cycle. One group of compounds binds to the highly conserved zinc-fingers of the HIV nucleocapsid and blocks packaging of the viral genomic RNA into the capsid. A second, and extremely promising, group of anti-HIV drugs are the **protease inhibitors.** The HIV protease enzyme is packaged into the viral capsid, along with the viral RNA, the reverse transcriptase, and the integrase. The protease becomes active during the late stage of viral assembly, and plays an essential role in processing the precursors of the nucleocapsid proteins and the viral enzymes. The protease inhibitors block the activity of the HIV protease, halting the process of viral maturation and rendering the new virions noninfectious. Protease inhibitors include drugs such as saquinavir, ritonavir, indinavir, nelfinavir, and their close relatives.

Several additional classes of anti-HIV drugs have joined the arsenal of therapeutic agents. The first of these are classified as attachment and fusion inhibitors, agents that protect cells from infection by HIV by preventing the virus from attaching to a new cell and penetrating the cell membrane. (AMD-3100, FP21399, PRO 542, T-20, and T-1249 all fall into this category.) Another new group of anti-HIV agents, the antisense drugs, represent a mirror image of part of the HIV genetic material, and, as such, are capable of hydrogen bonding to the single-stranded viral nucleic acid and blocking its functioning. One antisense drug, HGTV43, recently entered Phase II clinical trials.

Other therapeutic approaches are designed to boost the immune response to the AIDS virus. One more traditional type of approach that has been proposed is that of "recreating" the immune system, much as we do when we perform bone marrow transplants in cancer patients. The patient's damaged cells could be destroyed, and replaced with healthy cells from a donor. Scientists also hope that they will someday be able to develop a vaccine that would prevent HIV infection. One major stumbling block has been the high variability of the AIDS virus, which results from its high mutation rate. Recently, it has been discovered that CD8+ cells (the cytotoxic T cells that kill virus-infected cells in the body) produce several unique polypeptides (**chemokines**) that work together to suppress HIV replication. If these substances prove to be effective therapeutic agents, molecular biologists should be able to clone the genes encoding them and produce these chemokines in the laboratory, making them both readily available and affordable.

Current therapeutic protocols for treatment of newly diagnosed HIV infections recommend using a *combination of two different nucleoside analogues plus one of the following:* protease inhibitors indinavir or nelfinavir, *or* double protease combinations of ritonavir plus saquinavir, indinavir, or lopinavir, *or* the non-nucleoside reverse transcriptase inhibitor (NNRTI) efavirenz. If the treatment regimen proves effective, the patient should experience a 90% drop in viral load within 8 weeks. If viral resistance develops, a new therapeutic approach must be adopted immediately, often involving use of at least two new drugs, a combination of two protease inhibitors, or a protease inhibitor plus an NNRTI. Unfortunately, even with the availability of new therapeutic agents, the mutation rate of the HIV virus frequently outpaces the rate of development of new treatment options.

It is obvious that there is no simple solution to this complex AIDS problem. Its ultimate resolution will depend upon a combination of preventative measures designed to curtail new infections, and therapeutic measures designed to halt the transmission and progression of the disease in those already infected. It is certain to be a long battle, and one in which molecular biology will play a critical role.

Social and Ethical Issues

Where is this brave new world of genetic engineering leading us? Are we on our way to creating a perfect world, or is there a darker side to this intrusion of science into every aspect of our lives? Clearly, humanity will derive many benefits from the molecular revolution, but what are some of its potential drawbacks? Let's examine some of the ethical issues that are arising today.

First, some might ask whether scientists are "playing God" by altering the genomes of living organisms. Do we have the "right" to alter the genetic material of other organisms, thereby altering their evolutionary future? Although, from our human perspectives, we may be "improving" these organisms, it is difficult to say what impact this will have in the long run. Just as we learned what a devastating effect destruction of habitat, pollution, indiscriminate hunting, and use of toxic chemicals can have upon living organisms, so might we someday realize that our genetic experimentation has had unforeseen consequences.

Let us consider our genetically engineered plants. We can create plants that are resistant to viruses, bacteria, and fungi, but this does not guarantee that our future plants will be healthy. Just as human pathogens have developed resistance to antibiotics, so plant pathogens are likely to evolve in ways that will allow them to infect resistant species of plants. Ultimately we may end up with even deadlier, resistant strains of plant pathogens.

Second, what would happen if some of these recombinant plants cross-pollinate with their wild relatives? One area of great concern is the potential spread of herbicide resistance to wild plants, many of which are considered to be "weeds." If a wild relative of one domestic plant were to acquire herbicide resistance genes, it might become almost impossible to control. This is a real threat in areas where economically important crops grow side by side with related weed species, as in Mexico, where a close relative of corn (*teosinte*) may be found growing near cornfields. Even our own domestic crops could potentially become weeds. If crops were rotated in a particular field, herbicide resistant "stragglers" from the first crop might be difficult to eliminate, and might eventually begin to overgrow the second crop planted in the same area. Another point should be considered with respect to the development of herbicide resistant plants. Should we, in fact, be creating varieties of plants that will encourage the increased use of toxic herbicides? This is an important issue that should not be ignored.

Such a situation already exists in the case of herbicide-resistant soybeans. In 2001, genetically modified, herbicide-resistant soybeans accounted for at least 60% of the crop planted. Nevertheless, herbicide use actually increased during that same time period, largely due to two factors. First, herbicides can now be sprayed over an entire field, rather than being selectively applied to weed-containing areas. Second, the increased reliance on herbicides has resulted in an increase in the prevalence of herbicide-resistant weeds. This, in turn, has fueled a demand for development of new and different herbicides. In one particularly ironic turn of events, farmers who rotate herbicide-resistant corn and soybean crops in the same fields now find themselves in the position of having to deal with the problem of trying to eliminate stubborn, herbicide-resistant "volunteer" corn from their soybean fields.

There is also concern that some of the strategies being used to create virus-resistant plants might backfire. In certain instances, plants have been rendered virus-resistant by cloning viral genes into the plant, which, when expressed, interfere with some aspect of the viral replication process. It has been demonstrated that recombination can occasionally occur between the transgenic RNA of the plant and the genomic RNA of an infecting virus. Some scientists fear that this recombination could give rise to new and potentially dangerous plant viruses.

The safety of certain genetically engineered foods has also been questioned. One major concern is the potential for causing allergic reactions in sensitive individuals. Let's consider a specific case: one seed company was in the process of developing a recombinant soybean producing high levels of the amino acid, methionine. Unfortunately, the methionine-producing gene was derived from the Brazil nut, a food to which many individuals have serious, sometimes fatal, allergic reactions. When individuals allergic to Brazil nuts ate the recombinant soybeans, many of them produced antibodies to the soybeans. Since this suggested that there was a real risk of an allergic reaction to the high-methionine soybeans, their commercial production was halted. Other insect-resistant transgenic plants produce substances that act as inhibitors of insect digestive enzymes. Unfortunately, some of these substances also inhibit certain human enzymes, and might prove harmful if expressed in parts of the plant normally consumed by humans.

Genetic engineering is also having an impact upon our legal system, and we are seeing a number of lawyers specializing in the area of biotechnology. One of the interesting legal questions already being considered is who, if anyone, should have the right to patent a cloned human gene? Should it be the scientists who performed the actual cloning? They may have cloned the gene, but they certainly didn't create the original genome from which it was obtained. What about the individual from whom the gene was obtained? Or the university or corporation where the research was performed? Or the agency that funded the research? Or the biotechnology company that markets the sequence for research or diagnostic purposes? It might be argued by some that a person has no more right to patent a human gene sequence than he would have to patent an oak tree found growing in the forest. What do you think?

Let us now turn our attention to the direct impact this genetic engineering revolution is likely to have on our lives and our society.

Genetic Testing

In June of 2000, scientists announced that they had completed the initial sequencing of the human genome. Although this is certainly one of the most highly publicized goals of the Human Genome Project, it was not the only objective. According to the official U.S. Government Human Genome Project publications, the entire list of stated goals included: (1) identify all the approximate 30,000 genes in human DNA; (2) determine the sequences of the 3 billion chemical base pairs that make up human DNA; (3) store this information in databases; (4) improve tools for data analysis; (5) transfer related technologies to the private sector; and (6) address the ethical, legal, and social issues (ELSI) that may arise from the project. With the completion of the initial phase, we can expect to see an upsurge in the mapping, isolation, and cloning of large numbers of human genes. As we gain the ability to detect the presence of these genes in asymptomatic carriers, in individuals whose symptoms have not yet appeared, and in prenatal tests, we as a society can expect to be faced with a number of complex decisions.

First of all, who should have access to the information derived from these genetic tests? Only the individual? His or her spouse? Parents? Children? Siblings? What about his or her physician, employer, or insurance company? These are not simple questions. One might argue that only the individual should be informed.

Yet the person who inherited a gene linked to a deadly genetic disorder shares her genes with other members of her family. Do they also have the right to know that they may be the potential victims of a genetic defect? What if the individual prefers *not* to know whether he has inherited a particular gene; does he have the right *not* to know the results of genetic tests, or to refuse to have them performed?' What about employers or insurance companies? One might argue that they have a vested interest in the health of their employees or insurees. On the other hand, we have already seen numerous instances of individuals being denied employment or insurance because of medical conditions. Do we really want to expand this situation into the realm of *potential* medical conditions?

Inheritance of a particular gene does not always insure that the gene will be expressed in detrimental ways. Since the cloning of the cystic fibrosis gene, it has been discovered that not all individuals who inherit the CF alleles develop full-blown cystic fibrosis. Some individuals exhibit only mild, nondebilitating symptoms. Drug therapy, diet, exercise, medical intervention, avoidance of environmental risks— all of these can help to delay the onset or reduce the severity of the symptoms of certain genetic disorders. Should a person be denied insurance coverage because he *might* someday become ill? Should only the healthy have access to insurance? Will we see the development of a whole new class of disadvantaged citizens— the genetically disadvantaged?

Should genetic testing be limited to those conditions for which treatment is currently available, or in which access to test results would aid the individual in making informed decisions, such as the decision to have children? Some have argued that it is cruel to inform an individual that he has inherited a genetic disorder for which there is no treatment or cure, especially if its symptoms may not develop for another 20 or 30 years. On the other hand, doesn't the individual have the right to that information? Genetic counseling will clearly become more and more important as growing numbers of individuals are faced with such situations, and the number of genetic counselors in this country will need to increase to meet this demand.

Gene Therapy

Once we have isolated a gene associated with a genetic defect, cloned it, and begun to understand its mode of action and how it is regulated, the next logical step may be the development of gene therapy to treat its symptoms, or even cure it. As was the case with other genetic engineering endeavors, we may expect that some in our society will be opposed to "genetic meddling," and see it as a form of eugenics. Even if one is not opposed to gene therapy, it does raise a number of ethical issues. What types of traits should be altered, and who should decide? Hypothetically, we might all agree that cystic fibrosis is a disorder that should be treated, but what about nearsightedness, or premature baldness, or short stature? Do we really know what is best for the future of our species? Another important issue is *who* will have access to gene therapy. Will this be an option reserved for the wealthy? Will it be covered by our health insurance (if we are even eligible for health insurance)?

These issues become even more complex when we consider the possibilities of cloning humans and creating transgenic humans. In November of 2001, Advanced Cell Technology, Inc. (ACT) stunned the world with the announcement that it had cloned a human embryo for the first time. The "embryos" in question were the products of two different experimental methods. One technique involved activating human eggs to undergo parthenogenesis (activation and subsequent development without exposure to a sperm). Out of 22 eggs, 20 underwent cleavage, and five of those (30%) actually continued to divide for 5 days, going on to form a blastocoell cavity. A second technique involved nuclear transplantation: the nucleus from a human somatic cell (a fibroblast or a cumulus cell) was transplanted into an enucleated human donor egg. Although the eggs containing the transplanted fibroblast nucleus formed a pronucleus, they did not undergo further cleavage. In the case of the reconstructed eggs containing the cumulus cell nucleus, four out of eight developed a pronucleus, and three out of those four underwent one or two cell divisions. The stated goal of the ACT experiments was not the production of a human infant, but rather the production of human embryonic stem cells (from the inner mass of the blastocyst), to be used for therapeutic purposes. Within a matter of days, a second group, Clonaid, claimed that they had also produced a human embryo, and, in this instance, their ultimate goal was reproductive cloning. Once a human has been cloned, what is to prevent scientists from creating a genetically altered "designer" baby?

The mere suggestion of human cloning is sufficient to trigger impassioned debate. First and foremost, many would argue, cloning a human would be tantamount to "playing God." Second, religious and ethical arguments aside, science's track record in cloning mammals is less than stellar. In the case of Dolly, the first cloned sheep, it required 277 attempts to produce one viable lamb. Many of the embryos failed to develop, were malformed, or died before birth. Furthermore, even though Dolly survived to adulthood and produced four offspring, she has experienced some health and genetic problems that may be related to being the product to cloning. Dolly's telomeres are shorter than normal, suggesting that she may undergo premature aging. Dolly has experienced problems with obesity, and, more recently, arthritis uncommon in such a young sheep). Either or both of these problems might be cloning related.

In cloned cattle, 73% of the pregnancies ended in spontaneous abortion, and 20% of the calves that did survive until birth died shortly afterwards. Health problems noted among cloned calves included enlarged hearts, liver abnormalities, lungs that fail to develop normally, and "large calf syndrome," a fatal condition in which the pregnancy results in a huge calf (with an abnormally large placenta) that dies around the time of birth. Typically, even in species of mammals where cloning has worked well, success rates average 1–4%. In mice, for example, the success rate is 2–3%, and cloned mice experienced problems including extreme obesity and developmental delays.

Cloning primates has failed miserably. In monkeys, no one has yet succeeded in producing a clone from an adult cell. Nearly 500 cloning attempts resulted in grossly abnormal embryos, which either failed to develop, or died shortly after being transferred to the uterus of a female monkey, and even resulted in placentas without a fetus. In most instances, the reconstructed egg did not undergo more than a few cell divisions, a result reminiscent of the results of the ACT human nuclear transplantation cloning experiments.

What does this mean in terms of the prospects for successful human cloning? It suggests that we might expect a 96–98% *failure* rate, that the cloned human might suffer from any of a number of different physical abnormalities, and might die suddenly, or age prematurely. Is this an acceptable risk? Most would argue that it is not!

What about correcting genetic defects in embryos conceived in the traditional way, or in the germ cells of adults? Are we playing God? Once again, who is to decide which traits should be altered? In the end, one of the questions we need to ask ourselves is whether we have the right to alter the genetic future of those who will have no say in the matter—the generations to come. These are not simple questions, and there will certainly be no simple answers. Nevertheless, we as members of a society must educate ourselves so that we can contribute to the discussion of these issues and make informed decisions when the time comes. Ultimately, our future may lie in our own hands.

SUMMARY

Molecular biology is having a profound effect upon life in our society. Recombinant DNA technology is being used in research for a variety of purposes. It has been used to produce specific mutations within the genome of an organism, to alter the genotypes of organisms, to create transgenic organisms capable of synthesizing molecules of medical or economic importance, to supply DNA and RNA molecules for research purposes, and to map the genomes of both humans and research organisms. Techniques such as RFLP analysis and DNA fingerprinting, which were originally developed in the research labs, are being employed in areas ranging from forensics to preservation of endangered species. Genetic engineering has made it possible to produce important, and previously scarce, pharmaceutical products in the laboratory. Genes encoding substances ranging from human insulin to tissue plasminogen activator have been cloned, and recombinant vaccines are increasing in prominence.

Gene therapy is currently being tested as a treatment for several different types of genetic disorders, including severe combined immune deficiency, hypercholesterolemia, and cystic fibrosis. A number of practical problems remain to be resolved before gene therapy will be widely available.

Transgenic plants and animals may soon revolutionize agriculture. Plants have been engineered to exhibit resistance to pathogens, insect pests, and herbicides. Properties such as post-harvest quality, resistance to environmental factors, nutritional value, and palatability have all been improved, using the techniques of molecular biology. Animals, too, have been genetically modified. Current efforts are being directed toward the production of animals that mature more rapidly, utilize feed more efficiently, or produce leaner meat or more milk.

In industry, too, molecular biology is having an effect. Genetically altered organisms are being used to ferment novel substances, clean up oil spills, and even convert waste into usable food.

The rapid growth of our genetic engineering capabilities has raised a multitude of social and ethical issues, some of which were considered in the final section of this chapter.

DRILL QUESTIONS

1. Provide examples from recent press reports (e.g., national news magazines or newspapers) of applications of genetic engineering that reach large segments of society.

2. For this question, refer to Figure 16-1.
   (a) What is an embryonic stem cell (ES), and what is the advantage of using such cells in the production of transgenic mice?
   (b) How does the insertion of the *neo^r* gene into the *c-abl* gene alter the expression of the *c-abl* gene? What effects does this alteration have upon the phenotypic properties of the cells into which the modified gene is inserted?
   (c) How is the researcher able to identify ES cells that have been successfully transformed with this modified *c-abl* gene?
   (d) What is meant by "homologous recombination"? Why is it important to utilize cells in which the *c-abl* gene has undergone homologous recombination when attempting to produce transgenic animals?
   (e) Describe an experimental procedure that could be used to determine whether a particular clone of transformed ES cells had integrated the *c-abl* gene via homologous recombination.

3. (a) Describe some of the possible practical applications of the technology used to create transgenic animals.
   (b) What are some of the potential disadvantages and problems associated with the use of such technology?

4. (a) What is gene therapy and how does it differ from supportive (= replacement) therapy?
   (b) Why is it much more difficult to utilize gene therapy to treat a human genetic disorder than it is to produce a transgenic animal in the laboratory?

5. (a) What are RFLPs and how did they arise?
   (b) What are some of the advantages of using RFLPs to diagnose human genetic disorders? What are the disadvantages?

6. (a) What is a VNTR?
   (b) Explain the difference between a microsatellite repeat and a mini-satellite repeat. Which of the two is more useful in DNA fingerprinting? Why?
   (c) Describe the general procedure used to produce a DNA fingerprint.
   (d) What are some of the practical applications of DNA fingerprinting?

7. Suppose a researcher wished to clone the human growth hormone gene (hGH) and produce hGH in the laboratory.
   (a) What technical problems might be encountered when cloning and expressing a gene such as the *human* growth factor gene in a *bacterial* cell?
   (b) Describe two alternative experimental approaches that might be used to circumvent such potential problems.

PROBLEMS

1. For problems 1 and 2, refer to Figure 16-1.
   (a) Diagram the wild-type *c-abl* gene and show what this gene might look like relative to the wild-type gene, after the insertion of the *neo^r* gene. Also, indicate a region of the wild-type gene that could be used as a hybridization probe in the Southern blot screening procedure.

(b) Show possible homologous and nonhomologous recombination events that could lead to the production of G418-resistant cells.

(c) Assuming that the restriction sites used for the Southern blot analysis flank the *c-abl* gene (are situated in the chromosomal DNA upstream and downstream of the gene), show the chromosomal organization of transformed strains (G418 resistant) for both homologous and non-homologous recombination events. Indicate the approximate positions of the restriction sites and the hybridizing and nonhybridizing fragments. (Please remember that this is a diploid organism.)

2. What would be the outcome if the procedure used to create transgenic mice (as shown in Figure 16-1) were altered in the following ways? (Consider each situation separately.)

(a) The ES cells did not possess a gene responsible for visible phenotypic trait (such as agouti coat color) that differed from that of the cell in the recipient blastocyst.

(b) The inserted "disrupting" gene (i.e., *neo^r*) did not insert into our gene of interest (*c-abl*).

(c) Electroporation failed.

(d) The feeder layer allowed the ES cells to differentiate. (*Hint:* use the term "totipotent" in your answer.)

3. For this problem, refer to Figure 16-4.

(a) In 16-4(a), explain how you could determine which, if any, of the three suspects could be a prime suspect in this rape case?

(b) Can you determine which of the three suspects is innocent, based upon the DNA fingerprints presented? Can you state with absolute certainty which of the three suspects is guilty? Do you suppose it would be easier to prove guilt or innocence in a court of law, based on DNA fingerprints?

(c) In 16-4(b), which of the two men (F1 or F2) is most likely to be the father of the child (C)? Please explain how you arrived at your conclusion.

4. Consider the following hypothetical situation:

H gene

Normal German Shepherd dogs possess a thick coat of hair. The gene responsible for this normal coat (the **H gene**) is cut by *Eco* RI into three restriction fragments: 0.4, 0.7, and 0.9 Kb in size, each of which is capable of hybridizing with a radioactive probe produced from the cloned H gene. A dog breeder was shocked to discover that his newly acquired, normal-haired male and female German Shepherds produced both hairless and normal-haired puppies when mated. (About 25% of the puppies in the litter lacked hair.)

DNA was extracted from skin samples obtained from the hairless puppies, from their normal-haired littermates, and from the parents of the puppies. The DNA samples were digested with *Eco* RI and Southern blotted, using the H gene as a probe. It was found that the DNA samples obtained from the puppies produced three different patterns of bands:

(1) All (3) of hairless puppies: 0.9 and 1.1 Kb

(2) Three of the normal-haired puppies: 0.4, 0.7, and 0.9 Kb

(3) Six of the normal-haired puppies: 0.4, 0.7, 0.9, and 1.1 Kb.

(a) Draw a diagram of a restriction map of the normal H gene, showing all *Eco* RI cutting sites.

(b) Now draw a second restriction map that explains the occurrence of the hairless puppies. (It should be consistent with the data previously presented.)

(c) Draw a clearly labeled diagram of the Southern blots produced from the three different types of puppies (Lanes 1–3), and from the parents of these puppies (Lanes 4 and 5).

(d) How could this information be used to devise a screening test for normal-haired German Shepherds possessing the undesirable hairless gene?

CONCEPTUAL QUESTIONS

1. Consider the issue of gene therapy in the treatment of human diseases and the potential of creating a world in which diseases such as muscular dystrophy or hemophilia would be curable, and our elderly need not become senile. Given the time and difficulty involved in identifying mutant genes that cause disease and replacing the mutant gene with a wild-type one in homologous fashion, is it worth the time and effort spent for the good of only a small group of people, in many cases? Who takes responsibility if a gene is integrated incorrectly during treatment?

2. One concern about recombinant DNA technology is the danger of creating recombinant organisms hazardous to the environment, or to life on earth. Consider such dangers for yourself, versus the current and potential good for society. Which side should prevail, and how great a threat do you think recombinant DNA technology is to life as we know it? Imagine doing scientific research today without recombinant DNA technology. What types of findings would be impossible to obtain? Consider also the legal applications of recombinant DNA technology, such as using PCR- or RFLP-mapping to confirm relatedness, or to identify criminals from a few cells left at the scene of a crime. What are the potential benefits and dangers of using recombinant DNA technology in this manner?

3. The entire human genome is being sequenced. What types of information can be gained? With the potential for identifying a great number of genes, how will their functions be determined? In your opinion, would it be better to use the total sequencing approach, or more conventional gene isolation techniques to unravel the mysteries of the human genome?

Postscript to Your Review of Molecular Biology

Molecular Biology Is Enjoying a Golden Era!

Congratulations on completing your review of the essential features of the discipline of molecular biology! As you have journeyed through this learning experience, perhaps you have become aware of the rapid pace with which discoveries are being made in molecular biology. Newspapers, television programs, and Internet sites regularly publicize discoveries that represent extensions of the information base included in this textbook. No doubt your instructor mentioned several during your course of study. You will hopefully feel a sense of accomplishment at being able to understand those recent discoveries. Please try explaining some of them to your fellow students and friends. For example, relate to them some of the applications of molecular biology described in Chapter 16. In doing so, your working knowledge of this discipline will be enhanced, as will your self-esteem from being able to engage the attention of an audience of your peers.

This learning experience will necessarily, however, serve only as a starting point:

> Essentials of Molecular Biology *leads us up to the end of the beginning.*
> *The beginning of the end of molecular biology is far off in the future!*
> GMM (adapted from Winston Churchill, 1942)

Indeed, although the template for *information flow* in living cells was thought to be understood almost four decades ago, once the conversion of nucleotide sequence information via mRNA into protein was understood, significant modifications in that scenario have been uncovered in the past decade or two. A half dozen of the biggest surprises are listed below:

Structure of Macromolecules

Ribozymes: RNA-containing enzymes that catalyze reactions in much the same fashion that typical protein enzymes speed up the rate of chemical reactions. Originally, it was believed that only regular proteins were capable of that activity.

Split genes: Introns punctuate the typical eukaryotic gene, and can comprise up to 99% of the linear sequence of some human genes. Previously, it was thought that a direct linear relationship exists between nucleotide and amino acid sequence.

Function of Macromolecules

Reverse transcriptase: This is an RNA-dependent DNA polymerase (found in some viruses) that catalyzes the synthesis of DNA using RNA as a template. Initially, it was doubted that such an enzyme exists, for it would appear to represent a "reversal of normal information flow."

DNA repair enzymes: The importance during natural selection of such enzymes has been seriously underestimated. At present, it is believed that more than 100 different enzymes play one or another role in maintaining a correct nucleotide sequence.

Coordination of Macromolecular Function

Nitric oxide: A gas—synthesized in some neurons by the enzyme nitric oxide synthase—that is released as a consequence of neural activity and functions as a signal molecule (neurotransmitter). Its discovery erased some of our simplistic thinking about intercellular signal pathways.

Redundancy in embryonic processes: Knocking out the expression of a presumed regulatory protein (e.g., activin, which normally plays a role in embryonic muscle formation) fails to disrupt muscle-cell development. Multiple regulatory proteins, or alternative metabolic pathways, rather than the simple linear representations of gene control mechanisms, apparently exist.

You have, of course, learned about many of these surprises here in *Essentials of Molecular Biology.* You can expect, however, many more surprises. Several factors account for our amazement. With the advent of genome sequencing, our choice of research projects has expanded, and our conceptual approaches to building knowledge have been enhanced. In addition, record numbers of scientists are engaged in molecular biology research. Such research is being carried out at a variety of institutions, including traditional colleges and universities, research institutes, pharmaceutical companies, government laboratories, and biotechnology companies. Expanded research funding from government, private, and philanthropic sources has fostered increases in the rate of discovery. In addition, the abundant flow of information that has been facilitated by the Internet, and the cooperation between scientists, which often crosses international boundaries, have also served to bolster progress.

Finally, improved technologies (including several mentioned in Chapter 13) are providing opportunities for greatly expanding the rate at which nucleic acid and protein samples are processed, and in some instances are dramatically improving the sensitivity of various measuring devices.

Discovering the unexpected and postulating the unpredicted will, therefore, likely fascinate molecular biologists for decades to come. During just the past 2 or 3 years, as this 4th edition of *Essentials of Molecular Biology* was being prepared, there have been many additional discoveries that came as big surprises even to veteran molecular biologists. The following list contains a few examples of these recent surprises:

Transport of RNA between (plant) cells: Our paradigm stated that RNAs function within cells in well-defined intracellular locations. Now it appears that some RNAs move between cells, carrying information.

Gene silencing by RNAi: Its mere existence came as a surprise. Now we know that some regulation of gene expression is achieved by "antiexpression" (inhibitory) double-stranded RNAs.

Human gene count: Present estimates favor approximately 30,000 protein-coding genes, which is only about twice as many as the *Drosophila* gene count. It appears, therefore, that combinatorial gene expression patterns are responsible for the complexity exhibited by vertebrates.

Indeed, your instructor may be able to add a few more entries to this list, since every few months additional revelations appear in both scientific journals and the popular press.

Speculation—Let's Anticipate a Few Discoveries in Molecular Biology!

The results of the basic research endeavors of the type that generated most of the information content of this textbook are often characterized as being "unpredictable." As a consequence, surprises emerge that often generate amazement and astonishment, sometimes from the most unlikely directions. For example, the ribozymes previously mentioned were discovered when scientists searched for the proteins responsible for editing RNAs in the single-celled, pond-dwelling protozoan (*Tetrahymena*). After repeated attempts failed to isolate a protein enzyme, and only after discussion, which at the time appeared to confound the use of "strong inference" as a logic (see Chapter 1), attention turned to what was then considered an unlikely possibility—that nucleic acids might be capable of enzyme-like catalysis.

Examples such as that one account for why scientists strongly hold onto the concept that basic research results are difficult to predict. Here are some quotes which reinforce that notion:

You rarely find the most important things by deliberately looking for them.
Joshua Lederberg, 1995

The Error of Futurism: Prediction is too unreliable to provide the basis for any restrictions on research.
David Baltimore, 1979

Unpredictable results are the most interesting.

Masami Wakahara, 1984

We can see when some area of science is useful or is about to be useful, but we can't see that some area of science will be useless.

Ralph Gomory, 1994

That unpredictability feature usually makes it difficult to target basic research to solve specific problems. Most often it is curiosity driven research, for which the investigator displays individual passion, or, at the personal level, for which making discoveries fills a basic psychological need of the investigator that generates the types of fundamental discoveries that comprise the "essential features" of molecular biology.

Nevertheless, most scientists enjoy speculating about which research areas provide the highest probability for generating new knowledge. Several possible "breakthrough" projects follow (in alphabetical order) that molecular biologists have recently highlighted:

Archaebacteria information processing systems: They are difficult to culture in the laboratory and, therefore, it is uncertain how diverse is this collection of organisms. For example, many strains are believed to exist in extremely inhospitable (e.g., thermophilic) environments, such as natural hot springs. It is likely that researchers have underestimated both their range of types and total biomass. Once representative groups are ready for study, novel insights into molecular mechanisms and phylogenetic histories will likely be forthcoming. Recall, these organisms represent a "bridge" between typical prokaryotes and eukaryotes.

Bioengineering proteins: Introducing novel amino acids—synthesized in an organic chemistry laboratory so as to mimic the generic structure of a naturally occurring amino acid, but with slightly different properties—into proteins in place of normal amino acids will presumably confer new characteristics. Both site-specific insertion and general replacement could be used as strategies for generating proteins with novel therapeutic applications, or enhanced stability properties, for example.

Genome sequencing enterprises: Comparisons of human genome sequence data and sequence data from genomes of closely related vertebrates (e.g., chimpanzee and mouse) with data from puffer fish (*Fugu rubripes*) genome sequencing will probably yield insights into the minimal information required for a vertebrate lifestyle. The *Fugu* genome is comprised of only 400 million bp, whereas the human genome consists of almost 10x more DNA (approximately 3 billion bp).

Metabolic engineering: Previous successes with genetic engineering (e.g., "golden rice" that synthesizes β-carotene and reduces the incidence of vitamin A deficiency in the third world) are prompting more ambitious attempts at remodeling natural organisms for specific tasks. For example, as information from proteomics (see Chapter 13) accumulates, combinations of metabolic pathways, rather than only the single-pathway changes that have been achieved so far, will be genetically modified to generate ever more useful organisms.

Protein synthesis logistics: Although it is certain that the cytoplasm is the major site of protein synthesis in mammalian cells, recent reports reveal that coupled transcription and translation (as is observed in prokaryotes—Chapter 9) occurs in mammalian cell nuclei. It is possible to speculate that this process reflects phylogenetic history. Additional studies with "bridge" organisms such as *Archaebacteria*, previously mentioned, will be very informative for taking inventory of the myriad mechanisms that regulate eukaryotic gene expression.

Regenerative biology: Genomics and proteomics will likely soon begin to lead the way in explaining why the appendages (e.g., limbs) of salamanders are able to completely regenerate after amputation, whereas those of higher vertebrates (e.g., humans) are not. The extent to which the embryonic gene expression pattern is recapitulated during regeneration will provide insights for the design of suitable therapeutic strategies for treating human injury.

Transgenic mouse models for human diseases: Research progress in some of the most common human genetic diseases (e.g., Alzheimer's disease) is presently rate limited by the lack of availability of suitable laboratory animal model systems. Once genes that are believed to represent good candidates for causing a specific disease are discovered, isolated, and cloned, it will be possible to introduce them into laboratory mammals (see Chapter 16) in order to provide experimental material for routine tests. Once "cause/ effect" experiments can be carried out on a regular basis, breakthroughs will quickly follow.

Enhance Your Ability to Learn Molecular Biology!

During your journey through this textbook, you have probably encountered several phenomena that were difficult to fully comprehend, at first reading. For some students, this might have been Chapter 3's description of the dideoxynucleotide sequencing method, Chapter 7's analysis of the Okasaki pulse-chase data, or Chapter 11's review of the regulation of the tryptophan operon. Should you elect to continue studying molecular biology, you might like to expand your repertoire of learning techniques. Here is a list of 12 ways to better understand complex phenomena in molecular biology:

1. **Begin at the beginning:** Learn what your *learning style preference* is, and then develop strategies for specific tasks based on your learning *strengths.* That is, begin to *learn how you learn best!* Are you a *visual* learner? An *auditory* learner? A *sensory* learner? Why not do some Internet searches to collect information about (your) learning styles?

2. **Collaborate:** As undergraduates, you learn best when you learn together, with *study partners.* Get a group together to review the intricacies of the polymerase chain reaction!

3. **Learn through metaphors:** Review a problem or phenomenon in biology in terms that are familiar to you from daily life in order to understand its intricacies. The events at the DNA replication fork would be amenable to this learning aid. For example you might describe DNA pol III as a "one-way enzyme that works on a two-way street."

4. **Perform a context review:** Understand a concept/theory/phenomenon by reviewing its history—what preceded it, what exists now, and what it will probably be like in the future. Your understanding of the basis for believing that DNA is the genetic material might be good for this approach.

5. **Read alternative explanations:** Since different authors often explain the same phenomenon in different ways (use of symbolism/level of detail/illustrations/etc.), reading more than one explanation often enhances understanding. Your understanding of biological membrane structure may be enhanced with this learning method.

6. **Surf the web:** Your computer is a library! See what's been written about what you would like to learn. Nonsense suppressor tRNAs are diagrammed in various ways on several websites. You may want to search them out.

7. **Construct a "concept map":** Compile a list of key words, then generate a flow sheet that cartoons the relationships between the terms. Regulation of transcription in eukaryotes would be a good candidate for this strategy.

8. **Write it out:** As Lee Iacocca, a prominent industrialist, once said:

   In conversation you can get away with all kinds of vagueness and nonsense, often without even realizing it. But there's something about putting your thoughts on paper that forces you to get down to specifics. That way, it's harder to deceive yourself—or anybody else.

   Why not write out (without including diagrams) your own description of the details of protein synthesis?

9. **Prepare a road map:** Photocopy a set of illustrations, paste them on a large sheet of paper, then draw arrows to connect key features in a linear, progressive fashion. Learning how to clone fragments of DNA would likely be facilitated with this approach!

10. **Construct a three-dimensional model:** Representing the phenomenon (e.g., molecular entity) you are attempting to understand using materials you can work with your hands leads to "crystal-clear" images. The quaternary structure of a protein could be modeled with paper cutouts, and, thus, enhance your understanding of the roles shape and weak forces play in establishing a macromolecule's three-dimensional structure.

11. **Step outside yourself:** Disconnect yourself for a moment, and ask "what big picture am I dealing with," and then try to assemble the facts in a way that has meaning to you. Feedback inhibition of protein function could be favorably discussed in this fashion.

12. **Work backwards:** Begin with the final question, or answer to a problem, and work your way back, filling in knowledge gaps along the way. For example, initiate a discussion with classmates by posing the following question: How can transposon movement within a genome alter the regulation of gene expression?

For diverse learning tasks, please feel encouraged to avail yourself of opportunities to sample what for you might represent a new *way of learning*. Novel learning challenges often require us to broaden our learning methods. You might like to ask your instructor to explain some of the ways he/she has learned to grasp what was found—at first reading—to be a difficult description or concept.

Consider Becoming a Molecular Biologist!

Why not develop a career for yourself in molecular biology? Opportunities are available for all levels of education, beginning with an undergraduate degree in either biology or chemistry. Adventure, as well as satisfaction from knowing that you are participating in a "golden age" endeavor, await newcomers. Internet search engines can be used to survey the range of employment opportunities, as well as chances for advanced (graduate) study. Included will be (in alphabetical order): animal technologist, bioinformatics data manager, biotechnology production manager, cell culture technician, cytologist (molecular), electron microscopist, graduate student, journal editor, medical laboratory technician, microbiologist, pharmaceutical salesperson, postdoctoral research fellow, product representative, professor, research associate, research technician, scientific curator, staff scientist, virologist, and many more.

Please be encouraged to take this copy of *Essentials of Molecular Biology* along to your new job, and refer to it from time to time, as you encounter challenging ideas in molecular biology.

Appendix

Chemical Principles Important for Understanding Molecular Biology

The complex structures and functions of the molecules described in this text are a consequence of the chemistry of their component parts. To understand how these molecules function, we need to briefly review some of the fundamentals of chemistry. In particular, we need to understand the unique features of carbon, hydrogen, oxygen, nitrogen, phosphorus, and sulfur, which enable these elements to form the compounds essential for life. We will begin with an introduction to atomic structure and chemical bonds.

Structure of the Atom

Atoms are composed of a nucleus, which contains protons (positively charged particles), neutrons (particles with no charge), and electrons (negatively charged particles). The number of protons in the nucleus defines the atomic number and specifies the identity of an element. For example, an atom with six protons has an atomic number of six and has been given the name "carbon." (Refer to a Periodic Table of the Elements.) The electrons are added to shells around the nucleus.

Isotopes of the same element all have the same number of protons, but differ in the number of neutrons they contain. In an uncharged atom, the number of electrons and protons is equal.

Chemical Bonds

When individual atoms form chemical bonds between one another, they create molecules. Chemical bonds are important for the structure and properties of small molecules, such as water, and simple salts, acids, and bases; they are equally important for the structure and properties of large molecules, such as DNA and proteins (see Chapters 2–4). Chemical bonds can be of several different kinds, which are described in the sections following.

Covalent Bonds

Covalent bonds are formed when electrons are shared between two atoms. The electrons occupy shells that optimally contain the number of electrons shown in **Figure A-1**. Elements with completely filled outer shells do not readily form bonds with other atoms. For example, helium with two electrons is unreactive. These elements are found in the far right column of a Periodic Table of the Elements. In contrast, the elements found in biological compounds (hydrogen, carbon, oxygen, nitrogen, sulfur, and phosphorus) are uniquely suited to share electrons in covalent bonds, thereby forming very stable structures. A simple example of a covalent bond is shown in **Figure A-2**, where an atom of hydrogen (atomic number 1), with a single electron occupying its outer shell, combines with another atom of hydrogen to form molecular hydrogen. The two electrons are equally shared, giving each atomic nucleus a filled shell.

As shown in Figure A-1, the second shell can hold eight electrons. This means that oxygen (atomic number 8) has only six electrons in its outermost shell and needs to attract two electrons to share to fill its outer shell. Molecular oxygen, shown in **Figure A-3**, consists of two oxygen nuclei sharing two pairs or four electrons (dotted circle in Figure A-3). This is known as a double bond. In this way, each oxygen gains two electrons and satisfies its need for a total of eight electrons.

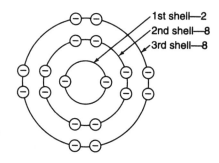

Figure A-1 Model of an atom showing electrons in shells around the nucleus.

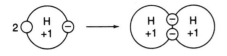

Figure A-2 Hydrogen atom and molecular hydrogen, H_2.

Oxygen can also share electrons with many other types of atomic nuclei, as can be seen in the molecules described throughout this text. When oxygen shares electrons to form covalent bonds with two hydrogen atoms, water is formed, as shown in **Figure A-4**. These covalent bonds are very strong, and they have unique properties, as described in the following section.

In **Table A-1**, we see the atomic numbers and the number of electrons required for a filled outer shell for each of the major elements found in biological molecules. The number of electrons needed to fill an atom's outermost shell reflects the number of bonds in which that atom can engage. Therefore, all the elements listed in Table A-1, except hydrogen, can form multiple bonds such that large, complex molecules are possible. Furthermore, the smaller an atom is, the easier it is to share electrons. Since the elements found in biological compounds all have relatively low atomic numbers, they form very strong covalent bonds. The most common bonds are single bonds (in which only one pair of electrons is shared), but double bonds (in which two pairs of electrons are shared) are also important for biological function. Carbon, oxygen, nitrogen, and phosphorus can all form a variety of double bonds. Carbon and nitrogen are also found in rare triple bonds (**Figure A-5**).

Ionic Bonds

In some compounds, the electrons are not equally shared, but are effectively transferred from one atom to the other. The result is two ionized species that associate due to their opposite charge. The salt that is formed from one atom of sodium and one atom of chlorine is an example of an ionic bond. In sodium chloride, an electron from sodium (atomic number 11) is transferred to chlorine (atomic number 17). The atomic nuclei in the resulting ionic compound each have a completely full outer shell of electrons and opposite charges (**Figure A-6**). The strength of ionic bonds varies more widely than does the strength of covalent bonds (**Table A-2**). In biological compounds, we will see many ionized species. In particular, oxygen, nitrogen, and phosphorus are frequently involved in ionic bonds.

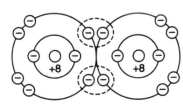

Figure A-3 Molecular oxygen, O_2.

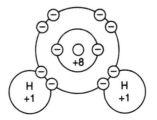

Figure A-4 Water, H_2O.

TABLE A-1	ATOMIC CONFIGURATION OF BIOLOGICALLY IMPORTANT ATOMS		
Element	Atomic Number	Electrons in Full Shell	Number Needed to Complete Shell
H	1	2	1
C	6	10	4
N	7	10	3
O	8	10	2
P	15	18	3
S	16	18	2

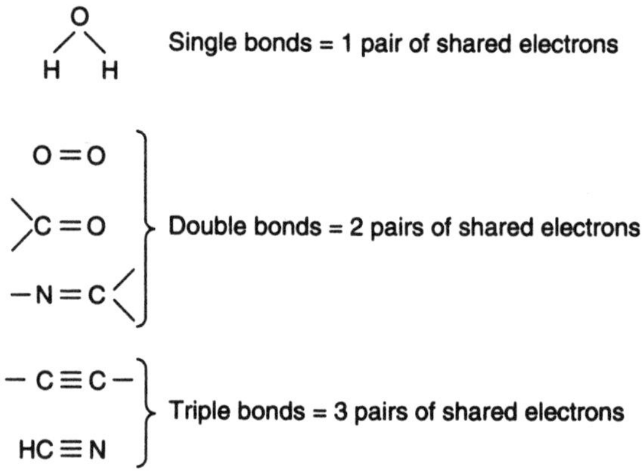

O
/ \
H   H   Single bonds = 1 pair of shared electrons

O = O

C = O   Double bonds = 2 pairs of shared electrons

— N = C

— C ≡ C —   Triple bonds = 3 pairs of shared electrons

HC ≡ N

Figure A-5 Some examples of single, double, and triple bonds.

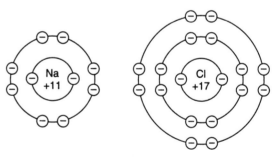

Figure A-6 The ionic bond of sodium chloride, NaCl.

Net charge +1   Net charge −1

TABLE A-2	STRENGTH OF VARIOUS TYPES OF CHEMICAL BONDS
Type of Bond	Force Required to Break (kcal/mole)
Covalent	50–100
Ionic	80–1
Hydrogen	3–6
Hydrophobic	0.5–3

Polar Covalent Bonds

Many bonds in biological compounds are intermediate between covalent and ionic bonds. This means that the electrons are neither equally shared nor completely transferred between atomic nuclei. For example, the hydrogen and oxygen in a water molecule are actually joined by polar covalent bonds; however, the shared electrons are held more tightly by the oxygen nucleus than by the smaller hydrogen nucleus. Since the bonds holding the two hydrogens to the oxygen form a 105° angle, there is a charge polarity to the water molecule (**Figure A-7**). The oxygen side of the molecule is more negatively charged than the hydrogen side, as indicated by the δ symbol, which stands for "partial."

Bonds between hydrogen, carbon, oxygen, nitrogen, and phosphorus are often polar covalent bonds. Hydrogen and nitrogen frequently are associated with a partial positive charge, while oxygen and phosphorus carry a partial negative charge. Carbon can carry either a partial positive or a partial negative charge (**Figure A-8**). When only carbon and hydrogen are present in a compound, the bonds are not polar, since carbon and hydrogen share electrons about equally. The four bonds formed by carbon show a tetrahedral symmetry (**Figure A-9**).

The existence of polar and nonpolar covalent bonds in biological compounds results in two additional classes of bonds that are very important for the structure of macromolecules. These are the hydrogen bonds and hydrophobic bonds described in the next two sections.

Hydrogen Bonds

When hydrogen is involved in a polar covalent bond with another atom, it carries a partial positive charge (Figure A-7). This partial positive charge will attract and associate with molecules (or parts of molecules) with a partial negative charge. For example, in water, the partially positively charged hydrogens will associate with the partially negatively charged oxygens of adjacent molecules. These hydrogen bonds are much weaker than a true covalent bond (Table A-2), but when present in large numbers, they can contribute substantially to the stability of a mixture of molecules. As shown in **Figure A-10**, liquid water is actually a highly ordered structure with each water molecule involved in hydrogen bonds with three other molecules. This stable network gives water many of its unique properties, including its high boiling point and low freezing point. In addition, the ability to form

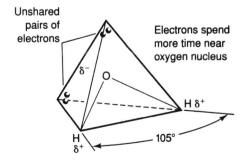

Figure A-7 Geometry and partial charge distribution of water.

$$\ce{>C=O}$$ with δ^+ δ^-

$$CH_3 - Li$$ with δ^- δ^+

Figure A-8 Partial positive and negative charges on carbon.

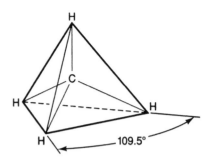

Figure A-9   Tetrahedral carbon.

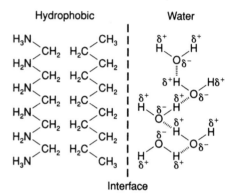

Figure A-10   Hydrogen-bonded structure of water.

hydrogen bonds with a variety of polar compounds makes water an extremely good solvent for most biological molecules.

Hydrogen bonds are also important for the structure of proteins and nucleic acids, as described in Chapters 3 and 4. Hydrogen bonds can be formed intramolecularly—that is, holding atoms in different parts of one molecule together, and intermolecularly—holding atoms in separate molecules together.

Hydrophobic Bonds

hydrophobic
hydrophilic

Nonpolar bonds, such as those found in compounds made up only of carbon and hydrogen, also have unique properties. These bonds are known as **hydrophobic** (literally "water fearing") because they do not readily associate with water. Remember that the structure of water is stabilized by the formation of hydrogen bonds. Nonpolar compounds cannot participate in hydrogen bond formation since there is no partial charge associated with their hydrogens. Mixing these compounds with water disrupts existing hydrogen bonds, a situation that is energetically unfavorable. Nonpolar compounds are, therefore, insoluble in water and tend to form separate phases or layers, as shown in **Figure A-11**. In complex molecules, we will see examples where part of a molecule is hydrophobic and part is polar or **hydrophilic** (which means "water-loving"). The structure of the molecule is stabilized by the energetics of hiding the hydrophobic ("water-hating") portions from water while exposing the hydrophilic portions for maximum interactions with water. These interactions are important for the structures of proteins, nucleic acids, and the macromolecules that make up cell membranes, as described in Chapters 3–5.

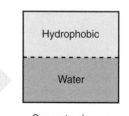

Separate phases

Figure A-11   Phase separation of hydrophobic molecules and water.

Resonance Bonds

One final class of bonds is found in biological molecules. These are the resonance bonds, which are in-between the single and double covalent bonds previously described. Resonance bonds were first described in synthetic molecules, such as benzene (**Figure A-12**), where it was determined that electrons were actually shared between a network of atomic nuclei. Resonance bonds are extremely stable and are found in many biological molecules, including the nucleotide bases, which are the building blocks of nucleic acids, and some of the amino acids, which make up proteins. Although many resonance-stabilized compounds are very hydrophobic (for example, benzene), sometimes resonance bonds are involved in stabilizing the components of an ionic bond, as shown in **Figure A-13**.

KEY CONCEPT

The Weak Forces → Shape Concept

Weak chemical interactions such as H-bonds and hydrophobic clustering are responsible for the overall shape of macromolecules.

The Ionization of Water—The pH Scale

Now we will review the interaction between water and the biologically important small molecules and macromolecules. This interaction is critical to both the structure and function of virtually all the molecules in the living cell. Water is, of course, the main component in a cell. All macromolecules reside in it, and all chemical reactions take place in it.

Water molecules ionize in the following way:

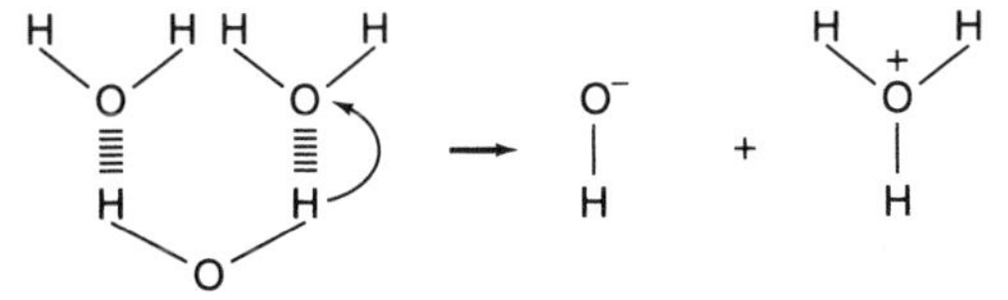

The polar covalent bonds between hydrogen and oxygen ionize forming hydroxyde ion (OH^-) and hydronium ion (H_3O^+). The hydronium ion may be treated simply as a hydrogen ion or proton (H^+), making the dissociation reaction

$$H_2O \rightleftharpoons H^+ + OH^-$$

The frequency of this dissociation is described by the following equation

$$K_{eq} = \frac{[H^+][OH^-]}{[H_2O]} \quad \text{or} \quad K_{eq}[H_2O] = [H^+][OH^-]$$

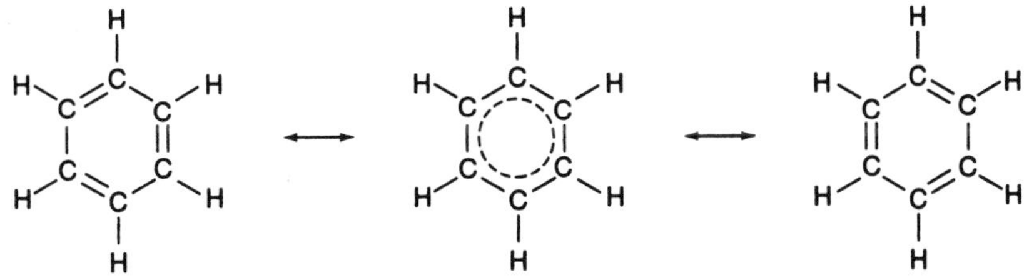

Equivalent structures

Figure A-12   Resonance forms of benzene.

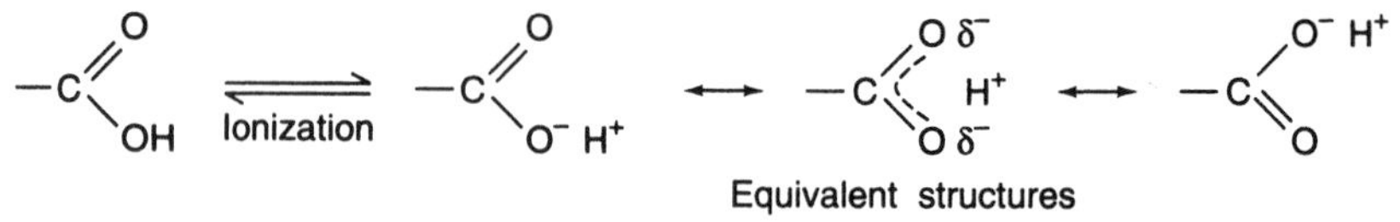

Figure A-13 Resonance forms of an ionic bond.

where K_{eq} is the equilibrium constant, and the terms in brackets are molar concentrations. The term $K_{eq}[H_2O]$ is a constant, such that in pure water

$$K_{eq}[H_2O] = K_w = [H^+][OH^-] = 1.0 \times 10^{-14}$$

where K_w is the dissociation constant or ion product of water. Thus, if the $[H^+]$ is known, the $[OH^-]$ can be calculated by solving the equation

$$[H^+][OH^-] = 1.0 \times 10^{-14}$$

In pure water, $[H^+] = [OH^-] = 1.0 \times 10^{-7}$ = neutral pH. Using such molar concentrations to express the $[H^+]$, however, would be difficult. Imagine if shampoo and soap makers had to say that their neutral products had a $[H^+]$ of 10^{-7}M. Instead, they say that the product is "pH balanced." The pH scale was devised to turn $[H^+]$ numbers into workable values. By definition

$$pH = -\log[H^+]$$

Therefore, neutral pH = 7, where $[H^+] = [OH^-] = 1.0 \times 10^{-7} = -\log_{10}[1.0 \times 10^{-7}]$. A pH scale follows. Note that it is a logarithmic scale such that the $[H^+]$ at pH 5 is 100 times greater than at pH 7; that is, a change of two pH units causes a $10^2 = 100$-fold change in $[H^+]$.

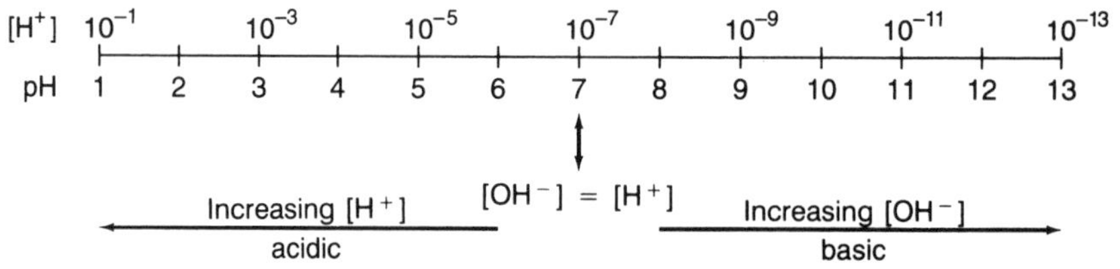

If a strong acid such as hydrochloric acid (HCl) is added to water, it will dissociate or ionize completely as follows

$$HCl \rightarrow H^+ + Cl$$

and the increase in $[H^+]$ causes the pH of the solution to drop (move to the left on the pH scale). If HCl is added to a final concentration of 0.1M (ignoring the very minor contribution of the 10^{-7}M H^+ of the water), then the pH = $-\log[10^{-1}] = 1$. Conversely, if a strong base such as sodium hydroxide (NaOH) is added to a final concentration of 0.1M, after complete dissociation, the $[OH^-] = 0.1$M or 10^{-1}M. By solving the equation $[H^+][OH^-] = 1.0 \times 10^{-14}$ for $[H^+]$, we get that $[H^+] = 10^{-13}$, since $(10^{-1})(10^{-13}) = 10^{-14}$ and the pH = $\log[10^{-13}] = 13$.

Most biological acids and bases are weak rather than strong, such as HCl and NaOH; therefore, they do not dissociate or ionize completely. An example is the carboxylic acid shown here.

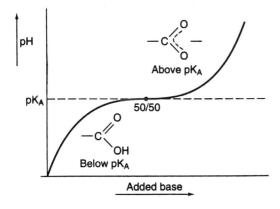

The dissociation of this carboxylic acid is described by the equation

$$K_A = \frac{[H^+][A^-]}{[HA]}$$

where A^- represents the deprotonated acid (often referred to as the **salt** or **conjugate base** of HA), and HA represents the protonated acid. Note the similarity between this equation and the earlier equation for the dissociation of water. If we rearrange the terms and solve for $[H^+]$, we get

$$[H^+] = K_A \frac{[HA]}{[A^-]}$$

If we then take the −log of both sides of the equation, we get

$$-\log[H^+] = -\log K_A - \log\frac{[HA]}{[A^-]}$$

This becomes

$$pH = pK_A + \log\frac{[A^-]}{[HA]}$$

by realizing that $-\log[H^+]$ is the definition of pH, and by analogy, $-\log K_A$ becomes pK_A. This is known as the Henderson-Hasselbach equation and permits us to calculate the pH of a solution if we know the ionization state of the constituents (the ratio of acid and conjugate base), as shown in **Figure A-14**.

When the $pH = pK_A$,

$$\log\frac{[A^-]}{[HA]} = 0 \quad \text{and} \quad \frac{[A^-]}{[HA]} = 1$$

salt

conjugate base

Figure A-14 Titration curve of a weak acid.

Therefore, the pK_A of an acid or base is the pH at which the ratio of protonated and deprotonated forms is $1:1$ (ionization is 50%), and is a constant for each acid or base. If we are dealing with a **weak acid,** then at a pH above its pK_A, most of the molecules will have lost their H^+ and be negatively charged. At a pH below its pK_A, most of the molecules will retain their H^+ and be neutral. If we have a **weak base** instead, then above its pK_A, most of the molecules will be deprotonated and, hence, neutral. Below its pK_A, almost every molecule will be protonated and, hence, carry a net positive charge as shown in the following.

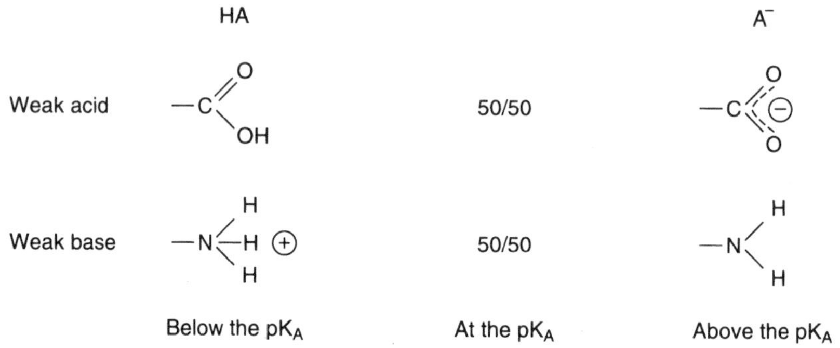

Remember that these relationships are logarithmic, such that at a pH one unit below the pK_A, the ratio of $[A^-]:[HA]$ will be $1:10$, while two pH units below the pK_A, the ratio would be $1:100$. The effects of pH on the rate of catalysis by an enzyme are dramatic, as illustrated earlier in Figure 4-16 in Chapter 4. This will also be briefly mentioned later.

Organic Chemistry

Now we turn our attention to a review of organic chemistry. Carbon is a special atom because it needs to share four electrons to fill its outer shell. It is very versatile because it can form **strong** bonds with many other elements, in particular, other carbon atoms, oxygen, hydrogen, phosphorus, nitrogen, and sulfur. Carbon is also versatile in that it can form single, double, or triple bonds as shown here.

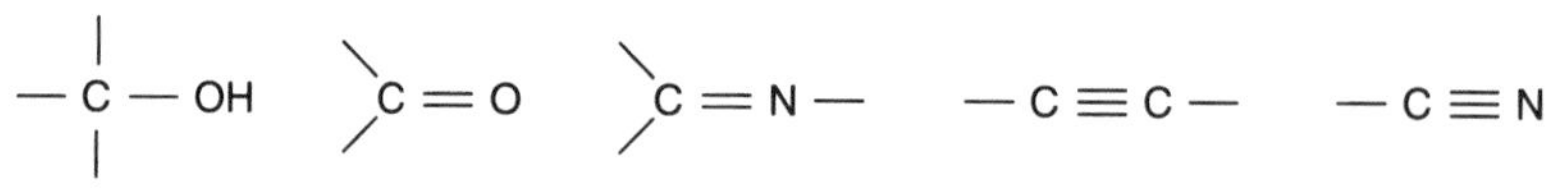

When four single bonds are formed, the shape of the molecule is a tetrahedron, as shown in Figure A-9, but for simplicity it is not always drawn as such. Several characteristic ways of drawing the tetrahedral methane molecule are shown in the following.

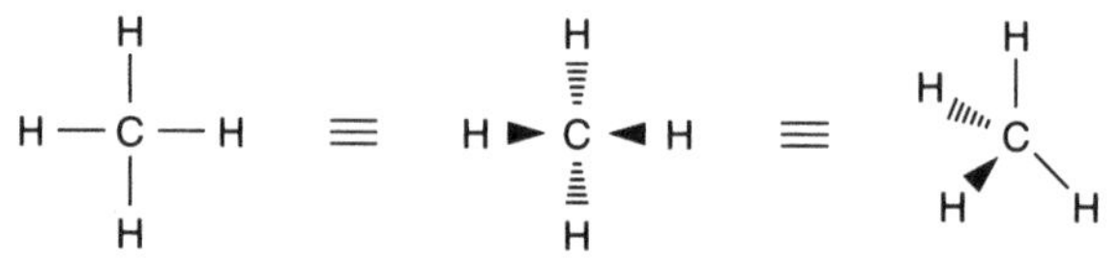

Note that the Fisher Projection (on the left) ignores the three-dimensional structure (it is merely assumed to be tetrahedral), while the other two drawings simulate the tetrahedral arrangement. Lines are in the plane of the page. Solid wedges project up out of the page, while dashed wedges project down into the page.

If there are four different substituents on one carbon atom, then this carbon is called a **chiral center,** and there are two possible **stereoisomers.** Stereoisomers are mirror images of each other and are not superimposable, rather like your left and right hands (see the following).

Any carbon that has four different groups attached to it is asymmetric. All living systems choose one version for each compound. Whereas ordinary chemical reactions cannot distinguish between stereoisomers, biological enzymes usually can act on only one stereoisomer, rather like a pair of gloves. One glove can fit the left hand, but not the right, and vice versa.

Next, we will discuss several classes of organic molecules that are important in molecular biology.

Alkanes

Alkanes are one class of **hydrocarbons,** molecules consisting only of carbon and hydrogen. In addition, they have only carbon–carbon single bonds, and have the general formula C_nH_{2n+2}. A series of alkanes containing from one to four carbons is shown in **Figure A-15**.

Methane, with one carbon, is the simplest alkane. For butane (four carbons) and all larger alkanes, **structural isomers** exist. Structural isomers have the same chemical formula, but differ in the exact set of chemical bonds they contain. The different carbon chain lengths and different structural shapes give the various alkanes slightly different properties.

Alkanes are fully **reduced** (as opposed to oxidized) and are said to be **saturated,** meaning that they contain the maximum number of hydrogens possible (only single bonds).

Many of the shorter alkanes are used as amino acid side chains. The longer forms of alkanes are a major component of gasoline. These molecules, plus methane (natural gas) and propane, are very high-energy compounds that can be burned or **oxidized** to produce CO_2 and H_2O (plus other by-products). Living cells obtain energy in a similar fashion. It is just a controlled burning of reduced carbon compounds to produce CO_2 and H_2O. However, alkanes do not work as a food source, since they are very hydrophobic and are, therefore, *not water soluble*. They are, in fact, toxic.

Alkenes

Another class of hydrocarbons are the alkenes, which, like alkanes, contain only carbon and hydrogen, but they differ from alkanes in that they have one or more double bonds between carbon atoms. A series of alkenes is shown in **Figure A-16**.

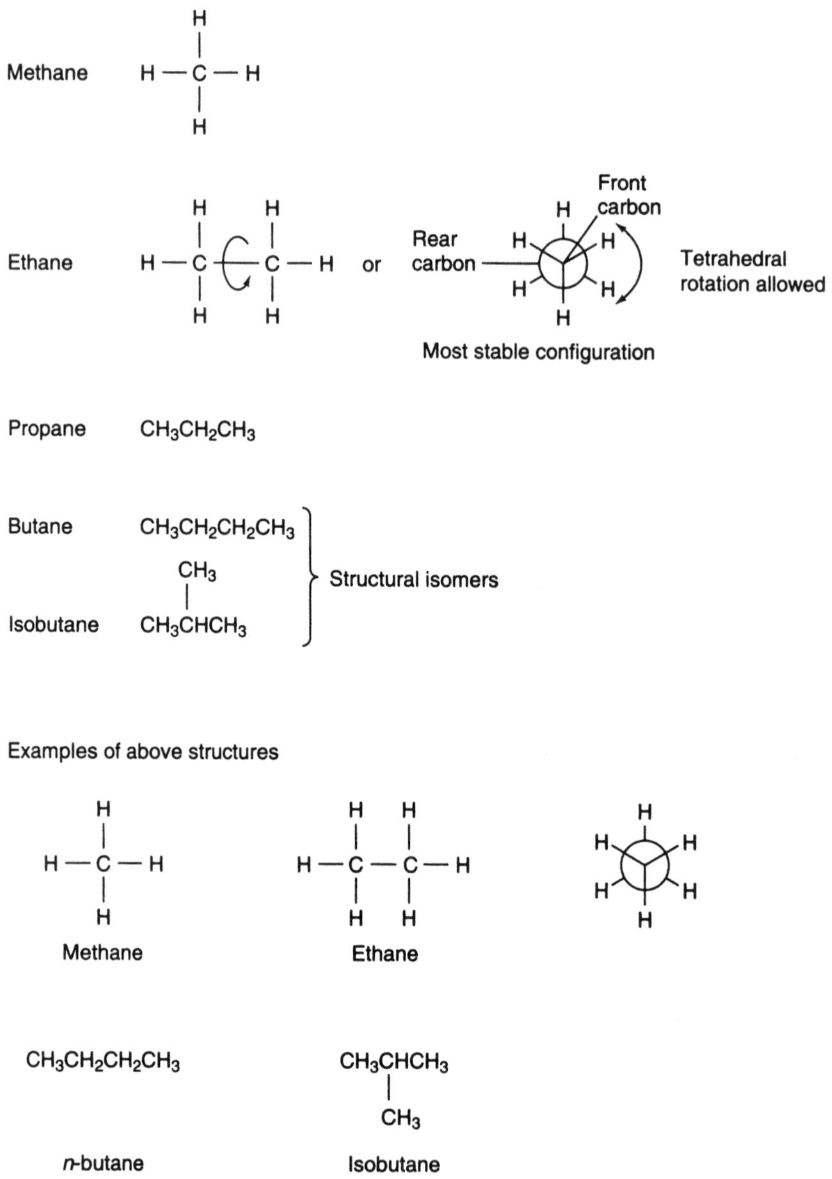

Figure A-15 Alkanes containing one to four carbons.

    The double bonds in alkenes are somewhat strained and are, therefore, more reactive than single bonds. Alkenes are said to be **unsaturated,** since they do not contain the maximum amount of hydrogen possible, and hydrogen or other atoms can be added to the carbons of a double bond. These highly reactive bonds are used as the starting material for many industrial polymers, including plastics, as well as for complex cellular compounds. For example, isoprene (Figure A-16) is the starting material for the synthesis of cholesterol and steroid hormones.

    The double bonds of alkenes also differ from the single bonds of alkanes in that the carbons on either side of the double bond cannot rotate freely around the double bond, causing the molecule to be planar rather than tetrahedral (compare

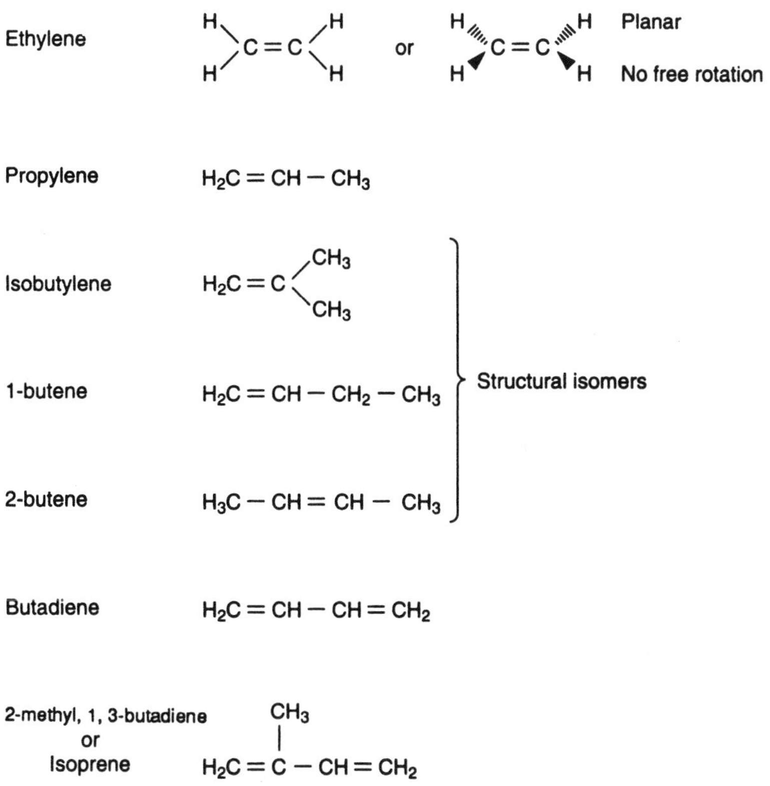

Ethylene	$H_2C=CH_2$ structure, or planar structure — Planar, No free rotation
Propylene	$H_2C = CH - CH_3$
Isobutylene	$H_2C = C(CH_3)(CH_3)$
1-butene	$H_2C = CH - CH_2 - CH_3$
2-butene	$H_3C - CH = CH - CH_3$
Butadiene	$H_2C = CH - CH = CH_2$
2-methyl, 1, 3-butadiene or Isoprene	$H_2C = C(CH_3) - CH = CH_2$

Structural isomers

Figure A-16 Some short chain alkenes.

ethane in Figure A-15 to ethylene in Figure A-16). Thus, in addition to the structural isomers that can form, **geometric isomers** are possible when different substituents are bonded to the carbons in an alkene as shown in the following.

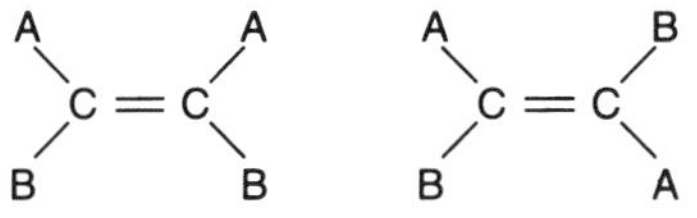

Cis isomer   Trans isomer

Geometric isomers have the same set of chemical bonds (unlike structural isomers), but the groups are arranged differently in space. In the **cis** isomer, the identical groups are on the same side of the double bond, whereas in the **trans** isomer, they are on opposite sides.

Alkynes

Alkynes are also hydrocarbons, but they contain one or more triple bonds between carbons. A simple alkyne, acetylene, which is the basis for the oxyacetylene torch, is shown here.

$$H-C \equiv C-H$$

Note that these molecules are linear around the carbon—carbon triple bond. The triple bonds of alkynes are very strained and, therefore, very reactive. Alkynes are even more unsaturated than alkenes. Carbon–carbon triple bonds are rare in biology.

Cycloalkanes and Cycloalkenes

So far, we have mentioned only straight chain molecules plus a few examples of branched structures. There are also cyclic forms of alkanes and alkenes. Cycloalkanes still contain only single bonds, but they are now in a ring structure with a general formula of C_hH_{2n}. Cyclopropane is the smallest cycloalkane possible, but it is unstable due to highly strained bonds. Cyclohexane is stable. Note in the following drawings how the hydrogens and carbons may be omitted from the drawing of the structure for simplicity. They are assumed to be there.

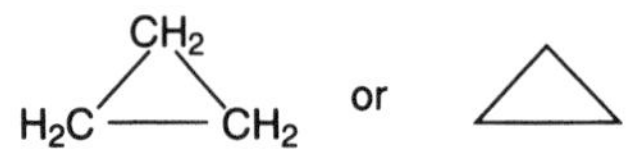

Cyclopropane 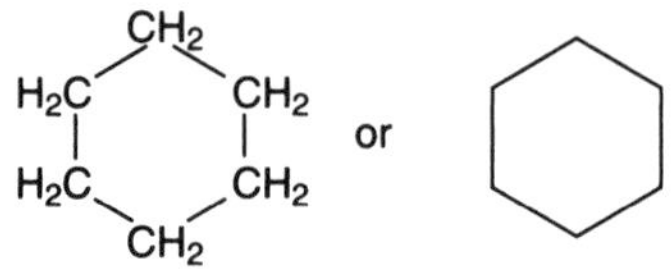   Cyclohexane

If a double bond is added, a cycloalkene is produced.

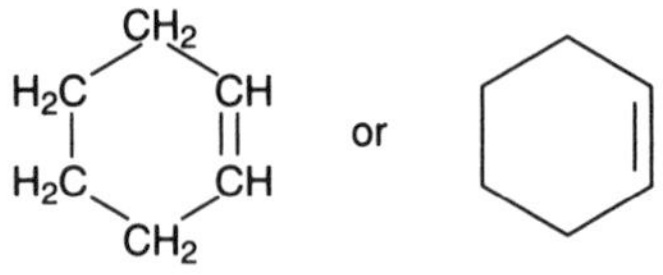

Cyclohexene

The alkanes, alkenes, alkynes, cycloalkanes, and cycloalkenes are all examples of **aliphatic** molecules, hydrocarbons that may be part of an open-chain or cyclic configuration, may have a straight chain or branched structure, and may contain single, double, or triple bonds.

Aromatic molecules are a special subclass of cycloalkenes that are not aliphatic. These molecules have the atoms arranged in rings, and the double and single bonds alternate in a resonance pattern that gives these molecules a high degree of stability. Many aromatic molecules are based on the structural backbone of benzene (shown in the following).

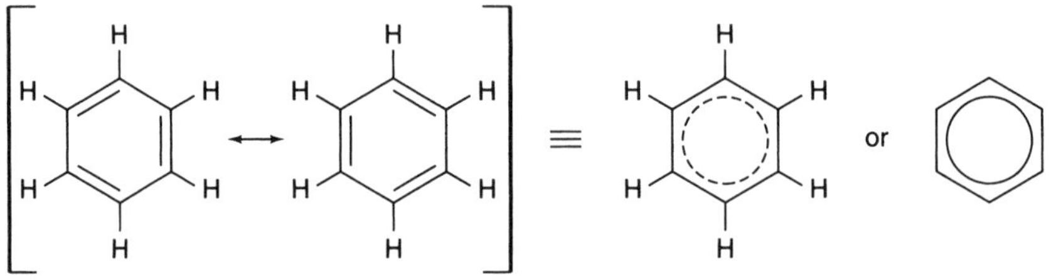

The actual structure of benzene (far right) is the instantaneous average of the two resonance structures shown at the left. Several amino acids have aromatic side

chains. We will also see further examples of resonance stabilization in the sections that follow.

So far, we have dealt only with carbon–carbon and carbon–hydrogen bonds. Now we will start adding other elements, such as oxygen, sulfur, and nitrogen.

Alcohols

Alcohols are compounds of the general formula ROH, where R is a hydrocarbon chain in which one hydrogen has been replaced with a hydroxyl group (OH). Alcohols are polar, hydrophilic molecules due to the partial charges on the OH group, as shown in the following.

Many alcohols are very soluble in water due to their ability to form hydrogen bonds with water molecules. A number of small alcohols are shown in **Figure A-17**. Note that, as with hydrocarbons, structural isomers of alcohols exist. However, since a functional group (the OH) is present, this leads to a distinction between **primary, secondary,** and **tertiary** alcohols. A primary alcohol is one in which only one carbon chain (or only H in the case of methanol) is attached to the carbon with the OH group, while secondary and tertiary alcohols have two and three carbon chains attached, respectively (see Figure A-17). Thus, although propanol and isopropanol are structural isomers, one is a primary alcohol, while the other is a secondary alcohol. Also note that, since the water solubility of alcohols is due to the OH group, as the hydrocarbon chain size increases, they become less soluble in water. Like other organic molecules, many alcohols are less dense than water.

KEY CONCEPT

The "Carbon Utility" Concept

Carbon's special properties (forms strong bonds with other carbon atoms; can form long chains and stable rings; forms single, double, and triple bonds) account for its widespread use in biological macromolecules.

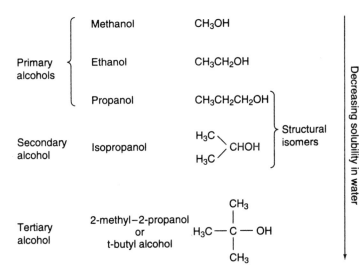

Figure A-17 Primary, secondary, and tertiary alcohols.

The alcohols depicted in Figure A-17 are all examples of aliphatic alcohols. The following is an example of an aromatic alcohol, phenol.

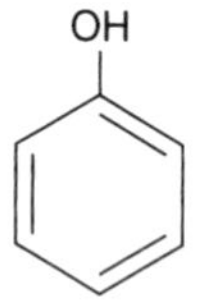

Many biological molecules are alcohols. Sugars, for example, are poly-alcohols. In fact, they are a special class of polyalcohols called **carbohydrates,** which have the general formula $(CH_2O)_n$. Glycerol is a simple three-carbon sugar and is shown here.

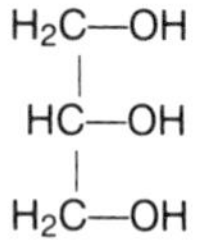

$$H_2C—OH$$
$$|$$
$$HC—OH$$
$$|$$
$$H_2C—OH$$

Notice that an **-ol** suffix following each of the compounds just discussed indicates that they are related to an alcohol.

Ethers

Ethers are another class of oxygen-containing compounds with the general formula R-O-R', where R and R' are identical or different hydrocarbon chains. Ethers may be thought of as the products of condensation reactions between two alcohols, specifically a **dehydration** reaction involving the loss of water as illustrated in the following.

$$ROH + R'OH \rightarrow ROR' + H_2O$$

Ethers are less polar than water or an alcohol, but more polar than an alkane due to the partial charges on the oxygen and carbon, shown here for dimethyl ether.

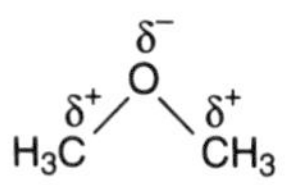

Ethers, in general, are not water soluble, but they are soluble in organic solvents and biological molecules, such as membrane lipids. Ethers are volatile, highly flammable, and frequently have strong odors. Some, such as diethyl ether (shown in the following), act as anesthetics at low doses by altering nerve transmission. Higher doses are toxic.

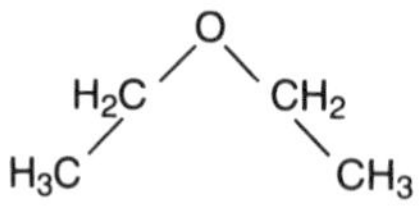

Aldehydes and Ketones

Aldehydes and ketones are molecules that contain a carbon atom double bonded to an oxygen. This is called a carbonyl group. Aldehydes and ketones may be thought of as the products of the oxidation (or **dehydrogenation**) reactions of primary and secondary alcohols, respectively.

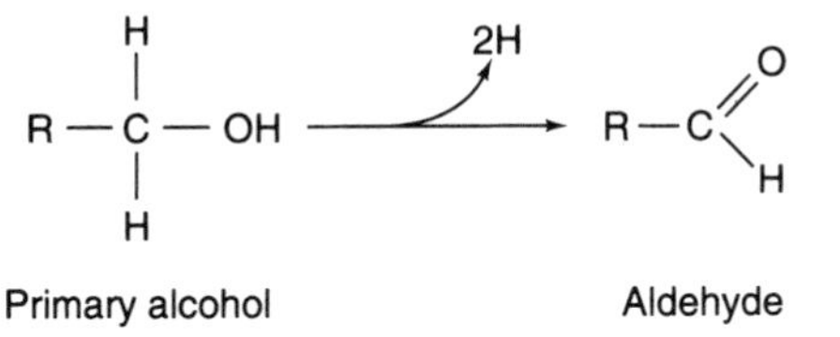

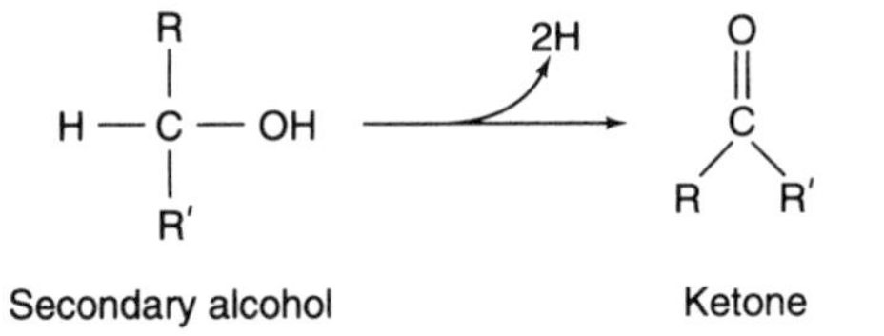

Primary alcohol   Aldehyde

Secondary alcohol   Ketone

Thus, aldehydes have the general formula RCHO, where one R chain is attached to the carbonyl, while ketones have the general formula RR′CO, where two R chains are attached to the carbonyl. As before, R or R′ refers to aliphatic or aromatic side chains. A commonly used aldehyde and a ketone are shown here.

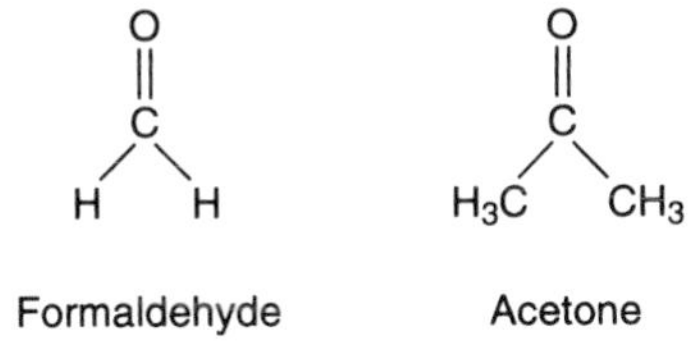

Formaldehyde   Acetone

Formaldehyde is used in biological laboratories as a fixative. Note that it has no carbon side chain, as it is derived from methanol (CH_3OH). Acetone is a laboratory solvent and also is used as nail-polish remover.

Aldehydes and ketones are somewhat polar, so they are somewhat water soluble. The extent of their solubility depends on the rest of the molecule (as with alcohols). Aldehyde and ketone functional groups may be found in amino acids and nucleotides, but are particularly evident in carbohydrates. An **aldose** or aldehyde containing sugar and a **ketose** or ketone containing sugar are shown in **Figure A-18**. In solution, sugars can form ring structures in which the aldehyde or ketone group is reduced to a hydroxyl (alcohol) and an ether linkage is formed. Due to the presence of the hydroxyl on one of the carbons adjacent to the ether linkage, it is referred to as a **hemiacetal** (Figure A-18).

Carboxylic Acids

Carboxylic acids are compounds that contain both a carbonyl and hydroxyl (alcohol) group bonded to the same carbon atom. They have the general structure shown in the following, where R, as usual, may be an aliphatic or aromatic side chain.

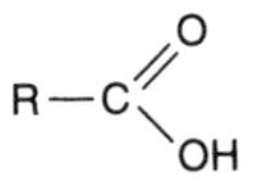

They may be thought of as the oxidized products of aldehydes, as shown later in the text.

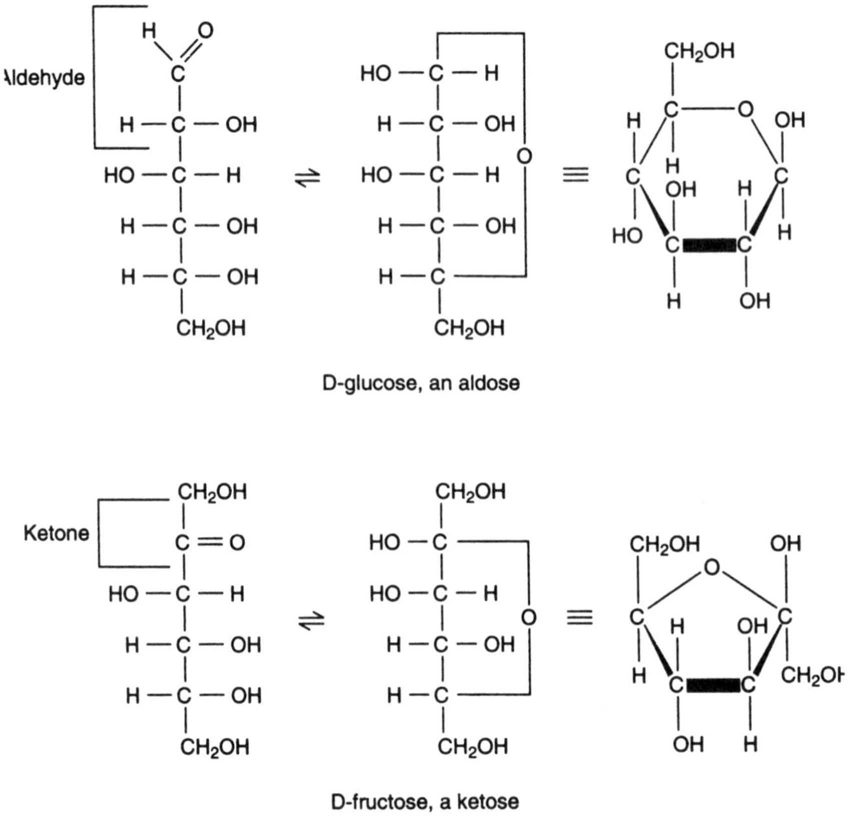

Figure A-18 The sugars, glucose and fructose, in open and cyclic configurations.

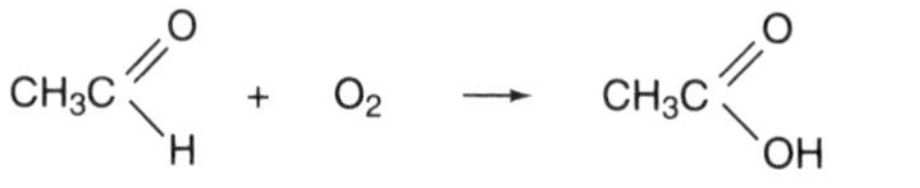

D-glucose, an aldose

D-fructose, a ketose

Acetaldehyde   Acetic acid, vinegar

For simplicity, the carbonyl functional group is often written as CO_2H or $COOH$.

Carboxylic acids are very important biological compounds. Many of the foods we eat derive their taste and smell from carboxylic acids. Carboxylic acids also give many animals their characteristic odors. Amino acids are carboxylic acids. In addition, carboxylic acids are involved in energy production (intermediates in the Krebs cycle, also referred to as the citric-acid cycle or tricarboxylic-acid cycle) and in energy storage (fatty acids). Some examples are given in **Figure A-19**.

Carboxylic acids tend to be water soluble, weak acids. As such, they can donate hydrogen ions (protons). This ionization is shown in the following:

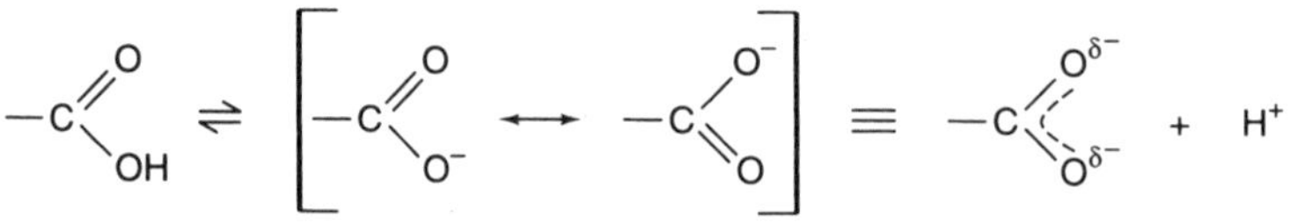

Resonance forms

Formic acid (methanoic acid)	HCOOH	Ant pheromone
Butyric acid (butanoic acid)	$CH_3CH_2CH_2COOH$	Rancid butter
Caproic acid (hexanoic acid)	$CH_3(CH_2)_4COOH$	Odor of goats
Stearic acid (ocatadecanoic acid)	$CH_3(CH_2)_{16}COOH$	A saturated fatty acid

Lactic acid

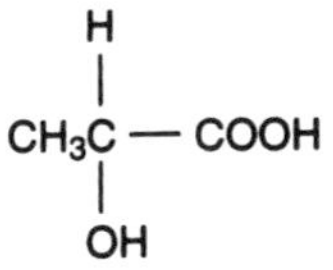

$$CH_3\overset{\overset{\displaystyle H}{|}}{\underset{\underset{\displaystyle OH}{|}}{C}} - COOH$$

Sour milk and overworked muscles

Citric acid

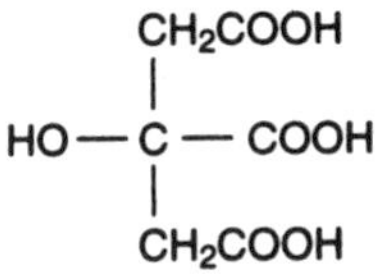

$$HO-\overset{\overset{\displaystyle CH_2COOH}{|}}{\underset{\underset{\displaystyle CH_2COOH}{|}}{C}} - COOH$$

Citrus fruits, Krebs cycle intermediate

Figure A-19 Several carboxylic acids and their biological significance.

For simplicity, carboxylate ions are usually written as only one of the resonance forms. The pK_A values for this ionization differ from one carboxylic acid to another but are in the range of 2 to 5.

Esters

A final class of oxygen-containing compounds are esters, which are the product of a dehydration reaction between an alcohol and a carboxylic acid, as shown in the following. Note that an ester contains a carbonyl group and an ether linkage.

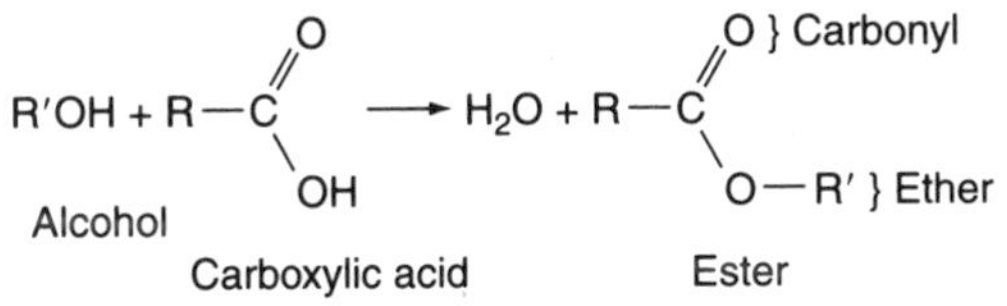

R'OH + R—C(=O)OH → H₂O + R—C(=O)O—R'

Alcohol / Carboxylic acid → Ester

Carbonyl } O
Ether } O—R'

Esters tend to be volatile and reactive. They frequently give various fruits and flowers their pleasant odors and flavors. **Figure A-20** shows several aliphatic and aromatic esters, as well as the **triglyceride** sterin, containing three stearic-acid molecules esterified to a glycerol backbone.

If one of the fatty acids in a triglyceride is replaced with a phosphate-containing group, a phosphoester linkage is formed and a class of molecules called **phospho-**

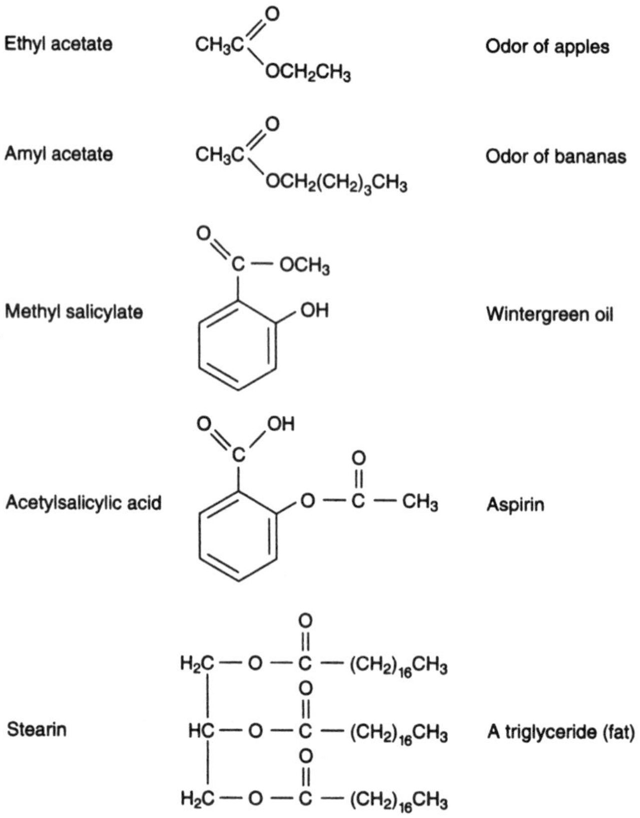

Ethyl acetate $CH_3C\overset{O}{\underset{OCH_2CH_3}{}}$ Odor of apples

Amyl acetate $CH_3C\overset{O}{\underset{OCH_2(CH_2)_3CH_3}{}}$ Odor of bananas

Methyl salicylate

Wintergreen oil

Acetylsalicylic acid

Aspirin

Stearin

A triglyceride (fat)

Figure A-20 Several esters and their biological significance.

lipids is generated. They comprise a major constituent of cellular membranes. Another example of the phosphoester is seen in the DNA backbone, where, in fact, a phosphodiester linkage occurs.

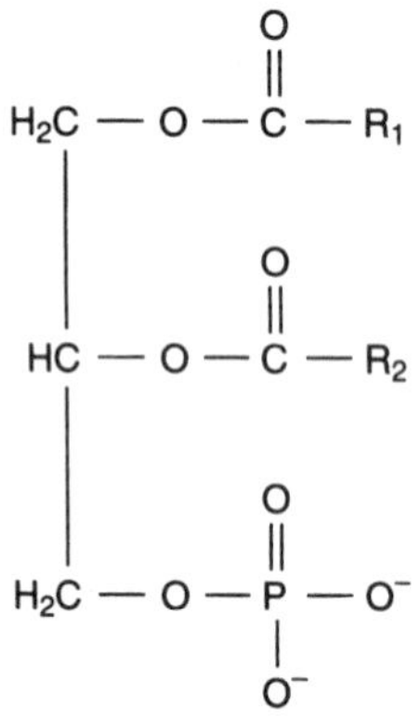

Phosphatidic acid
(precursor for phospholipids)

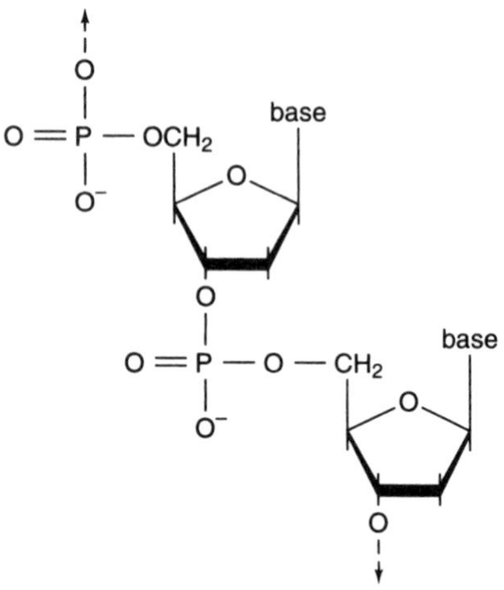

Strand of DNA

Note that phosphoesters are acidic and tend to be ionized at neutral pH.

Thiols, Thioethers, and Disulfides

Thiols (or mercaptans), thioethers, and disulfides may be thought of as the sulfur analogs of alcohols, ethers, and peroxides. They have the general formulas RSH, RSR′, and RSSR′, respectively. These sulfur-containing compounds generally have very bad odors. Some thiols and a thioether are shown in **Figure A-21**.

Two thiols can react with one another to form a disulfide. This is a common occurrence in proteins where two cysteine side chains react to form a disulfide bond as shown in the following.

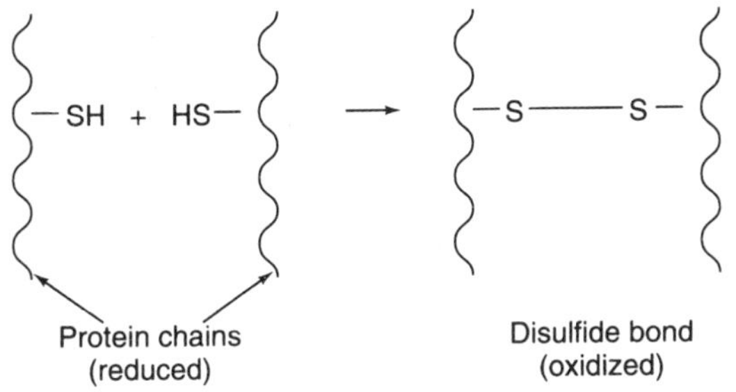

Protein chains
(reduced)

Disulfide bond
(oxidized)

These disulfide bonds, which can occur between different polypeptide chains or between different regions of the same chain, are involved in the structure and stability of proteins. At high concentrations, β-mercaptoethanol can reduce (break) disulfide bonds and destroy a protein's structure.

Amines, Imines, and Amides

Several classes of compounds contain carbon bonded to nitrogen. Amines are derivatives of ammonia (NH_3) in which one or more hydrogen atoms are replaced with carbon chains. A series of amines is shown in **Figure A-22**. Note that the amines may be primary, secondary, or tertiary, depending on whether one, two, or three of the ammonia hydrogens are replaced with carbon chains.

Methane thiol	CH_3SH	Rotten egg odor, cysteine side chain
Ethane thiol	CH_3CH_2SH	Sewer gas
Butane thiol	$CH_3CH_2CH_2CH_2SH$	Skunk secretion
β–mercaptoethanol	$HOCH_2CH_2SH$	Biochemical reagent, permanent wave solution
Thiophenol	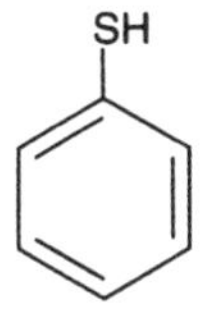	Odor of burnt rubber
Ethyl methyl thioether	$CH_3CH_2SCH_3$	Methionine side chain

Figure A-21 Thiols and a thioether with their characteristic odors and biological uses.

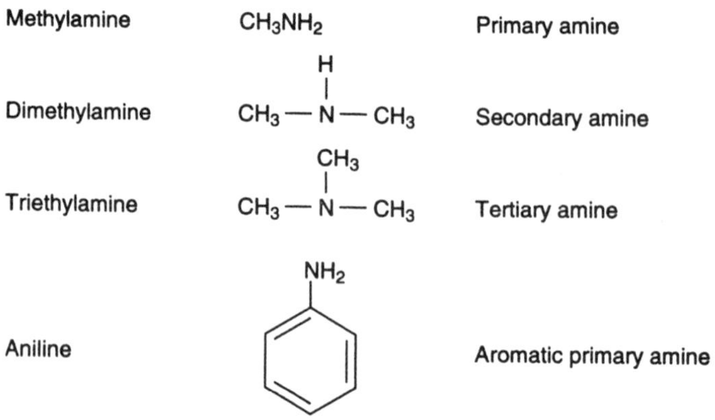

Methylamine CH_3NH_2 Primary amine

Dimethylamine $CH_3 - \overset{\overset{\displaystyle H}{|}}{N} - CH_3$ Secondary amine

Triethylamine $CH_3 - \overset{\overset{\displaystyle CH_3}{|}}{N} - CH_3$ Tertiary amine

Aniline   Aromatic primary amine

Figure A-22 Some simple primary, secondary, and tertiary amines.

In general, amines have fishy, ammonia-like odors. In contrast to carboxylic acids, which are weak acids, amines are weak bases and can be protonated as shown here.

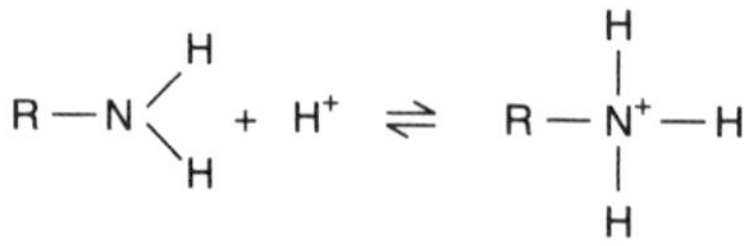

$$R - N\overset{H}{\underset{H}{\diagdown}} + H^+ \rightleftharpoons R - \overset{\overset{\displaystyle H}{|}}{\underset{\underset{\displaystyle H}{|}}{N^+}} - H$$

The pK_A values of amines range from 8 to 12. Therefore, at neutral pH, amines are protonated.

An imine is the nitrogen analog of a carbonyl and has the general form shown in the following.

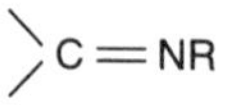

$$\diagup\hspace{-0.3em}\overset{\diagdown}{C} = NR$$

Imines can also be protonated, much like amines. They are often found in ring structures of biological importance, as shown in **Figure A-23**.

Primary, secondary, or tertiary amides are a combination of a carbonyl group and a primary, secondary, or tertiary amine, and have the general structures shown in the following.

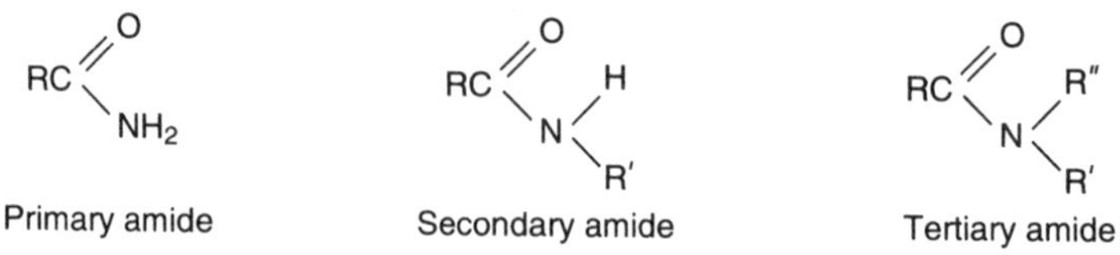

Primary amide   Secondary amide   Tertiary amide

A number of biological compounds contain one or more of these carbon–nitrogen linkages. Several are shown in Figure A-23.

Amino Acids

As we have seen, many molecules display more than one kind of functional group. One group of molecules that demonstrates a number of the functional

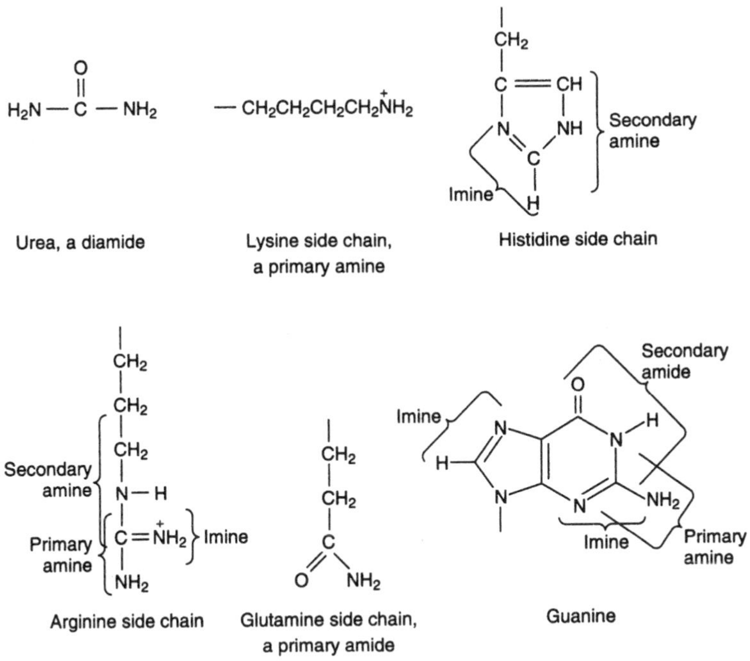

Figure A-23 Some important biological compounds containing carbon–nitrogen bonds.

groups and concepts previously discussed is the set of 20 naturally occurring amino acids. All amino acids have a carboxylic acid and amine group (both of which are ionizable) attached to a central carbon atom called the α-carbon. One hydrogen atom and a variable side chain (R) are also attached to the α-carbon. With the exception of glycine, in which R is a hydrogen, there are four different groups attached to the α-carbon, so it is a chiral center. Thus, all the amino acids except glycine have a specific stereochemistry. The side chain groups include aliphatic or aromatic carbon chains, primary or secondary alcohols, carboxylic acids, primary and secondary amines, primary amides, thiols, and thioethers.

Since amine and carboxylic acid groups have rather different pK_A values, amino acids have different charged states at different pH values. These charge states, as well as the L-isomer configuration of amino acids, are shown in **Figure A-24**. The pK_A values for the α-carboxyl and α-amine are about 2 and 9, respectively, differing only slightly from one amino acid to another. Keep in mind that some of the amino acids have ionizable carboxylic acid or amine side chains that have pK_A values of about 3.9 to 4.2 and 10.5 to 12.5, respectively. **Table A-3** illustrates the ionization of the side chains of several of these amino acids at various pH values.

Note that at neutral pH, most amino acids are **zwitterionic,** from the German meaning both positive and negative charge in a molecule with a net charge of 0. All amino acids are **amphoteric,** from the Greek meaning both acid and base. However, in proteins, the α-amine and α-carboxyl are involved

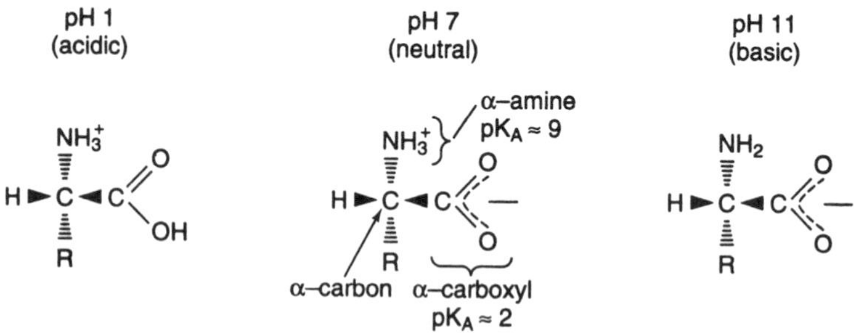

pH 1
(acidic)

pH 7
(neutral)

pH 11
(basic)

Figure A-24 A typical amino acid at acidic, neutral, and basic pH.

in peptide bonds linking one amino acid to the next and can no longer be ionized. Only the terminal carboxyl and amine groups, as well as the side chain carboxyls and amines, can be ionized and lead to the net charge of the protein. The structures of all 20 naturally occurring amino acids may be found in Chapter 2.

When studying enzyme function in a test tube, it is very important that the pH of the liquid solution be adjusted for an optimal reaction rate. This is achieved by setting a pH value so that the various amino acids that comprise the active site of the protein (see Figures 4-12 and 4-13 in Chapter 4) are appropriately ionized. An example of the effects of pH on enzyme activity was reviewed earlier in Chapter 4 (Figure 4-16).

Concluding Note

The chemistry of the complex macromolecules found in biological compounds is a simple extension of the rules described in this appendix. They contain a variety of the functional groups discussed previously. Remember that the types of bonds formed (covalent, polar covalent, ionic, hydrophobic, or hydrogen bonding) will determine how a molecule interacts with water and with other macromolecules. Also, many of the reactions we have reviewed (dehydration, dehydrogenation, reduction, oxidation, esterification, and the like) are common reactions during biochemical synthesis and degradation.

KEY CONCEPT

The "Surface Interaction" Concept

Interactions between surfaces, facilitated by a variety of complementary weak forces, permit macromolecules to serve as templates for ordering and/or synthesizing other cellular components.

TABLE A-3	EXAMPLES OF IONIZATION OF SIDE CHAINS OF AMINO ACIDS AT VARIOUS pH VALUES		
	Low pH (e.g., 2.0)	Physiological pH (e.g., 7.0)	High pH (e.g., 12.0)
Lysine's side chain amino group	+ Charge	+ Charge	No charge
Aspartic acid's side chain carboxyl group	No charge	– Charge	– Charge

List of Essential Concepts

Part I: The Structure of Proteins, Nucleic Acids, and Macromolecular Complexes

Chapter 2, Macromolecules

The "Bipolymer" concept: Polymers of small molecules such as amino acids (proteins) or nucleotides (nucleic acids) are extraordinarily versatile, and, hence, have evolved in complexity to generate the diverse functional repertoires of living cells. (p. 23)

Chapter 3, Nucleic Acids

The "Evolution of Structure" concept: Natural selection has favored the evolution of double-stranded DNA, for, with its tight helical structure and ease of unwinding, the bases are protected, but also available for replication. (p. 36)

The "RNA Enzyme" concept: Although most enzymes are proteins, some RNAs are catalytically active. (p. 48)

Chapter 4, The Physical Structure of Protein Molecules

The "Primary Sequence" concept: The primary sequence of building blocks (amino acids in proteins and nucleotides in nucleic acids) dictates the folding pattern and final shape, and, therefore, activity of a macromolecule. (p. 60)

The "Shape/Size" concept: The specific function that a macromolecule is best suited to perform is largely dictated by its size, shape, and charge. Those parameters set distinct limits on the number and types of interactions and relationships in which macromolecules can participate. (p. 64)

The "Shapes and Surfaces" concept: Subtle features of the shape and surface characteristics of a macromolecule allow it to interact with complementary surfaces on a neighboring macromolecule. (p. 68)

The "Enzyme Catalysis" concept: Enzymes lower the activation energies of the chemical groups that participate in a reaction, and thereby speed up the reaction. (p. 69)

Chapter 5, Macromolecular Interactions and the Structure of Complex Aggregates

The "Macromolecular Complex" concept: By acting in concert, the functional capacity and repertoires of proteins and nucleic acids are enhanced. (p. 87)

The "Connectedness" concept: The cell's need for connectivity and communication to the outside is achieved through the action of various membrane proteins. (p. 92)

Part II: Function of Macromolecules

Chapter 6, The Genetic Material

The "DNA-Genetic Material" concept: Double-stranded DNA has evolved as the genetic material because it is especially well-suited for replication, repair, change, and long-term stability. (p. 112)

The "Gene" concept: Genes contain all the information for the synthesis and functioning of cellular components. (p. 115)

Chapter 7, DNA Replication

The "Evolution of DNA Replication" concept: Due to the antiparallel nature of DNA strands, as well as the need for rapid replication and timely error correction, the discontinuous DNA replication mode has evolved. (p. 137)

Chapter 8, Transcription

The "Upstream/Downstream" concept: Transcription starts upstream of a gene's protein coding region and ends downstream, beyond the coding sequence. Information for regulating gene expression, thus, resides outside (upstream/ downstream) the protein coding region. (p. 154)

The "RNA Processing" concept: Eukaryotic mRNAs are extensively processed ("cut and shut") prior to functioning as template for protein synthesis. (p. 161)

Chapter 9, Translation

The "Universal Genetic Code" concept: The triplet code is universal (i.e., the same code is used by all organisms). It also has redundancies (i.e., multiple codons exist for most amino acids). (p. 173)

The "Prokaryotic Simplicity" concept: RNA synthesis and protein synthesis are coupled in prokaryotes, separate (due to the nuclear membrane) in eukaryotes. (p. 186)

The "Antibiotic Action" concept: The mechanism of action of most antibiotics is understood in terms of the antibiotic interfering with a specific step in protein synthesis, or a step in the synthesis of the cell wall, or other indispensable component of the bacterial cell. (p. 188)

Chapter 10, Mutations, Mutagenesis, and DNA Repair

The "Mutation → Variation" concept: Mutations, which represent changes in the nucleotide sequence of genes, generate the wide-ranging diversity that characterizes the biological kingdom. (p. 201)

The "DNA Repair" concept: Despite the inherent chemical stability of dsDNA, from time to time it sustains damage, and, hence, its continual repair is necessary for cells to function properly. (p. 216)

Part III: Coordination of Macromolecular Function in Cells

Chapter 11, Regulation of Gene Activity in Prokaryotes

The "Regulation" concept: A multitude of regulatory mechanisms function in cells to ensure that the supply of various macromolecules matches up with the cell's ever-changing needs. (p. 224)

The "Protein Levels" concept: The net amount of a protein in a cell represents a balance between the rate it is produced and the rate at which it is degraded or exported from the cell. (p. 243)

Chapter 12, Regulation of Gene Activity in Eukaryotes

The "Combinatorial Action" concept: The expression of eukaryotic genes is regulated by macromolecular complexes consisting of a variety of proteins that combine with each other and bind to specific nucleotide sequences (e.g., promoters and enhancers). (p. 257)

The "Mix and Match" Transcript concept: Many eukaryotic primary transcripts are "cut and shut" or "cut and pasted" to generate different mRNAs from a single long primary transcript. RNA editing can further change a reading frame. (p. 266)

The "mRNA Half-life" concept: The contribution an mRNA makes to cellular function is determined in large part by how quickly it is degraded. (p. 269)

The "Instability → Flexibility" concept: The inherent instability of many macromolecules (e.g., some proteins and mRNAs) in the living cell facilitates adapting to changing metabolic conditions. (p. 274)

The "Activation of a Transcription Factor" Concept: The activity of proteins that facilitate initiation of transcription (so-called "transcription factors") is often up or down regulated by the action of steroid hormones, small molecules, or even other proteins. (p. 275)

Chapter 13, Genomics and Proteomics Drive Information-Age Biology

The "Genome Size vs. Gene Number" concept: Organisms with very different genome sizes may nevertheless contain similar numbers of protein coding genes. **(p. 287)**

The "Defining a Gene" concept: The meaning of the term "gene" has become blurred by the discovery that a single stretch of nucleotide sequence can specify regions of amino acid sequence in more than one protein. **(p. 292)**

The "Protein Spectrum → Cell Function" concept: Identification of the proteins that are present and knowledge of how they interact with one another are required for understanding why a specific type of cell behaves the way it does. **(p. 298)**

Part IV: Experimental Manipulation of Macromolecules

Chapter 14, Transposons, Plasmids, and Bacteriophage

The "Changing Genome" concept: The sequence or arrangement of nucleotides in genes is constantly evolving, either by random mutation, by occasional recombination events, or from the action of transposable elements. **(p. 313)**

The "Simple Genome" concept: Viral genomes are extraordinarily simple, in part because viruses borrow functions from their host cells. **(p. 328)**

The "Levels of Sophistication" concept: Prokaryotes such as viruses and bacteria, under heavy natural selection pressure, have evolved more streamlined mechanisms for regulating gene expression than higher organisms (e.g., humans). **(p. 341)**

Chapter 15, Recombinant DNA and Genetic Engineering: Molecular Tailoring of Genes

The "Recombinant DNA" concept: DNA is DNA, regardless of the organism from which it is isolated. Hence, genes from diverse species can be recombined to construct novel genes, or, in certain instances, even new organisms. **(p. 366)**

Appendix

The "Weak Forces → Shape" concept: Weak chemical interactions such as H-bonds and hydrophobic clustering are responsible for the overall shape of macromolecules. **(p. 417)**

The "Carbon Utility" concept: Carbon's special properties (forms strong bonds with other carbon atoms; can form long chains and stable rings; forms single, double, and triple bonds) account for its widespread use in biological macromolecules. **(p. 425)**

The "Surface Interaction" concept: Interactions between surfaces, facilitated by a variety of complementary weak forces, permit macromolecules to serve as templates for ordering and/or synthesizing other cellular components. **(p. 434)**

Glossary

Activator - protein that acts to increase frequency of transcription of an operon.

Active site - region of an enzyme to which the substrate binds.

Adenine (A) - purine base that pairs with thymine (T) in DNA.

Agar - gelatinous substance prepared from refined seaweed extract providing a smooth, inert surface useful for growth of bacteria (when supplemented with nutrients).

A helix - double stranded DNA configuration featuring right-handed twisting and characteristic tilting of the helical axis.

AIDS - acquired immune deficiency syndrome (caused by virus).

Algorithm - formula for solving a specific problem based on a set of assumptions.

Alkylating agent - mutagenic compound that adds a methyl or ethyl group to a base (e.g., guanine).

Allele-specific oligonucleotides (ASO) - nucleotide polymers that are designed to be highly specific and are, therefore, capable of distinguishing, with hybridization procedures, between genes differing in only a single nucleotide.

Allosteric interaction - protein's functional activity is regulated by interaction of one part of protein with another molecule.

Alpha complementation - technique for recognizing bacterial colonies containing a hybrid plasmid (β-galactosidase enzyme is made functional and can be tested for).

Alpha helix - right-handed twist structure common to double-stranded DNA and portions of many proteins.

Alternative splicing - removal of introns and/or exons from a primary mRNA transcript in multiple patterns, generating more than one type of mRNA for translation.

Ames test - method for recognizing potential carcinogens, which uses the ability of test substances to generate mutations in bacteria as an indicator for potential carcinogen.

Amino acid - small molecule building block of protein.

Amino terminus - the end of a protein molecule displaying an amino group (also called "N-terminus").

Aminoacyl (A) site - binding site on larger ribosomal subunit serving as a docking site for incoming charged tRNA during protein synthesis.

Aminoacyl-tRNA synthetases - enzymes that covalently link amino acids to the 2′ or 3′ OH position of tRNA.

Annealing (of DNA) - reassociation to form a hybrid molecule of two complementary single-stranded DNA molecules.

Antibiotic - substance having the ability to inhibit the growth of bacteria, usually by interfering with specific metabolic process.

Antibody - protein (immunoglobulin) that recognizes a specific antigen.

Anticodon - three-base sequence in a tRNA that base pairs with a specific triplet codon in mRNA.

Antigen - molecule that provokes synthesis of an antibody (immunoglobulin).

Antiparallel - opposite polarities of two strands of a DNA double helix—one strand is 5′-3′ top to bottom and the other strand is 3′-5′.

Antisense mRNA - complementary copy of coding strand of bonafide mRNA, which usually will hybridize with the cognate mRNA.

Antiterminator - proteins associated with bacteriophage reproduction allowing RNA polymerase to transcribe beyond normal transcription termination sites.

Archaebacteria - procaryotic bacterial grouping (kingdom) that exhibits intermediate features between typical eubacterial types (e.g., *E. coli*) and lower eukaryotes.

Attenuation - regulatory mechanism for controlling frequency of transcription of a gene, which employs a nucleotide sequence that leads to premature transcription.

Autoradiography - method for detecting radioactively labeled molecules by the image they produce on photographic film.

Autoregulation - regulation system for gene expression in which the gene product serves to bind to the promoter region and, depending upon the specific gene, either enhances or diminishes transcription.

Auxotroph - organism that requires one or more growth supplements in the culture medium.

Bacteria - tiny, free-living microorganisms characterized by lack of nuclear membrane.

Bacteriophage - virus that infects and reproduces in bacteria (often denoted as "phage").

Base analog - chemically similar to one of the naturally occurring bases (A,T,G,C,U, etc.), except that one or more atoms are substituted or modified.

Base pair - pair of bases (either A-T or G-C) that are opposite one another in a double-stranded DNA molecule.

Base stacking - arrangement of nucleotide bases in a relatively rigid, strung-out fashion, due to hydrophobic interactions between ring structures of bases (see also "stacking of bases").

β-galactosidase - enzyme that hydrolyzes lactose into glucose and galactose.

Beta structure - extended region of a protein characterized by rigidity due to hydrogen bonds between adjacent polypeptides.

Bidirectional replication - DNA replication process that involves two replication forks that move in opposite directions.

Bioinformatics - study of the meaning (interpretation) of information contained in an organism's genome.

B lymphocytes (or B cells) - cells in which antibodies are synthesized.

bp - abbreviation for base pair.

cAMP - nucleotide that is similar to adenosine monophosphate, except that its phosphate group is bonded back onto the sugar.

Cap - a G nucleotide (sometimes methylated) that is added to the 5′ end of eukaryotic mRNA after transcription.

CAP - regulatory protein that is activated by cyclic AMP and is involved in transcription of the lac operon of *E. coli*.

Capsid - protein coat of virus.

Carboxyl terminus - end of a polypeptide that contains a free carboxyl group.

Carcinogen - agent (e.g., chemical or ultraviolet light) causing cells to express oncogenes (which leads to formation of a tumor or spreading cancer).

Catabolite repression - reduced expression of bacterial genes, which results from growth of excess glucose.

cDNA - single-stranded DNA with a nucleotide sequence, which is complementary to an RNA.

cDNA clone - double-stranded DNA sequence that is complementary to a specific RNA and inserted in a cloning vector such as a plasmid.

cDNA library - collection of cDNAs in a cloning vector, to be used for selecting clones containing specific sets of coding sequences.

Charging - enzymatic attachment of an amino acid to its appropriate tRNA.

Chromatin - macromolecular complex of DNA, protein, and RNA, which comprises chromosomes of eukaryotic cells.

Chromosome walk - method for sequentially isolating overlapping segments of DNA so that very large genes can be studied.

Cistron - region of DNA comprising coding sequence for one protein.

Clonal selection - biological process that sorts through antibody-producing cells and induces those combining with the challenging antigen to reproduce and thereby increase production of specific antibody.

Clone - large number of either molecules (e.g., DNA) or cells (e.g., *E. coli*) derived from a single progenitor.

Cloning vector - plasmid or phage into which a foreign DNA is inserted for replication.

Coding strand - DNA strand that is transcribed into mRNA.

Codon - triplet of nucleotides that codes for a specific amino acid or translation start-stop signal.

Cohesive ends - compatible (complementary) ends of double-stranded DNA generated by the action of restriction enzymes.

Cointegrate - intermediate in replicative transposition consisting of both donor and target molecules covalently linked as a large circular dsDNA molecule.

Colony hybridization - method for detecting bacteria that carry a vector with a desired inserted sequence.

Compatible ends - cohesive ends generated by the action of restriction enzymes on dsDNA.

Complementary DNA - cDNA.

Concatamer - series of repeats of a single nucleotide (DNA) sequence.

Conditional lethal mutation - kills the cell that harbors it, but only under certain (nonpermissive) conditions, such as elevated temperature.

Conjugation - "mating" between two bacteria, involving movement of genetic material from an F^+ to an F^- cell.

Consensus sequence - composite nucleotide sequence commonly believed to serve the same function in different genes (e.g., promoter sequences).

Constitutively expressed - gene that is expressed under all circumstances, regardless of the presence of inducing substance.

Core enzyme - portion of a larger enzyme complex containing catalytic activity, but which lacks regulatory subunits usually associated with "holoenzyme."

Core particle - enzyme digestion product of nucleosome that contains a histone octamer and approximately 150 bp of DNA.

Corepressor - small molecule combining with protein repressor to activate that protein for binding to an operon.

Covalent extension (DNA replication) - process of DNA replication that initiates with leading-strand synthesis extending from a preexisting 3'OH rather than from a newly synthesized RNA primer.

CpG island - sequences in eukaryotic DNA that are rich in CG bases, and usually exhibit promoter and/or enhancer activity.

Cruciform - loop-like structure formed when inverted complementary repeat nucleotide sequences within the same strand of double-stranded DNA match up.

Cytoplasm - collection of cell components that exist within the cell, but does not include nucleus.

Cytosine (C) - pyrimidine base that pairs with guanosine in DNA.

Cytoskeleton - fibrous network of structural proteins found in eukaryotic cell cytoplasm.

Deamination - loss of an amino group, usually by cytosine, causing a C-to-U conversion at that site.

Deletion - removal of gene region from chromosome.

Denaturation - for DNA or RNA, describes separation of a double-stranded molecule to a single-stranded state, usually by heating; for protein, describes change in physical shape, which usually renders it inactive.

Deoxyribose - five-carbon sugar found in DNA.

Diploid - the normal number of chromosomes (two copies of each—2N) in virtually all eukaryotes.

Direct repeats - multiple identical (or closely related) nucleotide sequences in the same orientation in a DNA molecule.

Directional cloning - use of recombinant DNA vectors that permit the inserted (foreign) DNA to fit into the cloning site in only one direction (vs. "bidirectional" cloning).

Discontinuous replication - synthesis of DNA in many short (Okazaki) fragments that are later joined to form a single, continuous strand.

Disulfide bond - covalent bond formed when two −SH groups in a protein(s) are oxidized, linking the two polypeptide regions together.

D loop - opening in double-stranded DNA formed when RNA primer for DNA synthesis is inserted into dsDNA.

DNA (deoxyribonucleic acid) - polynucleotide that contains deoxyribose sugar.

DNase - enzyme that cleaves phosphodiester bonds in DNA to break the molecule into pieces.

DNA fingerprint - individual gel pattern of DNA fragments produced when person's DNA is cut with restriction enzymes and sized on gel electrophoresis.

DNA ligase - enzyme that joins two double-stranded DNAs together, end to end.

DNA polymerase - enzyme that polymerizes nucleotides into double-stranded DNA using single-stranded DNA as a template.

Double-stranded helix - three-dimensional shape exhibited by two complementary base-paired DNA strands.

Downstream - direction away from transcription start site in DNA and toward the last section of the coding region to be transcribed.

Editing - a system of enzymes that monitors each nucleotide to ensure that it is correctly base paired during DNA replication.

Electrophoresis - separation technique that involves placing charged macromolecules in a voltage field and permitting them to migrate according to their individual net charges.

Electroporation - permeabilization of cells by electric shock for the purpose of inserting a foreign gene.

Elongation factors - regulatory proteins that associate with ribosomes to promote either the tRNA binding or the translocation step in the elongation phase of translation.

Endonuclease - enzyme that cleaves phosphodiester bonds within a nucleic acid.

Endoplasmic reticulum (ER) - folded membrane sheets to which polysomes (sites of protein synthesis) are anchored.

Enhancer - nucleotide sequence that increases the utilization of eukaryotic promoters; it can function in either an upstream or downstream location relative to the promoter.

ES cells - embryonic (stem) cells found in early mammalian embryos, which are capable, when transplanted into a host embryo, of differentiating into all types of tissues and organs.

ES complex - combination of an enzyme with its substrate, which results in catalysis of a chemical reaction to the substrate by the enzyme.

Escherichia coli (E. coli) - intestinal bacterium widely used because it can be conveniently cultured for prokaryotic genetics studies.

Excision repair - an enzyme system that removes a short, single-stranded sequence of double-stranded DNA, containing mispaired or damaged bases, and replaces it by synthesizing a sequence complementary to the remaining strand.

Exon - nucleotide sequence of a eukaryotic gene that is represented in the mRNA.

Exonuclease - enzyme that cleaves nucleotides one at a time off the end of a DNA or RNA molecule.

F′ - plasmid that has bacterial (host) genes incorporated into its genome.

F factor - bacterial plasmid that functions to generate sex or fertility characteristics.

F plasmid - fertility plasmid in *E. coli,* which permits a donor bacterial cell to conjugate (mate) with a recipient bacterial cell.

F1 generation - progeny of a cross between two parental types that differ in one or more genes.

Feedback inhibition - regulation mechanism for controlling activity of enzymes in a metabolic pathway—one of the metabolites inhibits one of the enzymes, which normally leads to its production (allosteric inhibition).

Filter hybridization - method for nucleic acid hybridization performed by soaking a denatured DNA preparation, immobilized on a nitrocellulose filter, in a solution of radioactively labeled RNA or DNA.

Fingerprint (DNA) - specific pattern of fragments of DNA on gel electrophoresis, generated by first treating DNA sample with endonucleases.

Fluid mosaic model - conceptualization of biological membranes in a manner that portrays some proteins within the membrane as moving laterally.

Frameshift mutation - insertion or deletion of either one or two bases in the coding region of a gene, which results in an altered downstream amino acid sequence.

Gene - basic unit of heredity consisting of the nucleotide sequences of DNA involved in synthesis of a protein.

Gene family - set of very similar genes derived by duplication of an ancestral gene and subsequent minor alteration in each gene in the family.

Gene therapy - introduction of a foreign gene into embryos or cells in an attempt to correct a genetic disease.

Genetic code - set of 64 codons (triplets) in DNA and the amino acids they represent.

Genetic engineering - use of recombinant DNA methods to insert foreign and/or manipulated genes into a host genome.

Genetic marker - mutant gene that is useful in genetic mapping studies for locating sites of other genes.

Genome - complete set of genetic information in a specific organism.

Genomic library - collection of cloning vectors (plasmids or phages) that contains fragments of chromosomal DNA.

Genomics - study of the physical structure and nucleotide sequence information contained in an organism's genome (DNA).

Genotype - genetic constitution of an individual organism.

Guanine (G) - purine base that pairs with cytosine in DNA.

Gyrase - topoisomerase (enzyme) that can introduce negative supercoils into DNA.

Haploid - chromosome number in the gametes of a species, symbolized by "N."

Helicase - enzyme that unwinds DNA double helix during DNA replication at the replication fork.

Heterochromatin - regions of chromosomes of eukaryotes that are highly condensed and not expressed.

Heteroduplexing - formation of double-stranded DNA by annealing two single-stranded DNAs differing in nucleotide sequence in one or more regions.

Heterogeneous nuclear (hn) RNA - transcripts of varying size synthesized by RNA polymerase II.

Heterozygote - diploid genotype with different alleles for a specific gene.

Hfr - strain of *E. coli* that exhibits an unusually high frequency of recombinations.

Histones - class of approximately five small, basic DNA binding proteins of eukaryotes that together with DNA and RNA comprise chromatin.

HIV - human immune-deficiency virus that causes AIDS.

Holoenzyme - complete enzyme complex, including catalytic subunits, as well as regulatory subunits.

Homeobox - nucleotide sequence, found in many genes, that codes for a polypeptide with DNA-binding properties.

Hotspot - region in DNA where mutations occur at exceptionally high frequency.

Human Genome Project - international program designed to obtain the complete nucleotide sequence of each and every base in the human genome.

Hybridization - formation of a double-stranded nucleic acid molecule from complementary single-stranded molecules.

Hydrophilic - describes molecules (e.g., some amino acids) that prefer to be in contact with water.

Hydrophobic - describes molecules that prefer to avoid direct contact with water (e.g., many lipids).

Hydroxylmethylcytosine (HMC) - cytosine base that contains a CH_2OH in place of a usual hydrogen. Found in the DNA of certain bacteriophages.

Hyperchromic - describes solution of macromolecule (e.g., DNA) that exhibits (relatively) increased absorbance of ultraviolet light (e.g., when denatured).

Immunity - for bacteriophages, the state that exists for the bacterial host when it contains a lysogenic phage and is thereby immune from further infection by phage of the same type as its lysogenic virus.

Immunoglobulin - a large antibody molecule consisting of four protein subunits.

Inducer - Small molecule (e.g., galactose) that causes the operon to be expressed.

Induction - expression of specific genes in response to the presence of substrates for the proteins (enzymes), coded by the induced genes. In the life cycle of a phage, this term means the initiation of the viral replication cycle.

Initiation factors - regulatory proteins that associate with the small subunit of the ribosome to facilitate initiation of translation of the mRNA.

Inosine (I) - nucleotide containing the base hypoxanthine, which base pairs with cytosine (C).

Insertion - placement of additional nucleotide pairs in a specific site in DNA.

Insertion sequence (IS) - DNA sequence in bacteria capable of relocating itself to a new location in the genome, using transposase encoded in its nucleotide sequence.

Insertional inactivation - cloning procedure involving insertion of foreign DNA into a plasmid at a site that disrupts (i.e., inactivates) the activity of an indicator gene (such as the β-galactosidase gene).

Intercalating agent - chemical that is approximately the same size as a normal DNA base pair and, when inserted into DNA between normal bases, can disrupt DNA replication and cause a mutation.

Intergenic suppression - canceling of the effects of an initial mutation by a second mutation in a different gene.

Intervening sequence - intron.

Intragenic suppression - canceling of the effects of an initial mutation by a second mutation in the same gene.

Intron - nucleotide sequence in a gene that is transcribed, but eventually removed from the mRNA by specific splicing enzymes.

Inverted repeats - symmetrical nucleotide sequence of DNA that is repeated in opposite orientations on same molecule.

in vitro - reaction or cell-culturing conditions "outside" of the natural environment (e.g., in a test tube).

Isopropylthiogalactoside - structural analog of lactose, which can induce expression of the *lac* operon without itself being metabolized.

Isozymes - multiple forms of enzymes that differ from one another in either catalytic or physical properties.

Joining region (J) - short segment of an antibody gene that is joined to a larger segment to constitute a complete variable region of an immunoglobulin gene.

kb - an abbreviation for 1,000 nucleotide base pairs of DNA or RNA.

Knockout mice - mice that have received a copy of a mutated foreign gene that inactivates the host's copies of the normal version of that inserted gene.

Lac **operon** - collection of genes in some bacteria that produce enzymes for metabolizing the milk sugar lactose.

Lactose permease - protein product of the *lac* operon that transports lactose into the bacterial cell.

Lagging strand (DNA) - replication of DNA employing short segments (Okazaki fragments).

Leader (sequence) - nontranslated sequence of mRNA, upstream of the initiation codon.

Leading strand (DNA) - replication of DNA as a continuous strand.

Library - collection of cloned fragments in a vector (plasmid or phage), usually "cDNA" or "genomic."

Ligation - enzymatically catalyzed formation of a phosphodiester bond that links two DNA molecules.

Linker (DNA) - DNA contained in a nucleosome that is not directly complexed with histones.

Lipid bilayer - model for cell membranes in which hydrophobic portions of two layers of lipid molecules face inward and hydrophilic portions face outward.

Long-terminal repeat (LTR) - nucleotide sequence that is repeated at the end of a DNA molecule.

Looping (of DNA) - double-stranded DNA bends back on itself, usually when in contact with proteins.

Lysis - the last step in the bacteriophage life cycle, which involves bursting of the bacterial cell membrane and release of multiple phage progeny.

Lysogenic - describes bacteriophage existing permanently in bacterium in the form of a "prophage" that is integrated into bacterial genome.

Lysozyme - enzyme that degrades bacterial cell wall during the lysis phase of the phage reproduction cycle.

Lytic - describes bacteriophage capable of undergoing typical reproductive life cycle.

Major groove - area of open space present in turns of the alpha-helix in the common staircase model of double-stranded DNA.

Melting temperature (T_m) - midpoint of heat denaturation curve for double-stranded DNA.

Membrane bilayer - for eukaryotic cell, a lipid bilayer that contains various proteins and surrounds the cytoplasm.

Messenger RNA (mRNA) - single-stranded template RNA that contains information for amino acid sequence of the protein.

Methylation - addition of methyl group by an enzyme to a preexisting base (usually cytosine).

Minimal medium - nutrient recipe for culturing cells, which contains fewest possible components required for growth, and thereby lacks supplements such as amino acids, vitamins, and the like.

Minor groove - area in double-stranded DNA between the antiparallel sugar-phosphate backbones, which includes the hydrogen-bonded nucleotide bases.

Mismatch repair - enzyme-mediated repair system for replacing bases in DNA that are incorrectly paired.

Missense mutation - change in DNA nucleotide sequence that leads to substitution of one amino acid for another in the protein.

Molecular biology - study of structure and function of the proteins, nucleic acids, other polymers, as well as related components, which comprise the working apparatuses of living cells.

Monocistronic - sequence of DNA that codes for a single protein.

Mutagen - an agent that causes changes in the nucleotide sequence of DNA.

Mutagenesis - process of producing nucleotide sequence changes, which are subsequently inherited (mutant).

Mutant - organism that carries a modified inherited gene.

Mutation - change in the nucleotide sequence of DNA that is inherited.

Mutator gene - gene (in bacteria) that, when mutated, leads to higher-than-usual number of mutations occurring.

Negative regulation - process resulting in the transcription of a specific gene or operon normally being repressed.

Negative supercoiling - twisting of double-stranded DNA in the opposite direction of the natural double helix twist.

Nick - gap in one strand of DNA of a double helix, caused by breakage of a phosphodiester bond.

Nick translation - method for radioactively labeling double-stranded DNA; DNA polymerase begins at a nick and replaces the original strand with a new one.

Nonsense mutation - mutation that changes a normal codon into a stop codon.

Nonsense suppressor tRNA - mutated tRNA containing an anticodon that can base pair with a stop codon, and thereby relieve the effects of the nonsense mutation.

Nuclease - enzyme that cleaves phosphodiester bonds in nucleic acids.

Nucleoside - chemical structure that includes either a purine or pyrimidine base and a five-carbon sugar (either ribose or deoxyribose).

Nucleosome - chromatin subunit that consists of DNA and a set of eight histone proteins.

Okazaki fragments - short run of nucleotide's synthesized lagging strand during DNA replication.

Oncogenes - genes that, when expressed, transform cells into a tumor-like state, characterized by rapid growth.

Operator - nucleotide sequence that binds repressor protein and thereby prevents transcription of adjacent gene.

Operon - set of genes in bacteria that is coordinately regulated and contributes to a single cellular function.

Ori - origin of replication in DNA.

Palindrome - DNA sequence consisting of adjacent inverted repeats that read the same in left direction on one strand as in right direction on the other strand.

P element - *Drosophila* transposon useful for generating mutants or transgenic flies for research purposes.

Peptide - polymer synthesized by forming a covalent bond between the alpha-amino group of one amino acid and the alpha-carboxyl group of another amino acid.

Peptidyl (P) site - binding site (similar to "A site") on larger ribosomal subunit serving as a docking site for incoming charged tRNA during protein synthesis.

Peptidyl transferase - enzyme that forms peptide bonds between amino acids during protein synthesis.

Photolyase - enzyme that breaks the thymine dimers in DNA produced by ultraviolet irradiation.

Photoreactivation - reversal of the effects of ultraviolet irradiation (thymine dimers) by the action of enzymes, which themselves require energization by visible light.

Pilus - extension of surface of male bacterial cell that connects to female cell for the purpose of transferring DNA from male (F^+) to female (F^-).

Poly A - sequence extension of polyadenylic acid at the 3′ end of a eukaryotic mRNA.

Polysome - macromolecular complex consisting of mRNA and several ribosomes, as well as associated components required for protein synthesis.

Positive regulation - process resulting in a specific gene or operon being transcribed, usually as the result of interaction of proteins with DNA.

Preinitiation (30S) complex - macromolecular complex consisting of mRNA, 30S ribosomal RNA subunit, fMet-tRNA, GTP, and initiation factors, which begins the process of protein synthesis by combining with a 50S ribosomal RNA subunit.

Prophase - early step in mitosis of eukaryotic cell in which chromosomes, although in a highly condensed state, have not interacted with the mitotic spindle.

Protease inhibitor - drug capable of inactivating enzymes that hydrolyze proteins (proteases), and is used in AIDS therapy.

Proteome - the sum total of all the functional proteins in a cell/tissue.

Proteomics - study that emphasizes the integration of structural, functional, and quantitative relationship between proteins in a cell/tissue.

Protoplast - cells that have been treated, usually with enzymes, to remove their cell walls.

R factor - bacterial plasmid that carries genes specifying resistance to one or another antibiotic.

Reading frame - the nucleotide sequence in an mRNA, beginning at a specific start codon, which is translated into a protein.

Recombinant DNA - DNA that has been experimentally manipulated in the laboratory with restriction enzymes/lygase so as to contain novel or modified genes.

Recombinase - enzyme that cuts and joins the V and J regions of genes that code for immunoglobulins.

Regulation - control of the activity of a gene or enzyme by mechanisms that increase or decrease that activity.

Regulatory gene - gene that codes for an RNA or protein used to control the expression of other separate genes.

Regulon - set of genes that are not necessarily located adjacent to one another, but that are regulated by a common mechanism.

Renaturation - process of a macromolecule returning to its native three-dimensional structure. For DNA, this involves the two strands base pairing; for protein, it involves folding to its active configuration. Also known as reannealing.

Replication fork - the Y-shaped region of DNA, which serves as the site of synthesis of new strands.

Repressor protein - a "regulatory" protein that binds to the operator locus on DNA and blocks transcription of the adjacent gene.

Resolution site (IRS) - nucleotide sequence in DNA recognized by resolvase enzyme for the recombination event used in replication of complex transposons.

Resolvase - enzyme that carries out recombination between two repeated transposons.

Restriction enzymes - enzymes that recognize specific, short nucleotide sequence DNA in double-stranded DNA and cleaves at that site.

Restriction fragment length polymorphism (RFLP) - individual DNA nucleotide sequence differences that lead to variation in lengths of fragments when DNA is digested with specific restriction endonuclease.

Restriction map - linear representation of DNA sequence that illustrates sites cut by specific restriction endonucleases.

Retrotransposon - DNA sequence present in eukaryotic genome that represents vestiges of earlier retrovirus integration.

Retrovirus - virus containing an RNA genome, but which replicates inside a cell by making a double-stranded DNA copy of itself as an intermediate for RNA duplication.

Reverse transcriptase - enzyme that synthesizes DNA from an RNA template.

Reversion - change in nucleotide sequence of DNA that reverses the effects of the original mutation.

Rho factor - *E. coli* protein that recognizes specific nucleotide sequences and causes RNA polymerase to terminate transcription.

Ribose - five-carbon sugar found in RNA.

Ribosomal RNA - RNA component of a ribosome.

Ribosome - component of protein synthesis machinery comprised of RNA and protein; it provides structural support for mRNA.

Ribosome binding site - Shine-Dalgarno sequence.

Ribozyme - RNA/protein complex that has catalytic activity of the type normally associated with protein enzymes.

RNA (ribonucleic acid) - polynucleotide that contains ribose sugar.

RNA editing - enzymatic modification of mature mRNA's nucleotide sequence so that the coding sequence is slightly changed.

RNA polymerase - enzyme that synthesizes RNA from ribonucleotide triphosphates; it requires a DNA template.

RNA splicing - process that removes segments of RNA (usually introns) from a long RNA molecule and joins remaining segments (usually exons).

rRNA - ribosomal RNA.

Rolling circle replication - process for replicating double-stranded circular DNA.

Sanger procedure - dideoxynucleotide DNA sequencing procedure (aka chain-termination method).

Satellite DNA - short nucleotide sequence repeated several times (in tandem) that, upon centrifugation, separates away from the main DNA fraction.

Scanning mode - movement of mRNA along the eukaryotic 30S ribosome in search of first AUG for initiation step in protein synthesis.

Secondary structure - the folding pattern of proteins or nucleic acids, which results from interaction of subunits.

Semiconservative replication - format for DNA replication that uses one old strand (conserved) as a template for the synthesis of a second, complementary new strand.

Sex factor - sex plasmid.

Sex plasmid - plasmid that enables host bacterium to conjugate (mate) with another bacterium.

Shine-Dalgarno sequence - the AGGAGG purine sequence on bacterial mRNA, which facilitates binding of the mRNA to a ribosome.

Sigma factor - subunit of RNA polymerase that facilitates binding to promoter sites on DNA.

Signal recognition particle (SRP) - macromolecular complex that recognizes signal sequence on proteins and helps transport them to the endoplasmic reticulum.

Silencer - nucleotide sequence having effects opposite those of enhancers—reduces rate of transcription of associated gene.

Site-specific mutagenesis - laboratory procedure that alters the nucleotide sequence of a cloned gene in a specific and deliberate fashion.

SOS response - DNA repair mechanism that is activated by ultraviolet light and permits error-prone DNA replication to proceed past thymine dimers.

Southern transfer (also known as Southern blotting) - method for transferring nucleic acids from an electrophoresis gel to nitrocellulose paper (for subsequent hybridization).

Spacer - nucleotide sequence of mRNA that is devoid of start/stop codons and, hence, is not translated.

Spliceosome - RNA/protein complex that carries out splicing of RNA.

Splicing - process that removes introns from RNA and joins exons (coding sequences) to form "translatable" mRNA.

SSB protein - protein that binds to single-stranded DNA during replication.

Stacking (of bases) - the tight packing of purine and pyrimidine base pairs in both single and double-stranded DNA due to hydrophobic interactions between the bases.

Substrate - the target molecule for an enzyme, which, when acted upon by the enzyme, is chemically altered.

Supercoiling - coiling of a large, circular, double-stranded DNA molecule back around itself.

Suppressor mutation - a mutation that acts to restore the original phenotype.

Tandem repeats - nucleotide sequence of DNA that is serially repeated many times.

TATA box - an A-T-rich sequence, usually consisting of 7 or 8 nucleotides, located approximately 10 base pairs upstream of the transcription start site. It serves as a binding site for RNA polymerase.

Temperate - describes a bacteriophage that integrates into the host genome and, although normally silent, can be induced to undergo a lytic life cycle.

Temperature sensitive - describes a mutant phenotype apparent at a relatively elevated temperature, but not recognizable at a "normal" temperature.

Template - nucleotide sequence that is "read" by one of the nucleotide polymerases and thereby serves to dictate the sequence of the newly synthesized (complementary) nucleic acid.

Tertiary structure - the native three-dimensional configuration of a macromolecule, which results after its subunit interactions are complete.

Theta replication - Configuration of prokaryotic cell's circular DNA replication apparatus, including parental and daughter strands, which, as drawn in two dimensions, resembles the Greek letter theta.

Thymine (T) - pyrimidine base found in DNA.

Thymine dimer - adjacent thymine residues in DNA that have been chemically linked, usually by the action of ultraviolet irradiation.

Tn element - bacterial transposable element.

Topoisomerase - enzyme that alters amount of supercoiling in large, double-stranded DNA molecules.

Totipotent - describes a cell that is capable of developing into virtually all the tissues and organs of an adult organism.

Trailer (sequence) - nucleotide sequence at the 3′ end of an mRNA that is not translated into part of the protein.

Transcription - synthesis of RNA from a complementary DNA template.

Transcription factors - proteins that bind to specific nucleotide sequences and facilitate the binding of RNA polymerase to the DNA template.

Transduction - transfer of bacterial gene from one bacterium to another by a bacteriophage.

Transfer RNA (tRNA) - small RNA that transfers amino acids to the mRNA template for protein synthesis.

Transforming principle - extract of cells (DNA) that, when added to bacterial cells, has the capacity to change their phenotype.

Transgenic animal - animal that has been altered with genetic engineering technology so as to contain a foreign gene.

Translation - synthesis of a protein from an mRNA template.

Translational frameshifting - reading of an mRNA template for protein synthesis by ribosomes, which shift back one nucleotide when a stop codon is encountered, and, hence, continue translation, yielding a longer-than-usual protein product.

Translocation signal - amino acid sequences on proteins, which are recognized by "signal recognition particles (SRPs)," and thereby get transported to specific locations in the cell cytoplasm.

Transposable element - transposon.

Transposase - enzyme that functions to insert a transposon into a new site in DNA.

Transposition - relocation of transposon to new location.

Transposon - nucleotide sequence that can insert itself, with the aid of an enzyme (transposase), into new locations in the genome.

Trans-splicing - primary transcript processing that involves exons originating in primary transcripts of different genes being ligated together.

Triplet - sequence of three nucleotides that serve, in DNA, to ultimately specify the position of a single amino acid in a protein.

Tumor inducing (Ti) plasmid - plasmid that, when it infects a plant cell, can integrate into the host cell's DNA and lead to production of a tumor.

Ubiquitin - small protein that binds to other proteins and thereby targets them for destruction by proteolytic enzymes.

Upstream - direction ahead of the start site for transcription or translation.

Uracil (U) - pyrimidine base found in RNA.

Uracil *N*-glycosylase - enzyme that cleaves the *N*-glycosylic bond and thereby releases uracil from its sugar (which is usually incorporated into the DNA's sugar-phosphate backbone).

van der Waals forces - weak attractive forces that result from transitory fluctuations in electron-charged densities in neighboring atoms.

Variable number tandem repeat (VNTR) - short nucleotide sequences that are repeated, in variable "repeat numbers," in the genomes of higher eukaryotes.

Vector - plasmid, phage, or other genome into which a foreign gene is inserted for cloning.

Virulent - bacteriophage capable only of lytic (vs. lysogenic) growth.

Virus - smallest "living" form that, although not capable of an independent existence, can reproduce itself.

Wobble hypothesis - explanation for ability of certain tRNAs to exhibit unusual base pairing with more than one triplet codon.

Yeast artificial chromosome - genetically engineered yeast chromosome that can serve as a cloning vector for large fragments of eukaryotic DNA.

Z-helix - double-stranded DNA existing in a left-handed helix rather than the more common right-handed format.

Answers to Questions and Problems

Answers to Drill Questions (Ch. 2)

1. Proteins—amino acids; nucleic acids—nucleotides; polysaccharides—sugars (most often glucose).
2. An amino group and a carboxyl group are linked to form a peptide linkage.
3. The termini of a protein molecule consist of an amino group and a carboxyl group. These groups are also contained in the side chains of some of the amino acids forming proteins. For example, both aspartic acid and glutamic acid have a free carboxyl group, and both asparagine and glutamine have a free amino group.
4. DNA—thymine; RNA—uracil.
5. The 5′-phosphate of one nucleotide and the 3′-OH of the adjacent nucleotide; a phosphodiester group.
6. A nucleotide is a nucleoside phosphate.
7. Two molecules that are poorly soluble, that is, interact poorly with water, can reduce contact with water by clustering, because clustering minimizes the ratio of surface to volume.
8. Strongest—ionic bond; weakest—van der Waals attraction.
9. All DNA molecules have a negative charge and, therefore, move in the same direction when electrophoresed. This is not true for protein molecules, however, because they may have a net positive or negative charge.
10. One molecule of SDS binds per amino acid, giving each protein the same charge per amino acid, and, thus, the same net charge. When SDS is bound and there are no disulfide bonds, all proteins unfold and have nearly the same shape.

Answers to Problems (Ch. 2)

1. (a) No, since differences in shape can affect mobility—that is, two proteins differing in both molecular weight and shape might have the same electrophoretic mobility.
   (b) Yes. All linear DNA fragments have the same basic shape and same charge-to-mass ratio. Therefore, if all bands seen were of equal intensity, the conclusion would be that five different DNA fragments were present.

2. Cysteine—disulfide bonds; lysine—ionic bonds and hydrogen bonds; isoleucine—hydrogen bonds; glutamic acid—ionic bonds and hydrogen bonds. All could participate in van der Waals attractions.

3. (a) The compactness probably results from an ionic bond(s) between unlike-charged groups.

   (b) The molecule probably contains many groups having like charges, and, in the lower ionic strength of 0.01 M NaCl, these groups repel one another.

4. Three. Other possible interactions involving amino acid side chains, such as hydrophobic interactions, ionic bonds, and van der Waals attractions, would determine the naturally occurring structure of the protein.

5. Electrophoretic mobility increases with charge, but is reduced by friction between the moving molecule and the solvent molecules. Since valine has a longer side chain than alanine, it generates greater friction and moves more slowly.

Answers to Drill Questions (Ch. 3)

1. The deep, wide groove made by the turns in double-stranded DNA constitutes the major groove. The shallow, narrow groove between the anti-parallel single strand of DNA makes up the minor groove.

2. The virus must have as its genome a single-stranded DNA molecule (at least one), because [A] + [G] is not equal to [T] + [C].

3. In a double-stranded molecule, [A] + [G] = [T] + [C]. The relationship between [A] + [T] and [G] + [C] varies from molecule to molecule.

4. The B helix is considered to be the common form of DNA, but DNA forms an A helix under conditions of dehydration. A Z helix is favored by segments with alternating Gs and Cs.

5. (a) No—there would be nothing to prevent the circle from untwisting.

   (b) Yes—this would form a supercoil.

   (c) G; No—this would merely form a relaxed circle.

6. Hydrophobic and hydrophilic properties of the bases and phosphate groups, respectively, and hydrogen bonding between the bases of a base pair.

7. The salt concentration must be high enough to neutralize the mutually repulsive negative charges of the phosphate groups, and the temperature must be sufficient to avoid extensive intrastrand base pairing.

8. With no cations to complex with the phosphate groups, the repulsion among them would be great enough to denature the DNA.

9. 480 turns of the helix are present (4,800 bases at 10 bases per turn). The molecule would be 1.632 μm long (480 turns $\times$ 34 Å/turn).

10. No, there would be no free ends to the molecule.

Answers to Problems (Ch. 3)

1. (a) Would be least likely to re-form the original structure, as it could have extensive intrastrand base pairing.

2. (a) Would require a higher temperature for reannealing (relative to its T_m) in order to overcome the intrastrand base pairing.

   (b) Would have the highest T_m, as it has the highest G^+C content.

3. 5′-ATCAAGGTA-3′

4. With too much ddT, chain termination would occur abruptly, and, thus, the fragments would be shorter and found toward the bottom of the gel.

5. Ribosomal RNA. It is complexed with several dozen proteins to form a compact structure that is not easily attacked by RNase.

6. The solution contained a mixture of copies of two different DNA molecules. One type had a considerably higher G+C content than the other.

Answers to Drill Questions (Ch. 4)

1. (a) The polar amino acids are arginine, asparagine, aspartic acid, cysteine, glutamic acid, glutamine, histidine, lysine, serine, threonine, and tyrosine. The nonpolar amino acids are alanine, glycine, isoleucine, leucine, methionine, phenylalanine, proline, tryptophan, and valine.

   (b) Isoleucine is more nonpolar than alanine, because it has a long, nonpolar side chain.

2. The peptide bond.

3. Cysteine—disulfide; arginine—ionic; valine—hydrophobic; aspartic acid—ionic and hydrogen bonds. All, of course, could be involved in van der Waals attractions.

4. Both will probably be near an amino acid with the opposite charge.

5. Set (c), in a hydrophobic cluster.

6. They would probably be scattered throughout the primary sequence, although there would probably also be a few in small clusters.

7. Primary structure is simply the linear sequence of amino acids. Secondary structure involves hydrogen bonding between peptide groups. Examples are the α helix and β structure. Tertiary structure is the folding of a polypeptide chain and is formed from secondary structure elements, plus several other interactions between amino acid side chains, including ionic bonds between oppositely charged side chains, various hydrogen bonds, hydrophobic clustering of nonpolar side chains, and the coordination of metal ions.

8. Since the α helix and β structure are both rigid, the former protein would be long, and thin or fibrous; the latter would be more flexible, and spherical or globular in shape.

9. Both the enzyme active site and the substrate change shape slightly upon binding, and the strain placed on the substrate by the shape change aids in the catalysis.

Answers to Problems (Ch. 4)

1. (a) Yes. A van der Waals attraction would favor aggregation. This would be aided by a hydrophobic interaction if the side chains on the surface were very nonpolar.

   (b) No. For geometric reasons, the hydrophobic regions probably cannot come into contact.

   (c) No. Charge repulsion by the lysines will effectively counteract any tendency to form a hydrophobic cluster.

   (d) Yes. If the geometry is appropriate, the alternation of unlike charges will allow unlike charges in the two molecules to attract one another.

2. No. In general, enzymes must be slightly flexible in order to adapt their shape to the substrate and carry out the required chemical reaction.

3. The enzyme could be a multisubunit protein in which one defective subunit is sufficient to eliminate the enzymatic activity.

4. When there is a single binding site, it is probably located at or near the junction of all the subunits. This location is certainly unlikely if there are several identical binding sites. If the number of binding sites equals the number of subunits, one might guess that the binding sites are far from all regions of contact between the subunits. If the number of binding sites is half the number of subunits, each site probably includes the contact region of two subunits.

5. If you looked at the protein structures with and without substrate bonds, there should be no change in the structure for a lock-and-key mechanism, but there will be a change in the structure for an induced-fit mechanism.

Answers to Drill Questions (Ch. 5)

1. An octameric disk is an aggregate of two each of histones H2A, H2B, H3, and H4. A core-particle consists of an octameric disk plus the 140 base pairs of DNA that encircle it. A nucleosome includes the core particle, linker DNA, and the histone H1.

2. A eukaryotic cell has more DNA than a prokaryotic cell, and that DNA must be contained within a nucleus.

3. During cell division, the DNA must be organized into a compact unit in order to be moved from the metaphase plate toward the poles of the cell. During other stages of the cell cycle (when replication and transcription occur), proteins must interact with specific base sequences of DNA, so the DNA must be less compact to permit these interactions.

4. The DNA would be most compact during mitosis, to facilitate movement of the chromosome as a unit. During S phase, DNA is being replicated, and so must be less compact.

5. 3.4 nm, the length of one turn of the helix.

6. The polar regions are on the "ends" of the protein that are outside of the membrane, while the nonpolar region is the part of the protein that is actually within the membrane.

7. Van der Waals attractions and hydrophobic interactions are the main factors stabilizing a lipid bilayer.

Answers to Problems (Ch. 5)

1. In this case, there are 200 base pairs per nucleosome. A DNA molecule 2.4 cm long contains about 7×10^7 base pairs (0.34 nm per base pair). Therefore, there would be about 3.5×10^4 nucleosomes.

2. Electrostatic interactions.

3. The unknown component is a covalently linked nucleotide-amino acid complex; the protein-DNA binding is, therefore, through a covalent bond.

4. The phosphate groups of the backbone of the helix.

5. Since the Cro protein binds stereospecifically, it is likely that the Cro-DNA complex would not form. Also, specific contact points might be blocked.

Answers to Drill Questions (Ch. 6)

1. The conversion of the genotype of a bacterium to the genotype of a second bacterium by exposure of the first bacterium to DNA from the second.
2. Many plant and animal viruses and some bacteriophages.
3. A codon.
4. ^{32}P for DNA and ^{35}S for proteins.
5. Mutation.
6. Uracil.
7. Thymine is 5-methyluracil.
8. Chemical alterations and replication errors.

Answers to Problems (Ch. 6)

1. The RNA in a virus is protected from the environment by a protein coat.
2. Progeny of a transformed cell all had the new character; in genetic terms, the cells bred true.
3. The chemical mechanism for hydrolysis of a phosphodiester bond requires participation of a free OH group, which is present on the 2′ carbon in RNA, but not in DNA.
4. It is widely believed that DNA was a simple tetranucleotide incapable of carrying the information required of a genetic material.
5. Transforming activity was not lost by reaction with either proteolytic enzymes or ribonucleases. Treatment with DNases inactivated the transforming principle.
6. During replication, an incorrect base or an extra base is inserted or deleted in the daughter molecule.

Answers to Drill Questions (Ch. 7)

1. Semiconservative.
2. Template.
3. All, ½, ¼, respectively.
4. Theta replication.
5. Rolling circle replication.
6. Positive supercoiling.
7. Negative supercoiling.
8. Polymerizing activity, 5′—3′ exonuclease, 3′—5′ exonuclease.
9. A 3′-OH group.
10. Polymerization, and the 5′—3′ exonuclease.

Answers to Problems (Ch. 7)

1. No. After one round of replication, it has hybrid density, but an ^{15}N strand always remains in the circle.
2. The 5′—3′ activity removes ribonucleotides from the 5′ termini of precursor fragments, and the 3′—5′ exonuclease removes a base that has been incorrectly added to the growing end of a DNA strand.

3. Pol III requires a helicase to unwind the DNA; pol I can unwind a helix without an accessory protein.

4. DNA polymerase joins a 5′-triphosphate to a 3′-OH group and in so doing removes two phosphates, so that the phosphodiester bond contains one phosphate; a ligase joins a 5′-monophosphate to a 3′-OH group.

5. One strand is copied from the 3′ end to the 5′ end, and the other strand is copied in the direction opposite to the movement of the replication fork by synthesis in short pieces (see Figures 7-16 and 7-17).

6. The RNA primer must be removed from the precursor fragment that was made first, because DNA ligase cannot join DNA to RNA.

Answers to Drill Questions (Ch. 8)

1. (a) Ribonucleoside 5′-triphosphates.
   (b) Double-stranded DNA.
   (c) RNA polymerase.
   (d) No.

2. The reactions are identical—reaction of a nucleoside 5′-triphosphate with a 3′-OH terminus of a nucleotide to form a 5′ to 3′-diester bond. The substrates differ in that DNA polymerase joins deoxynucleotides, while RNA polymerase joins ribonucleotides.

3. A 5′-triphosphate and a 3′-OH.

4. (a) An RNA molecule that is translated into one or more proteins.
   (b) A primary transcript is a complementary copy of a DNA strand. It could be a precursor to mRNA, tRNA, or rRNA and may be processed to form a functional RNA.
   (c) A cistron is a DNA segment between and including translation start-and-stop signals that contains the base sequence corresponding to one polypeptide chain. A polycistronic mRNA molecule contains sequences encoding two or more polypeptide chains.
   (d) Leaders, spacers, and the unnamed region following the last stop codon of a mRNA are not translated.

5. Five subunits; the sigma subunit is responsible for positioning.

6. (a) $^-10$ and $^-35$.
   (b) TATAAAA

7. (a) A terminal structure in which a methylated guanosine is in a 5′-to-5′-triphosphate linkage at the 5′ terminus of mRNA.
   (b) The 3′-OH end.
   (c) All mRNA molecules (except those of several viruses) are capped. Some mRNA molecules lack the poly(A) tail.

8. (a) Untranslated sequences that interrupt the coding sequence of a transcript, which are removed before translation begins.
   (b) Removal of introns and linking together of exon sequences.

Answers to Problems (Ch. 8)

1. (a) It is probably a sequence that existed at very early times and from which others were derived by mutation. Furthermore, it indicates that the biochemical system that used the sequence existed very long ago.

(b) It is essential for some stage or promotion—in particular, binding of RNA polymerase or initiating polymerization.

(c) The rates of initiation may differ slightly. It is difficult to be certain of this point, though, because the rate of initiation is mainly determined by the −35 region.

2. Since the triphosphate of the primary transcript is absent, the molecule has been processed. The processing could be extensive, or as simple as triphosphate hydrolysis, but there is no way of knowing from the information given.

3. Eukaryotic mRNA molecules contain a terminal poly(A) that can hydrogen-bond to the dT in the column under renaturing conditions. The mRNA, thus, will be retained by the column, and other RNA species will pass through. The mRNA can be eluted from the column simply by heating.

4. The number of times each of the four bases appears at each position is noted. The maximum value (as a percentage) is indicated as a subscript in the consensus sequence that follows:

$A_{80}T_{70}G_{80}C_{90}A_{80}C_{70}$.

5. 5′—AGCUGCAAUG—3′ and 5′—CAUUGCAGCU—3′

6. *E. coli* is a prokaryote and its RNA polymerase does not recognize all eukaryotic promoters. The eukaryote yeast, on the other hand, provides not only a suitable eukaryotic RNA polymerase, but also various transcription factors that act to facilitate transcription.

Answers to Drill Questions (Ch. 9)

1. (a), (b), and (c) are true. (d) Amino acyl tRNA synthetases are required (see question 2). (e) Rho is involved in the termination of transcription, not translation. However, if the anticodon of a tRNA has been altered, it can sometimes read a termination codon.

2. (c), (d), and (e).

3. Prokaryotic: 70S and 30S and 50S subunits containing 5S, 16S, and 23S RNA.
   Eukaryotic: 80S with 40S and 60S subunits containing 5S, 5.8S, 18S, and 28S RNA.

4. (1) Prokaryotes initiate translation at an AUG codon just downstream of the Shine-Dalgarno sequence (ribosome-binding site) AGGAGGU, which base pairs with a section of the 16S rRNA near its 3′ end. The eukaryotic ribosomal small subunit binds at the 5′ cap of mRNA, and the first AUG encountered is used.

   (2) Prokaryotes initiate translation with formylmethionine, whereas eukaryotes initiate with an unmodified methionine.

   (3) Prokaryotic mRNAs may be polycistronic, since the ribosome may reinitiate translation at a second AUG following termination of translation of the first reading frame. Eukaryotic ribosomes do not reinitiate translation without first dissociating.

5. Steps (a), (b), and (d) are true. Step (c) is false. Amino acids are coupled to tRNA molecules by aminoacyl tRNA synthetases.

6. A reading frame is a string of sequential, nonoverlapping codons beginning with an initiation codon (AUG), ending with a termination codon (UAA, UAG, or UGA), and containing codons for amino acids in between. In addition to a reading frame, most mRNAs have 5′ leader and 3′ tail sequences. A 5′ cap and poly A tail are present in eukaryotic mRNAs. Prokaryotic mRNAs can have several cistrons separated by short spacer regions.

7. Translation occurs along the mRNA in the 5′-to-3′ direction, the same direction as the synthesis of the mRNA itself. Thus, protein synthesis can occur while the mRNA is being copied from the DNA. In the reverse polarity, protein synthesis would have to await the completion of the molecule of mRNA. In the existing system, protein synthesis, thus, can start earlier than would be possible with reverse polarity, and the mRNA is relatively resistant to nuclease attack. This is only true in prokaryotes, of course.

8. (b).

9. Formation of the 70S initiation complex and translocation.

Answers to Problems (Ch. 9)

1. Met Pro Leu Ile Ser Ala Ser.

2. There are two families of codons for arginine, the CGX family and the AG_G^A family. Single base changes in the first two bases of the CGX family in any of the bases of the AG_G^A family could yield the following amino acid replacements: histidine, glutamine, cysteine, tryptophan, serine, glycine, leucine, proline, isoleucine, threonine, lysine, or methionine.

3. For UAG, the amino acids are Tyr, Leu, Trp, Ser, Lys, Glu, and Gln. For UAA, they are Tyr, Lys, Glu, Gln, Leu, and Ser.

4. The Arg-2 codon must have been AGG (the only arginine codon that can become AUG in one step), and Arg-2 could, therefore, also be replaced by Gly, Trp, Lys, Thr, or Ser. The Arg-3 codon must have been AGA, and Arg-3 could also be replaced by Ser, Lys, Thr, and Gly. The Arg-1 codon could have been any one of the six arginine codons, and cannot be unambiguously identified.

5. Val-Cys-Val-Cys-Val-Cys . . . , and peptides of various sizes starting with Val or Cys.

6. (a) Threonine, i.e., 5′-ACX-3′. Remember that the codon-anticodon pairing is antiparallel.

   (b) ACU, ACC, and ACA, since I in the wobble position can pair with U, C, or A.

   (c) CGU, since only ACG remains to be read, and C in the wobble position will only pair with G. U in the wobble position would read G, but also A, so that both ACG and ACA codons would be read, and ACA is already read by the first tRNA discussed.

Answers to Drill Questions (Ch. 10)

1. BU is a base analogue mutagen. It is a thymine analogue and is incorporated into the DNA during replication by base pairing with adenine. Due to the bromine atom, a shift in the keto-enol equilibrium occurs, such that the enol form is more prevalent than it would be with thymine. The enol form may base pair with guanine in future rounds of replication, leading to A.T → G.C or T.A → C.G transitions after one more round of replication.

2. Many amino acid changes will not yield a mutant phenotype, and many mutations will be chain-termination mutants.

3. There are two possibilities: There are two genes for the original tRNA species, and only one of these has been mutated, or natural chain-termination sequences might usually consist of two or more different termination codons. The nonsense suppressor would suppress only one, and chain termination would still occur. Both possibilities occur.

4. They clearly interact. Since a charge sign change yields a mutant and a second sign change in another amino acid yields a revertant, amino acids 28 and 76 are probably held together by an ionic bond.

5. Depurination and deamination of cytosine.

6. One explanation is that the T4 possesses its own repair system.

7. (d) DNA polymerase I.

Answers to Problems (Ch. 10)

1. This mutant could not be isolated because there is no temperature at which it could grow.
   (a), (c), (g), and (h), as they involve either changes in polarity, charge sign, or chemical properties.

2. No. If the original rate of mutation is one in 10^5 for a sequence 1,000 base-pairs long, this gives the approximate probability of hitting any base in the sequence and causing a detectable mutation. Since the reversion rate is also one in 10^5, this probably represents other random changes in the sequence. The frequency of an exact replacement should be the probability of a change × the number of sites that can be changed. For this problem, the answer would be one in ($10^5 \times 1,000$), which is one in 10^8, about 1,000 times less than the initial mutation rate. However, since not all changes at the original site would regenerate the original amino acid, the actual rate of exact reversal would still be somewhat lower.

3. The original mutation is probably a frameshift mutation. The mutagens listed cause a variety of transitions and transversions, some of which would almost certainly suppress a missense or nonsense mutation. To suppress a frameshift mutation would require an intercalating agent such as proflavine or acridine orange, mutagens that themselves cause frameshifts. The original mutation could also be caused by a transposable element, which would not be suppressed by the mutagens available.

4. (a) X is not inducible because the enzymes were present before protein synthesis was blocked by chloramphenicol.

(b) X would probably be considered to be inducible. The residual 5% could be due to either a second, noninducible system, or to a small amount of synthesis of X proteins when chloramphenicol is present.

Answers to Drill Questions (Ch. 11)

1. In negative regulation, an inhibitor bound to the DNA must be removed before transcription can occur. In positive regulation, an activator molecule must bind to the DNA. Negative and positive autoregulation is the same, except the gene product is its own inhibitor or activator.

2. A repressor protein binds to operator on DNA to prevent transcription. A corepressor is a small molecule that binds to a regulatory protein, an aporessor, to prevent transcription.

3. Derepressed state is the same as induced when describing the normal state of a gene; it has the same meaning as constitutive in describing the effect of mutation.

4. F′ O^c $lacZ^+/O^+$ $lacZ^-$; constitutive expression of lacZ.
   F′ O^c $lacZ^-/O^+$ $lacZ^+$; inducible expression of lacZ.

5. In the absence of arabinose, AraC binds to operator and prevents transcription. In the presence of arabinose, arabinose-AraC complex activates transcription from the *ara* promoter.

6. In eukaryotes, transcription and translation are not coupled.

7. Frequency of transcription initiation.
   Frequency of transcription termination (attenuation).
   mRNA ability.

8. mRNA stability.
   Frequency of translation initiation.
   Protein stability.

Answers to Problems (Ch. 11)

1. (b), and probably also (c).

2. In all cases, *lacZ* will not be expressed.

3. Attenuation is controlled by the levels of charged $tRNA^{trp}$ inside the cell. If there is no tryptophan present, the concentration of charged $tRNA^{trp}$ becomes inadequate, and a translating ribosome stalls at the tryptophan codons of the leader peptide, preventing the terminator from forming, and, hence, preventing premature transcription termination. However, other mutations (RNase P) that decrease the levels of charged tRNAs in the cell will not induce transcription termination.

4. There might be a mutation in the gene for adenylate cyclase, or for the cyclic AMP receptor protein. Another possibility is a membrane transport mutant, if these sugars were to utilize the same transport system.

Answers to Drill Questions (Ch. 12)

1. RNA polymerase I transcribes genes encoding ribosomal 18S, 5.8S, and 28S RNA. RNA polymerase III transcribes genes encoding transfer RNA, ribosomal 5S RNA, and U6 small nuclear RNA (U6 snRNA). RNA polymerase II transcribes protein-encoding genes, and most of the snRNA types.

2. The position of eukaryotic regulatory sequences with regard to the genes they regulate is often of great importance. However, in contrast to such regulatory sequences in prokaryotic genomes, far more difference in type and exact position of these regulatory sequences is seen in eukaryotes. In fact, some of these sequences (known as enhancers) can be experimentally moved, often hundreds of nucleotide pairs upstream or downstream, without affecting regulatory activity.

3. Variable region; A V segment and a J segment are joined to the C region of the gene. The C region encodes the constant region.

4. Increase in the stability, and thus, lifetime, of the mRNA. Activation of a strong promoter by a transcription factor. Activation of transcription through binding of a hormone to its receptor.

5. Active chromatin is at least partially devoid of histones and hypomethylated. To define such active regions, chromatin can be treated with DNase to locate hypersensitive sites.

6. Both cis- and trans-splicing can occur in the processing of eukaryotic mRNA. With cis-splicing, exons from the same transcript are joined. In trans-splicing, an exon from one transcript is spliced to an exon located on a different mRNA.

7. The gene encoding the protein could have a unique promoter. An effector molecule (e.g., transcription factor) could cause synthesis of two new proteins, an inhibitor of the cellular RNA polymerase, and a new RNA polymerase that could recognize only that promoter. There are also other possibilities.

8. An embryonic gene contains a large number of different base sequences that constitute the coding sequences for all antibodies. These sequences are contiguous in the DNA. In the course of development, a genetic recombinational event removes large blocks of DNA that include many adjacent sequences. There are many blocks that can be removed, so that many different coding sequences can remain after this recombinational event occurs. One event occurs in a particular cell leaving that cell with a unique coding sequence that enables it to make a particular antibody.

Answers to Problems (Ch. 12)

1. (a) Increase the number of J genes.
   (b) $(150)(12)(3)(5000) = 2.7 \times 10^7$

2. Very likely, unless gene B was somehow separated from some element needed for its activity.

3. Some upstream sites at which regulatory proteins seem to act are too far from the start site for a polymerase to bind both to a regulatory protein and the start site. Many upstream sites can, furthermore, be moved somewhat without significantly affecting the rate of transcription, and enhancers can even be moved downstream of the genes with which they are associated.

4. E somehow regulates excision of a single, 30-nucleotide intron near the 5′ terminus of the RNA. The 5′ initiation AUG is located either slightly upstream of the intron, or is contained within the intron. With the intron present, there is also an in-frame stop codon that may or may not be in the intron. Thus, without splicing, only a short product can be made due to the early stop caused by the in-frame stop codon. Once the intron is excised, and the downstream exon placed in the correct frame, the Q product can be translated utilizing either the upstream exon's AUG if present, or the first AUG contained in the downstream exon. Translation continues until the proper in-frame stop codon in the downstream exon is encountered.

5. (a) The promoters for both transcription units probably have a common sequence acted on by either the effector itself, another factor requiring the effector for activity, or a negative regulator that is inactivated by the effector.

   (b) Both primary transcripts have a common sequence acted on by an element that prevents some stage of processing. This element is inactivated by the effector.

   (c) Both processed mRNA molecules have a common sequence involved in ribosome binding. The effector may remove a protein bound to this sequence, or it may denature a double-stranded region containing the ribosome binding site.

6. Probably the simplest explanation would be that the two mRNAs had different lifetimes. More complicated explanations include post-translational modification of one of the proteins decreasing its stability, or some inhibition of translation of one of the mRNAs.

Answers to Drill Questions (Ch. 13)

1. Genomics focuses on genome size and sequence, whereas bioinformatics interprets that sequence information, searching for patterns of sequence organization, gene arrangement, noncoding sequences, etc. Proteomics focuses on the diversity of proteins in a cell, attempting to inventory each and every one, and study their functions and interactions.

2. The C-paradox refers to the seemingly strange fact that some relatively simple organisms contain large genomes, while other (often closely related) species contain relatively small genomes. One explanation for genome expansion is the accumulation, over evolutionary time, of noncoding DNA in an organism's genome. Organisms with relatively low genome sizes are considered, in contrast, to have evolved mechanisms that eliminated much of that noncoding DNA.

3. Tallying up the total number of protein-coding genes in an organism's genome. Gene counts are valuable for estimating the minimal number of different types of proteins an organism might express during its life cycle.

4. Approximately 5:1. Perhaps even approaching 10:1. Each protein is, in principle, a candidate for posttranslational modifications, which may include shortening, addition of carbohydrate residues, phosphorylation of some amino acid side chains (e.g., serine), binding to other proteins, etc. Combinations of

those modifications would substantially increase the total number of different functional "proteins."

5. The expression of large numbers (e.g., a few thousand) genes can be monitored simultaneously.

6. Splicing patterns of primary transcripts and posttranslational modification (e.g., shortening) of proteins generate—from one gene—multiple possible functional proteins, rather than "one," as originally proposed by microbial geneticists during the earliest years of the discipline of molecular biology.

7. Figures 13-3 and 13-5 illustrate "infer function" steps. That is, once relevant data is collected, inference, intuition, and experience (common sense) come into play.

8. Data mining involves sorting through information, such as nucleotide sequences for patterns, anomalies, evolutionary trends, etc., and then developing unifying explanations to account for facts that are collected. In contrast, hypothesis-driven studies begin with an idea, model, or principle, and then collect evidence (information) that either supports or negates that hypothesis. If the data negates the hypothesis, another idea or principle is substituted, and supporting evidence is once again searched for.

Answers to Problems (Ch. 13)

1. Genome size, organism's place on the evolutionary tree, opportunities for performing genetic manipulations, applications to basic and applied research projects, and relevance to interpreting the human genome.

2. Define what comprises a "gene" (protein-coding only?); factor in considerations such as alternative use of promoters, exons, and termination sites (which increase the complexity of the protein complement of an organism); devise a method for accommodating antibody gene rearrangement; recognize "pseudogenes" (which resemble normal genes, but are never expressed); count only those sequences for which a corresponding protein can be detected in one or another cell, etc. Compared with model organisms such as *E. coli, Caenorhabditis elegans,* and *Drosophila,* human genes are relatively enormous, due to the presence of large introns, which complicates recognizing individual genes.

3. Disperse throughout the 95% of the genome, which is comprised of noncoding sequences, a wide array of sequences, including regulatory regions, pseudogenes (which resemble normal genes, but are incomplete and, thus, never transcribed), long repetitive sequences, short tandemly repeated sequences, transposable elements (see Chapter 14), "structural" regions (which allow for bending of dsDNA), etc. In the remaining ~5% that codes for proteins, include large numbers of introns, for more than 90% of a human "gene" is comprised of intron sequences.

4. Include in your essay the following general theme: "complex features of morphology and behavior are best understood in terms of gene expression networks that overlap or act in combination." Discuss the fact that gene counts for nematodes, fruit flies, plants (e.g., *Arabidopsis*), and mammals differ by only a factor of 2 or 3, yet their biological complexities are vastly different.

Explain how alternative splicing of transcripts, as well as trans-splicing (Chapter 12), increase the potential protein coding capacity of genes. Include mention of the likelihood that since mammals contain many more types of cells (e.g., >200) than lower organisms, regulatory mechanisms (e.g., transcription factors) are probably more intricate in more highly evolved species. Also, explain the possibility that regulatory circuits acting in combination may be more common in mammals than in lower organisms. Finally, discuss the likelihood that the proteome of a mammal is more diverse than the proteome of a nematode or fruit fly.

5. Inability to obtain sufficient quantities of "low abundance" proteins for structural analysis; some types of proteins (e.g., membrane proteins) are difficult to extract from cells; lack of suitable comparative structures in data banks; inability of present-day NMR spectroscopy methods to recognize minor structural differences in protein fragments (which may have major functional significance); high costs of technology (esp. NMR instrumentation); etc.

6. Initially, the natural substrate of the protease (aspartyl residues in a protein) was established. Then structural analogs were chemically synthesized and tested for their ability to block the activity of the enzyme. Finally, animal testing was performed to determine which of the analogs were highly specific for the HIV (vs. human host) protease. Inhibitors that are highly specific for the HIV version of the protease are given top priority for clinical trials. In addition, drugs that have minimal side effects on the patient's normal metabolism are chosen for further development. Finally, ease of synthesis (and, thus, lower production costs) is taken into consideration.

Answers to Drill Questions (Ch. 14)

1. Direct repeat: ABCD . . . ABCD, in which the dots represent non-repeated bases. Inverted repeat: ABCD . . . D′C′B′A′, in which X′ is the complement of X.

2. Conservative transposition involves no replication of the original element, only its movement to a new location. Replicative transposition involves the duplication of the transposon, such that one copy remains in the original location and a new copy is produced in a new location. In either case, the target sequence is duplicated.

3. Synthesis in the donor provides the single-strand that is copied. Synthesis in the recipient converts the transferred strand to double-stranded DNA.

4. (a) The various genes of the *tra* operon.
   (b) A pilus, or conjugational bridge.
   (c) Rolling circle replication.
   (d) An endonucleolytic nick at *ori*T producing 5′-PO_4^- and 3′-OH ends.

5. (a) An Hfr or high frequency of recombination strain.
   (b) One of its several IS sequences.
   (c) Conjugation is usually interrupted before the entire chromosome can be transferred, though it is theoretically possible.
   (d) An excised F plasmid that contains some chromosomal DNA.

6. Tailed icosahedra, nontailed icosahedra, filaments.

7. T4, T7, λ: linear, double-stranded DNA. M13: circular, single-stranded DNA. MS-2: linear, single-stranded RNA.

Answers to Problems (Ch. 14)

1. (a)   transposon
   5′----- GATCCAGCAATGGCA---------- TGCCATTGCGATCCA-----3′
   3′----- CTAGGTCGTTACCGT----------- ACGGTAACGCTAGGT----5′
   (b) 5′------TTAGCA-3′   5′–TTAGCA-----3′
   3′----- AATCGT-5′   3′–AATCGT-----5′
   double-stranded break in donor molecules
   (c)   transposon
   5′----- TTAGCAGCAATGGCA---------- TGCCATTGCTTAGCA-----3′
   3′----- AATCGTCGTTACCGT----------- ACGGTAACGAATCGT----5′

2. (a) If the mutation affects only the transposase repressor function, trans-position frequency would go up, and the products would be an unaltered chromosome and a plasmid with the Tn3 element. If the mutation affects only the recombination function, transposition frequency would remain the same, but the product would be a cointegrate, that is, the plasmid would be integrated into the chromosome. If the mutation affects both functions, transposition frequency would go up, but only cointegrates would form.
   (b) Very little to no resolvase is made, so transposition frequency goes up but only cointegrates form, since transposase is not repressed, and res-olution of the cointegrate cannot take place.
   (c) Transposition would be abolished either because the transposon exci-sion or target DNA cutting functions are impaired. Either no products would be formed, or the donor or recipient molecules might be lost due to cleavage of only one of these molecules.
   (d) Transposition would be abolished since little to no transposase is made, and no products or intermediates would be detected.
   (e) There would be no effect on transposition or resolution, but screening for transposition becomes impossible since the selectable market has been knocked out.
   (f) Transposition frequency would remain the same, but only cointegrates would form since resolvase cannot recognize the *res* sites.

3. (a) Lac⁻.
   (b) The F′ (ts) *lac*⁺ has integrated into the bacterial chromosome.
   (c) Integration has occurred inside a *gal* gene.
4. No. Phage can only develop in a metabolizing bacterium.
5. It must be a host enzyme.
6. A double-stranded circular molecule.

Answers to Drill Questions (Ch. 15)

1. (a) A restriction enzyme is an endonuclease that makes cuts in DNA at one particular base sequence.
   (b) The biological function is to destroy foreign DNA.

(c) Scientists use restriction enzymes to make defined, reproducible DNA fragments from DNA, and for cloning in vitro.

(d) Each sequence is a palindrome. That is, it has rotational (dyad) symmetry. In other words, it reads the same in both directions.

2. The single-strand breaks may be directly opposite one another at the center of symmetry of the recognition site, or they may be staggered (several nucleotides apart, but symmetric around the line of symmetry of the recognition site). Breaks directly opposite one another produce blunt ends, while staggered cuts produce either 5′ or 3′ overhangs. Overhanging breaks are called cohesive ends because they can stick to similar overhangs by base pairing.

3. They must both recognize the same base sequence.

4. An ideal cloning vector must have an origin of replication. Two selectable marker genes are helpful—one conferring antibiotic resistance or complementing a genetic mutation in the host, such that only transformed cells live on a selecting medium, and another that distinguishes recombinant molecules from nonrecombinant ones by being inactivated when a fragment is cloned into it. An MCS containing numerous, unique cloning sites would make the vector more versatile. If the fragment is eukaryotic and is to be expressed, a promoter sequence is necessary.

5. Insertional inactivation is the process of cloning a DNA fragment into a site within a vector gene such that the gene is no longer active. Frequently, insertion into the *lacZ* gene is employed.

6. If the DNA fragment is available, it may be radiolabeled and used for colony hybridization or DNA hybridization of membrane lifts taken from plates containing isolated plasmids or phage. If the fragment has a known restriction map, then a number of plasmid or phage DNA fragments can be isolated, cut with one or more restriction enzymes, and analyzed by gel electrophoresis. The resulting patterns can be compared to the predicted ones. Ultimately, DNA from the plasmids or phage can be sequenced to verify the insert.

7. (a) Kanamycin.

   (b) Kan-r Amp-r, Kan-r Amp-s.

   (c) Kan-r Amp-s; colonies resistant to kanamycin can be replicated on agar containing ampicillin. Those that do not grow are Amp-s.

8. Since DNA ligase requires a 5′-phosphate and a 3′-hydroxyl for ligation, removal by a phosphatase of the 5′-phosphate from the cut vector molecules will prevent relegation of the vector alone. Another method is to use directional cloning involving cutting the vector and using an insert with two different restriction enzymes. As long as the ends generated are not compatible, the plasmid cannot religate. The small vector DNA fragment between the cut sites must, of course, be removed to prevent its favored relegation into the plasmid over the desired insert.

9. A genomic library is made by cloning pieces of genomic (chromosomal) DNA into an appropriate vector. A cDNA library contains DNA fragments derived from cellular mRNA by reverse transcriptase. A genomic clone could contain promoter and regulatory elements as well as introns. The advantage here is that the gene may be studied as it exists in the organism with

control elements intact. A cDNA clone would have no introns or control elements, but if supplied with a bacterial promoter, it can still be expressed.

10. Site-specific mutagenesis is the alteration of a cloned gene in vitro either by synthesizing the entire gene using short, overlapping single-stranded oligomers, containing one or more specific changes from the wild-type sequence, or by replacing a wild-type region of a cloned gene with a mutant region. The advantages over conventional mutagenesis are twofold. First, the changes made are *specific;* they are the exact ones that the experimenter wishes to make. Second, the changes occur only in the gene of interest, and the time-consuming search for mutants in the gene and the complication of second-site mutations are eliminated. Several disadvantages are that the gene of interest often must be properly reintroduced into the organism, and since the experimenter chooses the changes that are made, some interesting mutations may be missed. Information gained from using site-specific mutagenesis includes understanding the role of specific amino-acid residues in enzyme structure and function, the interaction of multiple proteins, and the location within an enzyme of its active site. Likewise, mutagenesis within promoter and regulatory regions can lead to information on the regulation of gene expression.

Answers to Problems (Ch. 15)

1. (a) The EcoRI digest will generate 11 fragments from the linear molecule (n + 1 fragments, where n is the number of sites) and 10 fragments (n fragments) from the circular molecule.

   (b) Only nine of the fragments (n − 1) from the linear molecule will circularize, since the two end fragments will only have one EcoRI cohesive end. All of the fragments from the circular molecule can circularize. This assumes that all fragments are long enough to circularize.

2. (a) BamHI and BglII
   plasmid:

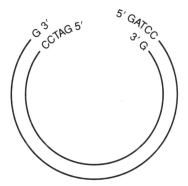

   gene fragment:

   gene
   →
   5′ GATCT ——————— A 3′
   3′ A ——————— TCTAG 5′

Yes, they are compatible as follows:

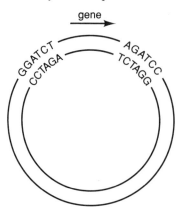

Neither site is regenerated. Note that the opposite cloning orientation is also possible.

(b) BamHI and Sau3A
plasmid:

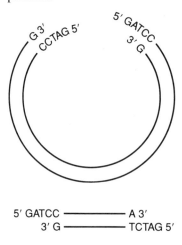

These ends are also compatible as follows:

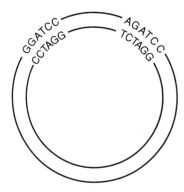

Again both cloning orientations are possible. In all cases, the Sau3A sites are regenerated. In this orientation, the left BamHI site is regenerated, but the right one is not.

c. BamHI and DpnI
plasmid: same as in (a) and (b)
fragment:

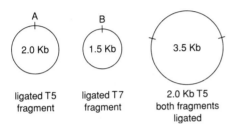

5′TCC ——————— AGA3′

3′AGG ——————— TCT5′

These are not compatible. Although DpnI recognizes the same sequence as Sau3A, the cut position is different. DpnI produces blunt ends which are not compatible with the BamHI cohesive ends.

3. The following circles would be produced. In addition, double T5 and double T7 circles would probably form.

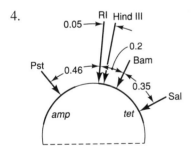

A	B	
2.0 Kb	1.5 Kb	3.5 Kb
ligated T5 fragment	ligated T7 fragment	2.0 Kb T5 both fragments ligated

Since low concentrations of DNA ligase allow the double fragment circle to form, the enzymes probably produce cohesive ends with either 5′ or 3′ overhangs. Blunt end ligations require high concentrations of DNA ligase. Since both enzymes can cut the double circle, the enzymes probably recognize identical sequences.

4.

RI Hind III
0.05 →

0.2
Bam

Pst
0.46

0.35
Sal

amp   tet

5. Chromosome walking using a human genomic clone bank is required. The first step is to create a genomic bank. Human genomic DNA can be partially digested and the fragments cloned into a phage vector. A partial digest is necessary to produce the overlapping fragments required for chromosome walking.

An initial genomic clone can be isolated by probing the library with a radiolabeled probe from one of the cDNA clones by plaque hybridization (similar to colony hybridization but on phage plaques). Radioactive probes corresponding to both ends of this genomic clone should be prepared by radiolabeling short restriction fragments or synthesized oligomers. These may be used to screen the genomic library for clones on either side of the first fragment. The walk must proceed in both directions, since the orientation of the two genes on the chromosome is not known. The new fragments can be properly oriented by restriction mapping and/or nucleotide sequencing. These new fragments should be probed with a radioactive fragment from the second gene by Southern or phage plaque hybridization to see if the walk has proceeded into the second gene. If the second gene is not found within a reasonable distance in both directions from the first gene, they are probably not linked.

Answers to Drill Questions (Ch. 16)

1. Genetically modified foods (e.g., corn and rice), swine with deactivated tissue antigens for potential use in human tissue/organ transplantation, dairy cattle with enhanced milk production, etc.

2. (a) ES cells are undifferentiated, totipotent cells found in the blastocyst of an early-stage mammalian embryo. Such cells can be removed from the embryo, genetically altered and introduced into another blastocyst, where they have the capability of being incorporated into any and all of the tissues of the developing embryo.

   (b) Insertion of the promoterless neo^r gene into the c-abl gene inactivates the latter gene. The cells into which this inactivated gene is inserted are unable to express this copy of the c-abl gene, and simultaneously become neomycin-resistant.

   (c) After the researcher can identify ES cells which have been successfully transformed with this modified gene by growing them on medium containing G418, a neomycin-like compound. Only cells that are neomycin-resistant (and therefore are likely to contain the neo^r gene, linked to the c-abl gene), will be able to grow.

   (d) Homologous recombination involves genetic exchange between two DNA molecules having substantial DNA base-sequence homology—in this case, between the wild-type c-abl gene found in the ES cell and the foreign, mutated c-able gene introduced into the ES cell via electroporation. Incorporation of the c-abl gene at a site other that its normal chromosomal location could result in its altered or abnormal expression, and in possible alteration, inactivation, or inappropriate expression of other genes in the ES cell.

   (e) Neomycin-resistant colonies of transformed ES cells could be isolated and their DNA analyzed by Southern blotting, using a probe which was derived from the c-abl gene. Since the parent cell line contained its own wild-type c-abl gene, once criterion from homologous recombination would be the presence of two distinct bands on the Southern blot: one at a position representing the normal size of the gene, and a second at a position slightly larger (representing the $cable$ gene containing the neo^r gene insert). If nonhomologous hybridization had occurred, only the normal-sized band would appear.

3. (a) The technology used to create transgenic animals could be used to introduce new and sometimes novel genes into various organisms, to in-

activate genes in order to learn more about their normal function in embryonic development, tissue differentiation, development of cancer, or functioning of the immune system; and a variation on this technique can be employed in gene therapy.

(b) If the engineered gene is essential for normal embryonic development, homozygous offspring may die before birth. Due to host-specific differences, the modified cells may fail to be incorporated into the developing embryo, and/or to be transmitted to subsequent generations of offspring. Nonhomologous recombination occurs more often than homologous recombination, and could result in the consequences described in 2(d).

4. (a) **In gene therapy,** a normal, functional copy of a gene is introduced into the tissues of a patient who is suffering from a genetic defect as a result of the presence of a specific defective gene, thus correcting the genetic defect in those tissues and allowing the patient to lead a more normal life. **Supportive (=replacement) therapy** involves the replacement of malfunctioning cells with normal cells, such as occurs in bone marrow transplants.

(b) It is difficult to utilize gene therapy to treat many disorders because, in order to do so, the functional gene must be introduced into the correct tissue, at the proper site, in a complex multicellular organism. Many tissues cannot be grown in culture, and cells cannot be removed from certain parts of the body in order to genetically modify them. In addition, viral vectors must be available, which selectively infect the proper types of cells and insert their DNA at an appropriate location. Unless these conditions are met, gene therapy can be ineffective or even harmful.

5. (a) RFLPs are restriction fragment length polymorphisms. They represent variations in DNA base-sequences from one individual to another, which are manifested in terms of variations in restriction enzyme cutting sites. RFLPs can arise if a deletion, insertion, or point mutation alters a restriction site within or near a gene, or if such a change occurs in a noncoding region between genes.

(b) By using RFLPs, it is often possible to diagnose human genetic disorders prior to the onset of their symptoms, in fetuses before birth, in ova prior to artificial insemination, and in individuals who may be carriers of detrimental genes and wish to make an informed decision prior to starting a family. Several possible disadvantages can be associated with the use of RFLPs in this manner. The gene associated with the genetic disorder must be identified, sequenced, and cloned before such tests are possible. An RFLP pattern difference must be detectable between normal and affected individuals. In some cases recombination may occur between the gene associated with the genetic disorder and the adjacent polymorphic DNA region associated with the RFLP, and some individuals may inherit the disease gene without inheriting the altered restriction site, or may inherit the altered restriction site but not possess the defective gene.

6. (a) **VNTRs (=Variable Number Tandem Repeats)** are short base-sequences occurring in variable numbers within tandemly repeated clusters throughout the genome.

 (b) **Microsatellite repeats** consist of 2–5 nucleotide-long repeats occurring at many locations throughout the entire genome. **Minisatellite repeats** consist of longer repeating units (30–35 bp) containing a shorter, common-core sequence, and tend to occur at fewer sites within the genome. Minisatellite repeats are more useful in DNA fingerprinting because their base sequences are more unique and do not occur so frequently within the genome that they are difficult to separate and distinguish electrophoretically.

 (c) Genomic DNA from the individual is isolated and cut with a restriction enzyme that does not cut the VNTR array itself, but which cuts to include the entire repeat section. (If the quantity of DNA available in the sample is small, it may first be amplified via PCR.) The restriction fragments produced are separated via gel electrophoresis. Southern blotting is carried out using a radioactive probe that hybridizes with the basic repeated sequence. Autoradiography will reveal the presence of a large number of DNA fragments of different sizes that hybridize with the probe, giving rise to a highly individual pattern of fragments (the DNA fingerprint).

 (d) DNA fingerprinting has been used in forensics to establish the paternity of a child, in cases of wildlife poaching, in captive breeding programs of endangered animals, and in a number of other instances.

7. (a) The technical problems encountered when attempting to clone and express the *human* growth factor gene in a bacterial cell might include the following: (1) The eukaryotic promoters might not be recognized by the bacterial RNA polymerases; (2) The mRNA transcribed from the eukaryotic gene might not be translatable on bacterial ribosomes; (3) Introns may be present in the eukaryotic gene, and bacteria lack the capability to excise introns and splice the exons together; (4) The protein itself might require processing, and the bacterium might be unable to respond appropriately to the eukaryotic processing signals; and (5) Eukaryotic proteins might be perceived as "foreign" material by bacterial proteases and degraded.

 (b) One alternative approach might involve cloning a eukaryotic gene into an *eukaryotic* cloning vector such as a yeast artificial chromosome (YAC), and utilizing yeast cells as the host for the vector. A second approach might involve using transgenic animals such as sheep or goats to produce the gene product, which would then be secreted in the animal's milk.

Answers to Problems (Ch. 16)

1. (a)
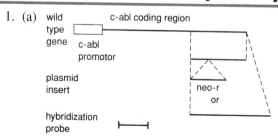

(b) Homologous

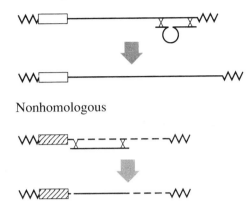

Nonhomologous

Note that the open box again represents the *c-abl* promoter, while
the hatched box represents a different cellular promoter. Sawtooth lines
represent flanking chromosomes, and the dashed line represents the
coding region downstream of the hatched promoter.

(c) Homologous

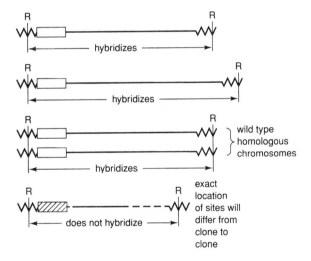

2. (a) It would not be possible to distinguish the transgenic offspring from the
nontransgenic offspring on the basis of physical appearance, making
the subsequent steps of the experiment much more difficult.

(b) Our gene of interest would not be disrupted (and, hence, not inactivated).
In addition, cells that had been transformed with the *c-abl* gene would not
become neomycin resistant, and would fail to grow on the G418 medium.

(c) The recombinant gene would not enter the cell, and no transgenic ES
cells would be produced.

(d) If the ES cells began to differentiate, they would no longer be totipotent, and, hence, we might fail to observe integration of the transformed ES cells into certain tissues, possibly including the germline tissues and/or the tissues exhibiting the visible phenotype on the basis of
which we were identifying chimeric animals.

3. (a) The DNA fingerprint pattern obtained from suspect #1 most closely matches that found in the semen specimen. The DNA fingerprints of the other two suspects (#2 and #3) do not match that of the semen specimen at all.

(b) Suspects #2 and #3 are innocent, since their DNA fingerprints *do not* match that of the semen specimen. We cannot say with absolute certainty that suspect #1 is guilty without knowing how often this particular DNA fingerprint pattern occurs in the general population or in the ethnic group to which suspect #1 belongs. (We also don't know whether suspect #1 has an identical twin brother, whose DNA fingerprints would be expected to be identical to his!) It would obviously be easier to prove innocence, based upon the *failure* of a DNA fingerprint to match that found at a crime scene, than it would be to prove guilt based upon a single apparent match.

(c) F2 is most likely to be the father of the child. The child's DNA fingerprint pattern is a combination of the bands found in the mother's pattern and in F2's pattern. F1's DNA pattern is much less consistent with that of the child.

4. (a)

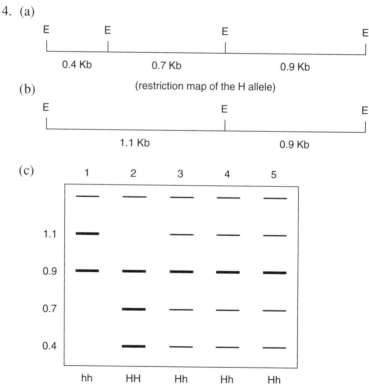

(d) Restriction analysis of DNA from normal-haired German shepherds would enable you to distinguish between (i) dogs that possessed two copies of the normal (H) allele of the gene (HH) and thus exhibited only the 0.4, 0.7, and 0.9 Kb bands, and (ii) those dogs that were heterozygous (Hh) and consequently exhibited a combination of the (0.4, 0.7, 0.9 Kb) and (0.9, 1.1 Kb) bands.

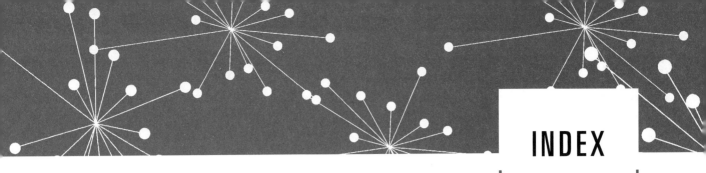

INDEX

A

A_{260}, 40, 43–44
Absorbance A, 40, 43–44
Ac, 312
Acrylamide, 27
Actin, in cytoskeleton, 92–93
Activator, 225
Active site, 69
Adenine
    methylation of, 206–208
    mismatch repair and, 206–208
    molar concentration in DNA, 33–34
    paired with thymine, 35
    structure of, 23
Affinity, between enzyme and substrate, 69
Affinity maturation, of antibodies, 280
Agar, 3
Agarose, 27
A helix, 36, 38
AIDS. See HIV
Alanine, 21, 25
Alcohols, 425–426
Aldehydes, 426–427
Aldose, 427
Aliphatic, 424
Alkanes, 421, 422
Alkenes, 421–423
Alkylating agents, as mutagens, 199
Alkynes, 423–424
Allele-specific oligonucleotides (ASO), 378
Allelic exclusion, 279
Allosteric effectors, 243
α helix, 60–61, 64
Alternative splicing, 260–261, 263–264
Ames test, 205–206
Amides, 431–432
Amines, 431–432
Amino acids
    acidic, 21
    basic, 21
    chemical properties of, 432–434
    chemical structure of, 20, 21
    codons for, 171–172
    hydrophilic, 21
    hydrophobic, 21
    sequencing, 8, 27
    tRNA attachment site, 173
Aminoacyl synthetases, 172–174
Amino terminus, 20. See also N terminus
Amphipathic molecule, 89
Amphoteric, 433
Antibiotics
    and DNA cross-linking, 197
    as protein synthesis inhibitors, 188
    resistance, transmission of, 309–310, 316
Antibodies
    and gene rearrangement, 275–280
    structure of, 66–68, 162
Anticodon, 169, 173, 175
Antigens, 276, 278, 279–280
    interaction with antibodies, 66–68
Antiparallel, 35, 62, 132, 148–149
Antisense, 163, 389
Antiterminator, 337
AP endonucleases, 208–209
Arabidopsis, 7
Arabinose operon, 235–236
ara operon, 235–236
Archaebacteria, 5
Arginine, 21, 81
Aromatic, 424
Asparagine, 21
Aspartic acid, 21
Atom, structure of, 411–412
Attenuation, 236–240
Attenuator, 155, 237–241
Autoregulation, 241
Auxotroph, 3
Avery, Oswald, 103–105
AZT, 144, 392

B

Bacillus thuringiensis (Bt), 387
Back mutation, 200
Bacteria, 2–4
    G+C content, 36
Bacteriophage, 2, 4–5
    and Blender Experiment, 106–107
    general, 324–325
    in genetic engineering, 341–342

lysogenic life cycle, 338–339
lytic life cycle, 326–328, 336–338
structure of, 4, 325
temperate, 326
virulent, 326
Base addition, 194, 195
Base analogue, 199, 208
Base composition, 33–34
Chargaff's rule, 105
Base deletion, 194, 195
Base pairing, 34–36
Bases, nucleic acid, chemical structures of, 23
Base stacking, 25, 33
Base substitution, 194
β structure, 61–62, 64
B helix, 34–38
Bioinformatics, 285, 289–294
Biosynthetic pathways, branched, 243–245
Blender experiment, 106–109
Blunt ends, 349
Bonds
    carbon, 420–421
    chemical, 412–417
    covalent, 412–415
    hydrogen. *See* Hydrogen bonds
    hydrophobic. *See* Hydrophobic bonds
    ionic. *See* Ionic bonds
    N–glycosylic. *See* *N*–glycosylic bond
peptide. *See* Peptide bond
phosphoester. *See* Phosphoester bond
polar, 415
resonance, 417
5-Bromouracil, structure of, 199
Broth, 3

c

cAMP, 232–234
Cap, 159, 187
CAP protein, 151
Capsid, 106, 325
Carboxylic acids, 427–429
Carboxyl terminus, 20
Carcinogen, 205–206
cDNA, 293, 362
cDNA library, 362
Cells
    behavior patterns, 1
    culture, 6

merodiploid, 322
    nucleus of, 99–100
Chain termination mutation, 194
    suppression of, 203–205
Chargaff, Erwin, 105
Chargaff's rule, 105
Chase, Martha, 106–109
Chemokines, 393
Chiral center, 421
Chromatin, 81–82, 250
    solenoidal model, 84
    and transcription, 158, 257–258
Chromatography, 8
Chromosome jumping, 353–355
Chromosomes, general, 79–85
Chromosome walking, 352–354
Cistron, 155, 177
Clonal selection, 279–280
Cloning, 162, 348
    of animals, 383
    directional, 358–361
    in genetic engineering, 358
    reverse transcriptase, 362
    using plasmids, 27–28
    vectors, 355
Codon
    chain termination, 183–184
    and mRNA synthesis, 111, 155, 169
    start, 172
    stop, 171
Cohesive ends, 336, 349–351
Cointegrate, 311
Complementary ends. *See* Cohesive ends
Complexity, of repeating sequence, 44
Computational biology, 289
Concatemer, 125
Concepts of molecular biology, essential, 437–440
Conceptual model building, 10
Conditional mutation, 195
Conformational changes, allosteric, 245
Conjugated base, 419
Conjugation, of plasmids, 317–322, 347
Constant region, in IgG, 67
Constitutive expression, 229
Core enzyme, 130
Core particle, 82–83
Corepressor, 236
Covalent bonds, 412–415
CpG islands, 258

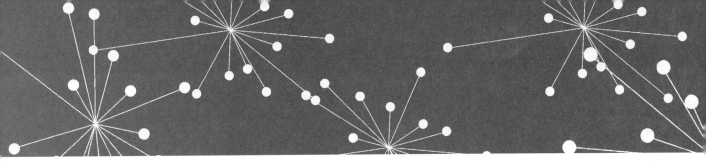

Crick, F.H.C., 2, 34
  and wobble hypothesis, 176
Critical thinking, 16
Cro protein, 86–87
CRP, 232–234
Cruciform structure, of DNA, 39–40
C terminus, 20
C-value paradox, 286
Cyclic AMP, 232–234
Cycloalkanes, 424–425
Cycloalkenes, 424–425
Cysteine, structure, 21
Cytosine
    deamination of, 113–114
    molar concentration in DNA, 33–34
    paired with guanine, 35
    structure of, 23
Cytoskeleton, 92–93
    intermediate filaments in, 93
    microtubules, 93
Cytosolic pathway, 272–273

D

Daughter strand, 120
Deamination, 113–114, 199
    repair mechanisms, 209–210
Dehydrogenation, 426
Denaturants, NaOH, 42
Denaturation, 40
Depurination, 197
    repair mechanisms, 208
Derepression, 231
Direct repeat, 307
Disulfide bonds, 431
    in IgG, 66–67
    and protein folding, 59–60
D loop, 125
DNA. See also DNA repair; DNA replication
    absorbance A of, 40–41
    alternate structures, 36–38
    antiparallel strands, 35, 132
    of bacterophages, 5
    base composition of, 33–34, 105
    base pairing in, 34–36
    base stacking in, 33
    B helix, 34–36
    Chargaff's rule, 105
    chemical synthesis of, 50–52
    circular, 38–39
    common defects in, 197

cruciform structures in, 39–40
C-value paradox, 286
databanks, 51, 53
denaturation of, 40–42
DNA-binding proteins, 85–87
DNA-DNA hybridization, 163
DNA-RNA hybridization, 162
of E. coli, 77–79
eukaryotic, organization in chromosomes, 80–85
fingerprinting, 379–381
four-stranded, 37–38
gyrase, 80, 122–123, 138
A helix, 36
hybridization, 45
hydrogen bonds in, 34, 35
hyperchromic, 41
inverted repeat in, 39–40
junk, 291
ligase, 136–137
linker, 83–84
looping, 236
melting temperature of, 41–42
microarray analysis of, 292–294
mitochondrial, 12
mutation of, 114–115. See also Mutations
native, 40
palindromic region in, 39–40
polymerase. See DNA polymerase
in prokaryotic chromosome, 77–80
properties of, 110–115
protein binding sites in, 40, 85–88
protein scaffold, 84–85
reannealing, 42–45
recombinant, 8, 28, 348
renaturation of, 42–45
repetitive sequences in, 44, 286, 291
replication of, 112
sequencing, 8
single-stranded, 37, 40, 41
stability of, 22, 112–114
steric constraints on structure, 176
storage and transmission of genetic
    information by, 111
supercoiled, 38–40
superhelical, 38–40
synthesis, 8
template, 127, 129, 152, 155
tetranucleotide model, 100, 101
as transforming principle, 103–105
triple-stranded, 37
viral, 2
Z helix, 36–37
DNA polymerase
    comparison with pol I, 131

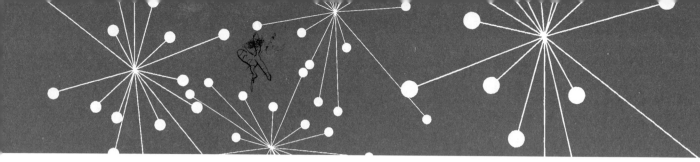

in covalent extension, 139–140
direction of chain growth, 131
and DNA replication, 120
exonuclease activity, 130–131
holoenzyme, 130–131
and nucleic acid sequencing, 48–50
    pol I, 126–130, 132
      comparison with pol III, 131
exonuclease activity, 129–130
pol III, 125–132, 138–139
polymerase α, 141, 143
polymerase δ, 143
and precursor fragment joining, 136–137
proofreading function of, 129
Sanger procedure, 48–50
and SOS response, 215

DNA repair
    AP endonucleases and, 208–209
    by direct reversal, 210
    excision repair, 210–212
    mismatch repair, 206–208
    postreplicational, 213
    recombinational repair, 212–214
    sister-strand exchange, 213
    SOS response, 214–215

DNA replication. *See also* DNA polymerase
    bidirectional, 140–141
    by covalent extension, 123–126, 139–140
    de novo initiation, 124
    discontinuous, 132–137
    editing function, 129
    elongation of new strands, 126–130
    error-prone, 214–215
    eukaryotic, 141–143
    initiation of, 124–126, 135–136
    lagging strand, 132–133, 138
    leading strand, 132–133, 138
    nick translation during, 130
    Okazaki fragments and, 132–135, 138
    ori, 123, 140–141
    origin of replication, 123
    overview, 137–140
    in phage M13, 333–335
    pol III and, 138–139
    postdimer initiation, 212–213
    precursor fragments and, 133–135
    primase and, 138, 140
    primosome in, 138, 140
    proofreading function of, 129
    rolling circle, 139–140, 335
      of F plasmid, 319
      in phage M13, 335
    semiconservative, 120
    sigma, 139–140

SOS response, 215
θ replication, 121–122, 138–139
and thymine dimers, 212
transdimer synthesis, 212–213
transposons as primers for, 308
unwinding of DNA for, 121–123, 126
DNase, 47–48
    and nucleic acid hydrolysis, 47–48
    use for purifying RNA, 162
Drososphila, DNA replication in, 141
Ds, in eukaryotic transposition, 312

E

E. coli
    chromosome, 77–79, 80
    and *de novo* initiation of DNA replication, 124
    insertion sequences in, 306
    nucleoid, 78
Editing function, 129
Efficiency argument, 9
Electron microscopy, 8, 27–28
Electrophoresis
    acrylamide, 27
    with ethidium bromide, 29
    Northern transfer, 164
    of nucleic acids, 27–28
    and nucleic acid sequencing, 44–50
    for protein purification, 26–27
    with SDS, 27
    Southern transfer, 162–164
    two-dimensional, 27–28
Electroporation, 373–374
Embryonic stem cells, 6, 372–374, 376
Endolysin, 328
Endonucleases, restriction, 48
Endoplasmic reticulum, 267, 273–274
End product inhibition, 243–245
Enhancers, 157, 254–255
Enterotoxin, 316
Enzymes. *See also* Restriction enzymes
    active site of, 69
    enzyme-substrate complex, 69–73
    general, 68–69, 68–72
Enzyme-substrate complex, 69–73, 244
ES cells, description of, 372–374
Escherichia coli. See E. coli
ES complex. *See* Enzyme-substrate complex
Esters, 429–430
Ethidium bromide, 29

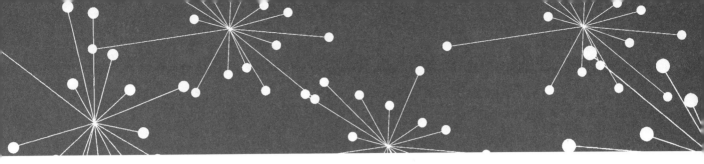

Eukaryotes, 2, 5–7
    chromosome replication in, 141–143
    genetic organization of, 249–251
    repetitive DNA sequences in, 44
    transcription in, 156–161
Excision repair, 210–212
Exon, 159
    splicing, 260–265
Exon cassette mode, 264
Exonucleases, 48, 129–130

F

F_{ab} fragment, 66–68
Familial hypercholesterolemia, 385
Feedback inhibition, 243–245
Ferritin, 269–270
5'-P terminus, 35
Fluid mosaic model, 90–91
Flush ends, 349, 351
Formaldehyde, 427
F plasmid
    and conjugation, 317
    for genetic experiments, 226
    in Hfr cells, 320, 321
    *ori*T, 318, 319
    semiconservative replication of, 319
    and transposition, 310
Fraenkel-Conrat, H., 110
Frameshift mutation, 196

G

Galactose operon, 233–235
    Tn elements in, 306
gal operon, 233–235
    Tn elements in, 306
G+C content, 36
    and DNA melting temperature, 41–42
Gel electrophoresis. *See* Electrophoresis
Genes
    cloning, 27–28
    constitutive, 241
    counting, 289, 291–292
    definition of, 291–292
    DNA restriction/modification, 317
    gene therapy, 6, 376, 384–386
        ethics of, 396–398
    housekeeping, 241, 252, 254, 258
    inducible, 229, 241
    knockouts, 8

    monocistronic, 250
    mutator, 198, 200
    number in humans, 266
    overlapping, 177–178
    plasmid-borne, 316–317
    polycistronic, 250–251
    rearrangements, 275–280
Genetic code, 170–172, 176
Genetic engineering, applications, 366
Genetic testing, 395–396
Genome, modeling, 289
Genomic library, 352
Genomics
    definition of, 285
    functional, 290, 292, 293
    general, 286–289
Glutamic acid, 21
Glutamine, 21
Glycine, 21
Griffith, Fred, 101–103
Growth media, 3
Guanine, 23, 33–35
Guide RNAs, 265–266
Gyrase, 80, 122–123, 138

H

H chain, 66–68
Headful mechanism, 330–331
Helicase, and DNA unwinding, 121–122, 126, 136, 138
    in bidirectional DNA replication, 140–141
Hemiacetal, 427
Henderson-Hasselbach equation, 419
Heredity, early observations on, 99–101
Hershey, Alfred, 106–109
Hershey-Chase blender experiment, 106–109
Heterochromatin, 258
Heterogeneous nuclear RNA, 160
Heterogenous nuclear ribonucleoprotein, 267–268
Hexokinase, 65
Hfr cells, 320–321
Histidine, 21
Histones
    and DNA replication, 141–143
    in eukaryotic chromosomes, 81–82
    introns lacking, 160
    mRNA stability and, 269
    in nucleosomes, 82
HIV, 313
    and AZT, 144

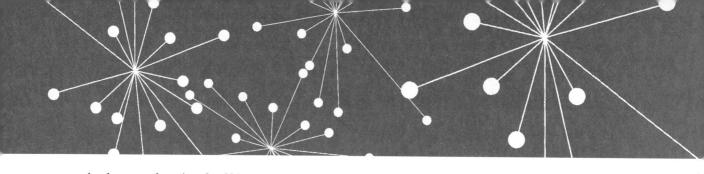

development of vaccines for, 384
gene therapy for, 376
protease inhibition and, 298
therapies for, 391–393
translational frameshifting by, 272
and unprocessed RNA, 268
hnRNA, 160
hnRNP, 267–268
Holistic approach, 11–12
Holoenzyme, 130–131, 149
Homologous recombination, 374–375
Hot spots, 209–210, 308
Human Genome Project, 298, 395
Hybridization, 45
    of allele-specific oligonucleotides, 378
    colony, 363–364
    DNA-DNA, 163
    DNA-RNA, 162
    microarray analysis, 292–294
    Northern, 45, 164
    of nucleic acids, 8
    probe, 162
    Southern blotting, 162–164, 292
Hydrocarbons, 421–425
Hydrogen bonds
    in β structures, 61–62
    description of, 415–416
    in DNA, 34–35
    and DNA denaturation, 41–42
    in α helix, 60–61
    in nucleic acids, 24
    and protein-DNA binding, 85–86
    in proteins, 24, 26, 59, 63–64
    strength of, 414
    types, 24
Hydrolysis, 47–48
Hydrophilic
    amino acids, 21
    definition of, 416
Hydrophobic amino acids, 21
Hydrophobic bonds
    description of, 416
    in polypeptide, 26
    in protein tertiary structure, 63–64
    strength of, 414
Hydrophobic interactions, 24–25
Hydroxylamine, 199
5–Hydroxymethylcytosine (HMC), 329
Hyperchromicity, 41

I

IgG. *See under* Immunoglobulins
Imines, 431–432
Immunoglobulins
    and gene rearrangements, 275–280
    IgG, 66–68, 276–277
    tertiary structure, 65
    V_2 domain structure, 65
Induced-fit model, 69–70
Inducer, 225
Induction, in phage, 338, 340
Inhibitor, 225
Insertion sequences, 306, 322
Integrase inhibitors, 392
Intercalating agents, 199
Interferon, 160
Intermediate filaments, 93
Internal resolution site (IRS), 309, 311
Intron retaining mode, 263–264
Introns
    comparison with retrotransposons, 314
    in mRNA, 156, 159–161
    and splicing, 159–161, 260–265
Inverted repeat, 39–40
Ionic bonds, 25
    description of, 413–414
    in polypeptides, 25–26, 26
    in protein tertiary structure, 63–64
Isoleucine, 21, 25
Isomers
    geometric, 423
    structural, 421
Isopropylthiogalactoside (IPTG), 229
Isozymes, 244–245
IS sequences. *See* Insertion sequences

K

Ketones, 426–427
Ketose, 427
Knockout mice, 372–375
Kornberg, Arthur, 127

L

lac operon, 226–233
Lactose operon, 226–233
Lactose permease, 226
Lagging strand, 132–133, 135–136

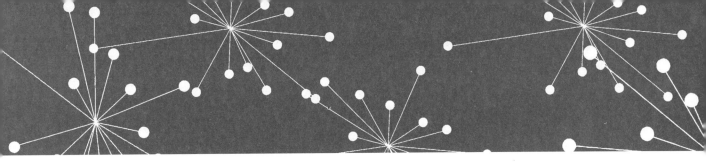

lariat product, 262
Layering approach, 16
L chain, 66–68
Leader sequence, of RNA, 155
Leading strand, 132–136
Leaky mutation, 196
Leucine, 21, 25
Leucine zipper, 87, 88
Levene, Phoebus, 100
LexA protein, 214–215, 240
Ligase, 136–137
Lipid bilayer, 89–92
Lock-and-key model, 69–70
Lysine, 21, 81
Lysogenic life cycle, 326, 338–339
Lysozyme, 70–71, 328
Lytic life cycle, 326–328, 336–338

M

Macleod, Colin, 103–105
Macromolecular complex concept, 87
Macromolecules, 19, 40
    Isolation and characterization, 26–29
Major groove, 34, 37
Malaria, cause of, 93
McCarty, Maclyn, 103–105
McClintock, Barbara, 312
Membranes, 87, 89–92
Mendel, Gregor, 100
Merodiploid cells, 322
Messenger RNA. *See* mRNA
Metabolic regulation, in bacteria, 3–4
Methionine, 21
Methods, 7–9
    nucleic acid isolation and characterization, 27–29
    protein isolation and characterization, 26–27
Methylation
    of cytosines, 258
    heterochromatin and, 258
    in mismatch DNA repair, 206–208
    regulation by, 257–260
    and transposition, 312
Microarray analysis, DNA, 292–294
Microsatellite repeats, 379
Microtubules, 93
Miescher, F., 2
    discovery of DNA, 99–100
Minimal medium, 3

Minisatellite repeats, 379
Minor groove, 34, 37
Mismatch repair, 206–208
Missense mutation, 194–195
    suppression of, 204–205
Mitochondria, 12
Models, conceptual, 10
Model systems, 2–7
Molecular biology
    concepts of, 12–13, 437–440
    early history, 1–2
    logic of, 9–11, 298–299
    rewards from studying, 16
Monocistronic, 155, 184, 250
Monoclonal antibody, 8
Morphogen, 256
Morphogenesis, of phage, 327
mRNA
    abundance, 46
    antisense, 389
    cap, 159, 187
    codon, 155, 169
    eukaryotic, 184, 187
    lifetime, 155
    maternal, 271
    monocistronic, 155, 184–185
    nucleocytoplasmic transport, 267–268
    poly(A), 187
    polycistronic, 155, 177, 184, 250–251
    Reading frames, 177–178
    role of, 111
    stability of, 242–243, 268–270
    synthesis of, 158–159
Mutagen, 193, 199, 205–206
Mutagenesis
    induced, 194
    mechanisms of, 196–197
    site-directed, 363–366
    site-specific, 363–366
    spontaneous, 194
    by transposition, 198
Mutations, 114–115
    biochemical basis of, 195–196
    chain termination, 203
    conditional, 195
    constitutive, 229
    frameshift, 196
    hot spots, 209–210
    induced, 198
    leaky, 196
    neutral, 194–195
    nonleaky, 196

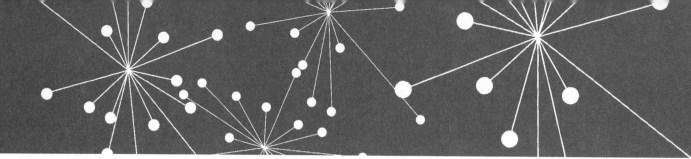

nonsense, 194–195
polar, 198
reversion, 200–201
types of, 193–195
Myohemerythrin, 65

N

Negative translational control, 271–272
N–glycosylic bond, 20, 22, 23, 47
Nicked, definition of, 38
Nick translation, 130
Nitrous acid, 199
Noncovalent interactions, 23–26, 58
Northern transfer, 45, 164
N terminus, 20, 273
Nuclear magnetic resonance, 296–297
Nuclear pore complexes, 267
Nucleases, 47–48
Nucleic acids, 20, 22
    hybridization of, 8
    hydrolysis of, 47–48
    isolation and characterization, 27–29
    sequencing of, 48–50
Nuclein, 99–100
Nucleocytoplasmic mRNA transport, 267–268
Nucleoid, 78
Nucleoproteins, 100
Nucleoside, 20
Nucleosomes, 82–85
Nucleotides, 8, 20, 22–23, 28

O

Okazaki, R., 132–135
Okazaki fragments, 133–135
Oncogenes, 374
Operator, 227, 230
Operons
    arabinose, 235–236
    galactose, 233–235
    lactose, 226–233
    tryptophan, 236–240
Optimism, 10–11
ori, 123, 140–141
    oriC, 124

P

Palindromes, 39–40, 349

Parallelism, 10
Parent strand, 120
Parthenogenesis, 260
PCR, 8, 355–357
P element, 313, 342
Peptide bond, 20, 59, 182–183
Peptidyl transferase, 182–183
p53 gene, 216–217
pH, 417–420
Phage. *See also* Bacteriophage
    lysate, 107
    transducing, 339–341
Phage G4, 178
Phage λ
    Cro-DNA binding in, 86–87
    and *de novo* initiation of DNA replication, 124, 125
    lysogenic life cycle in, 338–339
    lytic life cycle in, 336–338
Phage M13
    general, 333–335
    in genetic engineering, 342
    and site-specific mutagenesis, 363–366
Phage φX-174, 109, 178
Phage PM2, 39
Phage T2, 106–109
Phage T4
    circular permutation in, 331
    DNA polymerase of, 131
    electron micrograph of, 4
    metabolism in, 328–330
Phage T7, 330, 332–333
Phenylalanine, 21, 25
Phosphodiesterase, and deamination repair, 209
Phosphodiester group, 22
    breakage leading to mutagenesis, 197
    during DNA denaturation, 41
    and nucleic acid hydrolysis, 47
    rotation about, 23
Phosphoester bond, 20, 22, 429–430. *See also*
    Phosphodiester group
Phospholipids, 90, 429
Phosphorylation, 273, 275
Photolyase, 210
Photoreactivation, 198
Phylogenetic history, 9
Pilus, pili, 318, 348
Plasmids. *See also* F plasmid
    and antibiotic resistance, 316
    Col plasmids, 316
    conjugation, 317–322

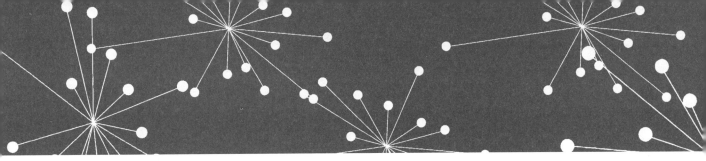

copy number, 323–324
DNA replication in, 322–323
F plasmid, 317, 318–322
in gene cloning, 27–28
general, 314–316
in genetic engineering, 341–342, 347–348
for genetic experiments, 226
relation to bacteriophage, 317
self-transmissible, 317
as vector, 27–28

Plasmodium, 93

Pluripotent, 376

Point mutation, 194

Polar mutations, 198

Pol I. *See under* DNA polymerase

Pol III. *See under* DNA polymerase

Poly(A) sequence, 159, 187, 271
and alternative splicing, 263–264

Polycistronic, 155, 177, 250–251

Polymerase chain reaction, 8, 355–357

Polymorphism, 379

Polypeptides, 19–20, 22, 25–26. *See also* Proteins
polyprotein, 185
synthesis of. *See* Protein synthesis

Polyribosome, 185–187

Polysaccharides, 22–23

Polysome, 185–187, 272

Postdimer initiation, 212–213

Posttranscriptional regulation, 241–243

Prealbumin, 65

Precursor fragments, 133–135

Primase, 138, 140
in bidirectional DNA replication, 140
and primer synthesis, 135–136

Primer, 127–129, 135–136, 138

Primosome, 136, 138, 140

Probe, 162

Problem solving, 16

Progression diagrams, 13–15

Prokaryotes, 2–5

Proline, 21

Promoters, 227
differential selection of, 263–264
eukaryotic, 157–158
in *gal* operon, 233–235
in insertion sequences, 306
open-promoter complex, 151
for RNA polymerase II transcription, 252–255
in transcription, 149–151, 153

Proofreading function, 129

Prophage, 338, 340

Protealysis, 243

Protease inhibitors, 392

Protein Data Bank, 299

Protein processing, 181, 185

Proteins. *See also* Enzymes, Polypeptides, Protein
synthesis
activator, 225
allosteric, 243
α carbon, 58–59
α helix, 60–61, 63, 64
β structure, 61–62, 64
Cro, 88
DNA-binding, 85–87
fibrous, 62–63
folding, 58–60, 296–297
general, 19–20
globular, 62, 64
histones, 81–82
integral membrane, 89, 91
isolation and characterization, 26–27
leucine zipper, 87, 88
membrane, 89–92
peripheral, 89
polymerases, 85
posttranslational modifications, 272–274
primary structure, 58
processing, 181, 185
regulatory, 85–86
secondary structure, 58, 60–62, 63
size, 58
stability of, 272–274
structural, 62–63
subunits in, 64–68
synthesis. *See* Protein synthesis
synthesis of. *See under* Protein synthesis
tertiary structure, 63–64
trafficking, 290
transmembrane, 90, 91
zinc finger, 87, 88

Protein synthesis
direction of, 170, 171
elongation, 179, 183
error frequency, 175
inhibitors, 188
initiation factors, 181, 182
of novel proteins, 188
preinitiation complex, 181, 182
in prokaryotes, 181–185
reading frame, 182
release factors, 184
ribosomes in, 179–181
termination of, 179, 183–184

translocation, 183
wobble hypothesis, 176–177
Proteome, 294
Proteomics
chemical, 297–298
definition of, 285
examples of projects, 290
expressional, 297
general, 294–298
structural, 295–297
Prototroph, 3
Provirus, 313
Pseudogenes, 314
Pseudoreversion, 200
Pulse-chase experiment, 133–134
Pulse-labeling experiment, 133–134
Purines, 20, 34
Pyrimidines, 20, 34
Pyruvate kinase, 65

Q

Quantitative assessment, 9

R

Radioisotope tracers, 8
Random coil, 24
Reading frame, 177–178, 182
Reannealing, of DNA, 42–45
RecA protein, 214–215
Recognition region, 174
Recombinant DNA technology
definition of, 348
detection of, 362–363
ethical issues, 393–395
with plasmid vector, 27–28
and production of pharmaceuticals, 381–384
uses in agriculture, 387–390
uses in medicine, 377–386
uses in research, 371–377
Recombinase, 278, 279
Recombinational repair, 212–214
Reductionist approach, 11
Redundancy, in genetic code, 171, 176
Regulation
coordinate, 224
global, 240
methylation, 257–260
and mRNA stability, 268–270

negative, 225
nucleocytoplasmic mRNA transport, 267–268
by nucleocytoplasmic transport, 267–268
positive, 225
posttranscriptional, 241–243
by protealysis, 243
of protein activity, 272–274
RNA editing, 265–266
specific, 240
splicing, 260–265
and transcription initiation, 251–252
transcriptional, 224–226
of translation, 270–272
Regulon, 240
Release factors, 184
Renaturation, 42–45
Repetitive sequences, 43–45
Replacement therapy, 384
Replication bubble, 122, 140
Replication fork, 132, 138
Replicon, 311
Replisome, 323
Repressor gene, 227
Repressor protein, 225, 230
Resolution, 311
Resolvase, 309
Resonance
of benzene, 417, 424
peptide bonds and, 59
Resonance bonds, 417
Restriction endonucleases. See Restriction enzymes
Restriction enzymes
chromosome jumping, 353–355
chromosome walks, 352–353, 354
and genetic engineering, 349
and restriction maps, 351–352
use in cloning, 358–361
Restriction fragment length polymorphism, 377–378, 379
Restriction map, 351–353
Retrotransposon, 286–287, 313–314
Retroviruses, 313
Reverse genetics, 385
Reverse mutation, 200
Reverse transcriptase, 313, 362, 392
Reverse transcription, 293
Reversion
mechanisms of, 202
and mutagen detection, 205–206
of mutations, 200–201
R factor, 316

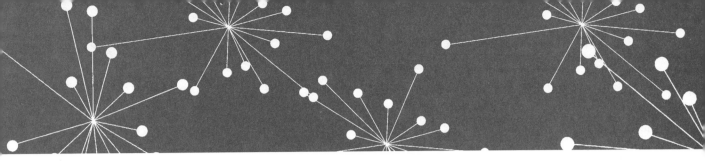

RFI, in phage M13, 334
RFLP, 377–378, 379
Ribonucleoprotein particle, 160
Ribose, 20
Ribosomal RNA. *See* rRNA
Ribosomes
    definition of, 155–156
    and protein synthesis, 170, 179–181
    structure of, 179–180
Ribozymes, 48, 157
RNA. *See also* mRNA, rRNA, tRNA
    base pairing in, 46–47
    classes of, 154–156
    double-stranded, 36
    editing, 265–266
    enzymes, 48
    as genetic material, 109–110, 115
    guide, 265–266
    hydrolysis of, 48
    microarray analyses, 293–294
    polymerase. *See* RNA polymerase
    primary structure, 46
    processing. *See* RNA processing
    secondary structure, 46, 47
    splicing, 159, 160–161, 260–265
    stability compared with DNA, 22
    structure, 46
    synthesis, 52. *See also* RNA synthesis, Transcription
    viral, 2
5S RNA, synthesis of, 158
RNA polymerase
    classes of, 158
    core enzyme, 149
    in eukaryotic RNA synthesis, 156, 251, 252
    holoenzyme, 149
    in prokaryotic RNA synthesis, 148–151
    sigma factor, 149, 153
RNA processing
    combinatorial, 266–267
    definition of, 156
    in *E. coli,* 157
    regulation of, 260–267
RNase, 48
RNA synthesis. *See also* RNA polymerase, Transcription
    basic features of, 147–149
    direction of, 171
    direction of growth, 148
    initiation of, 152
    stages of, 153
    termination of, 152–154
RNP, 160

rRNA
    abundance, 46
    and protein synthesis, 155–156
    species of, 180–181
    synthesis of, 158

S

Saccharomyces cerevisiae, 5
Salmonella typhimurium, 205–206
Salt, of acid, 419
Sanger procedure, 48–50
Scaffold, 78–79
Scanning mode, 181
SDS, 27
Second–site mutation, 200
Sedimentation coefficient, 133
Sequencing, 28, 48–50, 287–289
Serine, 21
Severe combined immunodeficiency (SCID), 385
Shine-Dalgarno sequence, 181–182
Short tandem repeats, 379
Shuttle vectors, 357–358
Side chain, 19–20
Sigma factor, 149, 153
Sigma replication, 139–140
Signal recognition particle, 272
Silencer sequence, 255
Silent mutation, 194–195
Singer, B., 110
Sister-strand exchange, 213
snRNP, 161, 261–262
Sodium dodecyl sulfate, 27
SOS response, 214–215
Southern blotting, 162–164, 292
Spacer, 155, 177
Spliceosome, 160–161, 261–262
Splicing, 159–161, 260–265
SRBs, 256
SRP, 272
Ssb, 126, 335
Stem–loop structures, 46
Stereoisomers, 421
Sticky ends. *See* Cohesive ends
Strong inference, 10
    utility of, 299
Substrate, 69
Supercoiling, 38, 122

Supportive therapy, 384
Suppression, 200–205
Suppressor mutant, 200, 203
SV40, 124, 125, 143, 178
s value, 133
Synthetases, 172–174

T

Taq polymerase, 355, 390
Target sequence, 307, 308
TATA box, 252–254, 256
Tautomerization, 197, 199, 207
Terminally inverted sequences, 307
Terminal redundancy, 330
Terminal repeat sequence, 308
Terminators, 152–154
Tetranucleotide DNA, 100, 101
θ replication, 121–122
Thioethers, 431
Thiols, 431
3'-OH terminus, 35
Threonine, 21
Thymine, 23, 33–34, 35
    dimers, 212–213
Tissue engineering, 280
T_m, of DNA, 41–42
Tn elements. *See* Transposable elements
Tobacco mosaic virus, 65, 109–110
Topoisomerases, 40, 122–123
Totipotent, 372
Transcription. *See also* RNA synthesis, Transcription
    factors
    activator for, 225
    couple with translation, 186–187
    DNA as template for, 111
    in eukaryotes, 156–161
    initiation in eukaryotes, 251–252
    in phage λ, 336–338
    in phage T7, 332–333
    posttranscriptional modification, 156
    primary transcripts, 156
    regulation of, 224–225
    repressors, eukaryotic, 255
    signals, 149–154
    template strand, 152
Transcription factors
    definition of, 157
    posttranslational regulation of, 274–275
    for RNA polymerase II, 253–254

structure of, 255–256
    and transcription initiation, 251–252
Transdimer synthesis, 212–213
Transducing particles, 339–341
Transferrin, 269–270
Transfer RNA. *See* tRNA
Transformation experiments, 101–103
Transforming principle, 103–105
Transgenic animals, 6
    knockout mice, 372
    for pharmaceutical production, 382–383
    production of, 365
    transgenic mouse production, 372–375
Transition mutation, 194
Translation. *See also* Protein synthesis
    complex units of, 185–187
    coupled with transcription, 186–187
    definition of, 169
    editing of, 175
    frameshifting, 272
    general, 169–170
    initiation in eukaryotes, 181, 184
    regulation of, 242, 270–272
Translational frameshifting, 272
Translocation, 183
    signals, 272
Transposable elements. *See also* Transposons
    Ac, 312
    and antibiotic resistance, 309–310
    autonomous, 312
    Ds, 312
    in eukaryotes, 312–313
    general, 306–314
    and mutagenesis, 198
    nonautonomous, 312
Transposase, 307, 312
Transposition
    conservative, 310
    definition of, 306
    frequency of, 311–312
    mutagenesis by, 198
    replicative, 311
Transposons. *See also* Transposable elements
    compared with proviruses, 313
    complex, 308–309
    definition of, 306
    in genetic engineering, 341
    as primers for DNA synthesis, 308
trans-Splicing, 265
Transversion mutation, 194
Triplet, mRNA, 111

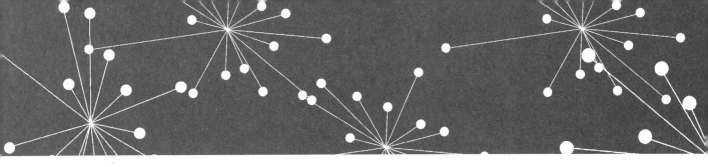

tRNA
    abundance, 46
    acylation, 174
    amino acid attachment site, 173
    and aminoacyl synthetases, 172–174
    anticodon, 169, 173
    attachment site, 173
    charged, 174
    mischarged, 174, 175
    nonsense suppressor, 203–205
    and protein synthesis, 156
    recognition region, 174
    structure of, 46, 173–174
    suppressor, 203–205
    synthesis of, 158
    uncharged, 174
Trp operon, 236–240
Tryptophan, 3, 21
Tryptophan operon, 236–240
Tubulin, 269–270
Tyrosine, 21

U

Ubiquitin, 272–274
Ultracentrifugation, 8, 27
Ultraviolet irradiation, 198, 200
Uracil, 23, 113–114
Uracil N-*glycosylase,* 209
UV reactivation, 215–216

V

Vaccines, 383–384
Valine, 21, 25
Van der Waals attraction, 25–26, 64
Variable number tandem repeats, 379–381
Variable region, 67
Vectors
    in genetic engineering, 348, 357–362
    plasmids as, 27–28
    shuttle, 357–358
    tumor inducing plasmids, 388
Viruses, 2, 37
    bacterial, 4–5
VNTRs, 379–381

W

Water, ionization of, 417–420
Watson, James, 2, 34

Weak acid, 420
Weak base, 420
Weaver, Warren, 1
Wobble hypothesis, 176–177

X

Xeroderma pigmentosum, and excision repair, 211–212
X-ray diffractometry, 8
    determination of protein structure, 27, 57, 296
    for nucleic acid isolation and characterization, 28, 33–34

Y

Yeast, 5
    hexokinase A, 70–73
Yeast artificial chromosome, 5, 382

Z

Z helix, 36–39
Zinc finger, 87–88
Zwitterionic, 433

A Typical Nucleotide

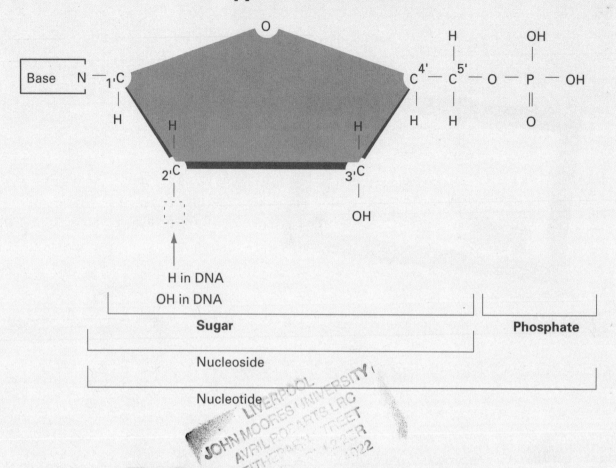

Base

H in DNA
OH in DNA

Sugar

Phosphate

Nucleoside

Nucleotide

The Bases Found in Nucleic Acids

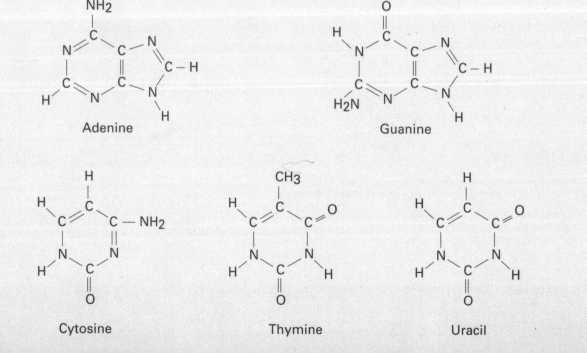

Adenine

Guanine

Cytosine

Thymine

Uracil